Lecture Notes in Computer Science

Lecture Notes in Artificial Intelligence 14839

Founding Editor

Jörg Siekmann

Series Editors

Randy Goebel, *University of Alberta, Edmonton, Canada*
Wolfgang Wahlster, *DFKI, Berlin, Germany*
Zhi-Hua Zhou, *Nanjing University, Nanjing, China*

The series Lecture Notes in Artificial Intelligence (LNAI) was established in 1988 as a topical subseries of LNCS devoted to artificial intelligence.

The series publishes state-of-the-art research results at a high level. As with the LNCS mother series, the mission of the series is to serve the international R & D community by providing an invaluable service, mainly focused on the publication of conference and workshop proceedings and postproceedings.

Mengjun Hu · Chris Cornelis · Yan Zhang ·
Pawan Lingras · Dominik Ślęzak · JingTao Yao
Editors

Rough Sets

International Joint Conference, IJCRS 2024
Halifax, NS, Canada, May 17–20, 2024
Proceedings, Part I

Editors
Mengjun Hu
Saint Mary's University
Halifax, NS, Canada

Yan Zhang
California State University
San Bernardino, CA, USA

Dominik Ślęzak
University of Warsaw
Warsaw, Poland

Chris Cornelis
Ghent University
Ghent, Belgium

Pawan Lingras
Saint Mary's University
Halifax, NS, Canada

JingTao Yao
University of Regina
Regina, SK, Canada

ISSN 0302-9743 ISSN 1611-3349 (electronic)
Lecture Notes in Artificial Intelligence
ISBN 978-3-031-65664-4 ISBN 978-3-031-65665-1 (eBook)
https://doi.org/10.1007/978-3-031-65665-1

LNCS Sublibrary: SL7 – Artificial Intelligence

© The Editor(s) (if applicable) and The Author(s), under exclusive license
to Springer Nature Switzerland AG 2024

This work is subject to copyright. All rights are solely and exclusively licensed by the Publisher, whether the whole or part of the material is concerned, specifically the rights of translation, reprinting, reuse of illustrations, recitation, broadcasting, reproduction on microfilms or in any other physical way, and transmission or information storage and retrieval, electronic adaptation, computer software, or by similar or dissimilar methodology now known or hereafter developed.
The use of general descriptive names, registered names, trademarks, service marks, etc. in this publication does not imply, even in the absence of a specific statement, that such names are exempt from the relevant protective laws and regulations and therefore free for general use.
The publisher, the authors and the editors are safe to assume that the advice and information in this book are believed to be true and accurate at the date of publication. Neither the publisher nor the authors or the editors give a warranty, expressed or implied, with respect to the material contained herein or for any errors or omissions that may have been made. The publisher remains neutral with regard to jurisdictional claims in published maps and institutional affiliations.

This Springer imprint is published by the registered company Springer Nature Switzerland AG
The registered company address is: Gewerbestrasse 11, 6330 Cham, Switzerland

If disposing of this product, please recycle the paper.

Preface

This volume comprises the papers selected for presentation at IJCRS 2024, the 2024 International Joint Conference on Rough Sets, held at Saint Mary's University in Halifax, Canada, on May 17–20, 2024. The annual IJCRS series conferences combine four distinct conferences linking rough sets to various paradigms: RSCTC (data analysis), RSFDGrC (granular computing), RSKT (knowledge technology), and RSEISP (intelligent systems). The first Joint Rough Set Symposium took place in Toronto, Canada, in 2007; followed by Symposiums in Chengdu, China (2012); Halifax, Canada (2013); Granada and Madrid, Spain (2014); Tianjin, China (2015), where the acronym IJCRS was proposed; and subsequent conferences IJCRS 2016 in Santiago, Chile; IJCRS 2017 in Olsztyn, Poland; IJCRS 2018 in Quy Nhon, Vietnam; IJCRS 2019 in Debrecen, Hungary; IJCRS 2020 in La Habana, Cuba (online); IJCRS 2021 in Bratislava, Slovakia (hybrid); IJCRS 2022 in Suzhou, China (hybrid); and IJCRS 2023 in Kraków, Poland (hybrid).

IJCRS 2024 continued to receive significant attention from researchers in the rough sets community. We received 56 full-length paper submissions, which went through a rigorous single-blind reviewing process. Each submission was reviewed by at least three domain experts. Some authors were requested to make revisions, which were further reviewed before the final decision was made. As a result, 43 top-quality submissions were accepted as full-length papers. The camera-ready versions underwent further review by the Program Committee Chairs and General Conference Chairs. The scientific discourse at IJCRS 2024 was complemented by ten extended abstracts, describing ongoing work or research published elsewhere in the past year. These extended abstracts were rigorously reviewed by the Program Committee Chairs and compiled into a Book of Abstracts edited by the Publication Chair, Xiaodong Yue, and his PhD student Zihao Li. The success of the conference owes much to the contributions of the authors, reviewers, and Program Committee Members.

The IJCRS 2024 program featured eight invited talks, including two presentations by former presidents of the International Rough Set Society, Duoqian Miao and Wojciech Ziarko, and six keynote talks by renowned researchers in the field, Lipika Dey, Jimmy X. Huang, Ryszard Janicki, Eric T. Matson, Jesús Medina, and Jarosław Wąs. We are grateful to all the invited speakers for their visionary talks on research related to rough sets. IJRCS 2024 also hosted two workshops on "Uncertainty, Three-Way Decision, and Explainable Artificial Intelligence" and "Applications of Deep Learning and Soft Computing" and two special sessions on "General Rough Set Perspectives on Foundations of AI and Machine Learning" and "Formal Concept Analysis, General Operators and Related Topics". Our gratitude is extended to all the workshop and special session chairs, Duoqian Miao, Jianfeng Xu, Chuanlei Zhang, Ying Yu, Hong Yu, Raavee Kadam, A. Mani, Stefania Boffa, Davide Ciucci, Jesús Medina, M. Eugenia Cornejo, and Eloísa Ramírez-Poussa.

The IJCRS 2024 program was further augmented by the Rough Set School and Tutorials. We are grateful to the chairs, Piotr Artiemjew and Zaineb Chelly Dagdia, and the tutorial speakers, Stefania Boffa, James F. Peters, Usman Qamar, Andrzej Skowron, Dominik Ślęzak, Arkadiusz Wojna, and Yiyu Yao. IJCRS 2024 also hosted a Data Mining Competition, sponsored by Southwest Properties. We would like to extend our thanks to the chairs Yasushi Akiyama and Andrzej Janusz, the judges Chris Cornelis, Dun Liu, Kanngi Mahajan, Dan Penny, Jiju Poovvancheri, Sanjeevi Ramachandran, Trishla Shah, and Yiyu Yao, as well as the participants.

We appreciate the sponsorship from Springer for the two Best Student Paper Awards. The awards were assigned based on a competitive process, considering scientific excellence and clarity of both articles and presentations. With a competition among 29 eligible papers, the two awards were presented to Qiaoyi Li, from the University of Regina in Canada under the supervision of Yiyu Yao, for the paper entitled "Granular Approximations of Partially-Known Concepts", and Hajime Okawa, from the Muroran Institute of Technology in Japan under the supervision of Yasuo Kudo, for the paper entitled "A Vector Is a Granule: A Novel Extension of the Variable Precision Rough Set Model". We are also grateful to Jimmy X. Huang and IEEE for their sponsorship through the IEEE TCII Fund.

IJCRS 2024 would not have been successful without the support of many people and organizations. We are grateful to the Program Committee Members for their effort and engagement in providing a rich and rigorous scientific program. We greatly appreciate the cooperation, support, and sponsorship from the MSc in Computing and Data Analytics (MCDA) program at Saint Mary's University and the International Rough Set Society. We acknowledge the use of the EasyChair conference system for paper submission and review. We are also grateful to Springer for publishing the proceedings as two volumes of LNCS/LNAI.

Lastly, thanks are extended to Raavee Kadam, Neelam Pal, Vrushali Prajapati, and other members in the local organizational team for their logistical, technical, and administrative support, without which IJCRS 2024 would not have been possible.

May 2024

Mengjun Hu
Chris Cornelis
Yan Zhang
Pawan Lingras
Dominik Ślęzak
JingTao Yao

Organization

Honorary Chairs

James F. Peters	University of Manitoba, Canada
Andrzej Skowron	Systems Research Institute of the Polish Academy of Sciences, Poland
Yiyu Yao	University of Regina, Canada

General Conference Chairs

Pawan Lingras	Saint Mary's University, Canada
Dominik Ślęzak	University of Warsaw, Poland
JingTao Yao	University of Regina, Canada

Program Committee Chairs

Mengjun Hu	Saint Mary's University, Canada
Chris Cornelis	Ghent University, Belgium
Yan Zhang	California State University San Bernardino, USA

Steering Committee Chairs

Witold Pedrycz	University of Alberta, Canada
Roman Słowiński	Poznań University of Technology, Poland
Guoyin Wang	Chongqing University of Posts and Telecommunications, China

Data Mining Competition Chairs

Yasushi Akiyama	Saint Mary's University, Canada
Andrzej Janusz	Queensland University of Technology, Australia

Rough Set School and Tutorial Chairs

Piotr Artiemjew   University of Warmia and Mazury in Olsztyn, Poland
Zaineb Chelly Dagdia   Université Paris-Saclay, UVSQ, France

Special Session and Workshop Chairs

Eloísa Ramírez-Poussa   University of Cádiz, Spain
Hong Yu   Chongqing University of Posts and Telecommunications, China

Publicity Chairs

A. Mani   Indian Statistical Institute, Kolkata, India
Stefania Boffa   University of Milano-Bicocca, Italy

Publication Chair

Xiaodong Yue   Shanghai University, China

Local Organization Chair

Raavee Kadam   Saint Mary's University, Canada

Administrative Support

Neelam Pal   Saint Mary's University, Canada

Webmaster

Vrushali Prajapati   Saint Mary's University, Canada

Program Committee Members

A. Mani	Indian Statistical Institute, India
Lubomir Antoni	Jozef Šafárik University in Košice, Slovakia
Roberto G. Aragón	University of Cádiz, Spain
Piotr Artiemjew	University of Warmia and Mazury, Poland
Nouman Azam	National University of Computer and Emerging Sciences, Pakistan
Jaume Baixeries	Universitat Politècnica de Catalunya, Spain
María José Benítez Caballero	University of Cádiz, Spain
Stefania Boffa	Universitá degli Studi di Milano-Bicocca, Italy
Henri Bollaert	Ghent University, Belgium
Nizar Bouguila	Concordia University, Canada
Andrea Campagner	Universitá degli Studi di Milano-Bicocca, Italy
Mihir Chakraborty	Jadavpur University, India
Shradha Chavan	Symbiosis International Deemed University, India
Yingxiao Chen	Beijing University of Posts and Telecommunications, China
Yumin Chen	Xiamen University of Technology, China
Zehua Chen	Taiyuan University of Technology, China
Mu-Chen Chen	National Yang Ming Chiao Tung University, Taiwan
Zehua Chen	Taiyuan University of Technology, China
Yumin Chen	Xiamen University of Technology, China
Davide Ciucci	Universitá degli Studi di Milano-Bicocca, Italy
M. Eugenia Cornejo	University of Cádiz, Spain
Chris Cornelis	Ghent University, Belgium
Zoltán Ernő Csajbók	University of Debrecen, Hungary
Jianhua Dai	Hunan Normal University, China
Dayong Deng	Zhejiang Normal University, China
Tingquan Deng	Harbin Engineering University, China
Murat Diker	Hacettepe University, Turkey
Pawel Drozda	University of Warmia and Mazury, Poland
Soma Dutta	University of Warmia and Mazury, Poland
Zied Elouedi	Tunis Higher Institute of Management, Tunisia
Luis Farinas	Centre National de Recherche Scientifique, France
Andrew Fisher	Saint Mary's University, Canada
Hamido Fujita	Iwate Prefectural University, Japan
Can Gao	Shenzhen University, China
Anna Gomolinska	University of Białystok, Poland
Rafal Gruszczynski	Nicolaus Copernicus University, Poland

Shen-Ming Gu	Zhejiang Ocean University, China
Quang-Thuy Ha	VNU-University of Engineering and Technology, Vietnam
Christopher Hinde	Loughborough University, UK
Mengjun Hu	Saint Mary's University, Canada
Feng Hu	Chongqing University of Posts and Telecommunications, China
Qinghua Hu	Tianjin University, China
Ryszard Janicki	McMaster University, Canada
Jouni Jarvinen	University of Turku, Finland
Richard Jensen	Aberystwyth University, UK
Xiuyi Jia	Nanjing University of Science and Technology, China
Chunmao Jiang	Fujian University of Technology, China
Shi Jinyuan	Tianjin University, China
Raavee Kadam	Saint Mary's University, Canada
Ting Ke	Tianjin University of Science and Technology, China
Michal Kepski	University of Rzeszow, Poland
Md. Aquil Khan	Indian Institute of Technology Indore, India
Marzena Kryszkiewicz	Warsaw University of Technology, Poland
Yasuo Kudo	Muroran Institute of Technology, Japan
Sergei O. Kuznetsov	National Research University Higher School of Economics, Russia
Tamás Kádek	University of Debrecen, Hungary
Guangming Lang	Changsha University of Science and Technology, China
Dajiang Lei	Chongqing University, China
Oliver Urs Lenz	Leiden University, Netherlands
Tianrui Li	Southwest Jiaotong University, China
Min Li	Tianjin University of Science and Technology, China
Jinhai Li	Kunming University of Science and Technology, China
Huaxiong Li	Nanjing University, China
Jiye Liang	Shanxi University, China
Churn-Jung Liau	Academia Sinica, Taiwan
Pawan Lingras	Saint Mary's University, Canada
Caihui Liu	Gannan Normal University, China
Guilong Liu	Beijing Language and Culture University, China
Sujuan Liu	Tianjin University of Science and Technology, China

Jiangtao Liu	Tianjin University of Science and Technology, China
Junfang Luo	Southwestern University of Finance and Economics, China
Domingo López-Rodríguez	University of Málaga, Spain
Vijay Mago	York University, Canada
Jesús Medina	University of Cádiz, Spain
Ernestina Menasalvas	Universidad Politécnica de Madrid, Spain
Claudio Meneses	Universidad Católica del Norte, Chile
Duoqian Miao	Tongji University, China
Marcin Michalak	Silesian University of Technology, Poland
Tamás Mihálydeák	University of Debrecen, Hungary
Sonajharia Minz	Jawaharlal Nehru University, India
Mikhail Moshkov	King Abdullah University of Science and Technology, Saudi Arabia
Michinori Nakata	Josai International University, Japan
Amedeo Napoli	Université de Lorraine, France
Nikita Neveditsin	Saint Mary's University, Canada
Thuc Nguyen	University of Science-VNUHCM, Vietnam
Hoang Son Nguyen	Hue University, Vietnam
Khuong Nguyen-An	Bach Khoa University, Vietnam
Agnieszka Nowak-Brzezinska	University of Silesia, Poland
Manuel Ojeda-Aciego	University of Málaga, Spain
Vladimir Parkhomenko	Peter the Great St. Petersburg Polytechnic University, Russia
Andrei Paun	University of Bucharest, Romania
Daniel Peralta	Ghent University, Belgium
Alberto Pettorossi	Università di Roma Tor Vergata, Italy
Małgorzata Przybyła-Kasperek	Uniwersytet Śląski w Katowicach, Poland
Jianjun Qi	Xidian University, China
Jin Qian	East China Jiaotong University, China
Sheela Ramanna	University of Winnipeg, Canada
Eloísa Ramírez-Poussa	University of Cádiz, Spain
Marek Reformat	University of Alberta, Canada
Mauricio Restrepo	Universidad Militar Nueva Granada, Colombia
Sergio Ribeiro	Pontificia Universidade Católica do Paraná, Brazil
Henryk Rybiński	Warsaw University of Technology, Poland
Hiroshi Sakai	Kyushu Institute of Technology, Japan
Ambuja Salgaonkar	University of Mumbai, India
B. Uma Shankar	Indian Statistical Institute, India
Marek Sikora	Silesian University of Technology, Poland

Andrzej Skowron	Systems Research Institute of the Polish Academy of Sciences, Poland
Dominik Slezak	University of Warsaw, Poland
Roman Slowinski	Poznań University of Technology, Poland
Łukasz Sosnowski	Systems Research Institute of the Polish Academy of Sciences, Poland
Urszula Stańczyk	Silesian University of Technology, Poland
John Stell	University of Leeds, UK
Lin Sun	Tianjin University of Science and Technology, China
Langwangqing Suo	Shaanxi Normal University, China
Marcin Szczuka	University of Warsaw, Poland
Ryszard Tadeusiewicz	AGH University of Science and Technology, Poland
Adnan Theerens	Ghent University, Belgium
Dan Thu Tran	University of Management and Technology, Vietnam
Bala Krushna Tripathy	Vellore Institute of Technology, India
Li-Shiang Tsay	North Carolina Agricultural and Technical Sate University, USA
Bay Vo	Ho Chi Minh City University of Technology, Vietnam
Zhen Wang	Weihai Institute of Beijing Jiaotong University, China
Ye Wang	Texas A&M University, USA
Guoyin Wang	Chongqing University of Posts and Telecommunications, China
Wei Wei	Shanxi University, China
Jun Xie	Taiyuan University of Technology, China
Jianfeng Xu	Nanchang University, China
Xin Yang	Southwestern University of Finance and Economics, China
Hailong Yang	Shaanxi Normal University, China
Jilin Yang	Sichuan Normal University, China
Tian Yang	Hunan Normal University, China
Yiyu Yao	University of Regina, Canada
Ning Yao	Tongji University, China
Minda Yao	China University of Mining and Technology, China
JingTao Yao	University of Regina, Canada
Hong Yu	Chongqing University of Posts and Telecommunications, China
Ying Yu	East China Jiaotong University, China

Xiaodong Yue	Shanghai University, China
Yanhui Zhai	Shanxi University, China
Jianming Zhan	Hubei Institute for Nationalities, China
Chuanlei Zhang	Tianjin University of Science and Technology, China
Yuanjian Zhang	Shanghai University, China
Yan Zhang	Cal State University San Bernardino, USA
Qinghua Zhang	Chongqing University of Posts and Telecommunications, China
Li Zhang	Soochow University, China
Hongyun Zhang	Tongji University, China
Shu Zhao	Anhui University, China
Huilai Zhi	Shanghai University, China
Bing Zhou	Sam Houston State University, USA
Jie Zhou	Shenzhen University, China
Wojciech Ziarko	University of Regina, Canada
Beata Zielosko	University of Silesia, Poland
Jianhang Yu	Chongqing University of Posts and Telecommunications, China

IRSS President Forum Talks

The Contributions of Rough Set Theory to Artificial Intelligence

Duoqian Miao

School of Electronic and Information Engineering, Tongji University, Shanghai, 201804 China
dqmiao@tongji.edu.cn

Fundamentals in artificial intelligence surround the concept of knowledge, such as knowledge representation, knowledge acquisition, and knowledge application. However, many books on artificial intelligence unfortunately fail to formally and clearly explain knowledge [1]. This talk will introduce the significant contributions of rough sets to AI in addressing these three pivotal aspects.

Rough set theory, introduced by Z. Pawlak in 1982 [2], provides at least two pathways to tackle the challenges, namely, the algebraic and information-theoretic approaches, as shown in Fig. 1 [1, 3, 4]. A formal knowledge representation may be achieved through equivalence relations [1], providing an algebraic approach based on set theory. Furthermore, attribute reduction is introduced to obtain significant features and reduce the complexity of knowledge representation [5, 6]. However, constructing reducts is only decidable and is an NP-hard problem. To address this challenge, we proposed a heuristic algorithm based on attribute significance [1, 4]. Skowron introduced the discernibility matrix for constructing attribute reducts [6, 7], which is also decidable. Thus, we developed an algorithm attuned to the attribute frequency within a discernibility matrix [1]. On the other hand, we introduced the information-theoretic representation of knowledge and defined the rough set approximations accordingly. A few heuristic methods based on information entropy, conditional entropy, and mutual information are also proposed [1, 12]. In a recent work, we introduced a zentropy-based uncertainty measure, which was applied to compute attribute reducts [12].

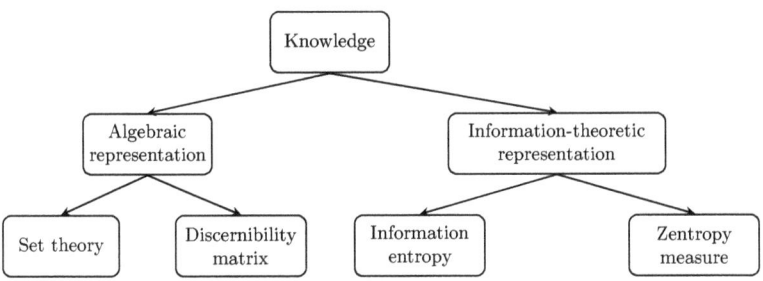

Fig. 1. Knowledge representation based on rough sets

Concept is the fundamental unit for knowledge acquisition [1, 8, 9]. The capability of handling imprecise concepts plays a vital role in artificial intelligence [9, 10]. Philosophical scholars contend that a concept is imprecise if its boundary region is nonempty [5]. Computers have a hard time in recognizing this semantic limitation. Rough set theory provides a mathematical framework for approximating an imprecise concept by a pair of lower and upper approximations, with uncertainty captured by their differences [1, 2, 11]. The positive, negative, and boundary regions are subsequently induced, associated with different knowledge applications. Inspired by rough sets and going beyond, three-way decision (3WD) is introduced by Y. Yao and has been successfully applied to various scenarios [9].

References

1. Miao, D.: Rough set theory and its applications in machine learning (Doctoral dissertation). Institute of Automation, Chinese Academy of Sciences (1997)
2. Pawlak, Z.: Rough sets. Int. J. Comput. Inf. Sci. **11**, 341–356 (1982)
3. Lingras, P., Chen, M., Miao, D.: Rough cluster quality index based on decision theory. IEEE Trans. Knowl. Data Eng. **21**, 1014–1026 (2009)
4. Miao, D., Zhao, Y., Yao, Li, H., Xu, F.: Relative reducts in consistent and inconsistent decision tables of the Pawlak rough set model. Inf. Sci. **179**, 4140–4150 (2009)
5. Ramsey, F.: Review of Ludwig Wittgenstein's "Tractatus Logico-Philosophicus". Mind **32**, 465–478 (1923)
6. Skowron, A., Cecylia, R.: The discernibility matrices and functions in information systems. In: Słowiński, R. (eds.) Intelligent Decision Support. TDLD, vol. 11, pp. 331–362. Springer, Dordrecht (1992). https://doi.org/10.1007/978-94-015-7975-9_21
7. Slezak, D., Glick, R., Betliski, P., Synak, P.: A new approximate query engine based on intelligent capture and fast transformations of granulated data summaries. J. Intell. Inf. Syst. **50**, 385–414 (2018)
8. Yao, Y.: Interval-set algebra for qualitative knowledge representation. In: Proceedings of the 5th International Conference on Computing and Information, pp. 370–375. IEEE (1993)
9. Yao, Y.: Three-way decisions with probabilistic rough sets. Inf. Sci **180**, 341–353 (2010)
10. Yao, J., Vasilakos, A., Pedrycz, W.: Granular computing: Perspectives and challenges. IEEE Trans. Cybern. **43**, 1977–1989 (2013)
11. Yuan, K., Miao, D., Yao, Y., Zhang, H., Zhao, X.: Feature selection using zentropy-based uncertainty measure. IEEE Trans. Fuzzy Syst. **32**, 2246–2260 (2024)
12. Yuan, K., Xu, W., Miao, D.: A local rough set method for feature selection by variable precision composite measure. Appl. Soft Comput. **155**, 111450 (2024)

From Deterministic to Probabilistic Rough Sets

Wojciech Ziarko

Department of Computer Science, University of Regina, Regina, Saskatchewan, S4S 0A2
Canada
ziarko@sasktel.net

The presentation reviews the evolution, based on author's and associates' research, of 'soft' models of rough sets. It starts with the fundamental deterministic model of rough sets, as introduced by Pawlak [2, 3], then is going over early models of variable precision approach to rough sets [1, 7], and ending with the review of probabilistic models [4, 5, 6, 8, 10]. In particular, the initial single control parameter variable precision model, based on the partial set inclusion relation is presented [7]. It is followed by the review of the variable precision models with symmetric and asymmetric bounds [1]. The most recent probabilistic and Bayesian models of rough sets are discussed next [4, 8, 10]. Basic ideas and some properties of the discussed models are presented and their advantages and disadvantages, commonalities and differences are reviewed. Classification tables, rough decision tables and probabilistic decision tables are introduced and used to illustrate the presented notions, and also to emphasize the practical aspects of the presented methodologies. The classification tables and decision tables acquired from data are considered the main building blocks of the rough set-based real-world applications. The optimization and analytical aspects of the rough decision tables and probabilistic decision tables are discussed in the context of classical rough set notions of attribute dependencies, such as, functional, partial functional and probabilistic. The optimization involves computation of locally minimal sets of attributes (attribute reducts), minimal rules (value reducts) and significance factors of attributes, which are also discussed [3, 9].

References

1. Katzberg, J.D., Ziarko, W.: Variable precision extension of rough sets. Fundam. Informaticae **27**, 155–168 (1996)
2. Pawlak, Z.: Rough sets. Int. J. Comput. Inf. Sci. **11**, 341–356 (1982)
3. Pawlak, Z.: Rough Sets - Theoretical Aspects of Reasoning About Data. Kluwer Academic Publishers (1991)
4. Slezak, D., Ziarko, W.: The investigation of the Bayesian rough set model. Int. J. Approximate Reasoning **40**, 81–91 (2005)
5. Yao, Y.Y.: Probabilistic rough set approximations. Int. J. Approximate Reasoning **49**, 255–271 (2008)
6. Yao, Y.Y.: Probabilistic approaches to rough sets. Expert Syst. **20**, 287–297 (2003)

7. Ziarko, W.: Variable precision rough sets model. J. Comput. Syst. Sci. **46**, 39–59 (1993)
8. Ziarko, W.: Probabilistic approach to rough sets. Int. J. Approximate Reasoning **49**, 272–284 (2008)
9. Ziarko, W.: Dependency analysis and attribute reduction in the probabilistic approach to rough sets. In: Stańczyk, U., Jain, L. (eds.) Feature Selection for Data and Pattern Recognition. SCI, vol. 584, pp. 93–111. Springer, Heidelberg (2014). https://doi.org/10.1007/978-3-662-45620-0_6
10. Ziarko, W.: Variable precision approximations of rough sets. In: Meyers, R.A., (eds.) Encyclopedia of Complexity and Systems Science, pp. 1–12. Springer, Heidelberg (2021). https://doi.org/10.1007/978-3-642-27737-5_719-1

Keynote Talks

Analysis of Healthcare Data - Explainability Using Rough Sets and LLMs

Lipika Dey

Department of Computer Science, Ashoka University, India
lipika.dey@ashoka.edu.in

The success of deep-neural networks for tasks like prediction as well as content generation, has raised expectations for their use in various application areas. The Healthcare sector, with an ever-increasing availability of data, is one of the most prolific areas focused on using these technologies for a multitude of applications. Healthcare data is available in many different forms which include structured reports as well as images and texts in the forms of test reports, nursing notes, prescriptions and so on. Healthcare analytics also make extensive use of bio-medical literature. A wide range of tasks are focused towards prediction of future events for hospital logistics management, like predicting length of stay of a patient in hospital or Intensive Care Unit, patient re-admission probability, predicting whether the patient would need a particular procedure and so on. Another line of research is focused towards personalised healthcare. These are aimed at deriving insights about individual patients or small patient cohorts, in order to deliver more targeted treatment as well as hospital experience to each patient. Predicting adverse effects of drugs for a particular patient is an important problem in the area. Designing personalised treatment plans, based on insights obtained from patient cohorts of similar patients, is also an important task.

While all the above areas have benefited with the application of deep-neural models, especially while working with unstructured data like text and images, one of the critical aspects essential for healthcare applications is explainability of the model. A significant amount of research is focused on interpretability of the models and generating post-hoc explanations, to make the models more acceptable [1, 2]. However, the problem is very complex, as applications need explainability from different perspectives, depending on the end-users.

In this talk we will be presenting the utility of model-agnostic explanation generation techniques like SHAP and LIME in explaining predictions [3, 5, 6]. We also show that using these in conjunction with multiple representations for the same data point, helps in generating alternative view-points about the same entity, which helps in interpretation and decision making. While we have worked primarily with text data in conjunction with a few physiological parameters, the core idea can be extended to include other forms of data also. Specifically, rough set concepts, which are ideally suited to interpret data that has inherent uncertainties and incompleteness, can add on to this capability [4]. Though at a very nascent stage, there are attempts to integrate rough-set concepts with deep neural networks to enhance the interpretability of the models. At the simplest level, rough sets have been applied for feature selection and dimensionality reduction before applying deep-learning models. However, the rough sets can be used at a much more complex level as techniques for approximations as well as for designing hybrid model architectures.

References

1. Allgaier, J., Mulansky, L., Draelos, R.L., Pryss, R.: How does the model make predictions? A systematic literature review on the explainability power of machine learning in healthcare. Artif. Intell. Med. **143**, 102616 (2023)
2. Bienefeld, N., et al.: Solving the explainable AI conundrum by bridging clinicians' needs and developers' goals. npj Digit. Med. **6**, 94 (2023)
3. Dey, L., Jana, S., Dasgupta, T., Gupta, T.: Deciphering clinical narratives - augmented intelligence for decision making in healthcare sector. In: Proceedings of the 18th Conference on Computer Science and Intelligence Systems (FedCSIS 2023), ACSIS, vol. 35, pp. 11–24. IEEE (2023)
4. Gaeta, A., Loia, V., Orciuoli, F.: An explainable prediction method based on fuzzy rough sets, TOPSIS and hexagons of opposition: applications to the analysis of information disorder. Inf. Sci. **659**, 120050 (2024)
5. Jana, S., Dasgupta, T., Dey, L.: Predicting medical events and ICU requirements using a multimodal multiobjective transformer network. Exp. Biol. Med. **247**, 1988–2002 (2022)
6. Krishnamoorthy, T.V., et al.: A novel NASNet model with LIME explanability for lung disease classification. Biomed. Signal Process. Control **93**, 106114 (2024)

Generative Information Retrieval: RAG and GAR

Jimmy X. Huang

Information Retrieval and Knowledge Management Research Lab, York University, Toronto, Canada

Abstract. Large language models (LLMs) are revolutionizing information retrieval (IR). Retrieval-Augmented Generation (RAG) and Generation-Augmented Retrieval (GAR) leverage LLMs to enhance IR. RAG improves response accuracy by integrating verified external information, addressing the "hallucinated" content. GAR refines search queries and enhances document representations, leading to more relevant results. The interaction between RAG and GAR creates a powerful synergy, improving both retrieval and generation processes. It is expected that rough set theory can be utilized to optimize the process in RAG and GAR via identifying the key features.

Keywords: Generative Information Retrieval · Generation-augmented Retrieval · Retrieval-augmented Generation

1 Introduction

The emergence of large language models (LLMs) has led to significant advancements in information retrieval (IR) technologies, resulting in the development of methods such as Retrieval-Augmented Generation (RAG) and Generation-Augmented Retrieval (GAR) [3]. These methods utilize the advanced generative abilities and deep semantic understanding of LLMs to increase the precision and effectiveness of information systems.

Retrieval-Augmented Generation (RAG) aims to improve the reliability of responses generated by LLMs [4]. By dynamically retrieving and integrating external information during the inference process, RAG attempts to anchor the model's responses in verified contents. This approach addresses the issue of "hallucinated" information—coherent yet factually incorrect content produced by LLMs. The success of RAG depends on the model's ability to effectively use the retrieved information, which relies on the quality and completeness of the external sources.

Conversely, Generation-Augmented Retrieval (GAR) seeks to improve search results by utilizing the generative capabilities of LLMs. GAR employs these models to extend and refine search queries or to enhance document representations [5][6], thus better aligning user queries with the document corpus. This method not only increases the relevance of search results but also broadens the range of contents accessible in response to complex queries.

Rough sets have been successfully applied for Web mining (e.g. Web usage mining and web page classification) [1][2]. We expect that rough set theory can also be utilized to optimize the process in RAG and GAR, ensuring that the retrieved contents are highly relevant and specific to the user's needs. This can be achieved by identifying the key features of the information that are most relevant to the generation task.

2 RAG and GAR

The interaction between RAG and GAR creates a powerful synergy where each component can inform and improve the other, forming a feedback loop that potentially enhances both retrieval and generation processes.

From RAG to GAR: The factual content retrieved and validated by RAG provides a reliable foundation for GAR tasks. By ensuring the accuracy of the information used to enhance queries or documents, RAG supports GAR in producing more focused and contextually relevant search enhancements. This leads to more effective and precise information retrieval, minimizing the occurrence of irrelevant results.

From GAR to RAG: In turn, the improved querying capabilities developed in GAR can substantially benefit RAG. Enhanced queries facilitate more effective and accurate information retrieval during the RAG process, ensuring that the external content integrated during generation is highly relevant and specific to the user's needs. This not only enhances the quality of the generated responses but also reduces the system's load by preventing the retrieval of unnecessary information.

The cyclical synergy between RAG and GAR presents a robust framework for leveraging the strengths of LLMs in both generating and retrieving information. Continuously refining these interactions is crucial for developing next-generation IR systems that are both intelligent and intuitive, capable of meeting the increasingly complex demands of users in the digital age. Further exploration and optimization of the dynamics between RAG and GAR are essential for achieving greater levels of accuracy and reliability in both generated content and search results.

References

1. An, A., Huang, Y., Huang, X., Cercone, N.: Feature selection with rough sets for web page classification. Trans. Rough Sets **2**, 1–13 (2004)
2. Huang, X., Cercone, N., An, A.: Comparison of interestingness functions for learning web usage patterns. In: Proceedings of 2002 ACM CIKM, pp. 617–620 (2002)
3. Huang, Y., Huang, J.X.: Exploring ChatGPT for next-generation information retrieval: opportunities and challenges. Web Intell. J. **22**, 31–44 (2024)
4. Huang, Y., Huang, J.X.: A survey on retrieval-augmented text generation for large language models. arXiv preprint arXiv:2404.10981 (2024)
5. Huang, J.X., Miao, J., He, B.: High performance query expansion using adaptive co-training. IPM **49**, 441–453 (2013)
6. Ye, Z., Huang, J.X., Lin, H.: Finding a good query-related topic for boosting pseudo-relevance feedback. JASIST **62**, 748–760 (2011)

On Optimal Approximations for Rough Sets

Ryszard Janicki

Department of Computing and Software, McMaster University, Hamilton, ON, L8S 4K1 Canada
janicki@mcmaster.ca

Classical Rough Sets [7, 8] consider only lower and upper approximations, however the concept of an approximation is not restricted only to lower and upper approximations. Consider the well known *linear least squares approximation* of points in the two dimensional plane. Here we know or assume that the points should be on a straight line and we are trying to find the line that fits the data best. However, this is not the case of an upper, or lower approximation in the sense of Rough Sets. The cases like the linear least squares approximation assume that there is a well defined concept of *similarity* (or *distance*) [1, 9], and there are some techniques for finding *maximal similarity* (*minimal distance*) between entities and their approximations.

In [4, 5] the concept of *optimal approximation* has been added to standard Rough Sets and discussed in some detail, and in [3] it was extended to the Rough Sets defined by coverings.

The word 'optimal' suggests some numerical calculations and comparisons; we need some proper definition of *similarity measure* for sets. We will discuss several such definitions, propose and justify some set of axioms for the similarity concept, as well the concept of *consistency*, that allows treating some similarities as equivalent.

An efficient greedy algorithm for finding optimal approximations for standard Rough Sets (i.e. induced by partitions) and *Marczewski-Steinhaus similarity measure* [6] (and all similarity measures consistent with it) will be discussed in detail. The algorithm is based on the properties of an index that quantifies the ratio of common to distinct elements of two given sets. We used the Marczewski-Steinhaus similarity measure as an engine of our algorithm because it has a very natural and regular definition and convenient mathematical properties. Moreover, the Marczewski-Steinhaus similarity measure is a generalization of a very popular Jaccard index [2], and it is also consistent (i.e. practically equivalent) to many other popular similarity measures.

In [3] a fairly natural transition from Rough Sets induced by coverings to standard Rough Sets (i.e. induced by partitions) have been proposed and discussed. We will use this transformation to extend our algorithm for Rough Sets induced by coverings.

References

1. Deza, M.M., Deza, E.: Encyclopedia of Distances. Springer, Berlin (2012). https://doi.org/10.1007/978-3-642-30958-8
2. Jaccard, P.: Étude comparative de la distribution florale dans une portion des Alpes et des Jura. Bull. Soc. Vaudoise Sci. Naturalles **37**, 547–549 (1901)

3. Janicki, R.: Yet another kind of rough sets induced by coverings. In: Polkowski, L., et al. (eds.) Rough Sets (IJCRS 2017). LNCS, vol. 10313, pp. 140–153. Springer, Cham (2017). https://doi.org/10.1007/978-3-319-60837-2_12
4. Janicki, R., Lenarčič, A.: Optimal approximations with rough sets. In: Lingras, P., et al. (eds.) Rough Sets and Knowledge Technology. RSKT 2013. LNCS, vol. 8171, pp. 87–98. Springer, Heidelberg (2013). https://doi.org/10.1007/978-3-642-41299-8_9
5. Janicki, R., Lenarčič, A.: Optimal approximations with rough sets and similarities in measure spaces. Int. J. Approximate Reasoning **71**, 1–14 (2016)
6. Marczewski, E., Steinhaus, H.: On a certain distance of sets and corresponding distance of functions. Colloquium Mathematicum **4**, 319–327 (1958)
7. Pawlak, Z.: Rought sets. Int. J. Comput. Inf. Sci. **34**, 557–590 (1982)
8. Pawlak, Z.: Rough Sets. Kluwer, Dordrecht (1991)
9. Tversky, A.: Features of similarity. Psychol. Rev. **84**(4), 327–352 (1977)

Self-organizing Cyber-physical Systems Development with Rough Sets

Eric T. Matson

Department of Computer and Information Technology, Purdue University, 401 North Grant Street, West Lafayette, IN, 47907 USA
ematson@purdue.edu

The future in the enhancement of cyber-physical system and robotic functionalities lies not only in the mechanical and electronic improvement of the robots' sensors, mobility, stability and kinematics., but also, if not mostly, in their ability to connect to other actors (human, agents, robots, machines, and sensors HARMS). The capability to communicate openly, to coordinate their goals, to optimize the division of labor, to share their intelligence, to be fully aware of the entire situation, and thus to optimize their fully coordinated actions will be necessary. Additionally, the ability for two actors to work together without preference for any specific type of actor, but simple from necessity of capability, is provided by a requirement of indistiguishability, similar to the discernment feature of rough sets.

Once all of these actors can effectively communicate, they can take on group rational decision making, such as choosing which action to take that optimizes a group's effectiveness or utility. Given group decision making, optimized capability-based organization can take place to enable human-like organizational behavior. Similar to human organizations, artificial collections with the capability to organize will exhibit emergent normative behavior. The complexity of this type of system does not necessarily fit into some of the logical conventions of computing. Modeling the vagueness of the real world task environments in which these systems operates calls for an extension of classical theory, especially in the area of indiscernibility and approximation which must occur in real time decision making.

Rough Set Theory and Concept Lattices to Solve Fuzzy Relation Equations

Jesús Medina

Department of Mathematics, University of Cádiz, Spain
jesus.medina@uca.es
https://www.mcis.uca.es

Property-oriented concept lattices and object-oriented concept lattices [5, 11, 14] arose as a generalization of rough set theory (RST) considering the philosophy of formal concept analysis (FCA) [13]. These frameworks were introduced in the multi-adjoint fuzzy setting [6] in the paper [17]. Specifically, the lower and upper approximation operators are generalized to be applied to a quantitative relationship between the objects and attributes (informative table) instead of its discernibility relation [8, 17]. The new (modal) operators are called necessity and possibility operators, respectively.

Fuzzy relation equations (FRE) are associated with the composition of fuzzy relations and is an appropriate tool for handling and modeling non-probabilistic forms of uncertainty. From their introduction in the eighties by Elie Sanchez [19], FRE have received great interest from their introduction due to the wide range of their applications in different areas of real-world, such as decision making [7], bipolarity [12], optimization [2] and image processing [1].

A relevant achievement was to characterize the solvability of FRE with the computation of concepts in a property-oriented concept lattice or an object-oriented concept lattice [9, 10]. Given a FRE, depending on whether the unknown relation is to the left side or to the right side, we have determined an associated property-oriented context or an object-oriented context, respectively, whose concepts provide solvable FREs related to the original one [9].

This relationship has allowed diverse advances in FRE. For example, it has been possible to introduce three different mechanisms of approximating unsolvable FRE. Two of them are based on the computation of closest concepts associated with a given fuzzy subset of attributes [7] and the third one is based on the theory of attribute reduction [3, 4, 18] developed in the mentioned concept lattices frameworks [16, 15]. This last theory has also allow the reduction of equations of the system associated with a given FRE [15], which is a novel procedure that had not been envisaged until now.

This contribution will be focused on these relations and achievements, and will present new challenges.

Partially supported by the projects PID2019-108991GB-I00 and PID2022-137620NB-I00/AEI/10.13039/501100011033/ FEDER, UE, and with grant TED2021-129748B-I00 funded by MCIN/AEI/10.13039/501100011033 and, as appropriate, by "ERDF A way of making Europe", by the "European Union" or by the "European Union NextGenerationEU/PRTR".

References

1. Alcalde, C., Burusco, A., Díaz-Moreno, J.C., Medina, J.: Fuzzy concept lattices and fuzzy relation equations in the retrieval processing of images and signals. Int. J. Uncertainty Fuzziness Knowl.-Based Syst. **25**, 99–120 (2017)
2. Aliannezhadi, S., Abbasi Molai, A.: A new algorithm for geometric optimization with a single-term exponent constrained by bipolar fuzzy relation equations. Iran. J. Fuzzy Syst. **18**(1), 137–150, (2021)
3. Antoni, L., Cornejo, M.E., Medina, J., Ramirez, E.: Attribute classification and reduct computation in multi-adjoint concept lattices. IEEE Trans. Fuzzy Syst. **29**, 1121–1132 (2021)
4. Benítez-Caballero, M.J., Medina, J., Ramírez-Poussa, E., Ślęzak, D.: Rough-set-driven approach for attribute reduction in fuzzy formal concept analysis. Fuzzy Sets Syst. **391**, 117–138 (2020)
5. Chen, Y., Yao, Y.: A multiview approach for intelligent data analysis based on data operators. Inf. Sci. **178**(1), 1–20 (2008)
6. Cornejo, M.E., Medina, J.: Impact Zadeh's theory to algebraic structures. multi-adjoint algebras. J. Pure Appl. Math. **12**, 126–141 (2021)
7. Cornejo, M.E., Medina, J., Ramírez-Poussa, E.: Attribute and size reduction mechanisms in multi-adjoint concept lattices. J. Comput. Appl. Math. **318**, 388–402 (2017)
8. Cornelis, C., Medina, J., Verbiest, N.: Multi-adjoint fuzzy rough sets: definition, properties and attribute selection. Int. J. Approximate Reasoning **55**, 412–426 (2014)
9. Díaz-Moreno, J.C., Medina, J.L Multi-adjoint relation equations: definition, properties and solutions using concept lattices. Inf. Sci. **253**, 100–109 (2013)
10. Díaz-Moreno, J.C., Medina, J.: Using concept lattice theory to obtain the set of solutions of multi-adjoint relation equations. Inf. Sci. **266**(0), 218–225 (2014)
11. Düntsch, I., Gediga, G.: Approximation operators in qualitative data analysis. In: Theory and Applications of Relational Structures as Knowledge Instruments, pp. 214–230 (2003)
12. Freson, S., De Baets, B., De Meyer, H.: Linear optimization with bipolar max-min constraints. Inf. Sci. **234**, 3–15 (2013). Fuzzy Relation Equations: New Trends and Applications
13. Ganter, B., Wille, R.: Formal Concept Analysis: Mathematical Foundation. Springer, Heidelberg, (1999). https://doi.org/10.1007/978-3-642-59830-2
14. Gediga, G., Düntsch, I.: Modal-style operators in qualitative data analysis. In: Proceedings of the IEEE Interenational Conference on Data Mining, pp. 155–162 (2002)
15. Lobo, D., López-Marchante, V., Medina, J.: Reducing fuzzy relation equations via concept lattices. Fuzzy Sets Syst. **463**, 108465 (2023)
16. Lobo, D., López-Marchante, V., Medina, J.: On measuring the solvability of a fuzzy relation equation. Stud. Comput. Intell. **1127**, 9–15 (2024)
17. Medina, J.: Multi-adjoint property-oriented and object-oriented concept lattices. Inf. Sci. **190**, 95–106 (2012)
18. Medina, J.: Relating attribute reduction in formal, object-oriented and property-oriented concept lattices. Comput. Math. Appl. **64**(6), 1992–2002 (2012)
19. Sanchez, E.: Resolution of composite fuzzy relation equations. Inf. Control **30**(1), 38–48, (1976)

Complex Collective Systems - Examples of Intelligent Particles Interactions

Jarosław Wąs

Department of Applied Computer Science, Faculty of Electrical Engineering, Automatics, IT and Biomedical Engineering, AGH University of Krakow, Mickiewicza 30, 30-059 Krakow, Poland

Keywords: Complex Collective Systems, Agent Based Modeling, Cellular Automata

1 Introduction

Complex systems are networks created using different components, that interact with each other in a non-linear fashion [7].

In complex systems representing living organisms (pedestrians, skiers, drivers - who are interpreted as intelligent agents [1]) we can observe various interesting collective behaviors. The complexity of their behavior and non-trivial forms of self-organization led to their being called intelligent particles, and this class of systems can be included in the category of Complex Collective Systems [9].

Two concepts are key from the point of view of describing the collective properties of complex systems: emergence in scale domain on the one hand and selforganization in time domain on the other [9]. Emergence means the spontaneous formation of qualitatively new forms and behaviors visible at the macroscopic level, from interactions between simpler elements at the microscopic level. Selforganization means that the elements of a complex system spontaneously create organized spatial structures.

2 Examples

There are many models of crowds (pedestrian) dynamics and behavior. The models include microscopic models in which the pedestrian is presented as a state of a cellular automaton with simple or more complex movement rules[2], as well as a particle whose movement is determined on the basis of the superposition of forces. Regardless of the model used, the applied rules of human movement at the microscopic level create complex macroscopic patterns. This approach is used, for example, in large-scale crowd models [10] or in data driven modelling approaches [6].

Downhill skiers represent a different type of movement. As this sport becomes popular, the number of skiers increases and safety issues become crucial [4]. Therefore, models are created which are most often based on the principle of superposition of forces. These models reflect movement and behavior patterns on the ski slope [3] and make it possible to create advanced safety analyses [4].

Vehicle simulations are of a completely different nature than the previous examples and are of great practical importance. Simple traffic models should be identified, starting from the classic Nagel-Schreckenberg model, through complex implementations in urban environments [5], to the movement of autonomous vehicles [8].

References

1. Abar, S., Theodoropoulos, G.K., Lemarinier, P., O'Hare, G.M.: Agent based modelling and simulation tools: a review of the state-of-art software. Comput. Sci. Rev. **24**, 13–33 (2017). https://doi.org/10.1016/j.cosrev.2017.03.001, https://www.sciencedirect.com/science/article/pii/S1574013716301198
2. Burstedde, C., Kirchner, A., Klauck, K., Schadschneider, A., Zittartz, J.: Cellular automaton approach to pedestrian dynamics – applications. In: Pedestrian and Evacuation Dynamics, pp. 87–98. Springer, New York (2002)
3. Holleczek, T., Tröster, G.: Particle-based model for skiing traffic. Phys. Rev. **E 85**, 056101 (2012). https://doi.org/10.1103/PhysRevE.85.056101, https://link.aps.org/doi/10.1103/PhysRevE.85.056101
4. Korecki, T., Pałka, D., Wąs, J.: Adaptation of social force model for simulation of downhill skiing. J. Comput. Sci. **16**, 29–42 (2016). https://doi.org/10.1016/j.jocs.2016.02.006, https://www.sciencedirect.com/science/article/pii/S1877750316300138
5. Małecki, K., Kaminski, M., Wąs, J.: A multi-cell cellular automata model of traffic flow with emergency vehicles: effect of a corridor of life and drivers' behaviour. J. Comput. Sci. **61**, 101628 (2022). https://doi.org/10.1016/j.jocs.2022.101628, https://www.sciencedirect.com/science/article/pii/S1877750322000515
6. Porzycki, J., Wąs, J.: Modeling spatial patterns in a moving crowd of people using data-driven approach—a concept of interplay floor field. Saf. Sci. **167**, 106266 (2023). https://doi.org/10.1016/j.ssci.2023.106266, https://www.sciencedirect.com/science/article/pii/S0925753523002084
7. Sayama, H.: Introduction to the Modeling and Analysis of Complex Systems. Milne Open Textbooks (2015), https://knightscholar.geneseo.edu/oer-ost/14
8. Szlachetka, M., Borkowski, D., Wąs, J.: The downselection of measurements used for free space determination in ADAS. J. Comput. Sci. **63**, 101762 (2022). https://doi.org/10.1016/j.jocs.2022.101762, https://www.sciencedirect.com/science/article/pii/S1877750322001454
9. Wąs, J.: Modeling and Simulation of Complex Collective Systems. CRC Press, Boca Raton (2023)
10. Wąs, J., Lubas, R.: Towards realistic and effective agent-based models of crowd dynamics. Neurocomputing **146**, 199–209 (2014). https://doi.org/10.1016/j.neucom.2014.04.057, https://www.sciencedirect.com/science/article/pii/S0925231214007838. bridging Machine learning and Evolutionary Computation (BMLEC) Computational Collective Intelligence

Contents – Part I

Rough Set Models and Foundations

Mapper-Based Rough Sets 3
    Mauricio Restrepo and Chris Cornelis

On Tolerance-Based Rough Set Operators and Their Covering
Generalizations 18
    Piotr Wasilewski and Dominik Ślęzak

Parametrized γ-Decision Valuation for Variable Precision Rough Set Model . . . 34
    Soma Dutta and Dominik Ślęzak

Rough Algebraic Semantics of Concepts in a Distributed Cognition
Perspective 50
    A. Mani

Description Logic for Rough Concepts 67
    *Krishna B. Manoorkar, Andrea De Domenico,
    and Alessandra Palmigiano*

Rule Induction and Machine Learning

Greedy Algorithm for Construction of Deterministic Decision Trees
for Conventional Decision Tables from Closed Classes 93
    Azimkhon Ostonov and Mikhail Moshkov

Study of Dependency Degree and Bayesian Networks for Conflict Scenarios . . . 105
    Małgorzata Przybyła-Kasperek and Rafał Deja

Consideration of Detecting Data and Functional Dependency in Tabular
Data with Missing Values by the Obtained Rules 120
    Hiroshi Sakai, Michinori Nakata, Dominik Ślęzak, and Junzo Watada

Distance-Based Fuzzy-Rough Sets and Their Application
to the Classification Problem 134
    Amrit Kumar and Niladri Chatterjee

Dealing with Missing Values Meaning Unknown in Probabilistic
Approximations 157
    Michinori Nakata, Hiroshi Sakai, and Takeshi Fujiwara

On Complexity of Deterministic and Nondeterministic Decision Trees
for Decision Tables with Many-Valued Decisions from Closed Classes 173
    Azimkhon Ostonov and Mikhail Moshkov

Simulating Functioning of Decision Trees for Tasks on Decision Rule
Systems 188
    Kerven Durdymyradov and Mikhail Moshkov

RIONIDA: A Novel Algorithm for Imbalanced Data Combining
Instance-Based Learning and Rule Induction 201
    Grzegorz Góra and Andrzej Skowron

Granular Computing

Information System in the Light of Interactive Granular Computing 223
    Soma Dutta and Andrzej Skowron

GBTWSVM: Granular-Ball Twin Support Vector Machine 238
    Lixi Zhao, Zhifei Zhang, Wenjun Liu, and Guangming Lang

Fuzzy Granular-Balls Based Spectral Clustering 252
    Yueyang Li, Siheng Chen, and Guangming Lang

A Vector Is a Granule: A Novel Extension of the Variable Precision Rough
Set Model 266
    Hajime Okawa, Yasuo Kudo, and Tetsuya Murai

Rough Set Applications

Cross-Weighting Knowledge Distillation for Object Detection 285
    Zhaoyi Li, Zihao Li, and Xiaodong Yue

A Method of Multi-USV Reward Design Using Fuzzy Control 300
    Jianfeng Xiao, Qun Liu, and Xin Huang

Hyp-DAN: Hyperbolic Distance-Aware Attention Networks 314
    Fuchuan Xiang, Jianhang Tang, Shaobo Li, Guoyin Wang, and Ji Xu

Optimizing Rough Set Flow Graph Inference 329
    Jun Wang and Cory J. Butz

Multimodal Propaganda Detection in Memes with Tolerance-Based Soft Computing Method 343
Siddharth Kelkar, Srinivasa Ravi, Sheela Ramanna, and Anand Kumar Madasamy

Author Index 353

Contents – Part II

Three-Way Decision and Rough Sets

The Visual Analysis of Three-Way Decision Based on Decision-Theoretic Rough Set: A Perspective of Fusing Two-Way Decision Pair 3
   Jing Tu, Hong Rao, Jianfeng Xu, Duoqian Miao, and Yuanjian Zhang

Triangular Fuzzy Number Intuitionistic Fuzzy Covering Rough Sets and Applications to Decision Making 14
   Zhongling Bai, Jiang Chen, and Xianyong Zhang

Three-Way Decision of Granular-Ball Rough Sets Based on Fuzziness 29
   Zhuangzhuang Liu, Taihua Xu, Jie Yang, and Shuyin Xia

Three-Way Decision in Data Analytics

Three-Way Cost-Performance Approximate Attribute Reduction 47
   Jialin Hou and Yiyu Yao

Granular Approximations of Partially-Known Concepts 59
   Qiaoyi Li, Chengjun Shi, Han Yang, and Yiyu Yao

Robust Online Satellite Video Object Tracking with Self-adoption Uncertainty 74
   Ziye Wang and Duoqian Miao

Three-Way Hybrid Sampling Using Granular Balls for Imbalanced Classification 86
   Qin Xie, Qinghua Zhang, Nanfang Luo, and Guoyin Wang

Identifying Important Concepts in the Concept Lattice Based on Concept Indices 103
   Kuo Pang, Zhen Wang, Li Zou, and Mingyu Lu

Result-Fusion-Based Temporal-Spatial Composite Sequential Three-Way Decisions 118
   Yi Xu, QiSheng Zhu, ZhengYue Pan, ZiHeng Qiu, and XiaoJun Sun

Three-Way Decision in Broad Senses

A Machine-People-Government Triangular Model of Smart Agriculture 135
   Chuanlei Zhang and Yiyu Yao

Three-Way Bibliometrics Analytics for Supporting Literature Review 149
   Langwangqing Suo, Hai-Long Yang, and Yiyu Yao

Approximate Criterion Reduction in Multi-criteria Trilevel Ranking
Analysis 165
   Chengjun Shi, Mengjun Hu, Qiaoyi Li, and Yiyu Yao

New Models of Three-Way Conflict Analysis Based on Decision-Theoretic
Rough Sets 181
   Ping Liu, Qimei Xiao, Huiying Yu, and Guangming Lang

Three-Way Conflict Analysis with Negative Feedback 196
   Yucong Yan and Xiaonan Li

Rental Market Data Mining

Estate 360: AI-Driven Centralized Real Estate Platform 213
   *Neeyati Mehta, Sameer Patel, Zaid Shaikh, Jerry Caleb,
   and Kritika Koirala*

Housing Rental Information Management and Prediction System Based
on CatBoost Algorithm - a Case Study of Halifax Region 230
   Shuangrun Shao, Bingxi Zhao, Xiangen Cui, Yihong Dai, and Beining Bao

A Novel Approach to Rental Market Analysis for Property Management
Firms Using Large Language Models and Machine Learning 247
   Raoof Naushad, Rakshit Gupta, Tejasvi Bhutiyal, and Vrushali Prajapati

Applications of Deep Learning and Soft Computing

Clinical Medical Test Decision-Making of Liver Disease Using
Granular-Ball Rough Set 265
   Fanxin Xu, Zuqiang Su, and Guoyin Wang

Lung Cancer Risk Prediction Model Trained with Multi-source Data 280
   Shijie Sun, Hanyue Liu, Ye Wang, and Hong Yu

Advancing ITS Applications with LLMs: A Survey on Traffic
Management, Transportation Safety, and Autonomous Driving 295
   Dingkai Zhang, Huanran Zheng, Wenjing Yue, and Xiaoling Wang

Automated Brain Tumor Classification with Deep Learning 310
   Venkata Sai Krishna Chaitanya Kandula and Yan Zhang

Author Index 325

Rough Set Models and Foundations

Mapper-Based Rough Sets

Mauricio Restrepo[1](✉) and Chris Cornelis[2]

[1] Department of Mathematics, Universidad Militar Nueva Granada,
Bogotá, Colombia
mauricio.restrepo@unimilitar.edu.co
[2] Department of Applied Mathematics, Computer Science and Statistics,
University of Ghent, Ghent, Belgium
Chris.Cornelis@UGent.be
https://www.unimilitar.edu.co, https://www.cwi.UGent.be

Abstract. This paper presents a new approach to analyzing numerical data sets using covering-based rough sets based on the Mapper algorithm, a fundamental tool of topological data analysis (TDA). Specifically, by varying the parameters of Mapper, our approach generates different coverings from a numerical dataset that can be used to define lower approximations, and the associated quality of classification, for covering-based rough sets. We discuss the fundamental ideas of how to integrate both theories, and explore possible lines of further work.

Keywords: topological data analysis · mapper algorithm · covering-based rough sets · approximation operators

1 Introduction

Topological data analysis (TDA) attempts to extract some geometric properties present in numerical datasets [2,6,10]. Concretely, algebraic topology is used to identify certain geometric regularities in the data. For example, using homology groups, it is possible to identify basic elements such as connected components, loops, voids, and n-dimensional voids.

From the point of view of topology, a cloud of points in an n-dimensional space is a totally disconnected set. If we use a notion of distance or similarity, it is possible to connect points, or clusters of points, to construct simplicial complexes. The Mapper algorithm, introduced by Singh, Mémoli and Carlsson [24], is a computational method for extracting descriptions of high-dimensional datasets from simplicial complexes. The Mapper algorithm has been used to reduce and capture topological and geometric information present in data.

Using this algorithm, relevant results have been obtained in various fields such as medicine [1,11,16,17,23], economics [7,8,15], sports [14] and chemical engineering [25]. For example, in [13] the Mapper algorithm was applied to determine a new subtype of type 2 diabetes.

On the other hand, rough sets were introduced by Pawlak [19,20] to provide lower and upper approximations of concepts in the presence of inconsistent information. At the heart of the theory is an equivalence relation that partitions the available data into sets of mutually indiscernible instances.

Several generalizations and variations of classical rough set theory were proposed to extend its range of possible applications. In this paper, we focus on covering-based rough sets (CBRSs), where the aforementioned partition of the data space is replaced by a more general covering [4,21,22,27,28,30,31]. Such a generalization is useful, for example, to deal with numerical and missing data.

Most of the existing work on CBRSs has been dedicated to the study of approximation operators and their algebraic properties, but little attention has been paid so far to practical methods for obtaining coverings of data. Therefore, this paper proposes the integration of CBRSs and the Mapper algorithm in such a way that it generates not only one, but a series of coverings. The suitability of these coverings is then evaluated using the quality of classification from rough sets on the one hand, and the number of elements in the covering on the other hand.

The remainder of the paper is organized as follows: Sect. 2 presents preliminary concepts regarding rough sets, covering-based rough sets, the definition of the particular lower approximation that we will use and the associated definition of the quality of classification. Section 3 recalls the Mapper algorithms and explains how to apply it to obtain different coverings from a numerical dataset. Section 4 analyzes the main effects of the different parameters of the Mapper algorithm on the type of coverings obtained on a number of popular benchmark data sets. Finally, in Sect. 5 we summarize the main conclusions of the paper and outline future work.

2 Preliminaries

Classical rough set analysis aims to analyze a decision system $(U, A \cup \{d\})$, consisting of a set U of instances, characterized by means of a set of conditional attributes A and a decision attribute d, all of which are assumed to take categorical (nominal) values. For any subset B of A, we may construct an equivalence relation R_B on U, to which a couple (u, v) of instances belongs if they share the same values for all attributes in B. Likewise, the decision attribute also induces a partition of the data into a finite number of decision classes.

The lower approximation of each decision class D is then defined as the union of equivalence classes fully contained in D, while the quality of classification [9] associated with R_B is computed as the fraction of elements in U that belong to the lower approximation of any decision class. In other words, the quality of classification expresses the percentage of instances for which we can predict the decision class unequivocally based on the attributes in B.

Later on, the idea of partition obtained from the equivalence relation was generalized to coverings, allowing instances to be connected to each other in different ways, not only based on the equality of attribute values.

Definition 1. *[31] Let $\mathbb{C} = \{K_i\}$ be a family of nonempty subsets of U. $\mathbb{C}$ is called a covering of U if $\bigcup K_i = U$.*

The generalization of partitions to coverings raises the question of how to define the rough set approximations in this case. A systematic investigation of covering-based rough sets was conducted in [27], while in [21,22] the order relations between different approximation operators was studied. While it is possible to use any kind of covering-based lower approximation, in this paper we consider the common granule-based definition:

$$\underline{apr}'_{\mathbb{C}}(A) = \bigcup\{K \in \mathbb{C} : K \subseteq A\} \quad (1)$$

The quality of classification for a covering $\mathbb{C}$ on U is defined as follows:

$$\gamma_{\mathbb{C}} = \frac{1}{|U|} \sum_{D} |\underline{apr}'_{\mathbb{C}}(D)|. \quad (2)$$

where D belongs to the partition obtained from the decision attribute, i.e., D is a decision class. Note that this definition matches the quality of classification in rough set theory when $\mathbb{C}$ is a partition.

3 Generating Coverings with the Mapper Algorithm

In this section we show how to obtain coverings from a dataset using the Mapper algorithm.

3.1 The Mapper Algorithm

A difficult and recurring question that arises when dealing with high-dimensional data is how to properly visualize them, in order to expose their inherent characteristics. Typically, such a visualization requires mapping the original data to a lower-dimensional structure, as is done for example in principal component analysis (PCA). Recently, various attempts have been made to use persistent homology in data visualization [5], resulting in the Mapper algorithm.

By means of this algorithm, it is possible to obtain a graph containing relevant information about a dataset. A particularity of this graph is that it depends on several parameters, which must be adjusted in such a way that the representation reveals the desired characteristics of the data.

In particular, let X be a numerical dataset in which each sample is described by m attributes or numerical values. The set X is thus represented by a point cloud in $\mathbb{R}^m$. The Mapper algorithm, schematically described by Algorithm 1 (adapted from [2]), transforms such a cloud of points into a graph.

An important parameter of Mapper is the lens function (also called projection function). Some frequently used lens functions include: projections on individual attributes, projections on principal components, and other functions based on distances or norms like L_2, density, centrality and eccentricity [14,18]. The lens function transforms the original dataset X to the lower-dimensional space $f(X)$, on which a covering $\{I_j\}$ is defined. In case $f(X) \subseteq \mathbb{R}$ (one-dimensional lens

---
Algorithm 1: The Mapper algorithm

Input: a dataset X with an associated metric, a *lens* function $f : X \to \mathbb{R}$ (or $\mathbb{R}^d$), a covering $\{I_j \mid j = 1, \ldots n\}$ of $f(X)$, and a clustering algorithm *clusters* on X.
for $j \in \{1, 2, \ldots, n\}$ **do**
$\quad A_j = \{A_{j1}, \ldots, A_{jk} \mid i = 1, \ldots k\} \leftarrow clusters(f^{-1}(I_j));$
end
Result: a graph that includes the following;
- a vertex v_{ji} for each cluster A_{ji} and;
- an edge between v_{ji} and v_{lk} if $A_{ji} \cap A_{lk} \neq \emptyset$.
---

function), the covering typically consists of a number of overlapping intervals of $f(X)$.

Subsequently, for each I_j inside the covering, the instances from X projected onto it by f (i.e., $f^{-1}(I_j)$) are partioned by means of a clustering algorithm, another parameter of the algorithm. Commonly, k-means clustering is applied.

Finally, a graph is created where each unique cluster A_{ji} of $f^{-1}(I_j)$ becomes a node of the graph. If A_{ji} and A_{lk} are two clusters (nodes) such that $A_{ji} \cap A_{lk} \neq \emptyset$, we create an edge between nodes. The data structure used to represent these graphs corresponds to a simplicial complex that has information on connected components (homology of dimension 0) and cycles (homology of dimension 1). This is why Mapper has been widely used as a method of data visualization.

An important byproduct of Mapper's operation is that it also generates a covering of the instances in X, defined as

$$\mathbb{C} = \{A_{ji} \mid A_{ji} \text{ is a cluster obtained by Mapper}\}$$

In this paper, we propose to use this covering to apply the concepts of lower approximation and quality of classification. To illustrate our approach, we will use the Python implementation[1] of the Mapper algorithm provided by Van Veen et al. [26].

3.2 A Simple Example

In order to understand how coverings are obtained from the Mapper algorithm, let us consider the dataset shown in Table 1. The cloud point is the set

$$X = \{x_0, x_1, \ldots, x_9\},$$

where each x_i is a point in $\mathbb{R}^6$.

As our lens function $f : X \to \mathbb{R}$, we will use the first projection π_1, so the projected values are $\pi_1(X) = \{1.0, 1.5, 2.0, 3.0\}$ and these values belong to the interval $[1, 3]$. First of all, this interval is scaled to $I = [0, 1]$ and, according to this new scale, the values of $\pi_1(x_i)$ become $\{0, 0.25, 0.5, 1\}$.

---
[1] https://kepler-mapper.scikit-tda.org/en/latest/index.html.

Table 1. A simple dataset with numerical data.

Object	A1	A2	A3	A4	A5	A6	Class
x_0	3.0	1	0	0	2	0.0	0
x_1	2.0	0	0	1	0	1.0	1
x_2	1.5	1	1	1	1	1.0	0
x_3	1.0	2	2	0	2	1.0	1
x_4	1.0	2	0	1	0	0.0	0
x_5	3.0	0	1	0	1	0.5	1
x_6	3.0	1	1	0	1	0.5	0
x_7	2.0	1	2	1	0	1.0	1
x_8	2.0	0	2	0	2	1.0	1
x_9	1.5	1	0	1	0	0.0	0

Next, we introduce a covering of the interval I with five closed intervals with respective centers:
$$\{0.1, 0.3, 0.5, 0.7, 0.9\}$$
Moreover, suppose that the overlap percentage among intervals equals 50%. Therefore, the five closed intervals become: $I_1 = [-0.1, 0.3]$, $I_2 = [0.1, 0.5]$, $I_3 = [0.3, 0.7]$, $I_4 = [0.5, 0.9]$, $I_5 = [0.7, 1.1]$. These intervals are shown in Fig. 1.

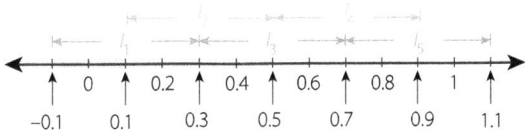

Fig. 1. Covering of $I = [0, 1]$ with five closed intervals and an overlapping of 50 %.

From the values in Table 1, we can see that $\pi_1^{-1}(I_1) = \{x_2, x_3, x_4, x_9\}$, because $\pi_1(x_2) = \pi_1(x_3) = \pi_1(x_4) = \pi_1(x_9) = 0 \in I_1$. Applying the k-means clustering algorithm with $k = 2$, we obtain:
$$clusters(\pi_1^{-1}(I_1)) = \{\{x_2, x_3\}, \{x_4, x_9\}\} \qquad (3)$$
Continuing in the same way for the remaining intervals I_2 through I_5, we obtain the clusters in Table 2.

The collection of clusters obtained by this process constitutes a covering of the set of instances. Removing the repeated clusters, we obtain the final covering:
$$\mathbb{C} = \{\{x_2, x_3\}, \{x_4, x_9\}, \{x_1, x_2, x_7, x_9\}, \{x_8\}, \{x_1, x_7\}, \{x_0\}, \{x_5, x_6\}\}$$
According to the decision attribute, we have a partition of X into two classes: $D_1 = \{x_0, x_2, x_4, x_6, x_9\}$ and $D_2 = \{x_1, x_3, x_5, x_7, x_8\}$. Using Equation (1), we

Table 2. Clusters of each interval I_j.

Interval I_j	$f^{-1}(I_j)$	Clusters of $f^{-1}(I_j)$
$I_1 = [-0.1, 0.3]$	$\{x_2, x_3, x_4, x_9\}$	$\{x_2, x_3\}, \{x_4, x_9\}$
$I_2 = [0.1, 0.5]$	$\{x_1, x_2, x_7, x_8, x_9\}$	$\{x_1, x_2, x_7, x_9\}, \{x_8\}$
$I_3 = [0.3, 0.7]$	$\{x_1, x_7, x_8\}$	$\{x_1, x_7\}, \{x_8\}$
$I_4 = [0.5, 0.9]$	$\{x_1, x_7, x_8\}$	$\{x_1, x_7\}, \{x_8\}$
$I_5 = [0.7, 1.1]$	$\{x_0, x_5, x_6\}$	$\{x_0\}, \{x_5, x_6\}$

have $\underline{apr}'_\mathbb{C}(D_1) = \{x_0, x_4, x_9\}$ and $\underline{apr}'_\mathbb{C}(D_2) = \{x_1, x_7, x_8\}$. Therefore, the quality of classification of X for the covering $\mathbb{C}$ equals $\frac{3+3}{10} = 0.6$.

Finally, Fig. 2 shows the graph obtained with Mapper. Its nodes correspond to the elements of the covering, before deleting the repeated sets.

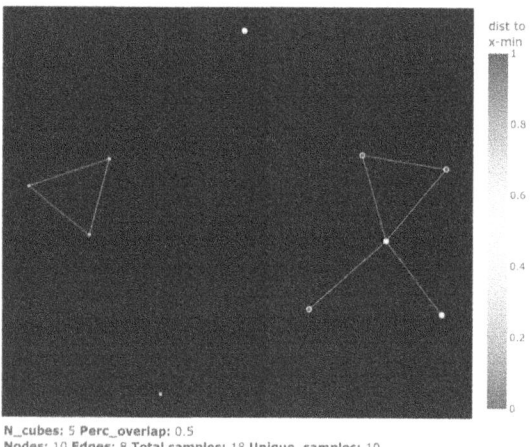

Fig. 2. Graph for data in Example 1.

The color scale shown on the right of the graph in Fig. 2 is constructed according to class membership. For the case of data in Table 1, it ranges from green (class = 0) to red (class = 1). If a node contains elements of only class 0 (pure node) it is green. Similarly, a node with only elements of class 1 will be colored red. Nodes with a different color contain elements of both classes.

By using different overlap percentage values, it is possible to obtain different coverings. For example, if we use the set of values for the overlap percentage p:

$$P = \{0.2, 0.4, 0.6, 0.8, 0.9\}$$

for the dataset in Table 1, we obtain the coverings shown in Table 3.

Table 3. Coverings for dataset in Table 1 obtained using the Mapper algorithm with different values of p.

Overlap	Covering: $\mathbb{C}$
0.2	$\mathbb{C} = \{\{x_4\}, \{x_3\}, \{x_9\}, \{x_2\}, \{x_8\}, \{x_1, x_7\}, \{x_5, x_6\}, \{x_0\}\}$
0.4	$\mathbb{C} = \{\{x_2, x_3\}, \{x_4, x_9\}, \{x_8\}, \{x_1, x_2, x_7, x_9\}, \{x_1, x_7\}, \{x_5, x_6\}, \{x_0\}\}$
0.6	$\mathbb{C} = \{\{x_2, x_3\}, \{x_4, x_9\}, \{x_9\}, \{x_2\}, \{x_8\}, \{x_1, x_7\}, \{x_5, x_6\}, \{x_0\}\}$
0.8	$\mathbb{C} = \{\{x_1, x_2, x_4, x_7, x_9\}, \{\{x_3, x_8\}\{x_0, x_3, x_5, x_6, x_8\}, \{x_0, x_5, x_6, x_8\},$ $\{x_1, x_7\}, \{x_1, x_2, x_7, x_9\}\}$
0.9	$\mathbb{C} = \{\{x_1, x_2, x_4, x_7, x_9\}, \{x_0, x_3, x_5, x_6, x_8\}\}$

The corresponding values of the quality of classification for the dataset in Table 1 are shown in Table 4. Note that the highest value 0.8 is obtained when the overlap is 0.2, but the associated covering also has the highest number of elements. Therefore, in general, a trade-off needs to be sought between optimizing the quality of classification and the granularity of the covering.

Table 4. Values for quality of classification in Table 1 with different values of p.

Overlap %	Quality of classification
0.2	0.8
0.4	0.7
0.6	0.6
0.8	0.2
0.9	0

3.3 Effect of the Lens Function

As mentioned, an important parameter in the Mapper algorithm is the lens function. Let us consider the dataset in Fig. 3 that consists of one hundred points belonging to two different classes. As two different scalar functions f, we use the projections $\pi_1(x, y) = x$ and $\pi_2(x, y) = y$. Furthermore, we construct a covering of $\pi_i(X)$ determined by ten closed intervals with an overlap percentage of 50%. Clustering is performed using the classical k-means clustering, with $k = 2$.

For each function π_i, the output of the algorithm is a graph in which each node represents a set of points obtained through the clustering process, and the edges represent a connection between nodes containing points in common, as shown in Fig. 3. Neither is better than the other; they are simply two different representations of the data (Fig. 4).

4 Application on Benchmark Datasets

A particular issue about the Mapper algorithm is related to the parameters necessary for its correct operation. As we know, each set of parameters in the

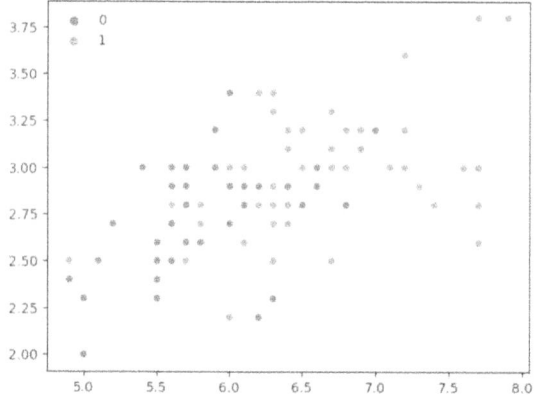

Fig. 3. Dataset in a two-dimensional space.

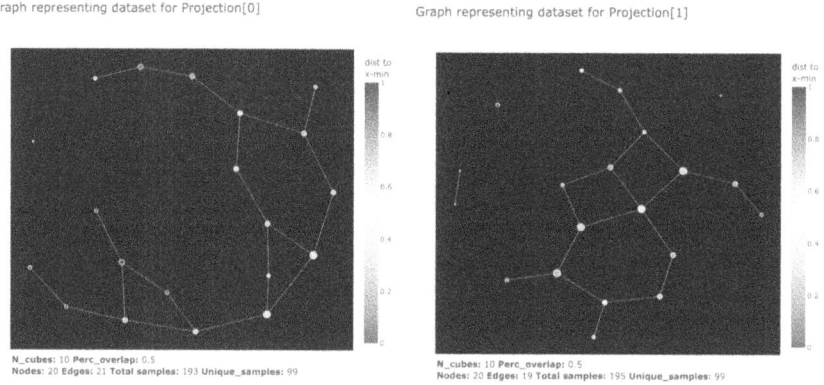

Fig. 4. Graphs obtained from data, using the first and second projections.

algorithm determines a different covering and therefore the graph and the quality of classification obtained from the lower approximation can change significantly.

In this section, we illustrate the operation of Mapper on three well-known benchmark datasets from the UCI repository [12]: iris, wine and breast cancer.

4.1 Iris Dataset

Iris is a well known dataset of about 150 flowers belonging to three different classes. Figure 5 below shows the values for quality of classification in the Iris dataset where different values of overlap percentage and different numbers of intervals (n-ranging from 10 to 29) are used. In this case, the number of k-means clusters equals three and the lens function is the projection on the first principal component of PCA.

Although it is possible to obtain very different graphs by varying the lens function, there are some characteristics that persist. For example, it is known in the Iris dataset that the "Setosa" class (green nodes in Fig. 5) is easily distinguishable from the other two.

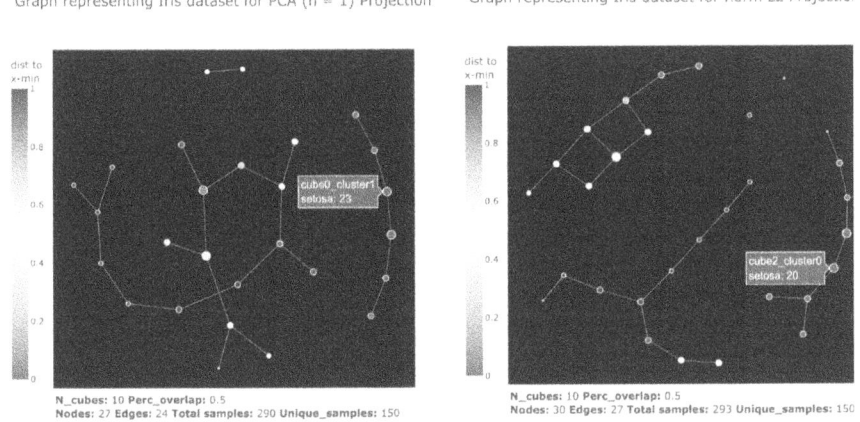

Fig. 5. Graphs for Iris dataset. In this data visualization, we can see that the Setosa class is easily distinguished from the other two classes. The green nodes contain only elements of this class. (Color figure online)

Ranging the overlap percentage and the number of intervals (in the Python implementation, they are called n-cubes), it is possible to obtain a heat map with the values of quality of classification as shown in Fig. 6. We can see that for $n = 28$ and $p = 50\%$, we obtain the maximum value for γ. In Table 5, we zoom in on this result by showing the number of elements in the lower approximation of each class.

Table 5. Number of elements in lower approximations of each class and quality of classification in the Iris dataset.

Class	% elements in lower approximation
Setosa	50/50
Virginica	49/50
Versi-color	44/50
Quality of classification	0.967

Cubes\Perc	0	0.1	0.2	0.3	0.4	0.5	0.6	0.7	0.8	0.9	0.95
10	0,773	0,780	0,847	0,813	0,813	0,727	0,807	0,867	0,893	0,547	0,333
11	0,820	0,833	0,833	0,780	0,827	0,827	0,847	0,827	0,760	0,660	0,333
12	0,820	0,807	0,827	0,840	0,853	0,893	0,773	0,853	0,873	0,787	0,333
13	0,893	0,867	0,860	0,793	0,793	0,813	0,853	0,807	0,833	0,687	0,547
14	0,827	0,873	0,873	0,853	0,867	0,893	0,860	0,873	0,820	0,813	0,547
15	0,840	0,873	0,880	0,860	0,893	0,840	0,880	0,827	0,880	0,687	0,547
16	0,953	0,913	0,833	0,833	0,833	0,867	0,867	0,847	0,900	0,820	0,547
17	0,847	0,880	0,887	0,893	0,907	0,900	0,873	0,820	0,893	0,873	0,693
18	0,867	0,887	0,907	0,900	0,827	0,847	0,867	0,860	0,873	0,867	0,547
19	0,887	0,920	0,900	0,900	0,853	0,900	0,880	0,853	0,900	0,900	0,727
20	0,873	0,920	0,907	0,920	0,880	0,907	0,893	0,893	0,833	0,887	0,827
21	0,867	0,847	0,920	0,907	0,907	0,873	0,893	0,893	0,860	0,900	0,800
22	0,893	0,900	0,900	0,920	0,927	0,867	0,880	0,907	0,873	0,873	0,840
23	0,900	0,907	0,913	0,913	0,907	0,907	0,940	0,867	0,860	0,873	0,687
24	0,873	0,880	0,927	0,900	0,907	0,893	0,873	0,913	0,840	0,867	0,833
25	0,873	0,887	0,900	0,913	0,933	0,947	0,867	0,860	0,860	0,887	0,840
26	0,867	0,867	0,893	0,920	0,893	0,907	0,933	0,867	0,833	0,873	0,827
27	0,853	0,887	0,907	0,927	0,913	0,927	0,913	0,907	0,860	0,847	0,840
28	0,847	0,873	0,887	0,900	0,953	0,967	0,853	0,887	0,880	0,940	0,853
29	0,887	0,887	0,893	0,873	0,887	0,920	0,933	0,893	0,867	0,887	0,833

Fig. 6. Heat map for quality of classification in the Iris dataset.

4.2 Wine Dataset

This dataset contains information about 178 wines belonging to three classes of wine produced by three different growers. In each analysis, 13 measurements on chemical and organoleptic properties of the wines were performed (Table 6).

Table 6. Quality of classification in the wine dataset.

Class	% elements in lower approximation
One	59/59
Two	69/71
Three	47/48
Quality of classification	0.9831

In Fig. 7, we can see that for a covering with 17 cubes, an overlap percentage of 0.8, the first principal component as projection, and $k = 3$ in k-means, a quality of classification of 0.9831 is obtained.

4.3 Breast Cancer Dataset

This dataset contains 569 records in two classes: tumor and benign. In a first approach, we used a scalar lens function, the projection on the first principal component. Figure 8 shows the γ values obtained in this case (a value of $k = 2$ was used in k-means), while Table 7 shows the number of elements in the lower approximation of each class corresponding to the optimal γ value we obtained ($\gamma = 0.721$, for 28 intervals and an overlap percentage of 70 %).

Cubes\Perc	0	0.1	0.2	0.3	0.4	0.5	0.6	0.7	0.8	0.9	0.95
10	0,6236	0,64045	0,80337	0,66292	0,64607	0,85955	0,86517	0,82022	0,84831	0,70225	0,36517
11	0,91011	0,77528	0,80337	0,75281	0,8427	0,86517	0,83708	0,89326	0,7809	0,67416	0,36517
12	0,79213	0,78652	0,6573	0,87079	0,88202	0,8764	0,76966	0,96067	0,83708	0,69663	0,37079
13	0,79213	0,80899	0,79775	0,76404	0,82584	0,83708	0,88202	0,92135	0,88764	0,79775	0,37079
14	0,76966	0,7809	0,87079	0,70225	0,92135	0,83146	0,83146	0,91011	0,93258	0,75281	0,38202
15	0,81461	0,76966	0,85955	0,79775	0,89888	0,86517	0,72472	0,90449	0,91573	0,80899	0,38202
16	0,83146	0,84831	0,83146	0,88202	0,80337	0,71348	0,90449	0,93258	0,92135	0,85955	0,38202
17	0,72472	0,83708	0,82022	0,82022	0,83146	0,85393	0,88764	0,91573	0,98315	0,85393	0,38764
18	0,83146	0,84831	0,88764	0,76966	0,88764	0,85393	0,88202	0,93258	0,91011	0,85393	0,38202
19	0,82022	0,8764	0,80337	0,88202	0,86517	0,87079	0,85955	0,91573	0,91573	0,87079	0,72472
20	0,76966	0,74157	0,76966	0,79213	0,80899	0,88764	0,87079	0,86517	0,9382	0,88202	0,70225
21	0,82584	0,80899	0,83146	0,79213	0,78652	0,92135	0,88202	0,93258	0,93258	0,86517	0,82022
22	0,85393	0,83146	0,86517	0,83708	0,92697	0,92697	0,85955	0,91573	0,91573	0,89888	0,73034
23	0,85393	0,80899	0,80899	0,88202	0,92697	0,91011	0,8764	0,92135	0,9382	0,87079	0,69663
24	0,92697	0,91573	0,8764	0,85393	0,8764	0,83146	0,85393	0,89888	0,93258	0,88202	0,87079
25	0,83708	0,85955	0,85955	0,82022	0,87079	0,88764	0,8764	0,91573	0,93258	0,89326	0,81461
26	0,84831	0,79775	0,79213	0,83146	0,88764	0,85393	0,89888	0,92697	0,95506	0,9382	0,85955
27	0,88764	0,89888	0,92697	0,94382	0,92697	0,88764	0,78652	0,94944	0,92697	0,92697	0,82022
28	0,83708	0,85955	0,84831	0,84831	0,8427	0,90449	0,93258	0,92135	0,93258	0,95506	0,88202
29	0,84831	0,85393	0,83708	0,83146	0,85393	0,92697	0,9382	0,88764	0,91573	0,94944	0,87079

Fig. 7. Heat map of quality of classification in the Wine dataset for the first component PCA.

Table 7. Quality of classification in the breast cancer dataset with a scalar function.

Class	% of elements in lower approximation
Malign	138/212
Benign	272/357
Quality of classification	0.721

Next, we proceed to examine what happens when we use a vectorial lens function. If we define f as the function that projects each point on $\mathbb{R}^2$, using the first and the second principal components, we can see that the classification quality values improve significantly, as shown in Fig. 9 and Table 8.

Table 8. Quality of classification in the breast cancer dataset with a vectorial function.

Class	% elements in lower approximation
Malign	191/212
Benign	344/357
Quality of classification	0.910

For cases in which more and more intervals are added or vector functions are used, it is possible to improve the quality of the classification, with the problem that the number of nodes in the graph is significantly increased, producing coverings with a very large number of elements.

This phenomenon is also illustrated in Fig. 10, where we show the graphs produced by Mapper for the breast cancer dataset, respectively using the scalar and the vectorial lens function introduced previously. In the first case, the number of

Cubes\Perc	0	0.1	0.2	0.3	0.4	0.5	0.6	0.7	0.8	0.9	0.95
10	0,357	0,387	0,561	0,608	0,650	0,612	0,204	0,450	0,220	0,216	0,000
11	0,308	0,304	0,464	0,373	0,547	0,612	0,464	0,401	0,220	0,218	0,000
12	0,390	0,322	0,348	0,381	0,408	0,457	0,647	0,241	0,494	0,223	0,000
13	0,622	0,466	0,453	0,518	0,429	0,547	0,599	0,227	0,485	0,227	0,185
14	0,606	0,562	0,476	0,457	0,490	0,540	0,585	0,671	0,452	0,244	0,220
15	0,589	0,576	0,606	0,518	0,641	0,550	0,617	0,619	0,445	0,230	0,258
16	0,524	0,598	0,606	0,641	0,406	0,531	0,592	0,610	0,482	0,230	0,218
17	0,483	0,490	0,534	0,571	0,496	0,457	0,522	0,627	0,492	0,232	0,218
18	0,441	0,462	0,482	0,580	0,582	0,640	0,503	0,624	0,471	0,243	0,223
19	0,460	0,483	0,522	0,562	0,622	0,645	0,664	0,552	0,455	0,239	0,228
20	0,580	0,571	0,441	0,524	0,555	0,627	0,664	0,397	0,670	0,236	0,244
21	0,608	0,550	0,638	0,671	0,550	0,638	0,634	0,689	0,677	0,241	0,258
22	0,634	0,663	0,677	0,663	0,682	0,627	0,557	0,671	0,659	0,482	0,228
23	0,627	0,518	0,587	0,615	0,629	0,518	0,594	0,649	0,627	0,489	0,228
24	0,513	0,524	0,594	0,559	0,631	0,573	0,606	0,492	0,640	0,485	0,232
25	0,599	0,543	0,536	0,594	0,633	0,687	0,589	0,631	0,622	0,457	0,239
26	0,582	0,603	0,533	0,594	0,555	0,643	0,643	0,659	0,692	0,490	0,243
27	0,554	0,634	0,654	0,578	0,612	0,636	0,694	0,677	0,650	0,531	0,230
28	0,650	0,656	0,619	0,634	0,685	0,582	0,634	0,721	0,684	0,529	0,236
29	0,698	0,650	0,654	0,654	0,661	0,707	0,675	0,689	0,661	0,552	0,236

Fig. 8. Heat map of quality of classification in the breast dataset for the first and second components PCA.

nodes is 50 and the number of elements in the covering equals 47, whereas the second graph has 646 nodes and the resulting covering contains 462 elements.

Therefore, with this type of projections, it is possible to improve the quality of classification, but the large number of elements in the covering is prone to data overfitting. Thus, we face a multi-objective problem in which the quality of classification is maximized and the number of sets in the covering is minimized.

To conclude this section, Table 9 presents a summary of the best values for the classification quality of the datasets, using different parameters in the Mapper algorithm. For a dataset such as Wine, a rather large classification quality can be achieved with a small number of elements in the covering, whereas obtaining a similar value in the breast cancer dataset requires a very large number of elements in the covering, as explained above.

Table 9. Parameters used for each dataset.

Dataset	Lens	Intervals	Elements in $\mathbb{C}$	Overlap	k-means clusters	Quality
Iris	L_2	21	57	0.7	3	0.960
Iris	$PCA = 1$	28	64	0.5	3	0.967
Wine	$PCA = 1$	9	27	0.6	3	0.938
Breast Cancer	$PCA = 1$	28	47	0.7	2	0.705
Breast Cancer	$PCA = 2$	27	462	0.7	2	0.910

Cubes\perc	0	0.1	0.2	0.3	0.4	0.5	0.6	0.7	0.8	0.9	0.95
0	0,670	0,752	0,736	0,710	0,770	0,692	0,531	0,666	0,582	0,311	0,000
1	0,659	0,622	0,612	0,747	0,735	0,838	0,670	0,564	0,666	0,506	0,000
2	0,689	0,652	0,638	0,692	0,756	0,800	0,786	0,740	0,587	0,350	0,000
3	0,775	0,777	0,729	0,743	0,726	0,798	0,796	0,779	0,684	0,494	0,202
4	0,722	0,750	0,736	0,761	0,726	0,757	0,845	0,822	0,680	0,591	0,253
5	0,740	0,814	0,822	0,775	0,794	0,822	0,766	0,851	0,656	0,641	0,267
6	0,764	0,775	0,784	0,819	0,812	0,826	0,807	0,837	0,815	0,557	0,292
7	0,754	0,735	0,763	0,756	0,810	0,824	0,838	0,796	0,798	0,707	0,309
8	0,787	0,773	0,800	0,837	0,817	0,805	0,842	0,817	0,861	0,622	0,308
9	0,770	0,801	0,831	0,814	0,840	0,856	0,840	0,861	0,861	0,714	0,302
10	0,779	0,789	0,789	0,812	0,807	0,858	0,861	0,831	0,847	0,754	0,311
11	0,757	0,819	0,828	0,819	0,835	0,828	0,851	0,863	0,831	0,670	0,529
12	0,789	0,812	0,847	0,865	0,849	0,842	0,849	0,880	0,882	0,786	0,371
13	0,777	0,789	0,796	0,828	0,865	0,856	0,849	0,859	0,868	0,773	0,536
14	0,777	0,817	0,830	0,858	0,863	0,863	0,859	0,859	0,886	0,752	0,583
15	0,754	0,803	0,805	0,826	0,840	0,882	0,868	0,868	0,884	0,814	0,629
16	0,747	0,800	0,826	0,872	0,858	0,865	0,854	0,873	0,900	0,798	0,677
17	0,777	0,824	0,835	0,861	0,861	0,859	0,879	0,910	0,884	0,754	0,615
18	0,764	0,784	0,824	0,842	0,873	0,872	0,886	0,882	0,889	0,793	0,645
19	0,740	0,784	0,817	0,851	0,865	0,880	0,882	0,893	0,888	0,882	0,694

Fig. 9. Heat map of quality of classification in the breast dataset for the first component of PCA.

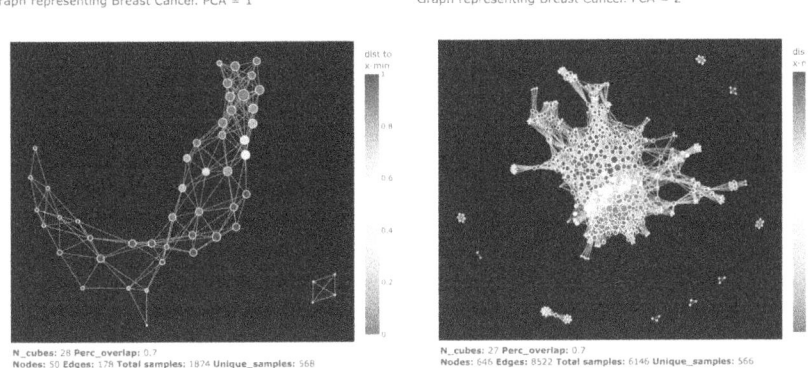

Fig. 10. Graphs obtained from Breast cancer dataset, using one and two PCA projections.

5 Conclusion and Future Work

In this paper, we showed how the Mapper algorithm can provide a covering of a numerical dataset that allows to evaluate the quality of classification γ for covering-based rough sets. By using different combinations of parameters in the algorithm, it is possible to improve the value of γ, but it is necessary to control the number of elements in the covering. Therefore it will be necessary to adopt a methodology that allows to get acceptable values γ, while maintaining adequate coverings. For future work, we plan to investigate the concepts of entropy of a partition, as well as its generalization to coverings [3,29], to control the number of elements in the covering.

Acknowledgements. This work was supported by Universidad Militar Nueva Granada's VICEIN Special Research Fund, under project CIAS 2548-2018.

References

1. Aljanobi, F.A., Lee, J.: Topological data analysis for classification of heart disease data. In: 2021 IEEE International Conference on Big Data and Smart Computing (BigComp), pp. 210–213 (2021)
2. Chazal, F., Michel, B.: An introduction to topological data analysis: fundamental and practical aspects for data scientists. Front. Artif. Intell. **4**, 667963 (2021)
3. Dai, J., Tian, H.: Entropy measures and granularity measures for set-valued information systems. Inf. Sci. **240**, 72–82 (2013)
4. D'eer, L., Cornelis, C., Yao, Y.: A semantically sound approach to Pawlak rough sets and covering-based rough sets. Int. J. Approximate Reasoning **78**, 62–72 (2016)
5. Dey, T.K., Memoli, F., Wang, Y.: Multiscale mapper: a framework for topological summarization of data and maps. arXiv:1504.03763 (2016)
6. Edelsbrunner, H., Letscher, D., Zomorodian, A.: Topological persistence and simplification. Discret. Comput. Geom. **28**, 511–533 (2002)
7. Gidea, M.: Topological data analysis of critical transitions in financial networks. In: Shmueli, E., Barzel, B., Puzis, R. (eds.) NetSci-X 2017. SPC, pp. 47–59. Springer, Cham (2017). https://doi.org/10.1007/978-3-319-55471-6_5
8. Goel, A., Pasricha, P., Mehra, A.: Topological data analysis in investment decisions. Expert Syst. Appl. **147**, 113222 (2020)
9. Greco, S., Matarazzo, B., Słowinski, R.: Rough sets theory for multicriteria decision analysis. Eur. J. Oper. Res. **129**, 1–47 (2001)
10. Gunnar, G.: Topology and data. Bulletin (New Series) AMS. **46**(2), 255–308 (2009)
11. Hwang, D., et al.: Topological data analysis of coronary plaques demonstrates the natural history of coronary artheroclerosis. Cardiovas. Imaging **14**(7), 1410–1421 (2021)
12. Kelly, M., Longjohn, R., Nottingham, K.: The UCI Machine learning repository. https://archive.ics.uci.edu
13. Li, L., et al.: Identification of type 2 diabetes subgroups through topological analysis of patient similarity**7**(311), 311ra174–311ra174 (2015)
14. Lum, P., et al.: Extracting insights from the shape of complex data using topology. Sci. Rep. **3**, 1236 (2013)
15. Majumdar, S., Laha, A.K.: Clustering and classification of time series using topological data analysis with applications to finance. Expert Syst. Appl. **162**, 113868 (2020)
16. Migdałek, G., Zelawski, M.: Measuring population-level plant gene flow with topological data analysis. Eco. Inform. **70**, 1–11 (2022)
17. Nicolau, M., Levine, A.J., Carlsson, G.: Topology based data analysis identifies a subgroup of breast cancers with a unique mutational profile and excellent survival. Proc. Nat. Acad. Sci. U.S.A. **108**(17), 7265–7270 (2011)
18. Offroy, M., Duponchel, L.: Topological data analysis: a promising big data exploration tool in biology, analytical chemistry and physical chemistry. Analytica Chimica Acta **910**, 1–11 (2016). ISSN 0003-2670
19. Pawlak, Z.: Information systems theoretical foundations. Inf. Syst. **6**(3), 205–218 (1981)
20. Pawlak, Z.: Rough sets. Int. J. Comput. Inform. Sci. **11**(5), 341–356 (1982)

21. Restrepo, M., Cornelis, C., Gómez, J.: Duality, conjugacy and adjointness of approximation operators in covering-based rough sets. Int. J. Approximate Reasoning **55**, 469–485 (2014)
22. Restrepo, M., Cornelis, C., Gómez, J.: Partial order relation for approximation operators in covering-based rough sets. Inf. Sci. **284**, 44–59 (2014)
23. Sasaki, K., Bruder, D., Hernandez-Vargas, E.A.: Topological data analysis to model the shape of immune responses during co-infections. Commun. Nonlinear Sci. Numer. Simul. **85**, 105228 (2020)
24. Singh, G., Mémoli, F., Carlsson, G.: Topological methods for the analysis of high dimensional data sets and 3D object recognition. Eurographics Symp. Point-Based Graph. **2**, 091–100 (2007)
25. Smith, A., Dlotko, P., Zavala, V.: Topological data analysis: concepts, computation, and applications in chemical engineering. Comput. Chem. Eng. **146**, 107202 (2021)
26. Van Veen, H.J., Saul, N., Eargle, D., Manghmam, S.W.: Kepler mapper: a flexible python implementation of the mapper algorithm. J. Open Source Softw. **4**(42), 1315 (2019)
27. Yao, Y.Y., Yao, B.: Covering based rough sets approximations. Inf. Sci. **200**, 91–107 (2012)
28. Zakowski, W.: Approximations in the space (u, π). Demonstratio Math. **16**, 761–769 (1983)
29. Zhaohao, W., Xiaoping, Z., Jianping, D.: The uncertainty measures for covering rough set models. Soft Comput. **24**, 11909–11929 (2020)
30. Zhu, W.: Basic concepts in covering-based rough sets. In: Proceedings of Third International Conference on Natural Computation, pp. 283–286 (2007)
31. Zhu, W.: Properties of the first type of covering-based rough sets. In: Proceedings of Sixth IEEE International Conference on Data Mining - Workshops, pp. 407–411 (2006)

On Tolerance-Based Rough Set Operators and Their Covering Generalizations

Piotr Wasilewski[1,2(✉)] and Dominik Ślęzak[3,4,5]

[1] Systems Research Institute, Polish Academy of Sciences,
ul. Newelska 6, 01-447 Warsaw, Poland
pwasilew@ibspan.waw.pl
[2] Faculty of Computer Science, Dalhousie University,
6050 University Avenue, Halifax, NS, Canada
[3] Institute of Informatics, University of Warsaw,
ul. Banacha 2, 02-097 Warsaw, Poland
slezak@mimuw.edu.pl
[4] QED Software Sp. z o.o., ul. Miedziana 3A m. 18, 00-814 Warsaw, Poland
[5] DeepSeas USA/Poland, ul. Aleje Jerozolimskie 123A, 02-017 Warsaw, Poland

Abstract. We investigate approximation operators determined by arbitrary families of granules which cover the space of objects and without posing any conditions on the nature of granules. We recall from literature some rough sets approximation operators: classical equivalential rough sets operators by Z. Pawlak and their generalizations for tolerance relations: tolerance approximation operators by A. Skowron and J. Stepaniuk and tolerance-granular approximation operators suggested by Z. Pawlak. Then we proposed granular generalization of tolerance approaches to tolerance-based rough sets by means of arbitrary covering of the object space proposed by Y.Y. Yao Then we propose a version of tolerance-granular operators suggested by Z. Pawlak with biting procedure proposed in our paper. We prove basic properties of recalled tolerance rough sets operators. We discuss which pair of operators possesses the property of mutual definability. Then we show generalizations of tolerance rough sets approximation operators which possess mutual definability property by granular covering operators from.

Keywords: rough sets · equivalential rough sets · tolerance relations · tolerance spaces · tolerance rough sets · tolerance-granular rough sets · biting procedure · coverings · granular sets · granules · granular computing

1 Introduction

Concepts modelling is one of the key issues in knowledge representation and it may be seen as one of the main subject of the study within the framework of the *granular computing* which is rapidly developing during last two decades [1,6,17]. Calls for granular computing came from necessity of dealing with incomplete information or reasoning with uncertainty. Therefore, granular computing is originated in various field approaching these challenges, including, among others, rough set theory (RST), fuzzy set theory, computing with words, Dempster-Shafer theory. In rough sets theory progenitors of information granules were Pawlak's *atoms* or *information atoms*, i.e. equivalence classes of indiscerniblity relations [12,13] derived from information systems [11]. Then the development of RST was determined by the nature of information atoms. In classical Pawlak's approach they were equivalence classes, then RST was generalized for tolerance relations, i.e. reflexive, symmetric and not necessary transitive relations and information atoms were tolerance neighbourhoods (Skowron and Stepaniuk [19]) or unions of tolerance neighbourhoods (Pawlak [14]). The next generalization of RST was made by Y.Y. Yao in [28], where information granules were sets of arbitrary coverings of the object space. In [20] by Ślęzak and Wasilewski one of the upper operators introduced in [28] were presented as unions of covering granules bited by its corresponding negative regions.

The rest of the paper is organized as follows. In Sect. 2 we present basic concepts of rough sets. Section 3 is devoted to presentation of tolerance rough sets proposed by Skowron and Stepaniuk in [19] while Sect. 4 is devoted to tolerance-granular rough sets suggested by Pawlak in [14]. Then in Sect. 4 we discuss tolerance-granular operators suggested by Z. Pawlak with biting procedure proposed in our paper [20]. In both Sects. 3 and 4 we compare all discussed operators with granular operators based on coverings proposed by Yao [28] concerning their mutual definability and their associations with definable sets. Results of this discussion are summarized in Conclusion section.

2 Elements of Rough Set Theory

In this section we are going to present basic concepts of rough set theory (RST) proposed by Zdzisław Pawlak [12,13] (see also [15]). RST is a kind of approach to concepts modelling and represents ideas of George Boole who proposed extensional formulation of concepts, identifying concepts with their extensions (denotations). In such an approach unions of concepts (extensions) as well as intersections of concepts (extensions) are also concepts. Moreover George Boole proposed the new operation on concepts, namely a complementation of a given concept.

Zdzisław Pawlak introduced rough sets [12,13] for the study of information systems [7,10,11]. Information systems represent information by means of objects and their attributes. In RST Pawlak understood *attributes* as functions of the form (we adopt here a slightly more general definition where an attribute takes a set of elements as its value, what represents uncertainty about its actual value):

$$a : U \longrightarrow \wp(Val_a)$$

where U is a set of objects, Val_a a set of values of attribute a and $\wp(Val_a)$ is the power set of Val_a. An *information system* is a triple of the form $\langle U, At, \{Val_a\}_{a \in At}\rangle$ where U is a set of objects, At is a set of attributes, and

$\{Val_a\}_{a \in At}$ is the family of value domains of attributes from the set At. Information system $\langle U, At, \{Val_a\}_{a \in At}\rangle$ is *complete* (also *total*) if $a(x) \neq \emptyset$ for every $x \in U$ and $a \in At$, otherwise $\langle U, At, \{Val_a\}_{a \in At}\rangle$ is *incomplete*. Originally Zdzisław Pawlak proposed *deterministic information systems* [11,12] (see also [7]) and after that Pawlak introduced *indeterministic information systems* [10]: information system $\langle U, At, \{Val_a\}_{a \in At}\rangle$ is *deterministic* if and only if $card(a(x)) = 1$ for every $x \in U$ and $a \in At$, otherwise $\langle U, At, \{Val_a\}_{a \in At}\rangle$ is *indeterministic*.

In RST Pawlak proposed *indiscernibility relations* for analysis of information represented in information systems. Let $\langle U, At, \{Val_a\}_{a \in At}\rangle$ be an information system and let $B \subseteq At$. An *indiscernibility relation* $\sim_B$ is defined as follows

$$(x, y) \in \sim_B \Leftrightarrow a(x) = a(y) \tag{1}$$

for any $a \in B$ and $x, y \in U$. One can note that $\sim_B$ is an equivalence relation on U. One of the main tools for calculating indiscernibility relations are discernibility matrices [18].

In rough set theory, knowledge is founded on the ability to discern between objects and, in an algebraic approach to RST, is represented by abstract structures, referred to as *approximation spaces* which are ordered pairs of the form (U, R), where U is a set of objects and R is an equivalence relation on U. Subsets of U which are unions of equivalence classes of relation R are called *definable sets* in approximation space (U, R) and subsets of U which are not definable are called *rough sets* what gave the name for whole theory. We denote the family of definable sets of the space (U, R) by $\mathsf{Def}_R(U)$. In [13] Pawlak called the equivalence classes of a relation R as *atoms*. One of the reasons for this is that the family $\mathsf{Def}_R(U)$ is closed for complements, arbitrary unions and intersections i.e. the algebra $(\mathsf{Def}_R(U); \cup, \cap, ^c, \emptyset, U)$, where $X^c := U \setminus X$, is a complete and atomic Boolean in which its atoms are exactly equivalence classes of the relation R. Below we use more convenient notation for the complement of a set X, where $X^c = X'$. The second reason was that using this term Pawlak underlined the intuition that equivalence classes are basic, indivisible, building blocks of knowledge which is achieved by abstraction from particular information about objects. This is due to the fact that every information system $\langle U, At, \{Val_a\}_{a \in At}\rangle$ determines an approximation space $(U, \sim_B)$, for any $B \subseteq At$. Thus Pawlak viewed indiscernibility relations as bridges from a "concrete" level of information to an abstract level of knowledge and Pawlak called subsets of U as *concepts* following George Boole's extensional view of *concepts*.

For any approximation space (U, R) and for $X \subseteq U$ two operators are defined as follows [12]:

$$R_*(X) = \{x \in U : x_{/R} \subseteq X\} \quad R^*(X) = \{x \in U : x_{/R} \cap X \neq \emptyset\},$$

where $x_{/R}$ are equivalence classes of an equivalence relation R, i.e. $x_{/R} := \{y \in U : (x, y) \in R\}$. Sets $R_*(X)$ and $R^*(X)$ are called *lower* and *upper approximations* of a set $X \subseteq U$ respectively. In this approach definable sets are defined by the equations of approximations: $R_*(X) = R^*(X)$ where X is a definable set.

Approximation operators based on equivalence relations can be also defined as unions of equivalence classes:

$$R_*(X) = \bigcup \{Y \in U_{/R} : Y \subseteq X\} \qquad R^*(X) = \bigcup \{Y \in U_{/R} : Y \cap X \neq \emptyset\}.$$

In this approach set $X \subseteq U$ is definable, when X is equal to some union of equivalence classes, otherwise it is indefinable. One can note that both definitions of operators presented above are equivalent when relation R is an equivalence relation on the space U. However these operators are not equivalent in the case of tolerance based relations [16] when tolerance relations are not transitive, like in [19] or in [14] - we discuss these approaches in Sects. 3 and 4 respectively. Both approaches were generalized in [28] by granular approximation operators based on arbitrary coverings and are presented below:

$$G_*^\forall(X) := \{a \in U : \forall A \in Gr(a), A \subseteq X\} \qquad G_\forall^*(X) := \{a \in U : \exists A \in Gr(a), A \cap X \neq \emptyset\}.$$

$$G_*^\exists(X) := \{a \in U : \exists A \in Gr(a), A \subseteq X\} \qquad G_\exists^*(X) := \{a \in U : \forall A \in Gr(a), A \cap X \neq \emptyset\}.$$

where $Gr(a) := \{A \in Gr(U) : a \in A\}$ and $Gr(U)$ is a family of granules covering space U. Then in [20] operator $G_\exists^*$ was defined by biting procedure.

3 Tolerance Rough Sets

In [19] Andrzej Skowron and Jarosław Stepaniuk proposed tolerance approximation spaces where tolerance approximation operators were not unions of granules but just collected centers of tolerance granules. They proved some key properties including duality resulting in mutual definability of lower and upper tolerance approximation operators and present also some applications of proposed operators for information analysis in data mining and knowledge discovery. Here we present a study of basic properties of Skowron and Stepaniuk operators showing also that operators $G_*^\forall$ and $G_\forall^*$ are generalizations of Skowron and Stepaniuk operators.

Definition 1. *Let (U, τ) be a tolerance space. We define operators $\tau_*, \tau^* : \wp(U) \longrightarrow \wp(U)$ for any set $X \subseteq U$ as follows*

$$\tau_*(X) := \{x \in U : \tau(x) \subseteq X\} \qquad \tau^*(X) := \{x \in U : \tau(x) \cap X \neq \emptyset\}.$$

Theorem 1. *Let (U, τ) be a tolerance space. Then for any sets $X, Y \subseteq U$ operators τ_* and τ^* have the following properties:*

1a. $\tau_*(X) \subseteq X$	1b. $X \subseteq \tau^*(X)$
2a. $X \subseteq Y \Rightarrow \tau_*(X) \subseteq \tau_*(Y)$	2b. $X \subseteq Y \Rightarrow \tau^*(X) \subseteq \tau^*(Y)$
3a. $\tau_*(\emptyset) = \emptyset$	3b. $\tau^*(\emptyset) = \emptyset$
4a. $\tau_*(U) = U$	4b. $\tau^*(U) = U$
5a. $\tau_*(\tau_*(X)) \subseteq \tau_*(X)$	5b. $\tau^*(X) \subseteq \tau^*(\tau^*(X))$
6a. $\tau_*(X \cap Y) = \tau_*(X) \cap \tau_*(Y)$	6b. $\tau^*(X \cap Y) \subseteq \tau^*(X) \cap \tau^*(Y)$
7a. $\tau_*(X) \cup \tau_*(Y) \subseteq \tau_*(X \cup Y)$	7b. $\tau^*(X) \cup \tau^*(Y) = \tau^*(X \cup Y)$
8a. $\tau_*(X) \subseteq \tau^*(\tau_*(X))$	8b. $\tau_*(\tau^*(X)) \subseteq \tau^*(X)$
9a. $\tau_*(X)' = \tau^*(X')$	9b. $\tau^*(X)' = \tau_*(X')$

Proof

Let (U, τ) be a tolerance space and let $X, Y \subseteq U$ in all proofs below.

(1a) $\tau_*(X) \subseteq X$

Let $a \in \tau_*(X)$, thus by Definition of operator τ_* we get that $\tau(a) \subseteq X$. Since the tolerance τ is reflexive relation, then $a \in \tau(a)$, so $a \in X$.

(1b) $X \subseteq \tau^*(X)$

Let $a \in X$. Since relation τ is reflexive, then $a \in \tau(a) \cap X$. Thus by Definition 1 $a \in \tau^*(X)$.

(2a) $X \subseteq Y \Rightarrow \tau_*(X) \subseteq \tau_*(Y)$

Let $X \subseteq Y$ and $a \in \tau_*(X)$. Thus by Definition 1 $\tau(a) \subseteq X$ and since $X \subseteq Y$, then $\tau(a) \subseteq Y$. Therefore by Definition 1 $a \in \tau_*(Y)$. Since $a \in \tau_*(X)$ was selected arbitrary, then we show that $\tau_*(X) \subseteq \tau_*(Y)$.

(2b) $X \subseteq Y \Rightarrow \tau^*(X) \subseteq \tau^*(Y)$

Let $X \subseteq Y$ and $a \in \tau^*(X)$. Then by Definition 1 $\tau(a) \cap X \neq \emptyset$. Since $X \subseteq Y$, then also $\tau(a) \cap Y \neq \emptyset$. Thus $a \in \tau^*(Y)$. Since $a \in \tau^*(X)$ was selected arbitrary, then we show that $\tau^*(X) \subseteq \tau^*(Y)$.

(3a) $\tau_*(\emptyset) = \emptyset$

For non direct proof assume that there is $a \in U$ such that $a \in \tau_*(\emptyset)$. Thus by Definition 1 $\tau(a) \subseteq \emptyset$. Since $a \in \tau(a)$, then $a \in \emptyset$ what leads to contradiction. Thus $a \notin \tau_*(\emptyset)$. Because $a \in U$ was selected arbitrary and $\emptyset \subseteq \tau_*(\emptyset)$ then $\tau_*(\emptyset) = \emptyset$.

(3b) $\tau^*(\emptyset) = \emptyset$

Note that for every $a \in U$, $\tau(a) \cap \emptyset = \emptyset$, thus $a \notin \tau^*(\emptyset)$ for each $a \in U$. Therefore $\tau^*(\emptyset) = \emptyset$.

(4a) $\tau_*(U) = U$

It is enough to prove inclusion $\supseteq$. Since by Definition 1 $\tau_*(U) \subseteq U$, it is enough to prove that $U \subseteq \tau_*(U)$. Let $a \in U$, and so by Definition 1 $\tau(a) \subseteq U$. Thus for every $a \in U$, $\tau(a) \subseteq U$ and so $\tau(a) \subseteq \bigcup_{a \in U} \tau(a)$. Therefore $U \subseteq \bigcup_{a \in U} \tau(a)$ and by Definition 1 $U \subseteq \tau_*(U)$.

(4b) $\tau^*(U) = U$

It is enough to prove inclusion $\supseteq$. Since by Definition 1 $\tau^*(U) \subseteq U$, we will show that $U \subseteq \tau^*(U)$. Let $a \in U$, because relation τ is reflexive, $a \in \tau(a)$ and then $\tau(a) \cap U \neq \emptyset$. Therefore by Definition 1 $a \in \tau^*(U)$. Because $a \in U$ was selected arbitrarily $U \subseteq \tau^*(U)$.

(5a) $\tau_* \tau_*(X) \subseteq \tau_*(X)$

Let $a \in \tau_* \tau_*(X)$. Thus by Definition 1 $\tau(a) \subseteq \tau_*(X)$. Since τ is reflexive and so $a \in \tau(a)$, then $a \in \tau_*(X)$. Because $a \in \tau_* \tau_*(X)$ was selected arbitrarily, then $\tau_* \tau_*(X)) \subseteq \tau_*(X)$.

(5b) $\tau^*(X) \subseteq \tau^*(\tau^*(X))$

Let $a \in \tau^*(X)$. Since τ is reflexive, then $a \in \tau(a)$ and so $\tau(a) \cap \tau^*(X) \neq \emptyset$. Thus $a \in \tau^*\tau^*(X)$.

(6a) $\tau_*(X \cap Y) = \tau_*(X) \cap \tau_*(Y)$

($\subseteq$) Let $a \in \tau_*(X \cap Y)$. Then by Definition 1 $\tau(a) \subseteq X \cap Y$. Thus $\tau(a) \subseteq X$ and $\tau(a) \subseteq Y$ and so $a \in \tau_*(X)$ and $a \in \tau_*(Y)$. Therefore $a \in \tau_*(X) \cap \tau_*(Y)$. Since $a \in \tau_*(X \cap Y)$ was selected arbitrarily, then $\tau_*(X \cap Y) \subseteq \tau_*(X) \cap \tau_*(Y)$.

($\supseteq$) Let $a \in \tau_*(X) \cap \tau_*(Y)$. Then $a \in \tau_*(X)$ and $a \in \tau_*(Y)$. Thus by Definition 1 $\tau(a) \subseteq X$ and $\tau(a) \subseteq Y$. Thus $\tau(a) \subseteq X \cap Y$ and so $a \in \tau_*(X \cap Y)$. Therefore $\tau_*(X) \cap \tau_*(Y) \subseteq \tau_*(X \cap Y)$.

(6b) $\tau^*(X \cap Y) \subseteq \tau^*(X) \cap \tau^*(Y)$

Let $a \in \tau^*(X \cap Y)$ thus $\tau(a) \cap (X \cap Y) \neq \emptyset$. Since $X \cap Y \subseteq X, Y$, then $\tau(a) \cap X \neq \emptyset$ and $\tau(a) \cap Y \neq \emptyset$. Thus $a \in \tau^*(X)$ and $a \in \tau^*(Y)$, and so $a \in \tau^*(X) \cap \tau^*(Y)$. Since $a \in \tau^*(X \cap Y)$ was selected arbitrarily, then $\tau^*(X \cap Y) \subseteq \tau^*(X) \cap \tau^*(Y)$.

(7a) $\tau_*(X) \cup \tau_*(Y) \subsetneq \tau_*(X \cup Y)$

Let $a \in \tau_*(X) \cup \tau_*(Y)$, thus $a \in \tau_*(X)$ or $a \in \tau_*(Y)$. Without loss of a generality of the proof we assume that $a \in \tau_*(X)$. Then $\tau(a) \subseteq X \subseteq X \cup Y$, so $a \in \tau_*(X \cup Y)$. Because $a \in \tau_*(X) \cup \tau_*(Y)$ was selected arbitrarily, thus $\tau_*(X) \cup \tau_*(Y) \subseteq \tau_*(X \cup Y)$.

(7b) $\tau^*(X) \cup \tau^*(Y) = \tau^*(X \cup Y)$

($\subseteq$) Let $a \in \tau^*(X) \cup \tau^*(Y)$, then $a \in \tau^*(X)$ or $a \in \tau^*(Y)$. Without loss of a generality of the proof we assume that $a \in \tau^*(X)$. Since $a \in \tau^*(X)$ then $\tau(a) \cap X \neq \emptyset$. Because $X \subseteq X \cup Y$ thus $\tau(a) \cap (X \cup Y) \neq \emptyset$ and so $a \in \tau^*(X \cup Y)$. Since $a \in \tau^*(X) \cup \tau^*(Y)$ was selected arbitrarily, therefore $\tau^*(X) \cup \tau^*(Y) \subseteq \tau^*(X \cup Y)$.

($\supseteq$) Let $a \in \tau^*(X \cup Y)$, then $\tau(a) \cap (X \cup Y) \neq \emptyset$ and so $\tau(a) \cap X \neq \emptyset$ or $\tau(a) \cap Y \neq \emptyset$. Thus $a \in \tau^*(X)$ or $a \in \tau^*(Y)$, then $a \in \tau^*(X) \cup \tau^*(Y)$. Because $a \in \tau^*(X \cup Y)$ was selected arbitrary, therefore $\tau^*(X \cup Y) \subseteq \tau^*(X) \cup \tau^*(Y)$.

(8a) $\tau_*(X) \subseteq \tau^*(\tau_*(X))$

Let $a \in \tau_*(X)$, then $\tau(a) \subseteq X$. Since τ is reflexive, then $a \in \tau(a)$, thus $a \in \tau(a) \cap \tau_*(X) \neq \emptyset$, so by Definition 1 $a \in \tau^*(\tau_*(X))$. Since $a \in \tau_*(X)$ was selected arbitrary, therefore $\tau_*(X) \subseteq \tau^*(\tau_*(X))$.

(8b) $\tau_*(\tau^*(X)) \subseteq \tau^*(X)$

Let $a \in \tau_*(\tau^*(X))$, then $\tau(a) \subseteq \tau^*(X)$. Since relation τ is transitive, then $a \in \tau(a)$ and $a \in \tau^*(X)$. Since $a \in \tau_*(\tau^*(X))$ was selected arbitrary, so $\tau_*(\tau^*(X)) \subseteq \tau^*(X)$.

(9a) $\tau_*(X)' = \tau^*(X')$

($\subseteq$) Let $a \in \tau_*(X)'$, then $a \notin \tau_*(X)$ so $\tau(a) \nsubseteq X$. Thus there is $b \in U$ such that $b \in \tau(a)$ and $b \notin X$. Then there is $b \in U$ such that $b \in \tau(a)$ and $b \in X'$ so $\tau(a) \cap X' \neq \emptyset$, thus $a \in \tau^*(X')$. Since $a \in \tau_*(X)'$ was selected arbitrary, therefore $\tau_*(X)' \subseteq \tau^*(X')$.

($\supseteq$) Let $a \in \tau^*(X')$, then $\tau(a) \cap X' \neq \emptyset$ so $\exists b \in U : b \in \tau(a)$ & $b \in X'$. Thus $\exists b \in U : b \in \tau(a)$ & $b \notin X$ so $\tau(a) \nsubseteq X$ and $a \notin \tau_*(X)$, then $a \in \tau_*(X)'$. Because $a \in \tau^*(X')$ was selected arbitrary, therefore $\tau^*(X') \subseteq \tau_*(X)'$.

(9b) $\tau^*(X)' = \tau_*(X')$

($\subseteq$) Let $a \in \tau^*(X)'$, then $a \notin \tau^*(X)$. Thus $\tau(a) \cap X = \emptyset$ then $\tau(a) \subseteq X'$ thus $a \in \tau_*(X')$. Since $a \in \tau^*(X)'$ was selected arbitrary, therefore $\tau^*(X)' \subseteq \tau_*(X')$.

($\supseteq$) Let $a \in \tau_*(X')$, then $a \in \tau(a) \subseteq X'$. Thus $\tau(a) \cap X = \emptyset$ and $a \notin \tau^*(X)$ so $a \in \tau^*(X)'$. Since $a \in \tau_*(X')$ was selected arbitrary, thus $\tau_*(X') \subseteq \tau^*(X)'$. □

Proposition 1. *Let (U, τ) be a tolerance space, then there is granular space $(U, Gr(U))$ such that the following equalities holds:*

(1) $G_*^\forall(X) = \tau_*(X)$,
(2) $G_\forall^*(X) = \tau^*(X)$.

Proof

Let (U, τ) be a tolerance space and let

$$\mathbb{S}_\tau(U) := \{\{x, y\} : (x, y) \in \tau\}.$$

Note that for every $a \in U$, $\{a\} \in \mathbb{S}_\tau(U)$, since tolerance relation τ is reflexive. For $a \in U$ let us define also

$$\mathbb{S}_\tau(a) := \{\{a, x\} : (a, x) \in \tau\}.$$

Note that $a \in \mathbb{S}_\tau(a)$, since τ is reflexive. Now one can prove the following lemma:

$$\bigcup \mathbb{S}_\tau(a) = \tau(a).$$

Let us prove now equalities 1 and 2.

(1) $G_*^\forall(X) = \tau_*(X)$

Let $a \in G_*^\forall(X)$, then the following steps are equivalent:

$$a \in G_*^\forall(X)$$
$$\forall A \in \mathbb{S}_\tau(a), A \subseteq X$$
$$\bigcup \mathbb{S}_\tau(a) \subseteq X$$
$$\tau(a) \subseteq X$$
$$a \in \tau_*(X).$$

Since $a \in G_*^\forall(X)$ and $X \subseteq U$ were selected arbitrary, then we prove that $G_*^\forall(X) = \tau_*(X)$ for any set $X \subseteq U$.

(2) $G_\forall^*(X) = \tau^*(X)$

($\subseteq$) Let $a \in G_\vee^*(X)$, thus $\exists A \in \mathbb{S}_\tau(a) : A \cap X \neq \emptyset$. Therefore there is $x \in U$ such that $\{a, x\} \in \mathbb{S}_\tau(a)$ and $\{a, x\} \cap X \neq \emptyset$ where $A = \{a, x\}$. Since $\{a, x\} \in \mathbb{S}_\tau(a)$, then $(a, x) \in \tau$ and so $x \in \tau(a)$. Therefore there is $x \in U$ such that $x \in \tau(a)$ and $\{a, x\} \cap X \neq \emptyset$. Since $\{a, x\} \subseteq \tau(a)$, then there is $x \in U$ such that $x \in \tau(a)$ and $\tau(a) \cap X \neq \emptyset$. Thus in particular $\tau(a) \cap X \neq \emptyset$ so $a \in \tau^*(X)$. Since $a \in G_\vee^*(X)$ was selected arbitrary, then we already prove that $G_\vee^*(X) \subseteq \tau^*(X)$.

($\supseteq$) Let $a \in \tau^*(X)$, thus $\tau(a) \cap X \neq \emptyset$. Thus there is $x \in U$ such that $x \in \tau(a) \cap X$, thus there is $x \in U$ such that $\{a.x\} \in \mathbb{S}_\tau(a)$ and $x \in X$. Since $x \in X$, then $\{a, x\} \cap X \neq \emptyset$, thus there is $x \in U$ such that $\{a, x\} \in \mathbb{S}_\tau(a)$ and $\{a, x\} \cap X \neq \emptyset$. Thus $\exists \in \mathbb{S}_\tau(a) : A \cap X \neq \emptyset$ and so $a \in G_\vee^*(X)$. Since $a \in \tau^*(X)$ was selected arbitrary, then we proved that $\tau^*(X) \subseteq G_\vee^*(X)$. Therefore we have shown that $G_\vee^*(X) = \tau^*(X)$. □

4 Tolerance-Granular Rough Sets

Rough approximation operators based on union of tolerance granules were mentioned by Zdzisław Pawlak in [14] without broader presentation. In [20] we formulated the properties of these operators in general setting for arbitrary coverings and in this paper we present the proof theory for those properties.

Definition 2. Let (U, τ) be a tolerance space. We define operators P_τ, $P^\tau : \wp(U) \longrightarrow \wp(U)$ for any set $X \subseteq U$ as follows

$$P_\tau(X) := \bigcup\{\tau(x) : \tau(x) \subseteq X\} \qquad P^\tau(X) := \bigcup\{\tau(x) : \tau(x) \cap X \neq \emptyset\}.$$

Let (U, τ) be a tolerance space and let $G_\tau(U) := \{\tau(x) : x \in U\}$. Since $G_\tau(U) \subseteq \wp(U)$ and $\bigcup G_\tau(U) = U$, then $(U, G_\tau(U))$ is a covering of the space U.

Definable sets in the tolerance space (U, τ) are unions of sets from the family $G_\tau(U)$. By $Def_{G_\tau}(U)$ we denote the family of definable sets in the tolerance space (U, τ). Thus $Def_{G_\tau}(U) := \{\bigcup \mathcal{B} : \mathcal{B} \subseteq G_\tau(U)\}$.

Theorem 2. Let (U, τ) be a tolerance space. Then for any sets $X, Y \subseteq U$ operators P_τ and P^τ have the following properties:

1a. $P_\tau(X) \subseteq X$	1b. $X \subseteq P^\tau(X)$
2a. $X \subseteq Y \Rightarrow P_\tau(X) \subseteq P_\tau(Y)$	2b. $X \subseteq Y \Rightarrow P^\tau(X) \subseteq P^\tau(Y)$
3a. $P_\tau(\emptyset) = \emptyset$	3b. $P^\tau(\emptyset) = \emptyset$
4a. $P_\tau(U) = U$	4b. $P^\tau(U) = U$
5a. $P_\tau(P_\tau(X)) = P_\tau(X)$	5b. $P^\tau(X) \subseteq P^\tau(P^\tau(X))$
6a. $P_\tau(X \cap Y) \subseteq P_\tau(X) \cap P_\tau(Y)$	6b. $P^\tau(X \cap Y) \subseteq P^\tau(X) \cap P^\tau(Y)$
7a. $P_\tau(X) \cup P_\tau(Y) \subseteq P_\tau(X \cup Y)$	7b. $P^\tau(X) \cup P^\tau(Y) = P^\tau(X \cup Y)$
8a. $P_\tau(X) \subseteq P^\tau(P_\tau(X))$	8b. $P_\tau(P^\tau(X)) = P^\tau(X)$
9a. $P_\tau(X)' \subseteq P^\tau(X')$	9b. $P^\tau(X)' \subseteq P_\tau(X')$
10. $X \in Def_P(U) \Leftrightarrow X = P_\tau(X)$	
11. $X, Y \in Def_P(U) \Rightarrow X \cup Y \in Def_P(U)$	
12. $X, Y \in Def_P(U) \Rightarrow P_\tau(X \cup Y) = P_\tau(X) \cup P_\tau(Y)$	

Proof

Let (U, τ) be a tolerance space and let $X, Y \subseteq U$ in all proofs below.

1a. $P_\tau(X) \subseteq X$

Let $a \in P_\tau(X)$, then by Definition 2 there is $x \in U$ such that $\tau(x) \subseteq X$ and $a \in \tau(x)$. Therefore $a \in X$.

1b. $X \subseteq P^\tau(X)$

Let $a \in X$. Since tolerance relation τ is reflexive, then $a \in \tau(a)$. Since $a \in X$, then $\tau(a) \cap X \neq \emptyset$, so by Definition 2 we get $a \in P^\tau(X)$.

2a. $X \subseteq Y \Rightarrow P_\tau(X) \subseteq P_\tau(Y)$

Let $X \subseteq Y$ and let $a \in P_\tau(X)$. Thus by Definition 2, $\tau(a) \subseteq X \subseteq Y$. Therefore $\tau(a) \subseteq Y$ so by Definition 2 and fact that $a \in \tau(a)$ we get $a \in P_\tau(Y)$.

2b. $X \subseteq Y \Rightarrow P^\tau(X) \subseteq P^\tau(Y)$

Let $X \subseteq Y$ and let $a \in P^\tau(X)$. Thus by Definition 2, there is $x \in U$, $a \in \tau(x)$ and $\tau(x) \cap X \neq \emptyset$. Since $X \subseteq Y$ thus $\tau(x) \cap Y \neq \emptyset$, and since $a \in \tau(x)$ we get by Definition 2 that $a \in P^\tau(Y)$. Since $a \in P^\tau(X)$ was selected arbitrary, then we have shown that $P^\tau(X) \subseteq P^\tau(Y)$.

3a. $P_\tau(\emptyset) = \emptyset$
3b. $P^\tau(\emptyset) = \emptyset$

Every tolerance relation is reflexive then every tolerance neighbourhood is non-empty and so by definitions of operators P_τ and P^τ as some unions of granules (tolerance neighbourhoods) one can directly show that properties 3a and 3b holds.

4a. $P_\tau(U) = U$
4b. $P^\tau(U) = U$

Since every element $x \in U$ determines its neighbourhood $\tau(x) \subseteq U$, then both properties 4a and 4b holds.

5a. $P_\tau(P_\tau(X)) = P_\tau(X)$

($\subseteq$) This inclusion follows from the facts 1a and 2a.
($\supseteq$) Let $a \in P_\tau(X)$. Then there is $x \in U$ such that $a \in \tau(x) \subseteq X$. Since $\tau(x) \subseteq X$ and $P_\tau(X) := \bigcup\{\tau(x) : \tau(x) \subseteq X\}$, then $a \in \tau(x) \subseteq P_\tau(X)$. Therefore by Definition 2, $a \in P_\tau(P_\tau(X))$.

5b. $P^\tau(X) \subseteq P^\tau(P^\tau(X))$

Inclusion 5b follows from the facts 1b and 2b.

6a. $P_\tau(X \cap Y) \subseteq P_\tau(X) \cap P_\tau(Y)$
6b. $P^\tau(X \cap Y) \subseteq P^\tau(X) \cap P^\tau(Y)$

6a and 6 b follows from 1a and 1b respectively and from the fact that $X \cap Y \subseteq X, Y$.

7a. $P_\tau(X) \cup P_\tau(Y) \subseteq P_\tau(X \cup Y)$

7a follows from 1a and the fact that $X, Y \subseteq X \cup Y$ for any $X, Y \subseteq U$.

7b. $P^\tau(X) \cup P^\tau(Y) = P^\tau(X \cup Y)$

($\subseteq$) This inclusion, analogically to 7a, follows from the fact 1a the fact that $X, Y \subseteq X \cup Y$ for any $X, Y \subseteq U$.
($\supseteq$) Let $a \in P^\tau(X \cup Y)$. Thus by Definition 2, there is $x \in U$ such that $a \in \tau(x) \cap (X \cup Y) \neq \emptyset$ then either $\tau(x) \cap X \neq \emptyset$ or $\tau(x) \cap Y \neq \emptyset$. Assume that $a \in \tau(x) \cap X \neq \emptyset$. Thus by 1b we get that $a \in \tau(x) \cap P^\tau(X) \neq \emptyset$. Therefore by Definition 2, $a \in P^\tau(X \cup Y)$. The proof goes analogically when $a \in \tau(x) \cap Y \neq \emptyset$.

8a. $P_\tau(X) \subseteq P^\tau(P_\tau(X))$

8a follows from 1b and from the fact that $P_\tau(X) \subseteq P_\tau(X)$.

8b. $P_\tau(P^\tau(X)) = P^\tau(X)$

($\subseteq$) This inclusion follows from the fact where for X we substitute $P^\tau(X)$.
($\supseteq$) Let $a \in P^\tau(X)$. Then there is $x \in U$ such that $a \in \tau(x) \cap U \neq \emptyset$. Since by Definition 2, $P^\tau(X)$ is a union of τ-neighborhoods which have non-empty intersections with set X, then $a \in \tau(x) \subseteq P^\tau(X)$. Thus by Definition 2, we get that $a \in P_\tau(P^\tau(X))$.

(9a) $P_\tau(X)' \subseteq P^\tau(X')$

($\subseteq$) Let $a \in U$ and $a \in P_\tau(X)'$, thus $a \notin P_\tau(X)$ what by the Definition 2 is equivalent to that there is no $A \in G_\tau(U)$ such that $a \in A$ and $A \subseteq X$. Now the following propositions are equivalent

$$\neg \exists A \in G_\tau(U) : a \in A \,\&\, A \subseteq X$$
$$\forall A \in G_\tau(U) : \neg(a \in A \,\&\, A \subseteq X)$$
$$\forall A \in G_\tau(U) : (\neg a \in A \lor A \not\subseteq X)$$
$$\forall A \in G_\tau(U) : (a \in A \Longrightarrow A \not\subseteq X)$$
$$\alpha := \forall A \in G_\tau(U) : (a \in A \Longrightarrow A \cap X' \neq \emptyset)$$

Since $a \in A$ and family $G_\tau(U)$ is a covering of space U, then there is $B \in G_\tau(U)$ such that $a \in B$. This and α imply that $B \cap X' \neq \emptyset$. Thus there is $B \in G_\tau(U)$, $a \in B$ and $B \cap X' \neq \emptyset$ what by Definition 2 is equivalent to $a \in P^\tau(X')$. Because $a \in P_\tau(X)'$ was selected arbitrary, therefore $P_\tau(X)' \subseteq P^\tau(X')$.

(9b) $P^\tau(X)' \subseteq P_\tau(X')$

($\subseteq$) Let $a \in P^\tau(X)'$ thus $a \notin P^\tau(X)$ so $a \notin \bigcup\{A \in G_\tau(U) : A \cap X \neq \emptyset\}$. Therefore $\forall A \in G_\tau(U) : A \cap X = \emptyset\}$ what is equivalent to $\alpha := \forall A \in G_\tau(U) : A \subseteq X'$. Since family $G_\tau(U)$ is a covering of space U, then $\exists B \in G_\tau(U) : a \in B$. With respect to that by α we get $\exists B \in G_\tau(U) : a \in B \,\&\, B \subseteq X'$, thus $a \in P_\tau(X')$. Because $a \in P^\tau(X)'$ was selected arbitrary, then $P^\tau(X)' \subseteq P_\tau(X')$.

(10) $X \in Def_{G_\tau}(U) \Leftrightarrow X = P_\tau(X)$

($\Rightarrow$) Because of (1a) it is enough to prove that $X \subseteq P_\tau(X)$

Assume that $X \in Def_{G_\tau}(U)$, thus X is a union of granules from $G_\tau(U)$. Let $x \in X$, since X is a union of granules from $G_\tau(U)$, then there is $A \in G_\tau(U)$ such that $x \in A$ and $A \subseteq X$, thus $x \in P_\tau(X)$. Therefore $X \subseteq P_\tau(X)$.

($\Leftarrow$) By definitions of definable sets and operator P_τ, for every set $X \subseteq U$: $P_\tau(X) \in Def_{G_\tau}(U)$. Thus since $X = P_\tau(X)$, then $X \in Def_{G_\tau}(U)$.

(11) $X, Y \in Def_{G_\tau}(U) \Rightarrow X \cup Y \in Def_{G_\tau}(U)$

($\Rightarrow$) Let $X, Y \in Def_{G_\tau}(U)$, then there is $\mathbb{A} \subseteq G_\tau(U) : X = \bigcup \mathbb{A}$ and there is $\mathbb{B} \subseteq G_\tau(U) : Y = \bigcup \mathbb{B}$. Let us note that $\bigcup \mathbb{A} \cup \bigcup \mathbb{B} = \bigcup(\mathbb{A} \cup \mathbb{B})$. Therefore $X \cup Y = \bigcup \mathbb{A} \cup \bigcup \mathbb{B} = \bigcup(\mathbb{A} \cup \mathbb{B})$. In addition note that if $\mathbb{A}, \mathbb{B} \subseteq G_\tau(U)$, then $\mathbb{A} \cup \mathbb{B} \subseteq G_\tau(U)$. Thus there is $\mathbb{D} \subseteq G_\tau(U)$ such that $X \cup Y = \bigcup \mathbb{D}$, namely $\mathbb{D} = \mathbb{A} \cup \mathbb{B}$, so $X \cup Y \in Def_{G_\tau}(U)$. Since $X, Y \in Def_{G_\tau}(U)$ were selected arbitrary that we prove that for any $X, Y \subseteq U$: $X, Y \in Def_{G_\tau}(U) \Rightarrow X \cup Y \in Def_{G_\tau}(U)$.

(12) $X, Y \in Def_{G_\tau}(U) \Rightarrow P_\tau(X) \cup P_\tau(Y) = P_\tau(X \cup Y)$

($\supseteq$) Let $X, Y \in Def_{G_\tau}(U)$. In the light of 2.7a it is enough to show inclusion $\supseteq$. Let $a \in G_*(X \cup Y)$. Since $P_\tau(X \cup Y) \subseteq X \cup Y$, then $a \in X \cup Y$. Thus $a \in X$ or $a \in Y$. Consider that $a \in X$. Since $X \in Def_{G_\tau}(U)$, then there is $A \in G_\tau(U)$ such that $A \subseteq X$ and $a \in A$. Thus by Definition 2 $a \in P_\tau(X)$. Analogously $a \in P_\tau(Y)$ when $a \in Y$. Thus we have shown that $a \in P_\tau(X)$ or $a \in P_\tau(Y)$, thus $a \in P_\tau(X) \cup P_\tau(Y)$. Since $a \in P_\tau(X \cup Y)$ was selected arbitrary, therefore $P_\tau(X \cup Y) \subseteq P_\tau(X) \cup P_\tau(Y)$. □

Example 1. The following conditions do not hold for arbitrary tolerance space:

(1) $P^\tau(X') \not\subseteq P_\tau(X)'$,
(2) $P_\tau(X') \not\subseteq P^\tau(X)'$.

These are counterexmples to opposite inclusions for properties (9a) and (9b) respectively and we present them below.

(1) Let (U, τ) be a tolerance space and where $U = \{a, b, c, d\}$, $\tau(a) = \{a, b\}$, $\tau(b) = \{a, b, c\}$, $\tau(c) = \{b, c, d\}$ and $\tau(d) = \{c, d\}$. Let $X = \{a, b\}$, note that $b \in G^*(X')$, because $b \tau c$ and $c \in X'$ so $\tau(b) \cap X' \neq \emptyset$. Since $b \in \tau(a) \subseteq X$, then $b \in P_\tau(X)$, therefore $b \notin P_\tau(X)'$. Since $b \in P^\tau(X')$ and $b \notin P_\tau(X)'$, then $P^\tau(X') \not\subseteq P_\tau(X)'$.

(2) Let (U, τ) be a tolerance space where $U = \{a, b, c, d\}$, $\tau(a) = \{a, b\}$, $\tau(b) = \{a, b, c\}$, $\tau(c) = \{b, c, d\}$ and $\tau(d) = \{c, d\}$. Let $X = \{c, d\}$. Note that $b \in \tau(a)$ and $\tau(a) \subseteq X'$. Thus by Definition 2 $b \in P_\tau(X')$. Note also that $b \in \tau(b)$ and $\tau(b) \cap X \neq \emptyset$, thus by Definition 2 $b \in P^\tau(X)$. Since $b \in P^\tau(X)$, then $b \notin P^\tau(X)'$. Therefore we get $b \in P_\tau(X')$ and $b \notin P^\tau(X)'$, thus $P_\tau(X') \not\subseteq P^\tau(X)'$.

In [20] we proposed biting procedure in order to avoid the overlapping regions paradox discussed in [20] which shows that if relation τ is not transitive then in approximations space (U, τ), there is a set $X \subseteq U$ such that $P^\tau(X) \cap P_\tau(X') \neq \emptyset$.

One of our motivations for introducing biting procedure was the fact that approximation operators based on tolerance granules proposed by Pawlak fallen into that paradox. However proposing new "bitted" upper approximation operators in the case of tolerance granules based on Pawlak's operators or other granular operators not only helped to avoid the overlapping region paradox but also offered duality or mutual definability for approximation operators. Below we present properties of such operators being the result of application of biting procedure to Pawlak tolerance granule based operators.

Definition 3. *Let* (U, τ) *be a tolerance space. For* $X \subseteq U$ *we define operator* P_b^τ

$$P_b^\tau(X) := P^\tau(X) \setminus P_\tau(X').$$

and we call it bitten upper approximation operator *because we use granules from negative region* $NEG_{P_\tau}(X) = P_\tau(X')$ *(were* $X' := U \setminus X$*) to bite the overlapping parts of granules from upper approximation* $P^\tau(X)$. *A bitten upper approximation operator* $P_b^\tau(X)$ *determines also a bitten boundary of a set* X *defined as follows:*

$$BND_{P_b}(X) := P_b^\tau(X) \setminus P_\tau(X).$$

Let us point out the direct formal consequence of biting procedure: it strengthens inclusions 9a and 9b from Theorem 2 to identity and gives a mutual definability of operators P_τ and P_b^τ. More exactly, the following proposition holds:

Proposition 2. *Let* (U, τ) *be a tolerance space, then for any* $X \subseteq U$ *the following conditions hold:*

(1) $P_\tau(X)' = P_b^\tau(X'),$
(2) $P_b^\tau(X)' = P_\tau(X').$

Proof
(1)
$$\begin{aligned}
P_b^\tau(X') &= P^\tau(X') \setminus NEG_{P_\tau}(X') \\
&= P^*(X') \cap NEG_{P_\tau}(X')' \\
&= P^\tau(X') \cap P_\tau(X'')' \\
&= P^\tau(X') \cap P_\tau(X)' \\
&= P_\tau(X)'
\end{aligned}$$

by Theorem 2.9a.

(2)
$$\begin{aligned}
P_b^\tau(X)' &= [P^\tau(X) \setminus NEG_\tau(X)]' \\
&= [P^\tau(X) \cap NEG_\tau(X)']' \\
&= [P^\tau(X) \cap P_\tau(X')']' \\
&= P^\tau(X)' \cup P_\tau(X')'' \\
&= P^\tau(X)' \cup P_\tau(X') \\
&= P_\tau(X')
\end{aligned}$$

by Theorem 2.9b. □

Corollary 1. Let (U, τ) be a tolerance space, then for any $X \subseteq U$ the following conditions hold:

(1) $P_\tau(X) = P_b^\tau(X')'$,
(2) $P_b^\tau(X) = P_\tau(X')'$.

Due to limited space, below we presented without proofs the basic properties of operator P_b^τ together with its dual operator P_τ

Theorem 3. Let (U, τ) be a tolerance space. Then for any sets $X, Y \subseteq U$ bitten upper approximation operator P_b^τ have the following properties (we set up them with already proved properties of lower approximation $P_\tau(X)$ keeping the same numeration):

1a. $P_\tau(X) \subseteq X$	1b. $X \subseteq P_b^\tau(X)$
2a. $X \subseteq Y \Rightarrow P_\tau(X) \subseteq P_\tau(Y)$	2b. $X \subseteq Y \Rightarrow P_b^\tau(X) \subseteq P_b^\tau(Y)$
3a. $P_\tau(\emptyset) = \emptyset$	3b. $P_b^\tau(\emptyset) = \emptyset$
4a. $P_\tau(U) = U$	4b. $P_b^\tau(U) = U$
5a. $P_\tau(P_\tau(X)) = P_\tau(X)$	5b. $P_b^\tau(X) = P_b^\tau(P_b^\tau(X))$
6a. $P_\tau(X \cap Y) \subseteq P_\tau(X) \cap P_\tau(Y)$	6b. $P_b^\tau(X \cap Y) \subseteq P_b^\tau(X) \cap P_b^\tau(Y)$
7a. $P_\tau(X) \cup P_\tau(Y) \subseteq P_\tau(X \cup Y)$	7b. $P_b^\tau(X) \cup P_b^\tau(Y) \subseteq P_b^\tau(X \cup Y)$
8a. $P_\tau(X) \subseteq P_b^\tau P_\tau((X))$	8b. $P_\tau(P_b^\tau(X)) \subseteq P_b^\tau X$
9a. $P_\tau(X)' = P_b^\tau(X')$	9b. $P_b^\tau(X)' = P_\tau(X')$
10. $X \in Def_P(U) \Leftrightarrow X = P_\tau(X)$	
11. $X, Y \in Def_P(U) \Rightarrow X \cup Y \in Def_P(U)$	
12. $X, Y \in Def_P(U) \Rightarrow P_\tau(X \cup Y) = P_\tau(X) \cup P_\tau(Y)$	

Finally, we present the fact that operators $(G_*^\exists, G_\exists^*)$ are generalizations of (P_τ, P_b^τ):

Proposition 3. Let (U, τ) be a tolerance space, where $Gr(U) = G_\tau(U)$ for operators $G_*^\exists$ and $G_\exists^*$, then for any set $X \subseteq U$ the following identities hold:

(1) $G_*^\exists(X) = P_\tau(X)$,
(2) $G_\exists^*(X) = P_b^\tau(X)$.

Proof Let $Gr(U) := G_\tau(U)$.

(1) $G_*^\exists(X) = P_\tau(X)$

$$a \in G_*^\exists(X)$$
$$\exists A \in Gr(a) : A \subseteq X,$$
$$\exists A : a \in A \ \& \ A \in Gr(U) \ \& \ A \subseteq X,$$
$$a \in \bigcup\{Y \in G_\tau(U) : Y \subseteq X\}$$
$$a \in P_\tau(X)$$

Let us note that conditions $a \in A$ and $A \in Gr(U)$ are equivalent to $A \in Gr(a)$. Thus the above implicational steps can be strengthened to equivalential steps, so $a \in G_*^{\exists}(X) \Leftrightarrow a \in P_\tau(X)$. Since $a \in G_*^{\exists}(X)$ was selected arbitrary then we have proven that $G_*^{\exists}(X) = P_\tau(X)$.

(2) $G_{\exists}^*(X) = G_b^\tau(X)$

Equality (1) of this proposition was proven for any set '$X \subseteq U$. Thus in particular it holds also for the complement of set X: $G_*^{\exists}(X') = P_\tau(X')$

$$G_{\exists}^*(X) = G_{\exists}^*(X)''$$
$$= G_*^{\exists}(X')'$$
$$= P_\tau(X')'$$
$$= P_b^\tau(X)''$$
$$= P_b^\tau(X).$$

Therefore $G_{\exists}^*(X) = P_b^\tau(X)$. □

5 Conclusions

We presented and proved all basic properties of operators $(\tau_*, \tau^*), (P_\tau, P^\tau)$, including the proof of mutual definability of (τ_*, τ^*) operators and the lack of this property for operators (P_τ, P^τ) proving those inclusions which hold and presenting counterexamples to inclusions 2.9a and 2.9b which does not hold for arbitrary tolerance spaces and show that operators (P_τ, P^τ) lack the mutual definability property. We presented all basic properties of operators (P_τ, P_b^τ) and due to limited scope of space we proved properties of mutual definability of operators (P_τ, P_b^τ). We showed with proofs that operators $(G_*^\forall, G_\forall^*)$ are generalizations of operators (τ_*, τ^*) and that operators $(G_*^\exists, G_\exists^*)$ are generalizations of operators (P_τ, P_b^τ).

Table 1 presents approximation operators possessing duality properties and those operators which do not possess duality properties.

Table 1. Pairs of dual approximation operators, i.e. operators fulfilling properties $Q_*(X)' = Q^*(X')$ and $Q^*(X)' = Q_*(X')$ vs. pairs of non-dual operators. We presented operators according to historical order of appearance.

Pairs of dual operators	Pairs of non-dual operators
(R_*, R^*)	
(τ_*, τ^*)	
$(G_*^\forall, G_\forall^*)$	
$(G_*^\exists, G_\exists^*)$	
(P_τ, P_b^τ)	(P_τ, P^τ)

Table 2. Pairs of approximation operators associated with definable sets in the sense that their lower approximation operator Q_* possess the following property $Q_*(X) = X$.

Operators associated with definable sets	Operators not associated with definable sets
(R_*, R^*)	
	(τ_*, τ^*)
	$(G_*^\forall, G_\forall^*)$
$(G_*^\exists, G_\exists^*)$	
(P_τ, P_b^τ)	(P_τ, P^τ)

Table 2 presents pairs of approximation operators associated with definable sets, i.e. pairs such that approximation operators are defined on the basis of definable sets, together with application of biting procedure in the case of upper approximation operators.

In the future work we are going to present a complete proof theory of basic properties of all operators discussed in this paper and their full comparison. In addition we are going to present topological operators generalizing rough sets operators proposed by T.Y. Lin. [3–5].

References

1. Bargiela, A., Pedrycz, W.: Granular Computing: An Introduction. Kluwer Academic Publishers, Dordrecht (2003)
2. Demri, S., Orłowska, E.: Incomplete Information: Structures. Springer-Verlag, Inference, Complexity (2002)
3. Lin, T.Y.: Topological and fuzzy rough sets. In: Słowiński, R. (ed.), Intelligent Decision Support, pp. 287–304, Springer, Dordrecht (1992)
4. Lin, T.Y.: Granular computing on binary relations I, II. In: Polkowski, L., Skowron, A. (eds.), Rough Sets in Knowledge Discovery, pp. 107-140. Physica-Verlag (1998)
5. Lin, T.Y.: A roadmap from rough set theory to granular computing. In: Wang, G.-Y., Peters, J.F., Skowron, A., Yao, Y. (eds.) RSKT 2006. LNCS (LNAI), vol. 4062, pp. 33–41. Springer, Heidelberg (2006). https://doi.org/10.1007/11795131_6
6. Lin, T.Y., Liau, C.J., Kacprzyk, J. (eds.): Granular, Fuzzy, and Soft Computing: A Volume in the Encyclopedia of Complexity and Systems Science Series 1st ed. Springer-Verlag (2023)
7. Lipski, W.: Informational systems with incomplete information. In: 3rd International Symposium on Automata, Languages and Programming, Edinburgh, Scotland, pp. 120–130 (1976)
8. Orłowska, E.: Semantics of vague concepts. applications of rough sets, Polish academy of sciences 469. In: Dorn, G., Weingartner, P. (eds.) Foundations of Logic and Linguistics, pp. 465–482. Plenum Press, Problems and Solutions (1985)
9. Orłowska, E.: Reasoning with incomplete information: rough set based information logics. In: Algar, V.S., Bergler, F.Q. Dong (eds.), Incompleteness and Uncertainty in Information Systems Workshop, pp. 16–33. Springer, London (1993). https://doi.org/10.1007/978-1-4471-3242-4_2

10. Orłowska, E., Pawlak, Z.: Representation of nondeterministic information. Theoret. Comput. Sci. **29**, 27–39 (1984)
11. Pawlak, Z.: Information systems - theoretical foundations. Inf. Syst. **6**, 205–218 (1981)
12. Pawlak, Z.: Rough sets. Int. J. Comput. Inf. Sci. **18**, 341–356 (1982)
13. Pawlak, Z.: Theoretical Aspects of Reasoning About Data. Rough sets. Kluwer Academic Publishers, Dordrecht (1991)
14. Pawlak, Z.: Some issues on rough sets. In: Peters, J.F., Skowron, A., Grzymała-Busse, J.W., Kostek, B., Świniarski, R.W., Szczuka, M.S. (eds.) Transactions on Rough Sets I. LNCS, vol. 3100, pp. 1–58. Springer, Heidelberg (2004). https://doi.org/10.1007/978-3-540-27794-1_1
15. Pawlak, Z., Skowron, A.: Rudiments of rough sets. Inf. Sci. **177**, 3–27 (2007)
16. Polkowski, L., Skowron, A., Zytkow, J.: Tolerance based rough sets. In: Lin, T.Y., Wildberger, M.A. (eds.) Soft Computing: Rough Sets, Fuzzy logic, Neural Networks, Uncertainty Management, pp. 55–58. Simulation Councils Inc, San Diego (1995)
17. Pedrycz, W., Skowron, A., Kreinovich, V. (eds.): Handbook on Granular Computing. Wiley, New York (2009)
18. Rauszer, C., Skowron, A.: The discernibility matrices and functions in information systems. In: R. Słowiński, (ed.) Intelligent decision support. Handbook of Applications and Advances in the Rough Set Theory. Kluwer, pp. 331–362 (1991)
19. Skowron, A., Stepaniuk, J.: Tolerance approximation spaces. Fund. Inform. **27**, 245–253 (1996)
20. Ślęzak, D., Wasilewski, P.: Granular sets – foundations and case study of tolerance spaces. In: An, A., Stefanowski, J., Ramanna, S., Butz, C.J., Pedrycz, W., Wang, G. (eds.) RSFDGrC 2007. LNCS (LNAI), vol. 4482, pp. 435–442. Springer, Heidelberg (2007). https://doi.org/10.1007/978-3-540-72530-5_52
21. Wasilewski, P.: Dependency and supervenience. In: L. Czaja (ed.) Proceedings Concurrence, Specification and Programming (CS&P'2003), vol. 2, pp. 550–560. University of Warsaw Press (2003)
22. Wasilewski, P.: On selected similarity relations and their applications into cognitive science (in Polish). Unpublished doctoral dissertation, Jagiellonian University: Department of Logic, Krakow, Poland (2004)
23. Wasilewski, P.: Concept lattices vs. approximation spaces. In: Ślęzak, D., Wang, G., Szczuka, M., Düntsch, I., Yao, Y. (eds.) RSFDGrC 2005. LNCS (LNAI), vol. 3641, pp. 114–123. Springer, Heidelberg (2005). https://doi.org/10.1007/11548669_12
24. Wasilewski, P.: Algebras of definable sets vs concept lattices. Fundamenta Informaticae **167**(3), 235–256 (2019)
25. Wasilewski, P., Ślęzak, D.: Foundations of rough sets from vagueness perspective. In: Hassanien, A.E., Suraj, Z., Ślęzak, D., Lingras, P. (eds.) Rough Computing. Theories, Technologies and Applications, pp. 1–37. Information Science Refer (2008)
26. Yao, Y.Y.: Relational interpretations of neighborhood operators and rough set approximation operators. Inf. Sci. **111**, 239–259 (1998)
27. Yao, Y. Y.: Information granulation and rough set approximation. Int. J. Intell. Syst. **16**(1), 87–104 (2001)
28. Yao, Y.Y.: On generalizing rough set theory. In: Wang, G., Liu, Q., Yao, Y., Skowron, A. (eds.) RSFDGrC 2003. LNCS (LNAI), vol. 2639, pp. 44–51. Springer, Heidelberg (2003). https://doi.org/10.1007/3-540-39205-X_6
29. Yao, Y.Y., Yao, B.: Covering based rough set approximations. Inf. Sci. **200**, 91–107 (2012)

Parametrized γ-Decision Valuation for Variable Precision Rough Set Model

Soma Dutta[1(✉)] and Dominik Ślęzak[2]

[1] University of Warmia and Mazury in Olsztyn, Słoneczna 54, 10-710 Olsztyn, Poland
soma.dutta@matman.uwm.edu.pl
[2] Institute of Informatics, University of Warsaw, ul. Banacha, 02-097 Warsaw, Poland
slezak@mimuw.edu.pl

Abstract. The aim of the paper is to look for a decision valuation which can encode the decision related information available in a decision system following the strategy of Variable Precision Rough Set (VPRS) model. The paper presents different possible decision valuations and discusses from what aspects they represent or lack to represent the VPRS model.

Keywords: VPRS model · Decision valuation · Decision reduct · Positive region

1 Introduction

Variable Precision Rough Set (VPRS) model [1] provides a generalization of the classical rough set model [2] by incorporating a probabilistic measure of the occurrence of an event[1] and designing the positive, negative and boundary region of that event (or concept) with respect to the given probabilistic measure.

Given an information system $\mathbb{A} = (U, A)$ [3], one can consider an ordering among the attributes of A. Thus, for a particular object the values corresponding to these attributes can be represented as a vector. For any $X \subseteq A$, we may consider a vector of dimension $|X|$ just by eliminating attributes from $A \setminus X$ and this is usually known as the information signature of an object respective to the set of attributes X. Consequently, the whole universe of objects can be partitioned into some clusters, each of which contains objects with the same (or similar in some sense) information signature. Such clusters of objects can be equivalence classes generated from an equivalence relation or in more general context it can be even called as information granules (or granules) generated from a similarity relation or different strategies of covering [4] the objects from U. A set of objects from U, or a concept, is called definable with respect to a set of attributes $X \subseteq A$ if it can be expressed as the union of some above mentioned clusters; otherwise using the classical rough set model one can approximately describe it in terms of positive, negative and boundary regions of the concept.

---
[1] An event may be considered as a set of objects from the considered universe of objects. In other words, an event may represent a concept.

However, in practice going beyond such approximations more finer tuning techniques are required so that some important data about the objects can be restored while creating such clusters or information granules [5]. For instance, majority of objects from a cluster belonging to the positive region and a very few belonging to the negative region of a concept may indicate presence of an anomaly in acquiring data. Those few negative examples may be regarded as outliers [6] which might contain important information. Thus, to make use of such objects in defining the base concept a further tuning is required. In literature, one can find many such tools based on tolerance relation, similarity relation or some fuzzy-rough hybridization methods. Considering a probabilistic measure as a threshold for tuning the inclusion of an object in the set of positive examples is one such tools which often comes in help in practice as without any additional apparatus the measure is directly obtained from the available data in $\mathbb{A}$. Thus, VPRS model gains its usefulness among researchers.

While making granulation of data through restoring significant information, the aspects of data reduction inevitably come up. One can find different perspectives as well as techniques of data reduction available in different fields of studies, starting from probability theory [7] to relational databases [8], theories of semigraphoid [9] etc. For an overview the readers are referred to [10,11].

Decision reduct [12] refers to one such domains of research which deals with the criteria of data reduction in the context of a decision system, an information system with a (set of) designated attribute(s) representing decision(s). Here the crucial aspect of data reduction is to restore the significant character of the decision classes (i.e., the clusters of objects having specific decision values) as much as possible encoded in the granules, created with respect to the information signatures of the objects. In this regard, the first step is to design a decision valuation in a way that the information about the decision classes available in a given decision system, say $(U, A \cup \{d\})$, can be encoded in that decision valuation, say ν_d, in a compressed way. Then the next step is to look for a subset of attributes which can preserve the information compressed in ν_d with respect to the whole set of attributes. Following this line of thought there can be many decision valuations focusing on different aspects of decision making and consequently different notions of decision reduct obtained [11] based on them.

In this paper, our interest lies on finding a suitable generalization for a particular decision valuation, known as γ-decision valuation. Let us briefly state the ground behind the importance of γ-decision valuation and the need for its generalization. Given a decision system $(U, A \cup \{d\})$ and $X \subseteq A$, let us consider the equivalence relation IND_X, which contains all pairs of objects (u, u') such that there is a vector of values $\vec{x}$ over X with $Inf_X(u) = Inf_X(u') = \vec{x}$, where $Inf_X(u)$ denotes the information signature of u with respect to X. Then IND_X partitions U in clusters of objects $[\vec{x}]$, which is an equivalence class of objects having information signature $\vec{x}$. Respective to X the positive region is defined as $POS_X = \cup\{[\vec{x}] : [\vec{x}] \subseteq D_k \text{ for some } k\}$, where $D_k \subseteq U$ represents the k-th decision class containing all objects with the decision value v_k. Using the notion of positive region the criterion for POS-decision reduct is defined as follows. If $POS_A = POS_X$, and there is no proper subset Y of X such that

$POS_A = POS_Y$, then X is a POS-decision reduct. That is, X is a smallest such set of attributes, which can effectively describe the decision classes of the given decision system as available with respect to the whole set A of attributes. So, here the strategy is to ignore the decisions of those objects which are inconsistent, i.e., the objects having same information signature but different decisions. This nature of information synthesis while reducing data is designed by a decision valuation, namely γ-decision valuation, which imposes that if a cluster of objects $[\vec{x}]$ is completely contained in a decision class, say D_k, then the respective value-vector $\vec{x}$ would receive the value v_k, and in other cases a dummy value ? is assigned to the value-vector. Our aim is to propose a parametrized version of γ-decision valuation ν_γ, so that it fits well with the three (positive, negative and boundary) regions of the VPRS model.

In this regard, Sect. 2 presents a summary of VPRS model. Section 3 introduces the notion of decision valuation and the respective notion of decision reduct with some examples. Section 4 presents different decision valuations addressing different aspects of VPRS model. Section 5 discusses some prevalent properties of decision valuations and their relations in the context of the proposed decision valuations for VPRS model. Section 6 ends with a conclusion.

2 VPRS Model

In this section we present a brief overview of VPRS model by introducing a probabilistic measure in the context of decision systems.

Let $\mathbb{A} = (U, A \cup \{d\})$ be a decision system where U can be infinite in general. Any subset E of U can be regarded as an event and P is a probabilistic measure over the σ-algebra $\mathcal{M}(U)$ of measurable subsets of U [1]. Given $X \subseteq A$ and the indiscernibility relation IND_X, we assume that the family of equivalence classes U/IND_X is such that for each $[\vec{x}] \in U/IND_X$, $0 < P([\vec{x}]) < 1$ and $\Sigma_{[\vec{x}] \in I/IND_X} P([\vec{x}]) = 1$ hold. The first constraint indicates that every indiscernibility class belonging to U/IND_X must occur with a positive probability and it is not a certain event. Moreover, U/IND_X is considered to be a countable family of equivalence classes. From the perspective of applications, both the constraints are meaningful as in practice we usually do not go beyond a countably infinite data and a data set having only one possible vector of values over a set of attributes X (i.e., a certain event) does not need any further analysis.

Now for any $E \subseteq U$ the conditional probability of E given $[\vec{x}]$ is defined by $P(E|[\vec{x}]) = \frac{P(E \cap [\vec{x}])}{P([\vec{x}])} = \frac{|E \cap [\vec{x}]|}{|[\vec{x}]|}$. So, based on the notion of classical rough set we can have the following approximate regions [1].

Definition 1. *For any $E \subseteq U$ and $X \subseteq \mathcal{A}$*

(i) $POSITIVE_X(E) = \cup \{[\vec{x}] : P(E|[\vec{x}]) = 1\}$.
(ii) $NEGATIVE_X(E) = \cup \{[\vec{x}] : P(E|[\vec{x}]) = 0\}$.
(iii) $BOUNDARY_X(E) = \cup \{[\vec{x}] : 0 < P(E|[\vec{x}]) < 1\}$.

It is to be noted that following the standard terminologies of rough set based approximations [2], $POSITIVE_X(E)$ is nothing but $LOW_X(E)$, the lower approximation of E with respect to X, $BOUNDARY_X(E) = UPP_X(E) \setminus LOW_X(E)$, the upper approximation of E excluding the lower approximation of E with respect to X, and $NEGATIVE_X(E)$ is the complement of the upper approximation of E with respect to X.

Given a decision system $\mathbb{A} = (U, A \cup \{d\})$ let us consider the set of all possible value-vectors as $\mathcal{V}_A = \{Inf_X(u) : u \in U, X \subseteq A\}$. Now, referring to the notion of POS_X introduced in the Introduction we can see the following relation.

Proposition 1. Let $\mathbb{A} = (U, A \cup \{d\})$ and $\vec{x} \in \mathcal{V}_A$.

(i) $[\vec{x}] \subseteq POS_X$ if and only if $[\vec{x}] \subseteq POSITIVE_X(D_k)$ for some D_k.
(ii) $[\vec{x}] \subseteq (POS_X)^c$ if and only if $[\vec{x}] \subseteq NEGATIVE_X(D_k) \cup BOUNDARY_X(D_k)$ for all D_k.

VPRS model introduces a possibility of tuning these crisp notions of positive, negative and boundary regions, presented in Definition 1, by incorporating two parameters for lower (l) and upper (u) thresholds such that $0 \leq l < u \leq 1$. Using these thresholds the notions of positive region and negative region are generalized as follows [1].

Definition 2. Given $\mathbb{A} = (U, A \cup \{d\})$ the u-positive region and l-negative region of $E \subseteq U$ with respect to $X \subseteq A$ are given by:

(i) $POSITIVE_X^u(E) = \cup \{[\vec{x}] : P(E|[\vec{x}]) \geq u\}$
(ii) $NEGATIVE_X^l(E) = \cup \{[\vec{x}] : P(E|[\vec{x}]) \leq l\}$

So, clearly $POSITIVE_X^u(E)$ allows more flexibility in considering an object as a positive example of a concept E as it includes those equivalence classes for which the conditional probability of E given the value vector $\vec{x}$ over X is at least u. Thus, $POSITIVE_X(E) \subseteq POSITIVE_X^u(E)$. Similarly, it is easy to check that $NEGATIVE_X(E) \subseteq NEGATIVE_X^l(E)$. Thus, boundary region $BOUNDARY_X^{l,u}(E) = \cup \{[\vec{x}] : l < P(E|[\vec{x}]) < u\}$ becomes more thinner than that of the classical context, and compare to classical context in VPRS we can gain more certainty in making decision for unseen cases.

Now, considering that for two sets C and D, 'C is included in D to the degree a', denoted as $C \subseteq_a D$, iff $\frac{|C \cap D|}{|C|} \geq a$, a graded POS-zone is introduced by defining $POS_X^u = \cup\{[\vec{x}] : [\vec{x}] \subseteq_u D_k \text{ for some } k\}$. Hence, Proposition 1 can be rephrased in the context of VPRS model as follows.

Proposition 2. For $\mathbb{A} = (U, A \cup \{d\})$ and $X \subseteq A$

(i) $[\vec{x}] \subseteq POS_X^u$ if and only if $[\vec{x}] \subseteq POSITIVE_X^u(D_k)$ for some k.
(ii) $[\vec{x}] \subseteq (POS_X^u)^c$ if and only if $[\vec{x}] \subseteq NEGATIVE_X^l(D_k) \cup BOUNDARY_X^{u,l}(D_k)$ for all k.

3 Decision Valuation and Decision Reduct

This section aims to present a brief overview of the notion of decision valuation and decision reduct [11]. A decision valuation is a function which assigns a value from a decision space, say $\mathcal{D}$, to the objects of a decision system based on their information signatures. The meaning of $\mathcal{D}$ may depend on the way of designing a decision model to infer about the relationships between particular information signatures and their respective decision values. Thus, $\mathcal{D}$ may contain several subspaces such as the subspace containing all decision values of the given decision system, the subspace containing all subsets of decision values, the subspace containing all probabilistic distributions over the decision values, some other auxiliary symbols etc. The formal definition can be presented as follows.

Definition 3. *[11] Given a decision system $\mathbb{A} = (U, A \cup \{d\})$, a decision valuation is a function $\nu : \mathcal{V}_A \mapsto \mathcal{D}$, where $\mathcal{D}$ is a decision space of all possible decision assignments.*

Let us also note that Definition 3 is formulated for a given decision table; so for $X, Y \subseteq A$ and the respective vectors of values $\overrightarrow{x}, \overrightarrow{y} \in \mathcal{V}_A$ the concatenated vector $\overrightarrow{xy}$ represents the value-vector over $X \cup Y$ while appearance of the common attributes is considered exactly once (maintaining the order of the attributes in A) if X and Y overlap. Then the specific assignments $\nu(\overrightarrow{x}), \nu(\overrightarrow{xy}) \in \mathcal{D}$ are determined based on the information gathered in $\mathbb{A} = (U, A \cup \{d\})$. Now given a decision valuation ν we can formulate the notion of decision reduct in the following manner.

Definition 4. *Given a decision system $\mathbb{A} = (U, A \cup \{d\})$ and a decision valuation $\nu : \mathcal{V}_A \mapsto \mathcal{D}$, a subset $X \subseteq A$ is said to be a ν-decision reduct if it is a minimal such subset for which $\nu(\overrightarrow{xy}) = \nu(\overrightarrow{x})$ for all $\overrightarrow{xy}$ over $X \cup Y = A$.*

In more precise notation the condition $\nu(\overrightarrow{xy}) = \nu(\overrightarrow{x})$ for all $\overrightarrow{xy}$ may be denoted as $I_\nu(d|X|Y)$ which means that under the decision valuation ν the decision assignment to d given (the value-vectors for) the set of attributes X is independent to (the value-vectors for) the set of attributes Y.

Let us present an example of a decision valuation based on a specific decision model in the context of a decision system. If in a decision system all objects from an equivalence class with respect to some $X \subseteq A$ belong to a single decision class, then the whole equivalence class is contained in the POS_X, and thus there is no ambiguity in describing such objects by a decision rule of the form $\wedge_{a \in X}(a = v_a) \Rightarrow (d = v_k)$ for some decision value v_k. However, if there are two objects having the same vector of values with respect to X, but fall into two different decision classes, then modeling such a class of objects is not straightforward. One very well known approach of decision modeling, known as generalized decision function, is to group such decision values together and attach this set of values to the concerned equivalence class of objects. Formally the example can be presented as below.

Example 1. Given $\mathbb{A} = (U, A \cup \{d\})$, the generalized decision valuation (or ∂-decision valuation) $\nu_\partial : \mathcal{V}_A \mapsto \mathcal{D}_\partial$ is defined as follows; for any vector $\vec{x} \in \mathcal{V}_A$ of values on the set of attributes $X \subseteq A$:

$$\nu_\partial(\vec{x}) = \{d(u') : Inf_X(u') = \vec{x}\} \quad (1)$$

where $\mathcal{D}_\partial$ denotes the space of all subsets of decision values[2].

For the above decision model the respective notion of decision reduct is as follows.

Definition 5. *Given $\mathbb{A} = (U, A \cup \{d\})$ and $X \subseteq A$, if $\nu_\partial(\vec{x}) = \nu_\partial(\vec{xy})$ holds for every $\vec{xy}$ over $X \cup Y = A$ then X is said to be a ∂-decision superreduct. Moreover, if X is an irreducible set satisfying the above condition, then it is said to be a ∂-decision reduct.*

Contrary to ∂-decision reduct the POS-decision reduct, discussed in the Introduction, looks for a reduced set of attributes focusing only on the consistent objects of a given decision system. Thus, the information regarding the inconsistent object of a data table is completely ignored while making decision. As in Example 1 the generalized decision valuation ν_∂ is presented, below we present the decision valuation corresponding to the POS-decision model.

Example 2. Given $\mathbb{A} = (U, A \cup \{d\})$, the POS-decision valuation (γ-decision valuation in short) $\nu_\gamma : \mathcal{V}_A \mapsto \mathcal{D}_\gamma$ is defined as:

$$\nu_\gamma(\vec{x}) = \begin{cases} v_k & \text{if } \nu_\partial(\vec{x}) = \{v_k\} \\ ? & \text{otherwise} \end{cases} \quad (2)$$

whereby $\mathcal{D}_\gamma$ is the set of all decision values with an additional dummy value "?".

Decision valuation ν_γ assigns a single decision value to a given vector of values $\vec{x}$, over a set of attributes X, if $[\vec{x}] \subseteq D_k$ for some k, and else it returns ?. This models a scenario in which decision can be made only in the case of full certainty/consistency (i.e. $[\vec{x}] \subseteq POS_X$). The value "?" may be compared to a missing value, and following [13] it can be interpreted as *do not care*; however, the name *boundary value* seems to have stronger interpretation within the theory of rough sets [14], as we can rephrase Proposition 1, in the following way.

Corollary 1. *Let $\mathbb{A} = (U, A \cup \{d\})$ and $\vec{x} \in \mathcal{V}_A$.*

(i) $[\vec{x}] \subseteq POS_X$ if and only if $\nu_\gamma(\vec{x}) = v_k$ for some k.
(ii) $[\vec{x}] \subseteq (POS_X)^c$ if and only if $\nu_\gamma(\vec{x}) = ?$.

The role of "?" can be related to *three-way decision making*, where decisions may not be made due to a lack of sufficient evidence [15]. Based on ν_γ we can rephrase the notion of POS-decision reduct, presented in the Introduction, as follows.

---

[2] It is to be noted that given any $\vec{x}$ the decision valuation ν_∂ only assigns a non-empty subset of $\mathcal{D}_\partial$ to it.

Definition 6. A set X of attributes is said to be a γ-decision reduct if X is a smallest such set that for any $\vec{x}, \vec{xy} \in \mathcal{V}_A$, $\nu_\gamma(\vec{x}) = \nu_\gamma(\vec{xy})$ where $X \cup Y = A$.

Another very popular way of aggregating decision related information for a given information signature $\vec{x}$ is to consider the probability distribution of the decision values corresponding to that particular information signature.

Example 3. Given $\mathbb{A} = (U, A \cup \{d\})$, the probabilistic decision valuation (or μ-decision valuation in short) $\nu_\mu : \mathcal{V}_A \mapsto \mathcal{D}_\mu$ is defined as:

$$\nu_\mu(\vec{x}) = \langle \nu_\mu(\vec{x})[1], \ldots, \nu_\mu(\vec{x})[r] \rangle \qquad (3)$$

where $\mathcal{D}_\mu$ is the space of all probability distributions of the decision values, $r = |V_d|$ is the number of decision classes in $\mathbb{A}$ and for each $k = 1, \ldots, r$:

$$\nu_\mu(\vec{x})[k] = \frac{|[\vec{x}] \cap D_k|}{|[\vec{x}]|} \qquad (4)$$

Now based on ν_μ we have the respective notion of μ-decision reduct as follows.

Definition 7. A set X of attributes is said to be a μ-decision reduct if X is a minimal such set that for any $\vec{x}, \vec{xy} \in \mathcal{V}_A$, $\nu_\mu(\vec{x}) = \nu_\mu(\vec{xy})$ where $X \cup Y = A$.

The above mentioned decision valuations and respective decision spaces $\mathcal{D}_\partial, \mathcal{D}_\mu$, and $\mathcal{D}_\gamma$ can be considered as the most basic examples of compressing decision related information by some decision assignments returning values, sets of values, probability distribution of values, and/or some additional symbols. More complex forms of compressing decision related information can be obtained mostly by combining or modifying these decision spaces. For more examples of decision valuations and their respective decision reducts readers are referred to [11].

4 VPRS Model Versus a Parametrized γ-Decision Valuation

The aim of this section is to look for a suitable decision valuation that corresponds to the variable precision rough set model [1]. As γ-decision valuation corresponds to rough set based positive region (and its complement), intuitively, it seems that a generalized version of γ-decision valuation with a parameter representing the upper limit of VPRS model may match to the VPRS based positive region. First we consider the following variable precision decision valuation ν_p.

Example 4. Given $\mathbb{A} = (U, A \cup \{d\})$ and $0.5 < u \leq 1$, the variable precision decision valuation (p-decision valuation in short) $\nu_p : \mathcal{V}_A \mapsto \mathcal{D}_\gamma$ is defined as:

$$\nu_p(\vec{x}) = \begin{cases} v_k & \text{if } \nu_\mu(\vec{x})[k] \geq u \text{ for some } k \\ ? & \text{otherwise} \end{cases} \qquad (5)$$

Here to be noted that ν_p for $u \leq 0.5$ cannot be defined in the given decision space $\mathcal{D}_\gamma$ as for a given $\vec{x}$ there may be more than one decision values with probability $u \leq 0.5$. Decision valuation ν_p is a less strict variant of ν_γ as contrary to ν_γ, ν_p allows a room for inconsistent cases in the process of decision making.

Let us review the connection of ν_p with VPRS model discussed in Sect. 2. Given $\mathbb{A} = (U, A \cup \{d\})$, let $X \subseteq A$ and $\{v_1, v_2, \ldots v_r\}$ be the set of decision values for the decision attribute d. Thus, respectively there are decision classes $D_1, D_2, \ldots, D_r$. Referring to Sect. 2, let us assume $l = 0$ and $u > 0.5$. So, for the k-th decision class we have the following relation.

$$POSITIVE_X^u(D_k) = \cup\{[\vec{x}] : P(D_k|[\vec{x}]) \geq u\} = \cup\{[\vec{x}] : \nu_p[\vec{x}]) = v_k\} \quad (6)$$

However, if $[\vec{x'}] \subseteq BOUNDARY_X^{u,\ 0}(D_k) = \cup\{[\vec{x}] : 0 < P(D_k|[\vec{x}]) < u\}$ that does not necessarily imply $[\vec{x'}] \subseteq \cup\{[\vec{x}] : \nu_p[\vec{x}]) = ?\}$ as $\nu_p(\vec{x'}) = ?$ only when $P(D_k|[\vec{x'}]) \geq u$ does not hold for any k. So, we have the following relation.

$$BOUNDARY_X^{u,\ 0}(D_k) \cup NEGATIVE_X^0(D_k) \supseteq \cup\{[\vec{x}] : \nu_p[\vec{x}]) = ?\} \quad (7)$$

Similar to Corollary 1 we get the following corollary in the context of ν_p.

Corollary 2. *For $\mathbb{A} = (U, A \cup \{d\})$ and $X \subseteq A$ we have following relations.*

(i) $[\vec{x}] \subseteq POS_X^u$ if and only if $\nu_p(\vec{x}) = v_k$.
(ii) $[\vec{x}] \subseteq (POS_X^u)^c$ if and only if $\nu_p(\vec{x}) = ?$.

Thus, ν_p seems to be quite close to the VPRS model given the restriction $u > 0.5$ and the lower limit $l = 0$. Intuitively, in VPRS the perspective of introducing an upper limit u is to tune the crisp notion of 'belonging to a concept (to the full extent)' by a relaxed notion of 'belonging to a concept to a sufficiently large degree'. More support to this ground is emphasized by the fact, that for $u \leq 0.5$ we may have a class $[\vec{x}]$ which is included in two different decision classes, say D_k and D_j. Consequently, POS_X^u is supposed to include a class of inconsistent objects. This is contrary to the idea of POS-region. From that angle ν_p defined for $u > 0.5$ seems meaningful. However, this is a limitation that the general constraint $0 \leq l < u \leq 1$ of VPRS model cannot be captured by ν_p. So, naturally question arises if we can replace the single value case in Example 4 by a set of values when $\nu_\mu(\vec{x})[k] \geq u$ for $u \leq 0.5$. For instance, for $u = 0.5$ if we have two decision classes D_k and D_j satisfying $\nu_\mu(\vec{x})[k] \geq u$ and $\nu_\mu(\vec{x})[j] \geq u$ then what should be assigned to $\nu_p(\vec{x})$. One quick impression can lead to the possibility of replacing the single value case of ν_p by the majority decision valuation [16,17] which selects the highest probable decision values corresponding to a given $\vec{x}$. So, let us consider the following modification of ν_p.

Example 5. Given $\mathbb{A} = (U, A \cup \{d\})$, the variable precision majority decision valuation $\nu_{p:m}^u : \mathcal{V}_A \mapsto \mathcal{D}_{\partial:\gamma}$, is defined as:

$$\nu_{p:m}^u(\vec{x}) = \begin{cases} \nu_m(\vec{x}) & \text{if } \nu_\mu(\vec{x})[k] \geq u \\ ? & \text{otherwise} \end{cases} \quad (8)$$

where $\mathcal{D}_{\partial:\gamma} = \mathcal{D}_\partial \cup \{?\}$ and $\nu_m(\vec{x}) = \{v_k : \nu_\mu(\vec{x})[k] = \max_l \nu_\mu(\vec{x})[l]\}$.

It is easy to observe that for $u > 0.5$, $\nu^u_{p:m} = \nu_p$. The case for $u = 0.5$ can also be handled nicely as there can be at most two decision values with probability of occurrence 0.5. Moreover, like in the context of ν_p we have the correspondence presented in Eq. 6, in the context of $\nu^u_{p:m}$ a natural expected translation of Eq. 6 can be presented as follows.

$POSITIVE^u_X(D_k) = \cup\{[\vec{x}] : P(D_k|[\vec{x}]) \geq u\} = \cup\{[\vec{x}] : v_k \in \nu^u_{p:m}(\vec{x})\}$.

However, this does not work as it may fail for $u < 0.5$. For example, let $\nu_\mu(\vec{x'}) = \langle 0, \frac{1}{3}, \frac{2}{3}\rangle$ represent respectively the probability distribution of the decision classes D_1, D_2, D_3 in $[\vec{x'}]$. So, for $u = \frac{1}{3}$, $\nu^{\frac{1}{3}}_{p:m}(\vec{x'}) = \{v_3\}$. That is, though $[\vec{x'}] \subseteq POSITIVE^{\frac{1}{3}}_X(D_2)$, $[\vec{x'}] \nsubseteq \cup\{[\vec{x}] : v_2 \in \nu^{\frac{1}{3}}_{p:m}(\vec{x})\}$. Hence, clearly use of ν_m in $\nu^u_{p:m}$ cannot completely capture the essence of VPRS model.

In Example 5, for the case $\nu_\mu(\vec{x})[k] \geq u$, let us replace $\nu_m(\vec{x})$ by another decision valuation $\nu_{\partial^u}(\vec{x}) = \{v_i : \nu_\mu(\vec{x})[i] \geq u\}$, which is the u-cut [18] of $\nu_\partial(\vec{x})$. Furthermore, keeping analogy to the parameters u and l of VPRS model, in the underlying decision valuation let us introduce the possibility of returning a pair of sets corresponding to the positive and negative examples of VPRS model.

Example 6. Given $\mathbb{A} = (U, A \cup \{d\})$, the variable precision (u, l)-decision valuation $\nu^{u,l}_{p:\partial} : \mathcal{V}_A \mapsto \mathcal{D}_{\partial^2:\gamma}$, is defined as:

$$\nu^{u,l}_{p:\partial}(\vec{x}) = \begin{cases} (\nu_{\partial^u}(\vec{x}), \nu_{\partial_l}(\vec{x})) & \text{if } \nu_{\partial^u}(\vec{x}) \neq \phi \text{ or } \nu_{\partial_l}(\vec{x}) \neq \phi \\ ? & \text{otherwise} \end{cases} \quad (9)$$

where $\mathcal{D}_{\partial^2:\gamma} = (\mathcal{D}_\partial \times \mathcal{D}_\partial) \cup \{?\}$ and $\nu_{\partial_l}(\vec{x}) = \{v_j : \nu_\mu(\vec{x})[j] \leq l\}$.

Let us assume that $\langle i \rangle \nu^{u,l}_{p:\partial}(\vec{x})$ denotes the i-th projection of $\nu^{u,l}_{p:\partial}(\vec{x})$ where i can be 1 or 2 representing respectively the case for $\nu_{\partial^u}(\vec{x})$ and the case for $\nu_{\partial_l}(\vec{x})$. Now we can have the following one to one correspondence between the VPRS model and the decision valuation $\nu^{u,l}_{p:\partial}$.

Proposition 3. *Given $\mathbb{A} = (U, A \cup \{d\})$ and respective $\nu^{u,l}_{p:\partial} : \mathcal{V}_A \mapsto \mathcal{D}_{\partial^2:\gamma}$*

(i) $POSITIVE^u_X(D_k) = \cup\{[\vec{x}] : P(D_k|[\vec{x}]) \geq u\} = \cup\{[\vec{x}] : v_k \in \langle 1 \rangle \nu^{u,l}_{p:\partial}(\vec{x})\}$

(ii) $NEGATIVE^l_X(D_k) = \cup\{[\vec{x}] : P(D_k|[\vec{x}]) \leq l\} = \cup\{[\vec{x}] : v_k \in \langle 2 \rangle \nu^{u,l}_{p:\partial}(\vec{x})\}$

(iii) $BOUNDARY^{u,l}_X(D_k) = \cup\{[\vec{x}] : l < P(D_k|[\vec{x}]) < u\}$
$= \cup\{[\vec{x}] : v_k \notin \nu_{\partial^u}(\vec{x}) \cup \nu_{\partial_l}(\vec{x})\} \supseteq \cup\{[\vec{x}] : \nu^{u,l}_{p:\partial}(\vec{x}) = ?\}$.

Moreover, Corollary 2 can be reformulated as follows.

Corollary 3. *For $\mathbb{A} = (U, A \cup \{d\})$ and $X \subseteq A$ we have following relations.*

(i) $[\vec{x}] \subseteq POS^u_X$ if and only if $\langle 1 \rangle \nu^{u,l}_{p:\partial}(\vec{x}) \neq \phi$.
(ii) $[\vec{x}] \subseteq (POS^u_X)^c$ if and only if $\langle 2 \rangle \nu^{u,l}_{p:\partial}(\vec{x}) \neq \phi$ or $\nu_p(\vec{x}) = ?$.

Before passing to the next section let us note that like ν_p is a generalized version of ν_γ this new decision valuation $\nu_{p:\partial}^{u,l}$ also can be considered as a generalized version of ν_γ. For $u = 1$ and $l = 0$ if $\nu_{\partial^u}(\overrightarrow{x}) \neq \phi$ then it would be a singleton set, say $\{v_k\}$ and $\nu_{\partial_l}(\overrightarrow{x})$ would be just the complement of $\{v_k\}$ over the set of decision values. On the other hand, if $\nu_{\partial^u}(\overrightarrow{x}) = \phi$, then there must be some j such that $0 < \nu_\mu(\overrightarrow{x})[j] < 1$ and thus $\nu_{p:\partial}^{1,0}(\overrightarrow{x}) = ?$. Hence, the reference of the second component of $\nu_{p:\partial}^{1,0}(\overrightarrow{x})$ will no more be needed; hence we can easily get ν_γ as a special case of $\nu_{p:\partial}^{u,l}$. Moreover, the respective notion of decision reduct for $\nu_{p:\partial}^{u,l}$ can be formulated just by following the definition of decision reduct generated based on a given decision valuation (see Definition 4).

5 Properties of Decision Valuation Representing VPRS Model

In Sect. 3 we presented a brief background for the notion of decision valuation and respective decision reduct with some examples. Let us now briefly present some of the properties of decision valuations that will be relevant for further discussion. In this regard, we present a few notations and definitions below.

Definition 8. *Let $\nu : \mathcal{V}_A \mapsto \mathcal{D}_\nu$ be a decision valuation with the decision space $\mathcal{D}_\nu$. For any decision values v_k, v_l,*

(i) $\mathcal{D}_\nu^{\setminus k}$ denotes a subspace of $\mathcal{D}_\nu$ where v_k is not represented and
(ii) $\mathcal{D}_\nu^{k \geq l}$ is a subspace of $\mathcal{D}_\nu$ where v_k is represented at least as strongly as v_l.

From now onwards we use $\nu : \mathcal{V}_A \mapsto \mathcal{D}_\nu$ to emphasize that for a decision value v_k being *not represented* in a decision space may have different interpretation based on the decision space $\mathcal{D}_\nu$ of a decision valuation ν. For instance, $\mathcal{D}_\partial^{\setminus k}$ denotes the set of subsets of decision values which do not include v_k, while $\mathcal{D}_\gamma^{\setminus k}$ simply does not contain v_k. On the other hand, $\mathcal{D}_\partial^{k \geq l}$ represents all subsets of decision values except those which include v_l but do not include v_k, and $\mathcal{D}_\gamma^{k \geq l} = \mathcal{D}_\gamma^{\setminus l}$.

Let us also assume that there is an injection of the set of decision values into the decision space $\mathcal{D}_\nu$, whereby each v_k is uniquely assigned with its representation denoted as $\nu_k \in \mathcal{D}_\nu$. For instance, the unique representation of a value v_k in $\mathcal{D}_\partial$ is $\{v_k\}$, denoted as ∂_k. On the other hand, the unique representation of v_k in $\mathcal{D}_\mu$ is a unit vector having 1 at the k-th position, denoted by μ_k. For ν_γ the unique representation of v_k in $\mathcal{D}_\gamma$, denoted as γ_k, refers to the value v_k.

We now present a few properties of decision valuations [11] reflecting different decision making aspects while aggregating the decision information from a given decision table. If for $\overrightarrow{x} \in \mathcal{V}_A$ the respective decision value is v_k, we write $\overrightarrow{xv_k} \in \mathcal{V}$.

(Exclusion) If $\overrightarrow{xv_k} \notin \mathcal{V}$, then $\nu(\overrightarrow{x}) \in \mathcal{D}_\nu^{\setminus k}$.
(Inclusion) If $\overrightarrow{xv_k} \in \mathcal{V}$, then $\nu(\overrightarrow{x}) \notin \mathcal{D}_\nu^{\setminus k}$.

(**Monotonicity**) If $|\overrightarrow{xv_k}| \geq |\overrightarrow{xv_l}|$, then $\nu(\overrightarrow{x}) \in \mathcal{D}_\nu^{k \geq l3}$.
(**Consistency**) The equality $\nu(\overrightarrow{x}) = v_k$ holds, if and only if v_k is the only decision value that occurs for $\overrightarrow{x}$.

The exclusion property seems to be the most basic requirement and all the decision valuations discussed above satisfy it. However, though ν_∂ and ν_μ satisfy the inclusion property, for ν_γ it does not hold. On the other hand, all these decision valuations satisfy the monotonicity property. The consistency property ensures that all the consistent cases that appear in $\mathbb{A}$ are uniquely represented by a given $\nu : \mathcal{V}_A \mapsto \mathcal{D}_\nu$. For example, the definition of ν_∂ yields that $\nu_\partial(\overrightarrow{x}) = \partial_k$, if and only if for some $\overrightarrow{x}$ there is only one v_k such that $\overrightarrow{xv_k} \in \mathcal{V}$. Similarly, one can check that the consistency property follows for ν_μ and ν_γ as well.

There is one more property, known as discernibility property, that influences the processes of attribute reduction. Intuitively, it ensures that a given attribute is redundant, if and only if its removal does not cause shortening of two different value-vectors with different decision values to the same vector.

(**Discernibility**) For any $\overrightarrow{x}$ on $X \subseteq A$, the following conditions are equivalent:
  (i) the equality $\nu(\overrightarrow{x}) = \nu(\overrightarrow{xy})$ holds for every extension $\overrightarrow{xy}$ of $\overrightarrow{x}$ on $X \cup Y$
  (ii) the equality $\nu(\overrightarrow{xy_1}) = \nu(\overrightarrow{xy_2})$ holds for every pair of extensions $\overrightarrow{xy_1} \neq \overrightarrow{xy_2}$ of $\overrightarrow{x}$ on $X \cup Y$

The discernibility property [12] does not hold for all decision valuations; however, all the decision valuations (ν_γ, ν_∂, and ν_μ), discussed in Sect. 3, satisfy it. Some properties of semigraphoid axioms [9] are also relevant in this context. Let us briefly present the formal representation of those properties below.

(**Weak union**) $I_\nu(d|X|Y \cup W) \Rightarrow I_\nu(d|X \cup W|Y)$.
(**Decomposition**) $I_\nu(d|X|Y \cup W) \Rightarrow I_\nu(d|X|Y) \wedge I_\nu(d|X|W)$.
(**Contraction**) $I_\nu(d|X \cup W|Y) \wedge I_\nu(d|X|W) \Rightarrow I_\nu(d|X|Y \cup W)$.

In [11] it is shown that for all decision valuations the contraction property holds. Moreover, it has also been proved that any decision valuation satisfying the discernibility property satisfies the weak union property and the decomposition property. The contraction property ensures that if d is independent of Y in presence of $X \cup W$ and W is also redundant for d in presence of X then d can be determined without $Y \cup W$ in presence of X. Decomposition property allows to split the original data table into two smaller data tables each of which is enough to represent the nature of decision making categorized in the original table. On the other hand, weak union property is in some sense connected to monotonicity of checking irreducibility of the reduced set of attributes. Thus, this property is very crucial from the angle of computational complexity of many known attribute reduction algorithms.

---
[3] The notations $|\overrightarrow{xv_k}|, |\overrightarrow{xv_l}|$ represent the supports of the corresponding vectors in $\mathbb{A}$, i.e. the number of objects $u \in U$ such that $Inf_{X \cup \{d\}}(u) = \overrightarrow{xv_k}$ and $Inf_{X \cup \{d\}}(u) = \overrightarrow{xv_l}$, respectively.

Among the decision valuations, discussed in Sect. 4 to capture the essence of VPRS model, ν_p satisfies all the above mentioned properties other than inclusion property. As for $\overrightarrow{xv_k} \notin \mathcal{V}$, $\nu_\mu(\overrightarrow{x})[k] = 0$, the exclusion property is immediate. The inclusion property does not hold as for some $\overrightarrow{xv_k} \in \mathcal{V}$ with $\nu_\mu(\overrightarrow{x})[k] < u$ the claim $\nu_p(\overrightarrow{x}) \notin \mathcal{D}_\gamma^{\setminus k}$ does not hold. For monotonicity if we assume $|\overrightarrow{xv_k}| \geq |\overrightarrow{xv_l}|$, then there are two cases, namely $\nu_\mu(\overrightarrow{x})[k] \geq u$ and $\nu_\mu(\overrightarrow{x})[k] < u$. In both the cases $\nu_p(\overrightarrow{x}) \in \mathcal{D}_\gamma^{\setminus l} = \mathcal{D}_\gamma^{k \geq l}$. Hence, ν_p satisfies the monotonicity property. Moreover, it can be checked that ν_p satisfies one direction of the consistency property, all the semigraphoid axioms and the discernibility property. The argument for such a claim is already proved in [11] for $u = 0.7$, and it works for any $u > 0.5$.

However, as presented in Sect. 4, $\nu_{p:\partial}^{u,l}$ seems to be the more plausible decision valuation for VPRS model, and for $\nu_{p:\partial}^{u,l}$ we need to restate the above mentioned properties with some modifications as the decision space of $\nu_{p:\partial}^{u,l}$ deals with pairs of sets of values. So, below we present some necessary changes to state all the properties discussed in Sect. 5 in the context of $\nu_{p:\partial}^{u,l}$ whose decision space is $\mathcal{D}_{\partial^2 : \gamma} = (\mathcal{D}_\partial \times \mathcal{D}_\partial) \cup \{?\}$. First we need to specify the subspaces of $\mathcal{D}_{\partial^2 : \gamma}$ that are relevant for the exclusion, inclusion and monotonicity properties.

Definition 9. *Given $\nu_{p:\partial}^{u,l} : \mathcal{V}_A \mapsto \mathcal{D}_{\partial^2:\gamma}$ the subspaces of $\mathcal{D}_{\partial^2:\gamma}$ having no representation of k-th decision value and having representation of the k-th decision value as strongly as that of the j-th decision value are respectively*

(i) $\mathcal{D}_{\partial^2:\gamma}^{\setminus k} = (\mathcal{D}_\partial^{\setminus k} \times \mathcal{D}_\partial) \cup \{?\}$ and
(ii) $\mathcal{D}_{\partial^2:\gamma}^{k \geq j} = (\mathcal{D}_\partial^{k \geq j} \times \mathcal{D}_\partial) \cup \{?\}$.

Moreover, to state the consistency property we need to also specify how an injection from the set of decision values to $\mathcal{D}_{\partial^2:\gamma}$ should be defined so that each value v_k is assigned to a unique representation in $\mathcal{D}_{\partial^2:\gamma}$. This new decision valuation $\nu_{p:\partial}^{u,l}$ being a generalization of ν_γ, we may simply try to follow the analogy of the injection from the set of decision values to $\mathcal{D}_\gamma$ in the context of ν_γ; so, in the context of $\mathcal{D}_{\partial^2:\gamma}$ a straightforward and easy translation can be assigning the pair $(\{v_k\}, \phi)$ to each value v_k. That is, the unique value for v_k in $\mathcal{D}_{\partial^2:\gamma}$, denoted by $(\partial^2 : \gamma)_k$, is $(\{v_k\}, \phi)$. With this background setting we are now in the state of checking the above mentioned properties for $\nu_{p:\partial}^{u,l}$.

Theorem 1. *$\nu_{p:\partial}^{u,l}$ satisfies all the properties except the inclusion property mentioned at the beginning of Sect. 5.*

Proof. If $\overrightarrow{xv_k} \notin \mathcal{V}$, then $\nu_\mu(\overrightarrow{x})[k] = 0 < u$ and thus, $\nu_{\partial^u}(\overrightarrow{x}) \in \mathcal{D}_\partial^{\setminus k}$. This clarifies $\nu_{p:\partial}^{u,l}(\overrightarrow{x}) \in \mathcal{D}_{\partial^2:\gamma}^{\setminus k}$. That is, the exclusion property holds.

The inclusion property does not follow as for some $\overrightarrow{xv_k} \in \mathcal{V}$ if $\nu_\mu(\overrightarrow{x})[k] < u$ then $\nu_{\partial^u}(\overrightarrow{x}) \in \mathcal{D}_\partial^{\setminus k}$, and thus $\nu_{p:\partial}^{u,l}(\overrightarrow{x}) \in \mathcal{D}_{\partial^2:\gamma}^{\setminus k}$.

As $\mathcal{D}_{\partial^2:\gamma}^{k\geq j} = (\mathcal{D}_{\partial}^{k\geq j} \times \mathcal{D}_{\partial}) \cup \{?\}$ the monotonicity property of $\nu_{p:\partial}^{u,l}$ depends on the monotonicity property of ν_{∂^u}. Given $|\overrightarrow{xv_k}| \geq |\overrightarrow{xv_j}|$ there can be two cases: (i) $|\overrightarrow{xv_j}| \geq u$ and (ii) $|\overrightarrow{xv_j}| < u$. For case (i) $\nu_{\partial^u}(\overrightarrow{x}) \subseteq \{v_k, v_j\}$ and for case (ii) $\nu_{\partial^u}(\overrightarrow{x})$ either may contain v_k but not v_j or does not contain either of v_k, v_j. So, clearly $\nu_{\partial^u}(\overrightarrow{x}) \in \mathcal{D}_{\partial}^{k\geq j}$ and thus, $\nu_{p:\partial}^{u,l}(\overrightarrow{x}) \in \mathcal{D}_{\partial^2:\gamma}^{k\geq j}$.

One direction of the consistency property is straightforward as for $\overrightarrow{xv_k} \in \mathcal{V}$ where v_k is the only possible value, $\nu_\mu(\overrightarrow{x})[k] = 1 \geq u$ and $\nu_{\partial_l}(\overrightarrow{x}) = \phi$. So, $\nu_{p:\partial}^{u,l}(\overrightarrow{x}) = (\{v_k\}, \phi)$.
However, given $\nu_{p:\partial}^{u,l}(\overrightarrow{x}) = (\{v_k\}, \phi)$ it does not mean that v_k is the only possible value satisfying $\overrightarrow{xv_k} \in \mathcal{V}$.

For the weak union property let $\nu_{p:\partial}^{u,l}(\overrightarrow{xyw}) = \nu_{p:\partial}^{u,l}(\overrightarrow{x})$ for all $\overrightarrow{xyw}$. So, two cases arise: (i) $\nu_{p:\partial}^{u,l}(\overrightarrow{xyw}) = \nu_{p:\partial}^{u,l}(\overrightarrow{x}) = (\nu_{\partial^u}(\overrightarrow{x}), \nu_{\partial_l}(\overrightarrow{x}))$ or (ii) $\nu_{p:\partial}^{u,l}(\overrightarrow{xyw}) = \nu_{p:\partial}^{u,l}(\overrightarrow{x}) = ?$.
Case (i) yields that for all $\overrightarrow{xyw}$, and v_k, v_j, $\nu_\mu(\overrightarrow{xyw})[k] \geq u$ if and only if $\nu_\mu(\overrightarrow{x})[k] \geq u$ and $\nu_\mu(\overrightarrow{xyw})[j] \leq l$ if and only if $\nu_\mu(\overrightarrow{x})[j] \leq l$.
Now as $[\overrightarrow{xw}] = \cup_{\overrightarrow{y}} [\overrightarrow{xyw}]$ for each $\overrightarrow{y}$, $\nu_\mu(\overrightarrow{xyw})[k] \geq u$ if and only if $\nu_\mu(\overrightarrow{xw})[k] \geq u$ and $\nu_\mu(\overrightarrow{xyw})[j] \leq l$ if and only if $\nu_\mu(\overrightarrow{xw})[j] \leq l$.
So, for all $\overrightarrow{xyw}$, $\nu_{p:\partial}^{u,l}(\overrightarrow{xyw}) = (\nu_{\partial^u}(\overrightarrow{x}), \nu_{\partial_l}(\overrightarrow{x}))$ implies $\nu_{p:\partial}^{u,l}(\overrightarrow{xw}) = (\nu_{\partial^u}(\overrightarrow{x}), \nu_{\partial_l}(\overrightarrow{x}))$.
For case (ii) the argument follows similarly. That is, $\nu_{p:\partial}^{u,l}(\overrightarrow{xyw}) = \nu_{p:\partial}^{u,l}(\overrightarrow{xw})$ for all $\overrightarrow{xyw}$.

As the contraction property holds for all decision valuations we do not need to prove it.

The decomposition property can be proved as from the weak union property we already know $\nu_{p:\partial}^{u,l}(\overrightarrow{xyw}) = \nu_{p:\partial}^{u,l}(\overrightarrow{x})$ for all $\overrightarrow{xyw}$ implies $\nu_{p:\partial}^{u,l}(\overrightarrow{xyw}) = \nu_{p:\partial}^{u,l}(\overrightarrow{xw})$ for all $\overrightarrow{xyw}$. So, we have $\nu_{p:\partial}^{u,l}(\overrightarrow{x}) = \nu_{p:\partial}^{u,l}(\overrightarrow{xw})$ for all $\overrightarrow{xw}$. That is, from $I_{p:\partial^{u,l}}(d|X|Y \cup W)$, $I_{p:\partial^{u,l}}(d|X|W)$ is established. In similar fashion, $I_{p:\partial^{u,l}}(d|X|Y)$ also can be established.

For discernibility property condition (i) implies condition (ii) always holds. Let us assume for any $\overrightarrow{y_1} \neq \overrightarrow{y_2}$, $\nu_{p:\partial}^{u,l}(\overrightarrow{xy_1}) = \nu_{p:\partial}^{u,l}(\overrightarrow{xy_2})$ holds. So, we can conclude that for all extensions $\overrightarrow{xy}$ of $\overrightarrow{x}$, $\nu_{p:\partial}^{u,l}(\overrightarrow{xy})$ is the same.

We have two cases: (1) $\nu_{p:\partial}^{u,l}(\overrightarrow{xy}) = (\nu_{\partial^u}(\overrightarrow{xy}), \nu_{\partial_l}(\overrightarrow{xy}))$ and (2) $\nu_{p:\partial}^{u,l}(\overrightarrow{xy}) = ?$.
For case (1), $\nu_{\partial^u}(\overrightarrow{xy}) \neq \phi$ or $\nu_{\partial_l}(\overrightarrow{xy}) \neq \phi$. That is, for all $\overrightarrow{xy}$, $\nu_\mu(\overrightarrow{xy})[k] \geq u$ or $\nu_\mu(\overrightarrow{xy})[j] \leq l$ for some v_k, v_j.
As $[\overrightarrow{x}] = \cup_{\overrightarrow{y}} [\overrightarrow{xy}]$, for all $\overrightarrow{xy}$, $\nu_\mu(\overrightarrow{xy})[k] \geq u$ or $\nu_\mu(\overrightarrow{xy})[j] \leq l$ implies $\nu_\mu(\overrightarrow{x})[k] \geq u$ or $\nu_\mu(\overrightarrow{x})[j] \leq l$. So, we can claim that $\nu_{\partial^u}(\overrightarrow{xy}) \subseteq \nu_{\partial^u}(\overrightarrow{x})$ and $\nu_{\partial_l}(\overrightarrow{xy}) \subseteq \nu_{\partial_l}(\overrightarrow{x})$.
On the other hand, if $v_i \in \nu_{\partial^u}(\overrightarrow{x})$ then $\nu_\mu(\overrightarrow{x})[i] \geq u$. Then claim is that $\nu_\mu(\overrightarrow{xy})[i] \geq u$ for all extension $\overrightarrow{xy}$ of $\overrightarrow{x}$. Because, if not, there would be some $\overrightarrow{xy'}$ such that $\nu_\mu(\overrightarrow{xy'})[i] < u$. Then by assumption, for all extensions $\overrightarrow{xy} \neq \overrightarrow{xy'}$

we must have $\nu_\mu(\overrightarrow{xy})[i] < u$, which contradicts that $\nu_\mu(\overrightarrow{x})[i] \geq u$. Thus, for all $\overrightarrow{xy}$, $\nu_\mu(\overrightarrow{xy})[i] \geq u$. Hence, $\nu_{\partial^u}(\overrightarrow{x}) = \nu_{\partial^u}(\overrightarrow{xy})$, and similarly we can show $\nu_{\partial_l}(\overrightarrow{xy}) = \nu_{\partial_l}(\overrightarrow{x})$. That is, case (1) for $\nu_{p:\partial}^{u,l}(\overrightarrow{xy}) = \nu_{p:\partial}^{u,l}(\overrightarrow{x})$ for all $\overrightarrow{xy}$ is proved.

Case (2) also follows similarly. □

6 Conclusion

In this paper, our attempt has been to look for a suitable decision valuation that can represent the underlying strategy for synthesizing decision related information of an arbitrary value-vector in the context of VPRS model. In this regard, possibilities of different decision valuations, such as ν_p, $\nu_{p:m}^u$, and $\nu_{p:\partial}^{u,l}$ are proposed, and their properties and limitations are discussed. Among these discussed decision valuations, $\nu_{p:\partial}^{u,l}$ seems to be more suitable for VPRS model. However, there is still a room for other possible generalizations.

For instance, in contrary to other decision valuations $\nu_{p:\partial}^{u,l}$ accommodates the possibility of reflecting the decision classes that could be considered as the negative zone (or counterexamples) for a given information signature $\overrightarrow{x}$. However, formulation of all the discussed properties of decision valuations mostly focuses on analyzing the relationships between different subspaces of a particular decision space (for a given decision valuation) based on the information signatures falling into certain decision classes, or more precisely information signatures that are positive examples of certain decision classes. As an example, we can refer to the subspaces introduced in Definition 9. Clearly, while defining $\mathcal{D}_{\partial^2:\gamma}^{\backslash k}$ we only stick to the subspace where k-th decision value has not been represented in the set of positive examples ($\nu_{\partial^u}(\overrightarrow{x})$) for a given information signature $\overrightarrow{x}$. Similarly, in $\mathcal{D}_{\partial^2:\gamma}^{k \geq j}$ the only focus has been to ensure that in the set of positive examples of an information signature $\overrightarrow{x}$ the representation of the k-th decision class is at least as strong as that of the j-th decision class. In neither of the subspaces, we bother about the decision classes which are negative examples of $\overrightarrow{x}$. Such stipulations fit fine to the present statements of the properties of decision valuations. However, a need for further exploration by modifying the properties as well as subspaces focusing on both positive and negative examples of an information signature is quite intriguing.

In this regard, let us consider an example. Suppose we have a decision system $\mathbb{A} = (U, A \cup \{d\})$ where $X \cup Y = A$. We assume that there is only one information signature $\overrightarrow{x}$ over X; whereas the equivalence class $[\overrightarrow{x}]$ is split into two equivalence classes $[\overrightarrow{xy}]$ and $[\overrightarrow{xy'}]$ over $X \cup Y$. There are 150 objects with information signature $\overrightarrow{xy}$ among which 100 fall into the decision class corresponding to v_1 and 50 fall into the decision class for v_3. On the other hand, in $[\overrightarrow{xy'}]$ there are 100 objects among which 50 have decision value v_1, 40 have decision value v_2, and 10 have decision value v_3. So, for $u = \frac{1}{2}$ and $l = \frac{1}{3}$ we have the decision assignments, namely $\nu_{p:\partial}^{\frac{1}{2},\frac{1}{3}}(\overrightarrow{xy}) = (\{v_1\}, \{v_3\})$, $\nu_{p:\partial}^{\frac{1}{2},\frac{1}{3}}(\overrightarrow{xy'}) = (\{v_1\}, \{v_3\})$, and $\nu_{p:\partial}^{\frac{1}{2},\frac{1}{3}}(\overrightarrow{x}) = (\{v_1\}, \{v_2, v_3\})$. Following the definition of $\nu_{p:\partial}^{u,l}$ decision reduct we can conclude that X is not a reduct as $\nu_{p:\partial}^{\frac{1}{2},\frac{1}{3}}(\overrightarrow{x}) \neq \nu_{p:\partial}^{\frac{1}{2},\frac{1}{3}}(\overrightarrow{xy'})$. It is to be noted, that if we consider only the positive examples of $\overrightarrow{x}$ and all its extensions $\overrightarrow{xy}$, $\overrightarrow{xy'}$, then all have the same value $\{v_1\}$. However, on negative examples $\overrightarrow{x}$ and $\overrightarrow{xy'}$ differ and in order to be a decision reduct X should preserve the decision values for both positive and negative examples of all its extension over A. On the other hand, it seems to be an anomaly that while checking monotonicity we only try to ensure that if $|\overrightarrow{xv_k}| \geq |\overrightarrow{xv_j}|$ then only for positive examples v_k should be as strongly represented as v_j.

From another angle, if we consider a similar concept like $POSITIVE_X$ of classical rough set, we know if $[\overrightarrow{x}] \subseteq POSITIVE_X(D_k)$ then $[\overrightarrow{xy}]$ being a subset of $[\overrightarrow{x}]$ also is included in $POSITIVE_X(D_k)$. Now when we focus on both positive examples and negative examples of a given $\overrightarrow{x}$ is it not natural to think about similar relationship for $NEGATIVE_X(D_k)$. For instance, in the above example we see negative examples of both $\overrightarrow{xy}$ and $\overrightarrow{xy'}$ are preserved in $\overrightarrow{x}$.

All this discussion above points out to a need for further exploration - either by modifying the notion of decision reduct or by modifying statements of the some of the properties of decision valuations when such functions return as output both positive and negative instances of decision values for a given information signature.

References

1. Ziarko, W.: Variable Precision Rough Set Model. J. Comput. Syst. Sci. **46**(1), 39–59 (1993)
2. Pawlak, Z., Skowron, A.: Rudiments of Rough Sets. Inf. Sci. **177**(1), 3–27 (2007)
3. Skowron, A., Ślęzak, D.: Rough sets turn 40: from information systems to intelligent systems. Proc. FedCSIS **2022**, 23–34 (2022)
4. Chakraborty, M.K., Sardar, M.R.: Algebras and logics emerging out of rough sets. J. Comb. Inf. Syst. Sci. **46**(1-4), 231–313 (2021)
5. Pedrycz, W.: Information granules and their use in schemes of knowledge management. Scientia Iranica **18**(3), 601–610 (2011)
6. Agarwal, C.C.: Outlier Analysis. Springer, Cham (2017). https://doi.org/10.1007/978-3-319-47578-3
7. Dawid, A.P.: Conditional independence in statistical theory (with Discussion). J. Roy. Stat. Soc. B **41**(1), 1–31 (1979)
8. Wong, S.K.M.: Testing implication of probabilistic dependencies. Proc. UAI **1996**, 545–553 (1996)
9. Pearl, J., Paz, A.: Graphoids: graph-based logic for reasoning about relevance relations. In Boulay, B.D., Hogg, D., Steele, L. (eds.): Advances in Artificial Intelligence-II. North-Holland, pp. 357–367 (1987)
10. Ślęzak, D.: On generalized decision functions: reducts, networks and ensembles. Proc. RSFDGrC **2015**, 13–23 (2015)
11. Dutta, S., Ślęzak, D.: Nature of decision valuations in elimination of redundant attributes. Int. J. Approximate Reas. **165**(3), 109091 (2024)
12. Ślęzak, D., Dutta, S.: Dynamic and discernibility characteristics of different attribute reduction criteria. Proc. IJCRS **2018**, 628–643 (2018)

13. Clark, P.G., Gao, C., Grzymała-Busse, J.W.: A comparison of mining incomplete and inconsistent data. Inf. Technol. Control **46**(2), 183–193 (2017)
14. Pawlak, Z.: Rough sets. Int. J. Comput. Inform. Sci. **11**(5), 341–356 (1982)
15. Yao, Y.: Three-way decisions with probabilistic rough sets. Inf. Sci. **180**(3), 341–353 (2010)
16. Ślęzak, D.: Normalized decision functions and measures for inconsistent decision tables analysis. Fund. Inform. **44**(3), 291–319 (2000)
17. Stawicki, S., Ślęzak, D., Janusz, A., Widz, S.: Decision bireducts and decision reducts - a comparison. Int. J. Approximate Reasoning **84**, 75–109 (2017)
18. Klir, G.J., Yuan, B.: Fuzzy Sets and Fuzzy Logic: Theory and Applications. Prentics Hall PTR, Hoboken (1995)

Rough Algebraic Semantics of Concepts in a Distributed Cognition Perspective

A. Mani

Machine Intelligence Unit, Indian Statistical Institute,
203, B. T. Road, Kolkata 700108, India
a.mani.cms@gmail.com, amani.rough@isical.ac.in
https://www.logicamani.in

Abstract. Up-directed rough sets are introduced and studied by the present author in earlier papers. This is extended by her in two different granular directions in this research, with a surprising algebraic semantics. The granules are based on ideas of generalized closure under up-directedness that may be read as a form of weak consequence. This yields approximation operators that satisfy cautious monotony, while pi-groupoidal approximations (that additionally involve strategic choice and algebraic operators) have nicer properties. The study is primarily motivated by possible structure of concepts in distributed cognition perspectives, real or virtual classroom learning contexts, and student-centric teaching. This study thus provides directions for building AI models of distributed cognition, and related decision-making in general.

Keywords: Up Directed Relations · Granular Computing · Groupoidal Semantics · Directed Rough Sets · Concept Organization · Knowledge Representation · Aggregation Operations

1 Introduction

In general rough sets granular, pointwise or abstract approximations and related rough objects are studied [23]. These approximations can be derived from information tables or may be the result of abstractions from human (or machine) reasoning. A *general approximation space* or a *frame* is a pair of the form $S = \langle \underline{S}, R \rangle$ with $\underline{S}$ being a set and R being a binary relation (S and $\underline{S}$ *will be used interchangeably throughout this paper*). S is typically interpreted as a set of attributes or concepts.

In human reasoning contexts, in particular, it happens that any two elements of S is associated with (R-related to) *another element.* The association may be determined by reasoning patterns, preference, or through decision-making guided by an external mechanism. The property of up-directedness is essentially this. In [17], the general not-necessarily relational rough set context enhanced with such a directedness is studied by the present author and applied to specific

---

This research is supported by Woman Scientist Grant No. WOS-A/PM-22/2019 of DST, India.

problems in concept learning. *It should be noted that these approximation spaces are not necessarily derived from information tables, and a wider interpretation is admissible.* Among the many ways of representing knowledge, concepts, and higher order properties in these spaces, two specific granular approaches that capture forms of consequence and algebraic closure are studied in this research and usefully applied.

From a purely mathematical perspective, the property of up-directedness (often referred to as directedness) of generalized orders is widely used and is additionally used for studying *ideals of binary relations* [5]. *Algebras with a single 2-place operation (groupoids)* can be used to study them [3].

From the perspective of pedagogic content knowledge in classroom [1,8,19,21] or student centered teaching contexts [6,18], it is very important to dynamically represent the concepts generated from multiple sources. Knowledge representation in the groupoidal approach of the present author in [17], and [18] address aspects of this, and to improve the scope they are extended with entirely different granules (and weaker assumptions) in this research. These are better suited for modeling in the context as the granules have improved closure-related properties (as up-directedness is related to a form of closure under consequence of concepts). Additionally, the property is involved at a first order level.

In this research, general approximation spaces, in which the relation R is up-directed, are studied in relation to closure under up-directedness and choice functions in some detail by the present author. Related approximations are characterized, and illustrated with examples. It is additionally shown that the algebraic semantics of the latter is very distinct from those in which R is a partial or quasi-order ([7,15]). More specifically, it is proven that an algebraic model of pi-groupoidal approximations is an algebraic lattice with additional operations. The semantics is compatible with definable aggregation operations. Applications to human learning contexts (specifically, real or virtual classroom teaching), are additionally proposed. Missing proofs may be found in the extended version of this paper.

2 Some Background

Information tables or descriptive or knowledge representation systems [15,18] are not absolutely essential for general rough sets, however, they are useful (if available). *In this research, prefix or Polish notation is preferred for relations and functions defined on a set. So instances of a binary relation σ are denoted by σab.* If-then relations (or logical implications) in a model are written in infix form with $\longrightarrow$. For lattice theory, the reader is referred to [24], and for partial algebras to [11].

The following concepts were studied in earlier papers [17]. In a general approximation space $S = \langle \underline{S}, R \rangle$, let $U_R(a,b) = \{x : Rax \,\&\, Rbx\}$, and $U_R(S) = \{U_R(a,b) : a, b \in S\}$. Consider the following conditions:

$$(\forall a,b)(\exists c) Rac \,\&\, Rbc \qquad \text{(up-dir)}$$

$$(\forall a) Raa \qquad \text{(reflexivity)}$$

$$(\forall a,b)(Rab \,\&\, Rba \longrightarrow a = b) \qquad \text{(anti-sym)}$$

If S satisfies up-dir (or equivalently, the condition $(\forall a, b) \neg U_R(a, b) = \emptyset$), then it will be said to be an *upper directed approximation space*. If it satisfies all three conditions then it will be said to be an *up-directed parthood space*. For any element $a \in S$, the neighborhood $[a]$ generated by it will be the set $[a] = \{x : Rxa\}$. $[a]$ is the set of things that relate to a. The inverse neighborhood $[a]_i = \{x : Rax\}$ is the set of things that element a relates to. These are investigated by the present author and others in a separate joint paper.

Proposition 1. *For any $e, f, a \in S$,*

$$\text{If } e, f \in [a] \text{ then } a \in U_R(e, f), \text{ and conversely} \tag{nu1}$$
$$\text{If } (\forall a) \#([a]) \geq 2 \text{ then } (\forall a, b) \neg U_R(a, b) = \emptyset \text{ and conversely.} \tag{nu2}$$

For any subset $A \subseteq S$, the following approximations can be defined:

$$A^l = \bigcup \{[a] : [a] \subseteq A\} \tag{lower}$$
$$A^u = \bigcup \{[a] : \exists z \in [a] \cap A\} \tag{upper}$$

Theorem 1. *In an up-directed approximation space S, the following properties hold for any elements $a, b \in \wp(S)$:*

$$a^{ll} = a^l \subseteq a;\ a^u \subseteq a^{uu};\ a^l \subseteq a^{lu} \subseteq a^u \tag{l-id0, u-wid0, lu-inc}$$
$$(a \subseteq b \longrightarrow a^l \subseteq b^l);\ (a \subseteq b \longrightarrow a^u \subseteq b^u) \tag{l-mo,u-mo}$$
$$S^l = S^u \subseteq S\ \&\ \emptyset^l = \emptyset = \emptyset^u \tag{bnd0}$$
$$(a \cup b)^u = a^u \cup b^u;\ a^l \cup b^l \subseteq (a \cup b)^l \tag{u-cup l-cup}$$
$$(a \cap b)^l \subseteq a^l \cap b^l;\ (a \cap b)^u \subseteq a^u \cap b^u \tag{l-cap, u-cap}$$

Suppose a set $\underline{S}$ of concepts relating to a classroom lesson are given, and that some of these are vague. For any two concepts a and b, assume that a concept c that apparently contains the two exists – this type of search for a c amounts to taking decisions. Let this concept of apparent parthood be denoted by R. Depending on the context, the relation R may be an up-directed, reflexive and antisymmetric relation. *Apparent parthood* relation has been considered by the present author in [16].

For two concepts a and b, $ab = b$ may mean that b fulfils the functions of a in some sense, or is a preference as in *b is preferable over a*. If, on the other hand, $ab \in U_R(a, b)$ then there is a implicit reference to a choice function in the search for a concept that fulfils the role of both a and b. The idea of ab being preferable over a and b is also admissible. However, note that transitivity is not assumed in all this. For a subset of concepts A, the lower approximation is an aggregation of directed granules that are included in A. It may additionally be read as the collection of *relatively definite concepts* that are attainable from A (using common sense methods or through common knowledge).

It should be noted that up-directedness is not essential for a relation to be represented by groupoidal operations. Partial and quasi-orders lead to other groupoidal operations [13,15] in the context of knowledge generated by approximation spaces.

Under certain conditions, groupoidal operations can be inter-definable with binary relations on a set. The problem of rewriting the semantic content of binary relations of different kinds using total or partial operations has been of much interest in algebra (for example [25]). Results on using partial operations for the purpose are of more recent origin [2,3]. All binary relations can be read as partial groupoidal operations in a perspective ([3]) and therefore all general approximation spaces can be transformed into partial groupoids [17]. In this subsection known results for groupoids are stated for convenience.

Definition 1. *If a relational system is up-directed, then a number of groupoid operations can be defined on S under the constraint*

$$(\forall a, b)\, ab = \begin{cases} b \text{ if } Rab \\ c\ c \in U_R(a,b)\ \&\ \neg Rab \end{cases} \quad \text{(updg1)}$$

These are studied in [2]. The collection of groupoids satisfying the above condition will be denoted by $\mathfrak{B}(S)$ and an arbitrary element of it will be denoted by $B(S)$. It may be noted that *up-directed sets* (partially ordered sets that are up-directed) and related constructions are well-known in topology and algebra, however, the specific association of up-directedness mentioned is new. *Join directoids* [9] are groupoids of the form S that admit of a partial order relation $\leq$ that satisfies $(\forall a,b)\, a, b \leq ab$ and if $\max\{a,b\}$ exists then $ab = \max\{a,b\}$. It may additionally be noted that lambda lattices (that are commutative join and meet directoids) are related special cases (see [15]).

It is proved [2] that an up-directed reflexive relational system S corresponds to a groupoid A if the latter satisfies the equations $aa = a\ \&\ a(ab) = b(ab) = ab$ and conversely.

Theorem 2. *If $S = \langle \underline{S}, R \rangle$ is an up-directed relational system, then all the following hold:*

- *R is reflexive if and only if $B(S) \models aa = a$.*
- *R is symmetric if and only if $B(S) \models (ab)a = a$.*
- *R is transitive if and only if $B(S) \models a((ab)c) = (ab)c$.*
- *If $B(S) \models ab = ba$ then R is antisymmetric.*
- *If $B(S) \models (ab)a = ab$ then R is antisymmetric.*
- *If $B(S) \models (ab)c = a(bc)$ then R is transitive.*

3 CUD-Approximations

The connection between up-directedness and approximations can be explored from a higher order or an extended pseudo-closure perspective. Additionally, the latter can be interpreted from a distributed cognition perspective [17,26]. This is because the neighborhood generated by the point reaches out to (relative to the relation R) represent enough common upper bounds for all other points.

Definition 2. *In an up-directed approximation space S, for a given $A \subset S$, the i-distributed cognitive neighborhood (idc-nbd) $\nu_A(x)$ of an element $x \in S$ relative to A will be the set $\nu_A(x) = \{z : (\exists h \in A) Rhz \;\&\; Rxz\}$, while the distributed cognitive neighborhood (dc-nbd) $\eta_A(x)$ of an element $x \in S$ relative to A will be the set $\eta_A(x) = \{z : (\exists h \in A) Rhx \;\&\; Rzx\}$.*

Proposition 2. *The following properties hold for idc-nbds in the context of Definition 2:*

$$(\forall x) \nu_S(x) \neq \emptyset \qquad \text{(idcn-ne)}$$
$$(\forall x)(A \subset B \subseteq S \longrightarrow \nu_A(x) \subseteq \nu_B(x)) \qquad \text{(idcn-mo)}$$
$$(\forall x) \nu_S(x) \subseteq [x]_{-i} \qquad \text{(dcn-mo)}$$

Proposition 3. *The following properties hold for dc-nbds in the context of Definition 2:*

$$(\forall x)(A \subset B \subseteq S \longrightarrow \eta_A(x) \subseteq \eta_B(x)) \qquad \text{(idcn-mo)}$$
$$(\forall x) \eta_S(x) \subseteq [x] \qquad \text{(idcn-mo)}$$

Definition 3. *A subset A of a general approximation space S will be said to be closed under up-directedness (CUD) if and only if it satisfies $(\forall a, b \in A \exists c \in A) Rac \;\&\; Rbc$*

It can be checked that in general, neighborhoods of the form $[x], [x]_i, \eta_A(x)$, and $\nu_A(x)$ are not CUD sets (even in an up-directed approximation space).

Definition 4. *If S is a general approximation space, then a partial map $\eth : \wp(S) \longmapsto \wp(S)$ will be said to be an up-directed partial closure operator provided it satisfies*

1. $\forall A \in \wp(S)\ A \subseteq \eth(A)$
2. *For each $A \in \wp(S)$, $\eth(A)$ is a minimal CUD set containing A with respect to set inclusion.*

Clearly, in any unbounded non-directed lattice, $\eth$ will not be a total operation.

Theorem 3. *If S is a directed general approximation space, then $\eth$ is a well-defined cautious closure operator. That is it satisfies:*

$$(\forall A \in \wp(S)) A \subseteq \eth(A) \qquad \text{(Inclusion)}$$
$$(\forall A, B \in \wp(S))(A \subset B \subseteq \eth(A) \longrightarrow \eth(A) \subseteq \eth(B)) \qquad \text{(cmo)}$$
$$(\forall A \in \wp(S)) \eth\eth(A) = \eth(A) \qquad \text{(Idempotence)}$$
$$\eth(\emptyset) = \emptyset;\ \eth(S) = S \qquad \text{(Bot; Top)}$$

Proof. If A and B are subsets of S, and $A \subset B \subseteq \eth(A)$, suppose that $\eth(B) \subset \eth(A)$. However, by definition $\eth(B)$ is a minimal CUD-set containing B, and it must additionally be a CUD-set containing A. This contradicts the fact that $\eth(A)$ is a minimal CUD-set containing A. Therefore, it must be that $\eth(A) \subseteq \eth(B)$. So $\eth$ is cautious monotonic (cmo).

Further $\eth$ is idempotent because a minimal (and the smallest) CUD-set containing a given CUD-set $\eth(A)$ must be the same CUD-set $\eth(A)$. In other words, $\eth\eth(A) = \eth(A)$.

Since $\emptyset$ and S are CUD-sets, $\eth(\emptyset) = \emptyset$ and $\eth(S) = S$. □

However, $\eth$ is neither additive nor does it preserve intersections in general. Additionally, if A and B are CUDS in the power set $\wp(S)$, then it need not be that $A \cup B$ or $A \cap B$ are CUD sets as well. CUD-sets are the most natural of possible granules or higher-order neighborhoods generated by an up-directed relation. Let the set of all CUD sets of S generated by a relation R be $\mathcal{C}_R(S)$.

Proposition 4. *If S is up-directed, then $\mathcal{C}_R(S)$ under the set inclusion order forms a bounded partially ordered set.*

Definition 5. *For any $A, B \in \mathcal{C}_R(S)$ let*

$$A \oplus B := \eth(A \cup B) \qquad \text{(oplus)}$$
$$A \odot B := \eth(A \cap B) \qquad \text{(odot)}$$

The algebraic system $\langle \mathcal{C}_R(S), \subseteq, \oplus, \odot, \emptyset, S \rangle$ will be called a CUD-*algebraic system* (CUDAS).

Theorem 4. *The CUDAS $\langle \mathcal{C}_R(S), \subseteq, \oplus, \odot, \emptyset, S \rangle$ is well-defined, $\odot$ and $\oplus$ are commutative, and idempotent operations, and the following hold (for any $A, B, C, E \in \mathcal{C}_R(S)$):*

$$(A \subseteq B\ \&\ C \subseteq E\ \&\ B \cup E \subseteq A \oplus C \longrightarrow A \oplus C \subseteq B \oplus E) \qquad \text{(cmo-plus)}$$
$$(A \subseteq B\ \&\ C \subseteq E\ \&\ B \cap E \subseteq A \odot C \longrightarrow A \odot C \subseteq B \odot E) \qquad \text{(cmo-dot)}$$
$$A \subseteq A \oplus B;\ A \odot B \subseteq A \qquad \text{(incl+, incldot)}$$

Proof. $\cup$ and $\cap$ are commutative and idempotent operations on the power set $\wp(S)$. Because $\eth\eth(A) = \eth(A)$ for an element A of $\mathcal{C}_R(S)$, the properties ic-dot and ic-oplus hold.

cmo-plus and cmo-dot: If A and C are CUD-sets, then $A \cup B \subseteq \eth(A \cup B)$. $A \oplus C$ includes A, C and elements that ensure the CUD property. Such elements should necessarily be contained in $B \oplus E$. □

The above machinery is useful for studying the following new approximations that take $\mathcal{C}_R(S)$ as the set of granules.

Definition 6. *In a directed approximation space S, the lower and upper CUD approximations (l_{cd} and u_{cd} respectively) of a subset A can be defined using the granulation $\mathcal{C}_R(S)$ as follows (Υ being the union of minimal elements of the collection under set inclusion):*

$$A^{l_{cd}} = \bigcup \{H : H \in \mathcal{C}_R(S) \ \& \ H \subseteq A\} \tag{lcd}$$

$$A^{u_{cd}} = \Upsilon \{H : H \in \mathcal{C}_R(S) \ \& \ H \cap A \neq \emptyset\} \tag{ucd}$$

The restricted union Υ is required for ensuring non triviality. The following properties can be deduced.

Theorem 5. *In the context of Definition 6, the approximations satisfy the following (for any two subsets A and B of S):*

$$A^{l_{cd}} \subseteq A \subseteq A^{u_{cd}} \tag{cdInclusion}$$

$$A^{l_{cd}l_{cd}} = A^{l_{cd}}; \ A^{u_{cd}} \subseteq A^{u_{cd}u_{cd}} \tag{lcdId. ucdpId}$$

$$A^{l_{cd}} \subseteq A^{l_{cd}u_{cd}}; \ A^{u_{cd}l_{cd}} = A^{u_{cd}} \tag{ulci, lucpi}$$

$$(A \subseteq B \longrightarrow A^{l_{cd}} \subseteq B^{l_{cd}}); \ (A \subseteq B \longrightarrow A^{u_{cd}} \subseteq B^{u_{cd}}) \tag{lcdmo, ucdmo}$$

$$A^{l_{cd}} \cup B^{l_{cd}} \subseteq (A \cup B)^{l_{cd}}; \ A^{u_{cd}} \cup B^{u_{cd}} \subseteq (A \cup B)^{u_{cd}} \tag{lcdsa, ucdsa}$$

$$(A \cap B)^{l_{cd}} \subseteq A^{l_{cd}} \cap B^{l_{cd}}; \ (A \cap B)^{u_{cd}} \subseteq A^{u_{cd}} \cap B^{u_{cd}} \tag{lcdsm, ucdsm}$$

$$\emptyset^{l_{cd}} = \emptyset = \emptyset^{u_{cd}}; \ S^{l_{cd}} = S = S^{u_{cd}} \tag{cdbot, cdtop}$$

Definition 7. *A subset $A \in \wp(S)$ will be said to be a cud-subset of $B \in \wp(S)$ (in symbols, $A \sqsubseteq_{cd} B$) if and only if*

$$A^{l_{cd}} \subseteq B^{l_{cd}} \ \& \ A^{u_{cd}} \subseteq B^{u_{cd}} \tag{cud-ss}$$

Further, they will be said to be cud-roughly equal (in symbols, $A \backsimeq_{cd} B$) if they satisfy the condition cd-ro eq::

$$A^{l_{cd}} = B^{l_{cd}} \ \& \ A^{u_{cd}} = B^{u_{cd}} \tag{cd-ro eq}$$

It is easy to see that

Theorem 6. *In the context of Definition 7, $\sqsubseteq_{cd}$ is a quasi order and $\backsimeq_{cd}$ is an equivalence relation on $\wp(S)$.*

4 Pi-Groupoids

The non-uniqueness of the groupoidal operation associated with an up-directed relation is addressed here.

For a given $a, b \in S$, let $U_R^m(a, b)$ be a maximal subset of $U_R(a, b)$ that satisfies

$$(\forall e, f \in U_R^m(a, b)) \exists g \in U_R(a, b) \setminus U_R^m(a, b)) \text{Reg \& Rfg}$$

$U_R^m(a, b)$ will be called a set of *pseudo joins* of a and b. $U_R^m(a, b)$ is possibly the set of potential values of the join of the elements a and b.

Proposition 5. *If R is a join semilattice order, then for every $a, b \in S$, $U_R^m(a, b)$ is a singleton.*

Definition 8. *A map $\pi : U_R(S) \longmapsto S$ will be called a upper choice function (UCF) if for all a and b in S $\pi(U_R(a, b)) \in U_R(a, b)$ and a pseudo-join choice function (PJCF) if it satisfies the condition $\pi(U_R(a, b)) \in U_R^m(a, b)$.*

Definition 9. *If a relational system is up-directed, then it corresponds to a Pi-groupoid defined by (for a UCF or PJCF π)*

$$(\forall a, b) \; ab = \begin{cases} b & \text{if } Rab \\ \pi(U_R(a, b)) & \text{if } \neg Rab \end{cases} \quad \text{(upidg)}$$

Pi-groupoids can be used to study many groupoidal operations or to algebraically model a specific perspective implicit in the generalities expressed by the up-directed relation. Properties ensured by PJCFs will be described in a separate paper.

4.1 Pi-Groupoidal Approximations

Pi-groupoids generated by a directed approximation space are groupoids that can additionally be provided with nice granulations for algebraically computing approximations. For the granulation, it is natural to use sub-groupoids and this additionally amounts to restricting the construction of CUD-approximations by special choices.

Definition 10. *On a pi-groupoid S, relative to the granulation $\mho = Su(S)$ (the algebraic lattice of subgroupoids of S), the pi-lower (l_π) and pi-upper (u_π) approximations of a subset $A \subseteq S$ will be as follows (the smallest subgroupoid of S containing a set A is $Sg(A)$):*

$$A^{l\pi} = \bigcup \{X : X \in \mho \ \& \ X \subseteq A\} \quad \text{(lower-}\pi\text{)}$$
$$A^{u\pi} = \mathbf{Sg}(A) \quad \text{(upper-}\pi\text{)}$$

The boundary of A in the above definition will be taken to be $A^{b\pi} = A^{u\pi} \setminus A^{l\pi}$. Because the granules are subgroupoids, it follows that

Proposition 6. *The pi-lower approximations are unions of subgroupoids, while the pi-upper approximations are subgroupoids.*

Pi-approximations satisfy substantially better properties in comparison to cd-approximations.

Theorem 7. *Pi-lower and pi-upper approximations satisfy the following (for any two subsets A and B of a pi-groupoid S*

$$\begin{aligned}
A^{l\pi} \subseteq A \subseteq A^{u\pi}; \ A^{l\pi} = A^{l\pi u\pi} & \quad \text{(piInclusion, lupipId)} \\
A^{l\pi l\pi} = A^{l\pi}; \ A^{u\pi} = A^{u\pi u\pi}; \ A^{u\pi l\pi} = A^{u\pi} & \quad \text{(lpiId, upipId, ulpi)} \\
(A \subseteq B \longrightarrow A^{l\pi} \subseteq B^{l\pi}); \ (A \subseteq B \longrightarrow A^{u\pi} \subseteq B^{u\pi}) & \quad \text{(lpimo, upimo)} \\
A^{l\pi} \cup B^{l\pi} \subseteq (A \cup B)^{l\pi}; \ A^{u\pi} \cup B^{u\pi} \subseteq (A \cup B)^{u\pi} & \quad \text{(lpiad, upiad)} \\
(A \cap B)^{l\pi} \subseteq A^{l\pi} \cap B^{l\pi}; \ (A \cap B)^{u\pi} \subseteq A^{u\pi} \cap B^{u\pi} & \quad \text{(lpmul, upmul)} \\
\emptyset^{l\pi} = \emptyset = \emptyset^{u\pi}; \ S^{l\pi} = S = S^{u\pi} & \quad \text{(pibottom, pitop)}
\end{aligned}$$

Proof. The proof of this theorem is dictated by the algebraic closure operator $\mathbf{Sg}$, though the points in Theorem 5 applies to these cases as well.

For example, the subgroupoid generated by a subgroupoid is the same subgroupoid. So properties upipId, luipId, and ulpi hold.

The properties upimo, upiad and upmul are essentially about subgroupoids generated by a subset of a groupoid through the algebraic closure operator $\mathbf{Sg}$. □

Theorem 8. *In the above context,*

$$A^{l\pi} \subseteq \mathbf{Sg}(A^{l\pi}) \subseteq \mathbf{Sg}(A) = A^{u\pi}$$

In general, $\mathbf{Sg}(A^{l\pi})$ is not a subset of A. So the lower approximation cannot be replaced by it. All the above suggests the next definition.

Definition 11. *By a pi-groupoidal rough tuple (pg-rough tuple) will be meant a 3-tuple of the form $(A^{l\pi}, \mathbf{Sg}(A^{l\pi}), A^{u\pi})$. An algebraically constructed pi-groupoidal rough tuple (acpg-rough tuple) will be pairs of the form $(\mathbf{Sg}(A^{l\pi}), A^{u\pi}))$. The collections of all such pg-rough and acpg-rough tuples will respectively be denoted by $\mathcal{PG}(S)$ and $\mathcal{ACP}(S)$.*

Definition 12. *Two subsets* $A, B \in \wp(S)$ *will be said to be* **pg-roughly equal** *(in symbols,* $A \simeq_{pg} B$*) or* **acpg-roughly equal** *(*$A \simeq_{acpg} B$*) as they respectively satisfy the conditions* **pg-ro eq** *or* **acpg-ro** *respectively:*

$$(A^{l\pi}, \mathbf{Sg}(A^{l\pi}), A^{u\pi}) = (B^{l\pi}, \mathbf{Sg}(B^{l\pi}), B^{u\pi}) \qquad \text{(pg-ro eq)}$$
$$(\mathbf{Sg}(A^{l\pi}), A^{u\pi}) = (\mathbf{Sg}(B^{l\pi}), B^{u\pi}) \qquad \text{(acpg-ro eq)}$$

Proposition 7. *Both* $\simeq_{pg}$ *and* $\simeq_{acpg}$ *are equivalence relations on* $\wp(S)$*, and*

$$\simeq_{pg} \subseteq \simeq_{acpg}$$

Problem: Algebraic systems definable over $\mathcal{PG}(S)$, $\mathcal{ACP}(S)$, $\wp(S) \mid \simeq_{pg}$, and $\wp(S) \mid \simeq_{acpg}$ are all of natural interest. What are the best possible ones?

$\mathcal{ACP}(S)$ is obviously a subset of $(\mathbf{Su}(S))^2$. So the algebraic lattice order on the latter induces an order on the former. This is explored in Section refpialg1. The upper approximation u_π is arguably not the best, and a good contender is defined next.

Definition 13. *In the context of Definition 10, define the* **anti-lower upper approximation** u_a *of a proper subset* A *of* S *by (the anti-lower upper approximation of* S *will be taken to be* S*)*

$$A^{u_a} = \Upsilon\{H : A \subset H \ \& \ H \in \mathbf{Su}(S)\} \qquad \text{(a-upper)}$$

It is easy to see that repeated a-upper approximations will properly contain the set being approximated unless it is S. In fact, it is provable that

Theorem 9. *In the context of Theorem 7, all the following hold (for any two subsets* A *and* B *of a pi-groupoid* S*):*

$$A^{l\pi} \subseteq A \subseteq A^{u\pi} \subseteq A^{u_a} \qquad \text{(+piInclusion)}$$
$$A^{u_a} \subseteq A^{u_a u_a}; \ A^{l\pi} \subseteq A^{l\pi u_a} \qquad \text{(uapIn, luaIn)}$$
$$A^{u_a l\pi} = A^{u_a}; \ A^{u_a} \cup B^{u_a} \subseteq (A \cup B)^{u_a} \qquad \text{(ulaId,uaadd)}$$
$$(A \subseteq B \longrightarrow A^{u_a} \subseteq B^{u_a}) \qquad \text{(uamo)}$$
$$\emptyset^{l\pi} = \emptyset \subseteq \emptyset^{u_a}; \ S = S^{u_a} \qquad \text{(abottom, atop)}$$

4.2 Abstract Example

An abstract example is constructed in this subsection for illustrating aspects of up-directed approximation spaces.

Let $\underline{S}$ be the set $\underline{S} = \{a, b, c, e, f\}$ and let $R = \{ac, bc, cc, af, ff, bf, ef, ca, cb, eb, cf, ea, fa, fb\}$ be a binary relation on it (ac means the ordered pair (a, c) and so on for other elements). In Fig. 1, the general approximation space $S = \langle \underline{S}, R \rangle$ is depicted. An arrow from e to f is drawn because Ref holds. c and **f** are in bold because they are also related to themselves.

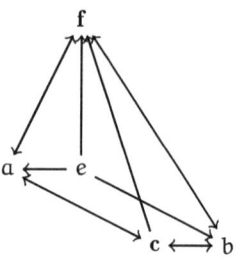

The up-directed approximation space $S = \langle \underline{S}, R \rangle$ is not reflexive and R is not antisymmetric. The set of R-upper bounds for pairs of elements are given in the symmetric Table 1 :

Table 1. Upper Bounds

	a	b	c	e	f
a	{c, f}	{c,f}	{c,f}	{f}	{f}
b	{c, f}	{c, f}	{c}	{f}	{f}
c	{c, f}	{c}	{a, b, c, f}	{a, f}	{a, b, f}
e	{f}	{f}	{a, f}	{a, b, f}	{a, b, f}
f	{ f}	{f}	{a, b, f}	{a, b, f}	{a, b, f}

Table 2. S Groupoid

/	a	b	c	e	f
a	c	f	c	f	f
b	f	f	c	f	f
c	a	b	c	a	f
e	a	b	a	b	f
f	a	b	f	a	f

A groupoid corresponding to S is given by Table 2 (hint: row names multiply column names) The subgroupoids of this groupoid are $\emptyset$, $\{c\}$, $\{f\}$, $\{a, c\}$, $\{b, f\}$, $\{c, f\}$, $\{e, f, b\}$, $\{a, c, f\}$, $\{b, c, f\}$, $\{a, b, c, f\}$ and S. The neighborhood granules corresponding to $\{a, b, c, e, f\}$ are respectively $\{e, f\}$, $\{c, e, f\}$, $\{a, b, c\}$, $\emptyset$, and S in order.

The set of subsets of S closed under up-directedness is

$$\mathcal{G} = \{\{c\}, \{f\}, \{a, c\}, \{b, c\}, \{f, c\}, \{b, f\}, \{e, f\}, \{a, f\},$$
$$\{a, c, f\}, \{b, c, f\}, \{c, e, f\}, \{b, e, f\}, \{a, e, f\}, \{a, b, f\}, \{a, b, c\},$$
$$\{a, b, c, f\}, \{a, b, e, f\}, \{a, c, e, f\}, \{b, c, e, f\}, S\} \quad (1)$$

Since $\wp(S)$ has 32 elements, approximations of specific subsets are alone considered next.

Let $A = \{e, b, c\}$, then its approximations are as below:

$$A^l = \emptyset \text{ and } A^u = S \qquad \text{(nbd)}$$
$$A^{lcd} = \{b, c\} \text{ and } A^{ucd} = \{b, c, e, f\} \qquad \text{(cd)}$$
$$A^{l\pi} = \{c\} \text{ and } A^{u\pi} = S \qquad \text{(pi)}$$

Note however that if $B = \{b\}$, then $B^{l\pi} = \emptyset$, $B^{u\pi} = \{b, c, f\} = B^{ua}$, and $A^{ua} = S$.

5 Pi-Groupoidal Algebraic Semantics

Universal algebraic properties motivate the following approach. If A and B are two subsets of S, then let $A * B = \cup\{H : H \subseteq A \cap B \ \& \ H \in \mathbf{Su}(S)\}$, and $A^\flat = \cup\{B : B \in \mathbf{Su}(S) \ \& \ B \subseteq S \setminus A\}$

Definition 14. *On $\mathcal{ACP}(S)$, let the operations $\sqcup, \sqcap, \widehat{\sqcap}, \Cup,$ and $\neg$ be defined as follows for any $A, B \in \mathcal{ACP}(S)$ ($\wedge$ and $\vee$ are the lattice operations on $\mathbf{Su}(S)$, and the subscripts indicate the first and second components)*

$$A \sqcup B = (\mathbf{Sg}(A_1 \cup B_1), A_2 \vee B_2)) \qquad \text{(acp-sum)}$$
$$A \sqcap B = (\mathbf{Sg}(A_1 * B_1), A_2 \wedge B_2) \qquad \text{(acp-pro)}$$
$$\neg A = (\mathbf{Sg}(A_2^\flat), \mathbf{Sg}(A_1^\flat)) \qquad \text{(acp-neg)}$$
$$\coprod(A) = (A_1^{\Cup \pi}, A_2^{\Cup \pi}) \qquad \text{(acp-u)}$$

The algebra $\langle \mathcal{ACP}(S), \sqcup, \sqcap, \neg, \bot, \top \rangle$ with $\bot = (\emptyset, \emptyset)$ and $\top = (S, S)$ will be called the *ACP-rough algebra* over S. It will be endowed with the induced order given by

$$(A_1, A_2) \trianglelefteq (B_1, B_2) \leftrightarrow A_1 \subseteq B_1 \ \& \ A_2 \subseteq B_2$$

Note that the last defining condition in the above actually requires that A_1 be a subgroupoid of B_1 and A_2 be a subgroupoid of B_2.

Proposition 8. *In Definition 14, all the operations are well-defined.*

Proof. It suffices to show that the expressions on the right are in $\mathcal{ACP}(S)$ as they are unique.

$\mathbf{Sg}(A_1 \cup B_1)$ is the groupoid generated by $A_1 \cup B_1$. So it is uniquely defined. $A_2 \vee B_2$ is a subgroupoid that contains the subgroupoid $\mathbf{Sg}(A_1 \cup B_1)$ because by definition the first components are subgroupoids of the second components. Therefore, $A \sqcup B$ is well-defined.

A_1 is a subgroupoid of A_2, yields $A_2^\flat$ is a subset of $A_1^\flat$. So $\neg A$ is well-defined. □

Theorem 10. *In the context of Definition 14, the following hold:*

$$\langle \mathcal{ACP}(S), \sqcup, \sqcap, \bot, \top \rangle \text{ is an algebraic lattice.} \qquad \text{(A1)}$$
$$A \trianglelefteq \neg\neg A \qquad \text{(A2)}$$
$$A \trianglelefteq B \longrightarrow \coprod A \trianglelefteq \coprod B \qquad \text{(A3)}$$
$$A \trianglelefteq \coprod A \qquad \text{(A4)}$$
$$A \trianglelefteq B \longrightarrow \neg B \trianglelefteq \neg A \qquad \text{(A5)}$$
$$\neg \coprod \neg A \trianglelefteq \coprod A \qquad \text{(A6)}$$

Proof. The direct product of two algebraic lattices is an algebraic lattice and a sublattice of the latter is also an algebraic lattice. Therefore $\langle \mathcal{ACP}(S), \sqcup, \sqcap, \bot, \top \rangle$ is an algebraic lattice.

¬A by definition is $(\mathbf{Sg}(A_2^{\flat}), \mathbf{Sg}(A_1^{\flat}))$. The subgroupoid generated by the union of subgroupoids in $S \setminus A_2$ is the first component. Because $S \setminus A_2$ is a subset of $S \setminus A_1$, the subgroupoid generated by the former is contained in that generated by the latter. Now a second negation of A would result in the form $\mathbf{Sg}((\mathbf{Sg}(A_1^{\flat}))^{\flat}), \mathbf{Sg}((\mathbf{Sg}(A_2^{\flat}))^{\flat}))$. It can be seen that $A_i \subseteq \mathbf{Sg}((\mathbf{Sg}(A_i^{\flat}))^{\flat})$ for $i = 1, 2$ because $X \subseteq \mathbf{Sg}(X)$ for any subset X. The reader may represent this proof visually.

Suppose $A \trianglelefteq B$, then $A_1 \subseteq B_1$ and $A_2 \subseteq B_2$. u_π is a monotonically increasing operator. Therefore, $A_1^{u_\pi} \subseteq B_1^{u_\pi}$ and $A_2^{u_\pi} \subseteq B_2^{u_\pi}$. From this it follows that $\bigsqcup A \trianglelefteq \bigsqcup B$.

A4 follows from the definition and basic properties of u_π.

If $A \trianglelefteq B$, then ¬B is the pair of subgroupoids generated by the union of subgroupoids of the form $S \setminus B_i$ (for $i = 2, 1$). $S \setminus A_i$ is a superset of $S \setminus B_i$ (for each i). So the union of subgroupoids contained in the former would also be a superset of the union of subgroupoids contained in the latter. This ensures A5. □

If S is finite, then the lattices are also atomistic (or even when the lattices are atomistic) and the decomposition into atomistic algebraic lattices are of interest [10]. This may lead to interesting representations in specific application contexts.

Problem. What additional assumptions ensure a duality result in the context?

6 Interpretation and Directions

In general, the groupoid operation can be read in at least three ways. The operation obviously adds information to the general approximation space – *this addition can be read as a decision because it involves choice among alternatives*. In fact, the collection of all possible groupoidal operations can be used to generate a decision space. Though these are always present in the original relation, the main issue here is of an accessible abstraction towards reasoning with the same. As such this aspect can be investigated in the given form or by taking the exact region to which the result of the operation belongs relatively. For the latter perspective, the groupoidal operation over $\wp(S)$ can be read as a combination of operations that are relatively better behaved relative to the approximations, aggregation and commonality operations. This permits easier interpretation, and semantics. In addition to these it is possible to interpret the up-directed aspect as a description of the *organization of a set of concepts*.

Specialization of these aspects to the specific approximations considered earlier, and pi-groupoids is motivated by many application contexts, and leads to distinct interpretations. In all cases, the following definition concerning regions determined by a pair of subsets can be reinterpreted in relation to associated granules. The semantic operations studied in previous sections can represent related decision regions in a better way.

Definition 15. *For any* $A, B \in \wp(S)$, *the following operations can be defined:*

$$n(A, B) = \{b : (\exists a \in A \exists b \in B)\, ab = b\} \quad \text{(normal)}$$
$$o_1(A, B) = \{c : (\exists a \in A \exists b \in B)\, ab = c \in U_R(a, b) \setminus A\} \quad \text{(outer-1)}$$
$$o_2(A, B) = \{c : (\exists a \in A \exists b \in B)\, ab = c \in U_R(a, b) \setminus B\} \quad \text{(outer-2)}$$
$$i_1(A, B) = \{c : (\exists a \in A \exists b \in B)\, ab = c \in U_R(a, b) \cap A\} \quad \text{(inner-1)}$$
$$i_2(A, B) = \{c : (\exists a \in A \exists b \in B)\, ab = c \in U_R(a, b) \cap B\} \quad \text{(inner-2)}$$
$$o(A, B) = o_1(A, B) \cap o_2(A, B) \quad \text{(outer)}$$

In the above definition, the global groupoid operation has been split into multiple operations based on the relative values assumed. For any two sets $A, B \in \wp(S)$, $n(A, B)$ is the set of things in B that have some part or approximate part in A, $o_1(A, B)$ is the set of things in the outer core determined by elements of $A \times B$ that are not in A, and $i_1(A, B)$ is the set of things determined by elements of $A \times B$ that are in A. All the six types lead to interesting operations on the decisions and are described in the extended version of this research paper.

When concepts are derived within a system that are closed under consequence or when they satisfy a number of mutual consistency properties then they may be representable within lattice-theoretical frameworks of general rough sets and formal concept analysis. Additionally, logics associated with such systems are known. In practice, such classification is often not feasible, and mutual inconsistency may be present. Circular reasoning happens even in philosophical contexts, where practitioners demonstrate high levels of engagement with their discourse. In situations where people are constrained by time to limit their engagement with information from multiple sources, and relative to their own perspective/biases about the nature of information in the constraint, clear order structures may not emerge. The connections are complicated additionally by the influence of the environment on the information generated by the sources in question – this is an essential part of the distributed cognition paradigm.

In the context of time-constrained evaluation by experts or teachers, it is necessary for expressions generated by non-experts or students to be approximated. Additionally, multiple sets of expressions by non-experts or students may be compared in the functional language of the expert or teacher (as opposed to the object-level language of the content) [1,8,19]. These acts result in up-directed approximation spaces or specialized variants like pi-groupoids. In comparison to the neighborhoods, and granules used in previous papers by the present author [17,18], *the ones introduced and studied here are more appropriate because their closure properties are more tightly coupled with consequence.* Closure under up-directedness of a granule ensures that apparent higher-level concepts are all within the granule – this helps in discerning things that are not subsumed by the concepts.

Further, it should be noted that the granular knowledge axioms of [14] and earlier work are not related to the groupoidal axioms directly. *This is because both CUD- and pi-groupoidal operations provide additional layers of decision-making that need be to integrated with existing work.* The application contexts in [17,18,22] and in previous sections support this claim.

Remarks

CUD-approximations are guided by closure under up-directedness – a feature of sets of concepts that are relatively closed under consequence in a perspective. These operators turn out to be cautious monotonic, and therefore is associated with non-monotonic reasoning (see [12]). These are studied, and applied to meaningful approximate organization of concepts in pedagogic content knowledge (and related modeling). It is also shown that CUD-approximations satisfy weaker properties in comparison to pi-groupoidal approximations. Modal connections are harder because complementation does not help. Other negations may however be considered. In the study on rough sets over partial and quasi orders, a similar closure has not been considered, and can be useful area of research. As the minimal assumptions on pi-groupoidal semantics restrict additive and multiplicative properties, extension to generalized orders such as λ-lattices, graphs, hyper-graphs, and partial lattices are of much interest (see [15]). Additionally, the aggregations similar to those studied in [20] are of interest in pi-groupoidal semantics.

A simplistic Boolean perspective is employed on hypergraphs (that are covering spaces in a perspective) to regard them as Boolean information tables in [4]. It extends earlier work of the authors on the adjacency matrix of finite simple graphs (viewed as information tables in which the set of objects and attributes are taken as the set of vertices). While these studies are of limited value in knowledge representation, graphs are related to up-directed approximation spaces, and related connections will be explored separately.

Appendix

Proof of Theorem 5

Proof. By definition, $A^{lcd} \subseteq A$. Now every subset of A is contained in some subset of S that is closed under directedness. Because the union of all granules is S, it follows that $A \subseteq A^{ucd}$.

$A^{lcdlcd} = A^{lcd}$ follows from the fact that A^{lcd} is a union of granules (that are closed under directedness).

$A^{ucdlcd} = A^{ucd}$ follows from the fact that A^{ucd} is a union of granules (that are closed under directedness), and the cd-lower approximation of such a set must be the same.

If $x \in A^{ucd}$, then it is in a granule $H \in \mathcal{C}_R(S)$ and $H \subseteq A^{ucd}$. But this means $x \in H \subseteq A^{ucducd}$. Therefore, $A^{ucd} \subseteq A^{ucducd}$ holds. In general, the converse inclusion will not hold. The proof of $A^{lcd} \subseteq A^{lcducd}$ is similar.

If X is an element of $\mathcal{C}_R(S)$, and $X \subseteq A \subseteq B$, then X is a subset of the cd-lower approximation of A and B. So lcdmo follows.

If X is an element of $\mathcal{C}_R(S)$, and $X \cap A \neq \emptyset$ and $A \subseteq B$, then $X \cap B \neq \emptyset$. Note that the restricted union Υ is at least as big for B. This ensures ucdmo.

If $x \in A^{lcd} \cup B^{lcd}$, then there exists a $X \in \mathcal{C}_R(S)$ such that $x \in X \subseteq A$ or $x \in X \subseteq B$. In either case $x \in X \subseteq A \cup B$ holds. Therefore, lcdsa holds.

If $x \in A^{ucd} \cup B^{ucd}$, then there exists a $X \in \mathcal{C}_R(S)$ such that $x \in X$ such that $X \cap A \neq \emptyset$ or $X \cap B \neq \emptyset$. In either case, $X \cap (A \cup B) \neq \emptyset$ holds. Therefore, ucdsa holds.

If $x \in (A \cap B)^{lcd}$, then there exists a $X \in \mathcal{C}_R(S)$ such that $x \in X \subseteq A \cap B$. This means $X \subseteq A$ and $X \subseteq B$. Therefore, lcdsm holds.

The rest of the proof is left to the reader. □

Proof of Theorem 8

Proof. $A^{l\pi}$ is a subset of the subgroupoid generated by itself. Therefore, $A^{l\pi} \subseteq Sg(A^{l\pi})$. $Sg(A^{l\pi}) \subseteq Sg(A)$ holds because Sg is an algebraic closure operator on $\wp(S)$. □

References

1. Ball, D., Thames, M., Phelps, G.: Content knowledge for teaching: what makes it special? J. Teach. Educ. **59**(5), 389–407 (2008)
2. Chajda, I., Langer, H.: Groupoids assigned to relational systems. Math Bohemica **138**, 15–23 (2013)
3. Chajda, I., Langer, H., Sevcik, P.: An algebraic approach to binary relations. Asian Eur. J. Math **8**(2), 1–13 (2015)
4. Chiaselotti, G., Ciucci, D., Gentile, T., Infusino, F.G.: Rough set theory and digraphs. Fund. Inform. **153**(4), 291–325 (2017)
5. Duda, J., Chajda, I.: Ideals of binary relational systems. Casopis pro pestovani matematiki **102**(3), 280–291 (1977). http://dml.cz/handle/10338.dmlcz/108456
6. Jacobs, G.M., Renandya, W.A., Power, M.: Simple, Powerful Strategies for Student Centered Learning. SE, Springer, Cham (2016). https://doi.org/10.1007/978-3-319-25712-9
7. Järvinen, J., Pagliani, P., Radeleczki, S.: Information completeness in nelson algebras of rough sets induced by Quasiorders. Studia Logica **101**, 1–20 (2012). https://doi.org/10.1007/s11225-012-9421-z
8. Jayasree, S., Subramaniam, K., Ramanujam, R.: Coherent formalisability as acceptability criterion for students' mathematical discourse. Res. Math. Educ. (2022). https://doi.org/10.1080/14794802.2022.2041469
9. Jezek, J., Quakenbush, R.: Directoids: algebraic models of up-directed sets. Algebra Univers. **27**, 49–69 (1990)
10. Libkin, L.: Direct decompositions of atomistic algebraic lattices. Algebra Univers. **33**, 127–135 (1995)
11. Ljapin, E.S.: Partial Algebras and Their Applications. Academic, Kluwer (1996)
12. Makinson, D.: General Patterns in Nonmonotonic Reasoning, pp. 35–110. Oxford University Press, Oxford (1994)

13. Mani, A.: Towards logics of some rough perspectives of knowledge. In: Skowron, A., Suraj, Z. (eds.) Rough Sets and Intelligent Systems - Professor Zdzislaw Pawlak in Memoriam. Intelligent Systems Reference Library, vol. 43, pp. 419–444. Springer, Heidelberg (2013). https://doi.org/10.1007/978-3-642-30341-8_22
14. Mani, A.: Algebraic semantics of proto-transitive rough sets. Trans. Rough Sets **XX**, 51–108 (2016)
15. Mani, A.: Algebraic methods for granular rough sets. In: Mani, A., Düntsch, I., Cattaneo, G. (eds.) Algebraic methods in general rough sets, pp. 157–336. Trends in Mathematics, Birkhauser Basel (2018)
16. Mani, A.: Dialectical rough sets, parthood and figures of opposition-I. Trans. Rough Sets **XXI**, 96–141 (2018)
17. Mani, A.: Functional extensions of knowledge representation in general rough sets. In: Bello, R., Miao, D., Falcon, R., Nakata, M., Rosete, A., Ciucci, D. (eds.) IJCRS 2020. LNCS (LNAI), vol. 12179, pp. 19–34. Springer, Cham (2020). https://doi.org/10.1007/978-3-030-52705-1_2
18. Mani, A.: Towards student centric rough concept inventories. In: Bello, R., Miao, D., Falcon, R., Nakata, M., Rosete, A., Ciucci, D. (eds.) IJCRS 2020. LNCS (LNAI), vol. 12179, pp. 251–266. Springer, Cham (2020). https://doi.org/10.1007/978-3-030-52705-1_19
19. Mani, A.: Mereology for STEAM and education research. In: Chari, D., Gupta, A. (eds.) EpiSTEMe 9, vol. 9, pp. 122–129. TIFR, Mumbai (2022). https://www.researchgate.net/publication/359773579
20. Mani, A.: Algebraic models for qualified aggregation in general rough sets, and reasoning bias discovery. In: Campagner, A., et al. (eds.) Rough Sets (IJCRS 2023), LNAI, vol. 14481, pp. 137–153. Springer, Cham (2023). https://doi.org/10.1007/978-3-031-50959-9_10
21. Mani, A.: Mereological methods for education research, and school and college-level mathematics. ICME **15**, 1–8 (2023)
22. Mani, A.: Representing pedagogic content knowledge through rough sets. IJCRS **2024**, 1–36 (2024). http://arxiv.org/abs/2403.04772
23. Mani, A., Düntsch, I., Cattaneo, G. (eds.): Algebraic Methods in General Rough Sets. Trends in Mathematics, Birkhauser Basel (2018). https://doi.org/10.1007/978-3-030-01162-8
24. Nation, J.B.: Revised Notes on Lattice Theory. Hawaii University (2020). http://www.math.hawaii.edu/~jb/
25. Poschel, R.: Graph algebras and graph varieties. Algebra Univers. **27**, 559–577 (1990)
26. Werner, K.: Enactment and construction of the cognitive niche: toward an ontology of the mind-world connection. Synthese **197**, 1313–1341 (2020)

Description Logic for Rough Concepts

Krishna B. Manoorkar[1]([✉]) , Andrea De Domenico[1] ,
and Alessandra Palmigiano[1,2]

[1] Vrije Universiteit Amsterdam, Amsterdam, The Netherlands
krishna.manoorkar@mail.com
[2] Department of Mathematics and Applied Mathematics,
University of Johannesburg, Johannesburg, South Africa

Abstract. Rough concepts have been introduced in [7] in the context of a mathematical framework unifying Rough Set Theory (RST) and Formal Concept Analysis (FCA). Algebraically, the lower and upper approximation operators on a concept lattice have similar order-theoretic properties to the $\Box$ and $\Diamond$ operators in modal logic. Thus, the logic of rough concepts has been defined as a (non-distributive) lattice-based modal logic whose relational semantics consists of formal contexts enriched with relations (interpreting the modal operators) satisfying the axioms classically corresponding to the reflexivity, symmetry, and transitivity of the accessibility relations of Kripke frames.

Recently, the description logic LE-$\mathcal{ALC}$ was introduced for reasoning in the semantic environment of these enriched formal contexts, and a tableaux algorithm was developed for checking the consistency of knowledge bases with acyclic TBoxes [5]. In the present paper, we introduce the description logic of rough concepts LE-$\mathcal{ALCR}$, which extends LE-$\mathcal{ALC}$ with the (modal) axioms classically corresponding to reflexivity, symmetry, and transitivity, and develop its corresponding tableaux algorithm. We then introduce two extensions of LE-$\mathcal{ALCR}$: the first one (LE-$\mathcal{ALCRO}$) extending LE-$\mathcal{ALCR}$ with generated concepts, and the second one (LE-$\mathcal{ALCRN}$) extending LE-$\mathcal{ALCR}$ with feature-pair inconsistencies. The resulting description logic is a framework for modeling reasoning problems related to rough concepts, which is demonstrated through the case-study of a knowledge base for Whittaker's five kingdom classification of living things.

Keywords: Description logic · Tableaux algorithm · Formal Concept Analysis · Rough concepts · LE-logics

1 Introduction

Description Logic (DL) [1] is a class of logical formalisms, typically based on classical first-order logic, widely used in Knowledge Representation and Reasoning to describe and reason about relevant concepts and their relationships in a given application domain. The fields in which these formalisms have been

---

This paper is partially funded by the EU MSCA (grant No. 101007627). The second and third authors are partially funded by the NWO grant KIVI.2019.001.

widely applied include semantic web [2,17], ontologies [27], and software engineering [4]. In several applications, representation and reasoning with uncertain or imprecise knowledge is required. Hence, there has been substantial work developing versions of description logic capable of handling uncertain or imprecise knowledge: these include probabilistic description logics [15,21], fuzzy description logics [22,31], rough description logics [18,19], possibilistic description logics [16,26].

In [5], the two-sorted non-distributive description logic LE-$\mathcal{ALC}$ was developed, based on *non-distributive* modal logic and its semantics based on formal contexts [8,9]. LE-$\mathcal{ALC}$ is a natural means of reasoning about the formal concepts (or categories) arising from formal contexts in Formal Concept Analysis [12,13]. As is well known from FCA, when endowed with its natural order, the set of formal concepts associated with a formal context is a complete, possibly non-distributive lattice known as the *concept lattice* associated with the given formal context. The logic LE-$\mathcal{ALC}$ entertains the same relationship with non-distributive modal logic and its semantics based on formal contexts as the relationship existing between $\mathcal{ALC}$, classical normal modal logic, and its Kripke frames semantics. Namely, LE-$\mathcal{ALC}$ facilitates the description of *enriched formal contexts*, i.e., formal contexts endowed with additional relations, which give rise to concept lattices extended with normal modal operators. Hence, LE-$\mathcal{ALC}$ is also a natural framework for representing and reasoning about general (possibly non-distributive) complete lattices equipped with modal operators. Similarly to the classical modal operators, also the 'non-distributive' modal operators can be given different interpretations, such as the epistemic one [8], and the one, most applied in Rough Set Theory, which regards them as lower and upper approximation operators.

The latter interpretation was pursued in [7], where non-distributive modal logic was studied and applied as a framework for describing approximate formal concepts or categories (referred to as rough concepts) where the modal operators $\Box$ and $\Diamond$ do not approximate *propositions*, as in the classical RST setting, but rather *categories* or *concepts*. Hence, the framework of rough concepts [7] generalizes classical Rough Set Theory from propositions (states of affairs) to categories[1]. Algebraically, this generalization corresponds to moving from the setting of modal algebras based on (complete atomic) Boolean algebras to modal algebras based on (complete) lattices. This generalization on the algebraic side corresponds, via discrete Stone duality, to generalizing Kripke frames (regarded as *approximation spaces* [25]) to enriched formal contexts (regarded as *conceptual approximation spaces* [7, Definition 4.1]).

Building on the results described above, the present paper refines LE-$\mathcal{ALC}$ to reason about conceptual approximation spaces and their associated rough concepts. In particular, we focus on the classes of enriched formal contexts defined by the modal axioms classically corresponding to reflexivity, symmetry, and transitivity. These conceptual approximation spaces can be understood

---

[1] There has been a vast amount of research on integrating Rough Set Theory and Formal Concept Analysis [3,20,28,29]. This work provides yet another new approach to this integration.

as the counterparts, in a non-distributive setting, of Pawlak's approximation spaces (which, as is well known, coincide with the class of Kripke frames for the classical modal logic S5). To this end, we introduce the description logic LE-$\mathcal{ALCR}$, which extends LE-$\mathcal{ALC}$ with the (RBox) axioms corresponding to reflexivity, symmetry, and transitivity. LE-$\mathcal{ALCR}$ can be seen as the description logic counterpart of non-distributive modal logic S5, akin to how $\mathcal{ALC}$ and LE-$\mathcal{ALC}$ are seen as description logic counterparts of classical basic modal logic and basic non-distributive modal logic, respectively. From another perspective, it can also be seen as a non-distributive rough description logic.

After defining the new description logic LE-$\mathcal{ALCR}$, we introduce a sound and complete polynomial-time tableau algorithm for checking the knowledge base consistency in this language. We then define two extensions of LE-$\mathcal{ALCR}$ that allow us to define concepts generated by a specific set of objects and features, and also allow us to define inconsistent pairs of features, i.e. pairs of features that no object can have together, providing tableaux algorithms for these extensions. These extensions increase the expressivity of LE-$\mathcal{ALCR}$, which is necessary for its applications. Lastly, we give an example of a knowledge base for Whittaker's five kingdom classification of living things in this extended description logic and reasoning based on it.

Structure of the Paper. In Sect. 2, we collect preliminaries on lattice-based modal logics, rough concepts, and LE-$\mathcal{ALC}$. In Sect. 3, we introduce LE-$\mathcal{ALCR}$, provide a tableaux algorithm for it, and prove its termination, soundness, and completeness. We also define extensions of LE-$\mathcal{ALCR}$ with generated concepts (LE-$\mathcal{ALCRO}$), and feature-pair inconsistencies (LE-$\mathcal{ALCRON}$), and provide tableaux algorithms for these extensions. In Sect. 7, we give an example of reasoning in LE-$\mathcal{ALCRON}$. In Sect. 8, we conclude and discuss some future research directions.

2 Preliminaries

The preliminaries collected in this section are based on [5,7,10,14].

2.1 Basic Lattice-Based Modal Logic and Its Polarity-Based Semantics

In this section, we introduce a basic lattice-based modal logic which is a member of a family of lattice-based logics, sometimes referred to as *LE-logics* [10], which have been studied in the context of a research program on the logical foundations of categorization theory [6–11,14]. Let Prop be a (countable) set of atomic propositions. The language $\mathcal{L} = \mathcal{L}(\mathcal{F}, \mathcal{G})$ is defined as follows:

$$\varphi := \bot \mid \top \mid p \mid \varphi \wedge \varphi \mid \varphi \vee \varphi \mid \Box \varphi \mid \Diamond \varphi,$$

where $p \in$ Prop, and $\Box \in \mathcal{G}$ and $\Diamond \in \mathcal{F}$ for finite sets $\mathcal{F}$ and $\mathcal{G}$ of unary $\Diamond$-type and $\Box$-type modal operators. The *basic*, or *minimal normal $\mathcal{L}$-logic* is a set **L** of sequents $\varphi \vdash \psi$, with $\varphi, \psi \in \mathcal{L}$, containing the following axioms:

$$p \vdash p \quad \bot \vdash p \quad p \vdash p \vee q \quad p \wedge q \vdash p \quad \top \vdash \Box\top \quad \Box p \wedge \Box q \vdash \Box(p \wedge q)$$
$$p \vdash \top \quad q \vdash p \vee q \quad p \wedge q \vdash q \quad \Diamond\bot \vdash \bot \quad \Diamond(p \vee q) \vdash \Diamond p \vee \Diamond q$$

for every $\Diamond \in \mathcal{F}$ and $\Box \in \mathcal{G}$, and closed under the following inference rules:

$$\frac{\varphi \vdash \chi \quad \chi \vdash \psi}{\varphi \vdash \psi} \quad \frac{\varphi \vdash \psi}{\varphi(\chi/p) \vdash \psi(\chi/p)} \quad \frac{\chi \vdash \varphi \quad \chi \vdash \psi}{\chi \vdash \varphi \wedge \psi} \quad \frac{\varphi \vdash \chi \quad \psi \vdash \chi}{\varphi \vee \psi \vdash \chi} \quad \frac{\varphi \vdash \psi}{\Box\varphi \vdash \Box\psi} \quad \frac{\varphi \vdash \psi}{\Diamond\varphi \vdash \Diamond\psi}$$

Unlike in classical modal logic, we do not assume that $\Box$ and $\Diamond$ are inter-definable.

Relational Semantics. The following notation, notions, and facts are from [7,11]. For any binary relation $T \subseteq U \times V$, and any $U' \subseteq U$ and $V' \subseteq V$, we let[2]

$$T^{(1)}[U'] := \{v \mid \forall u(u \in U' \Rightarrow uTv)\} \qquad T^{(0)}[V'] := \{u \mid \forall v(v \in V' \Rightarrow uTv)\}$$

In what follows, we fix two sets A and X, and use a, b (resp. x, y) for elements of A (resp. X), and B, C, A_j (resp. Y, W, X_j) for subsets of A (resp. X).

A *polarity* or *formal context* (cf. [13]) is a tuple $\mathbb{P} = (A, X, I)$, where A and X are sets, and $I \subseteq A \times X$ is a binary relation. Intuitively, formal contexts can be understood as abstract representations of databases [13], so that A and X represent collections of *objects* and *features*, and for any object a and feature x, the tuple (a, x) belongs to I exactly when the object a has the feature x.

As is well known, for every formal context $\mathbb{P} = (A, X, I)$, the pair of maps

$$(\cdot)^\uparrow : \mathcal{P}(A) \to \mathcal{P}(X) \quad \text{and} \quad (\cdot)^\downarrow : \mathcal{P}(X) \to \mathcal{P}(A),$$

defined by the assignments $B^\uparrow := I^{(1)}[B]$ and $Y^\downarrow := I^{(0)}[Y]$[3], form a Galois connection, and hence induce the closure operators $(\cdot)^{\uparrow\downarrow}$ and $(\cdot)^{\downarrow\uparrow}$ on $\mathcal{P}(A)$ and on $\mathcal{P}(X)$, respectively. The fixed points of $(\cdot)^{\uparrow\downarrow}$ and $(\cdot)^{\downarrow\uparrow}$ are the *Galois-stable* sets. A *formal concept* of a polarity $\mathbb{P} = (A, X, I)$ is a tuple $c = (B, Y)$ such that $B \subseteq A$ and $Y \subseteq X$, and $B = Y^\downarrow$ and $Y = B^\uparrow$. The subset B (resp. Y) is the *extension* (resp. the *intension*) of c and is denoted by $[\![c]\!]$ (resp. $([c])$). It is well known (cf. [13]) that the sets B and Y are Galois-stable, and that the set of formal concepts of a polarity $\mathbb{P}$, with the order defined by

$$c_1 \leq c_2 \quad \text{iff} \quad [\![c_1]\!] \subseteq [\![c_2]\!] \quad \text{iff} \quad ([c_2]) \subseteq ([c_1]),$$

forms a complete lattice $\mathbb{P}^+$, namely the *concept lattice* of $\mathbb{P}$. For any $a \in A$ (resp. $x \in X$) we use **a**, (resp. **x**) to denote the concept $(\{a\}^{\uparrow\downarrow}, \{a\}^\uparrow)$ (resp. $(\{x\}^\downarrow, \{x\}^{\downarrow\uparrow})$).

For any language $\mathcal{L} = \mathcal{L}(\mathcal{F}, \mathcal{G})$ as above, an *enriched formal $\mathcal{L}$-context* is a tuple $\mathbb{F} = (\mathbb{P}, \mathcal{R}_\mathcal{G}, \mathcal{R}_\mathcal{F})$, where $\mathcal{R}_\mathcal{G} = \{R_\Box \subseteq A \times X \mid \Box \in \mathcal{G}\}$ and $\mathcal{R}_\mathcal{F} = \{R_\Diamond \subseteq X \times A \mid \Diamond \in \mathcal{F}\}$ are sets of I-compatible relations, that is, for all $\Box \in \mathcal{G}$ and

---

[2] For any $u \in U$ (resp. $v \in V$) we will sometimes write $T^{(1)}[u]$ (resp. $T^{(0)}[v]$) in place of $T^{(1)}[\{u\}]$ (resp. $T^{(0)}[\{v\}]$).
[3] For any $a \in A$ (resp. $x \in X$) we will sometimes write $a^\uparrow$ and $a^{\uparrow\downarrow}$ (resp. $x^\downarrow$ and $x^{\downarrow\uparrow}$) in place of $\{a\}^\uparrow$ and $\{a\}^{\uparrow\downarrow}$ (resp. $\{x\}^\downarrow$ and $\{x\}^{\downarrow\uparrow}$), respectively.

any $\Diamond \in \mathcal{F}$, $a \in A$, and $x \in X$, the sets $R_\Box^{(0)}[x]$, $R_\Box^{(1)}[a]$, $R_\Diamond^{(0)}[a]$, $R_\Diamond^{(1)}[x]$ are Galois-stable in $\mathbb{P}$. For each $\Box \in \mathcal{G}$ and $\Diamond \in \mathcal{F}$, their associated relations $R_\Box$ and $R_\Diamond$ support their corresponding semantic interpretations as operations $[R_\Box]$ and $\langle R_\Diamond \rangle$ on the concept lattice $\mathbb{P}^+$ defined as follows: For any $c \in \mathbb{P}^+$,

$$[R_\Box]c = (R_\Box^{(0)}[([c])], I^{(1)}[R_\Box^{(0)}[([c])]]) \quad \text{and} \quad \langle R_\Diamond \rangle c = (I^{(0)}[R_\Diamond^{(0)}[[[c]]]], R_\Diamond^{(0)}[[[c]]]).$$

We refer to the algebra $\mathbb{F}^+ = (\mathbb{P}^+, \{[R_\Box]\}_{\Box \in \mathcal{G}}, \{\langle R_\Diamond \rangle\}_{\Diamond \in \mathcal{F}})$ as the *complex algebra* of $\mathbb{F}$. A *valuation* on such an $\mathbb{F}$ is a map $V: \text{Prop} \to \mathbb{P}^+$. For each $p \in \text{Prop}$, we let $[\![p]\!] := [\![V(p)]\!]$ (resp. $([p]) := ([V(p)])$) denote the extension (resp. intension) of the interpretation of p under V.

An $\mathcal{L}$-*model* is a tuple $\mathbb{M} = (\mathbb{F}, V)$, where $\mathbb{F} = (\mathbb{P}, \mathcal{R}_\mathcal{G}, \mathcal{R}_\mathcal{F})$ is an enriched formal context and V is a valuation on $\mathbb{F}$. For every $\varphi \in \mathcal{L}$, we let $[\![\varphi]\!]_\mathbb{M} := [\![V(\varphi)]\!]$ (resp. $([\varphi])_\mathbb{M} := ([V(\varphi)])$) denote the extension (resp. intension) of the interpretation of φ under the homomorphic extension of V. The following 'forcing' relations $\Vdash$ and $\succ$ can be recursively defined as follows:

$\mathbb{M}, a \Vdash p$ iff $a \in [\![p]\!]_\mathbb{M}$ $\mathbb{M}, x \succ p$ iff $x \in ([p])_\mathbb{M}$
$\mathbb{M}, a \Vdash \top$ always $\mathbb{M}, x \succ \top$ iff aIx for all $a \in A$
$\mathbb{M}, x \succ \bot$ always $\mathbb{M}, a \Vdash \bot$ iff aIx for all $x \in X$
$\mathbb{M}, a \Vdash \varphi \wedge \psi$ iff $\mathbb{M}, a \Vdash \varphi$ and $\mathbb{M}, a \Vdash \psi$ $\mathbb{M}, x \succ \varphi \wedge \psi$ iff $(\forall a \in A)(\mathbb{M}, a \Vdash \varphi \wedge \psi \Rightarrow aIx)$
$\mathbb{M}, x \succ \varphi \vee \psi$ iff $\mathbb{M}, x \succ \varphi$ and $\mathbb{M}, x \succ \psi$ $\mathbb{M}, a \Vdash \varphi \vee \psi$ iff $(\forall x \in X)(\mathbb{M}, x \succ \varphi \vee \psi \Rightarrow aIx)$.

As to the interpretation of modal operators, for every $\Box \in \mathcal{G}$ and $\Diamond \in \mathcal{F}$,

$\mathbb{M}, a \Vdash \Box\varphi$ iff $(\forall x \in X)(\mathbb{M}, x \succ \varphi \Rightarrow aR_\Box x)$ $\mathbb{M}, x \succ \Box\varphi$ iff $(\forall a \in A)(\mathbb{M}, a \Vdash \Box\varphi \Rightarrow aIx)$
$\mathbb{M}, x \succ \Diamond\varphi$ iff $(\forall a \in A)(\mathbb{M}, a \Vdash \varphi \Rightarrow xR_\Diamond a)$ $\mathbb{M}, a \Vdash \Diamond\varphi$ iff $(\forall x \in X)(\mathbb{M}, x \succ \Diamond\varphi \Rightarrow aIx)$

The definition above ensures that, for any $\mathcal{L}$-formula φ,
$$\mathbb{M}, a \Vdash \varphi \text{ iff } a \in [\![\varphi]\!]_\mathbb{M}, \quad \text{and} \quad \mathbb{M}, x \succ \varphi \text{ iff } x \in ([\varphi])_\mathbb{M}.$$
$$\mathbb{M} \models \varphi \vdash \psi \quad \text{iff} \quad [\![\varphi]\!]_\mathbb{M} \subseteq [\![\psi]\!]_\mathbb{M} \quad \text{iff} \quad ([\psi])_\mathbb{M} \subseteq ([\varphi])_\mathbb{M}.$$

The interpretation of the propositional connectives $\vee$ and $\wedge$ in the framework described above reproduces the standard notion of join and meet of formal concepts used in FCA. Note that in this semantics disjunction is interpreted in terms of the *intersection* of intensions. Therefore, $\vee$ does not produce branching for the same reason why $\wedge$ does not in the classical setting. This results in the tableaux rules for this logic (cf. Section 3.1) not having any branching as there is no real disjunctive i.e. "or" clause in the semantics. The interpretation of $\Box$ and $\Diamond$ operators is motivated by algebraic properties and duality theory for normal modal operators on lattices (cf. [11, Section 3] for an expanded discussion).

2.2 Rough Concepts

In [7], enriched formal contexts were employed to formalize approximate concepts, with the operations $\Box$ and $\Diamond$ interpreted as lower and upper approximation operations on a concept lattice, respectively[4]. The minimum requirement for

---
[4] This is motivated by the vast work in the classical rough set theory which relates Rough Set Theory and modal logic by interpreting $\Box$ and $\Diamond$ operators in classical modal logic as approximation operators in rough set theory [23,24].

the operators $\Box$ and $\Diamond$ to be interpreted as the lower and the upper approximations is given by $\Box c \leq \Diamond c$, indicating that the lower approximation of concept c is smaller than its upper approximation. Concept lattices with modal operators $\Box$ and $\Diamond$ satisfying this condition are called *conceptual rough algebras* (cf. [7, Section 6]). Various sub-varieties of conceptual rough algebras satisfying additional axioms were defined in [7, Section 6], including *conceptual tqBa-5*, which are conceptual rough algebras satisfying the modal axioms, classically corresponding to reflexivity, symmetry and transitivity, given in the following table.

Table 1. Algebraic axioms

reflexivity	$\Box p \leq p$	$p \leq \Diamond p$
symmetry	$p \leq \Box \Diamond p$	$\Diamond \Box p \leq p$
transitivity	$\Box p \leq \Box \Box p$	$\Diamond \Diamond p \leq \Diamond p$

Note that, unlike the classical case, the two axioms in each row of Table 1 are not equivalent in the lattice setting. Conceptual tqBa-5 algebras can be regarded as the non-distributive counterparts of tqBa-5 algebras (or S5 modal algebras) in classical RST. For any enriched formal context $\mathbb{F} = (\mathbb{P}, \{R_\Box\}_{\Box \in \mathcal{G}}, \{R_\Diamond\}_{\Diamond \in \mathcal{F}})$, the relations $R_\blacklozenge \subseteq X \times A$ and $R_\blacksquare \subseteq A \times X$ are defined as follows: $xR_\blacklozenge a$ iff $aR_\Box x$, and $aR_\blacksquare x$ iff $xR_\Diamond a$. The operators $\blacksquare$ and $\blacklozenge$ are the box and diamond operators defined by the relations $R_\blacksquare$ and $R_\blacklozenge$, respectively. Moreover, for all relations $R, S \subseteq A \times X$, we define $R\,;_I S \subseteq A \times X$ as follows: $a(R\,;_I S)x$ if $a \in R^{(0)}[I^{(1)}[S^{(0)}[x]]]$. Similarly, for all relations $R, S \subseteq X \times A$, we define $R\,;_I S \subseteq X \times A$ as follows: $x(R\,;_I S)a$ if $x \in R^{(0)}[I^{(0)}[S^{(0)}[a]]]$. The following theorem lists the first-order conditions on enriched formal contexts corresponding to the axioms listed in Table 1 on enriched formal contexts.

Theorem 1. *[7, Proposition 4.3] For any enriched formal context $\mathbb{F}$,*

$\mathbb{F}^+ \models \Box p \leq p$ *iff* $R_\Box \subseteq I$ $\quad\quad$ $\mathbb{F}^+ \models p \leq \Diamond p$ *iff* $I \subseteq R_\Diamond$
$\mathbb{F}^+ \models p \leq \Box \Diamond p$ *iff* $R_\Diamond \subseteq R_\blacklozenge$ $\quad\quad$ $\mathbb{F}^+ \models \Diamond \Box p \leq p$ *iff* $R_\blacklozenge \subseteq R_\Diamond$
$\mathbb{F}^+ \models \Box p \leq \Box \Box p$ *iff* $R_\Box \subseteq R_\Box\,;_I R_\Box$ $\quad$ $\mathbb{F}^+ \models \Diamond \Diamond p \leq \Diamond p$ *iff* $R_\Diamond \subseteq R_\Diamond\,;_I R_\Diamond$

Thus, the class of enriched formal contexts satisfying the aforementioned relational conditions corresponds to the variety of conceptual tqBa-5 algebras. As these enriched formal contexts satisfy the reflexivity, symmetry, and transitivity axioms, they can be seen as the non-distributive (conceptual) counterparts of Pawlak's approximation spaces in classical rough set theory. In this paper, we use the term *logic of rough concepts* to denote the logic of enriched formal contexts that satisfy these axioms.

2.3 Description Logic LE-$\mathcal{ALC}$

The language of LE-$\mathcal{ALC}$ contains individuals of two types, usually interpreted as the *objects* and *features* of the given database or categorization. Let OBJ and FEAT be disjoint sets of individual names for objects and features.

The set $\mathcal{R}$ of the role names for LE-$\mathcal{ALC}$ is the union of three types of relations: (1) A unique relation $I \subseteq \mathsf{OBJ} \times \mathsf{FEAT}$; (2) a set of relations $\mathcal{R}_\Box$ of the form $R_\Box \subseteq \mathsf{OBJ} \times \mathsf{FEAT}$; (3) a set of relations $\mathcal{R}_\Diamond$ of the form $R_\Diamond \subseteq \mathsf{FEAT} \times \mathsf{OBJ}$. While I is intended to be interpreted as the incidence relation of formal contexts and encodes information on which objects have which features, the relations in $\mathcal{R}_\Box$ and $\mathcal{R}_\Diamond$ encode additional relationships between objects and features (cf. [7] for an extended discussion).

For any set $\mathcal{C}$ of atomic concept names, the language of LE-$\mathcal{ALC}$ concepts is:

$$C := D \mid C \wedge C \mid C \vee C \mid \top \mid \bot \mid \langle R_\Diamond \rangle C \mid [R_\Box]C$$

where $D \in \mathcal{C}$, $R_\Box \in \mathcal{R}_\Box$ and $R_\Diamond \in \mathcal{R}_\Diamond$. This language matches the language of LE-logic and has an analogous intended interpretation on the complex algebras of enriched formal contexts (cf. Section 2.1). As usual, $\vee$ and $\wedge$ are to be interpreted as the smallest common superconcept and the greatest common subconcept, and the constants $\top$ and $\bot$ are to be interpreted as the largest and the smallest concept, respectively. We do not include $\neg C$ as a valid concept in our language, as there is no canonical and natural way to interpret negations in non-distributive (lattice-based) settings. The concepts $\langle R_\Diamond \rangle C$ and $[R_\Box]C$ in LE-$\mathcal{ALC}$ are intended to be interpreted as the operations $\langle R_\Diamond \rangle$ and $[R_\Box]$ as defined in Sect. 2.1, respectively, analogously to the way in which $\exists r$ and $\forall r$ in $\mathcal{ALC}$ are interpreted on Kripke frames as operations $\Diamond$ and $\Box$ of classical modal logic.

TBox *assertions* in LE-$\mathcal{ALC}$ are of the shape $C_1 \sqsubseteq C_2$, where C_1 and C_2 are concepts defined as above.[5] We use $C_1 \equiv C_2$ as shorthand for $C_1 \sqsubseteq C_2$ and $C_2 \sqsubseteq C_1$. ABox *assertions* are of one of the following forms:

$$aR_\Box x, \quad xR_\Diamond a, \quad aIx, \quad a:C, \quad x::C, \quad \neg\alpha,$$

where α is any of the first five ABox terms. We refer to the terms of the first three types as *relational terms*. We let $\mathcal{A}$ (resp $\mathcal{T}$) denote an arbitrary set of ABox (resp. TBox) assertions. The intended meaning of $a:C$ and $x::C$ is "object a is a member of concept C", and "feature x is in the description of concept C", respectively. Note that we add the negative terms to ABoxes explicitly, as the concepts in LE-$\mathcal{ALC}$ do not contain negations.

An *interpretation* for LE-$\mathcal{ALC}$ is a tuple $I = (\mathbb{F}, \cdot^I)$, where $\mathbb{F} = (\mathbb{P}, \mathcal{R}_\Box, \mathcal{R}_\Diamond)$ is an enriched formal context based on the polarity $\mathbb{P} = (A, X, I)$, and $\cdot^I$ maps:

1. individual names $a \in \mathsf{OBJ}$ (resp. $x \in \mathsf{FEAT}$), to some $a^I \in A$ (resp. $x^I \in X$);
2. relation names I, $R_\Box$ and $R_\Diamond$ to relations I^I, $R_\Box^I$ and $R_\Diamond^I$ in $\mathbb{F}$;
3. any primitive concept D to $D^I \in \mathbb{F}^+$, and other concepts as follows:

$$\bot^I = (X^\downarrow, X) \qquad \top^I = (A, A^\uparrow) \qquad (C_1 \wedge C_2)^I = C_1^I \wedge C_2^I$$
$$(C_1 \vee C_2)^I = C_1^I \vee C_2^I \quad ([R_\Box]C)^I = [R_\Box^I]C^I \quad (\langle R_\Diamond \rangle C)^I = \langle R_\Diamond^I \rangle C^I$$

where the operators $[R_\Box^I]$ and $\langle R_\Diamond^I \rangle$ are defined as in Sect. 2.1.

The satisfiability relation for an interpretation I is defined as follows:

---

[5] As is standard in DL (cf. [1] for more details), general concept inclusions of the form $C_1 \sqsubseteq C_2$ can be rewritten as $C_1 \equiv C_2 \wedge C_3$, where C_3 is a new atomic concept name.

1. $I \models C_1 \equiv C_2$ iff $[\![C_1^I]\!] = [\![C_2^I]\!]$ iff $([\![C_2^I]\!]) = ([\![C_1^I]\!])$.
2. $I \models a : C$ iff $a^I \in [\![C^I]\!]$ and $I \models x :: C$ iff $x^I \in ([\![C^I]\!])$.
3. $I \models aIx$ (resp. $aR_\Box x$, $xR_\Diamond a$) iff $a^I I^I x^I$ (resp. $a^I R_\Box^I x^I$, $x^I R_\Diamond^I a^I$).
4. $I \models \neg \alpha$, where α is any ABox term, iff $I \not\models \alpha$.

An interpretation I is a *model* for an LE-$\mathcal{ALC}$ knowledge base $(\mathcal{A}, \mathcal{T})$ if $I \models \alpha$ for every $\alpha \in \mathcal{A}$ (notation: $I \models \mathcal{A}$) and $I \models \beta$ for every $\beta \in \mathcal{T}$ (notation: $I \models \mathcal{T}$). A knowledge base $(\mathcal{A}, \mathcal{T})$ (resp. ABox $\mathcal{A}$, resp. TBox $\mathcal{T}$) is *inconsistent* if $(\mathcal{A}, \mathcal{T})$ (resp. $\mathcal{A}$, resp. $\mathcal{T}$) has no model.

We refer to [5] for details on polynomial-time tableaux algorithm for LE-$\mathcal{ALC}$.

3 Description Logic for Rough Concepts and Tableaux Algorithm for It

The description logic for rough concepts (referred to as LE-$\mathcal{ALCR}$) is an axiomatic extension of LE-$\mathcal{ALC}$ with the axioms reflexivity, symmetry, and transitivity reported in Table 1. Hence, any model of LE-$\mathcal{ALCR}$ must satisfy these axioms. LE-$\mathcal{ALCR}$ can be regarded as the description logic counterpart of the logic of rough concepts (cf. Sect. 2.2). In the present section, we introduce a sound and complete polynomial-time decision procedure for checking the consistency of LE-$\mathcal{ALCR}$ knowledge base.

3.1 Tableaux Algorithm for LE-$\mathcal{ALCR}$ ABoxes

In this section, we introduce the tableaux algorithm for checking the consistency of LE-$\mathcal{ALCR}$ ABoxes. An LE-$\mathcal{ALCR}$ ABox $\mathcal{A}$ contains a *clash* iff it contains both β and $\neg \beta$ for some relational term β. The expansion rules below are designed so that the expansion of $\mathcal{A}$ will contain a clash iff $\mathcal{A}$ is inconsistent. The set $sub(C)$ of sub-formulas of any LE-$\mathcal{ALCR}$ concept C is defined as usual.

A concept C' *occurs* in $\mathcal{A}$ ($C' \in \mathcal{A}$) if $C' \in sub(C)$ for some C such that one of the terms $a : C$, $x :: C$, $\neg a : C$, or $\neg x :: C$ is in $\mathcal{A}$. A constant b (resp. y) *occurs* in $\mathcal{A}$ ($b \in \mathcal{A}$, or $y \in \mathcal{A}$), iff some term containing b (resp. y) occurs in it.

The tableaux algorithm below constructs a model $(\mathbb{F}, \cdot^I)$ for every consistent $\mathcal{A}$, where $\mathbb{F} = (\mathbb{P}, \mathcal{R}_\Box, \mathcal{R}_\Diamond)$ is such that, for any $C \in \mathcal{A}$, some $a_{C'} \in A$ and $x_{C'} \in X$ exist such that, for any $a \in A$ (resp. any $x \in X$), $a \in [\![C^I]\!]$ (resp. $x \in ([\![C]\!]^I)$ iff $aIx_{C'}$ (resp. $a_{C'}Ix$) for some C' defined from C. We call a_C and x_C the *classifying object* and the *classifying feature* of C, respectively. To make our notation more easily readable, we will write $a_{\Box C}$, $x_{\Box C}$ (resp. $a_{\Diamond C}$, $x_{\Diamond C}$) instead of $a_{[R_\Box]C}$, $x_{[R_\Box]C}$ (resp. $a_{\langle R_\Diamond \rangle C}$, $x_{\langle R_\Diamond \rangle C}$).

Pre-processing. In pre-processing, we use the fact that the reflexivity and transitivity axioms together imply that a sequence of $\Box$ (resp. $\Diamond$) operators is equivalent to a single $\Box$ (resp. $\Diamond$) operator to replace any LE-$\mathcal{ALCR}$ concept with its *reflexive-transitive contraction*, obtained by replacing a sequence of $\Box$ (resp. $\Diamond$) operators with a single $\Box$ (resp. $\Diamond$) operator. This is similar to the strategy used in some tableaux algorithms for classical S5 modal logic, where all the

Algorithm 1. tableaux algorithm for checking LE-$\mathcal{ALC}$ ABox consistency

Input: An LE-$\mathcal{ALC}$ ABox $\mathcal{A}$.   **Output**: whether $\mathcal{A}$ is inconsistent.

1: Do **pre-processing** on $\mathcal{A}$.
2: **if** there is a clash in $\mathcal{A}$ **then return** "inconsistent".
3: **pick** any applicable expansion rule R, **apply** R to $\mathcal{A}$ and proceed recursively.
4: **if** no expansion rule is applicable **return** "consistent".

formulas are replaced by equivalent formulas of modal depth at most two. For any LE-$\mathcal{ALCR}$ concept C, its *reflexive-transitive contraction* is the concept C' obtained by replacing any sequence of more than one $[R_\Box]$ (resp. $\langle R_\Diamond \rangle$) operator with a single $[R_\Box]$ (resp. $\langle R_\Diamond \rangle$) operator. For example, the reflexive-transitive contractions of concepts $[R_\Box][R_\Box]C_1 \vee \langle R_\Diamond \rangle \langle R_\Diamond \rangle C_2$ and $\langle R_\Diamond \rangle \langle R_\Diamond \rangle [R_\Box] \langle R_\Diamond \rangle C_1$ are the concepts $[R_\Box]C_1 \vee \langle R_\Diamond \rangle C_2$ and $\langle R_\Diamond \rangle [R_\Box] \langle R_\Diamond \rangle C_1$, respectively. In the pre-processing step for LE-$\mathcal{ALCR}$ tableaux algorithm, we replace every concept occurring in ABox with its reflexive-transitive contraction. For example, if the terms $b : [R_\Box][R_\Box]C_1 \vee \langle R_\Diamond \rangle \langle R_\Diamond \rangle C_2$ and $y :: \langle R_\Diamond \rangle \langle R_\Diamond \rangle [R_\Box] \langle R_\Diamond \rangle C_1$ occur in an ABox, they are replaced with the terms $b : [R_\Box]C_1 \vee \langle R_\Diamond \rangle C_2$ and $y :: \langle R_\Diamond \rangle [R_\Box] \langle R_\Diamond \rangle C_1$, respectively. For any concept C, let $RT(C)$ denote its reflexive-transitive contraction. The following lemma follows straightforwardly from the definition.

Lemma 1. *For any concepts C_1 and C_2,*

1. $RT(C_1 \wedge C_2) = RT(C_1) \wedge RT(C_2)$ *and* $RT(C_1 \vee C_2) = RT(C_1) \vee RT(C_2)$.
2. $RT([R_\Box]C_1) = RT([R_\Box]RT(C_1))$ *and* $RT(\langle R_\Diamond \rangle C_1) = RT(\langle R_\Diamond \rangle RT(C_1))$.

The expansion rules for the tableaux algorithm are given below. The commas in each rule are metalinguistic conjunctions, hence every tableau is non-branching.

Creation rule

$$\text{create} \, \frac{\text{For any } C \in \mathcal{A}}{a_C : C, \quad x_C :: C}$$

Basic rule

$$I \, \frac{b : C, \quad y :: C}{bIy}$$

⊤ and ⊥ rule

$$\top \, \frac{}{b : \top} \quad \frac{}{y :: \bot} \, \bot$$

Rules for the logical connectives

$$\wedge_A \, \frac{b : C_1 \wedge C_2}{b : C_1, \quad b : C_2} \qquad \vee_X \, \frac{y :: C_1 \vee C_2}{y :: C_1, \quad y :: C_2} \qquad \frac{b : C_1 \vee C_2, \quad y :: C_1, \quad y :: C_2}{bIy} \, \vee_A$$

$$\Box \, \frac{b : [R_\Box]C, \quad y :: C}{bR_\Box y} \qquad \Diamond \, \frac{y :: \langle R_\Diamond \rangle C, \quad b : C}{yR_\Diamond b} \qquad \frac{y :: C_1 \wedge C_2, \quad b : C_1, \quad b : C_2}{bIy} \, \wedge_X$$

Adjunction rules

$$R_\Box \, \frac{bR_\Box y}{\Diamond bIy, \quad bI\Box y} \qquad \frac{yR_\Diamond b}{\Diamond bIy, \quad bI\Box y} \, R_\Diamond$$

Basic rules for negative assertions   **Appending rules**

$$\neg b \, \frac{\neg(b : C)}{\neg(bIx_C)} \qquad \frac{\neg(x :: C)}{\neg(a_C Ix)} \, \neg x \qquad x_C \, \frac{bIx_C}{b : C} \qquad \frac{a_C Iy}{y :: C} \, a_C$$

Rough concept rules					I-compatibility rules	
$Ref_\Box \dfrac{bR_\Box y}{bIy}$	$Ref_\Diamond \dfrac{yR_\Diamond b}{bIy}$	$\dfrac{bI\Box y}{bR_\Box \Box y} Trans_\Box$			$\dfrac{bI\Box y}{bR_\Box y} \Box y$	
$Trans_\Diamond \dfrac{\Diamond bIy}{yR_\Diamond \Diamond b}$	$Sym_\Diamond \dfrac{yR_\Diamond b}{bR_\Box y}$	$\dfrac{bR_\Box y}{yR_\Diamond b} Sym_\Box$			$\dfrac{\Diamond bIy}{yR_\Diamond b} \Diamond b$	

In rules $\top$ and $\bot$, b and y are any objects or features occurring in the tableau. In the adjunction rules $R_\Box$ and $R_\Diamond$ term $\Diamond bIy$ (resp. $bI\Box y$) is added only if b (resp. y) is not of the form $\Diamond d$ (resp. $\Box z$) for some object (resp. feature) name d (resp. z). When such terms are added, the individuals $\Diamond b$ and $\Box y$ are new and unique for each relation $R_\Box$ and $R_\Diamond$, except for $\Diamond a_C = a_{\Diamond C}$ and $\Box x_C = x_{\Box C}$.

The basic rule and the logical rules for the connectives encode the semantics of the logical connectives in LE-$\mathcal{ALC}$. The creation rule makes sure that, whenever successful, the algorithm outputs models with classifying object a_C and feature x_C for every concept $C \in \mathcal{A}$. The adjunction rules and I-compatibility rules ensure that all $R_\Box \in \mathcal{R}_\Box$ and $R_\Diamond \in \mathcal{R}_\Diamond$ are I-compatible. Appending and negative assertion rules encode the defining property of classifying objects and features of concepts. The rules $Ref_\Box$, $Ref_\Diamond$, $Sym_\Box$, $Sym_\Diamond$, $Trans_\Box$, and $Trans_\Diamond$ capture the reflexivity, symmetry, and transitivity axioms.

In the following sections, we will show that the above tableaux algorithm provides a sound and complete polynomial time decision procedure for checking the consistency of an LE-$\mathcal{ALCR}$ ABox.

Regarding termination, note that new constant names are added to the tableaux during the expansion phase only by the adjunction rules $R_\Box$ and $R_\Diamond$. The side conditions on these rules ensure that no constant name added occurs under the scope of more than one $\Box$ or $\Diamond$ operator. Moreover, note that only $\Box$ (resp. $\Diamond$) operators can occur in a feature (resp. an object) name. This ensures that the number of constant names that can be added during the expansion phase is linear in the size of the original ABox $\mathcal{A}$. As to concepts, note that new concepts are added to the tableau only by appending rules. Hence, no new propositional connective can be added to any concepts. By pre-processing and the previous argument, no constant name with more than one consecutive $\Diamond$ or $\Box$ operator in it appears in the tableaux expansion. Since the new concepts are added through their characterizing constants, this ensures that any newly added concept is obtained by the application of a single $\Box$ or $\Diamond$ operator to a concept already occurring in the original ABox $\mathcal{A}$. Thus, the number of new concepts, object names, and feature names added during the tableaux expansion are all linear in the size of the initial ABox $\mathcal{A}$. Hence, the total number of terms which can occur in a tableaux expansion is polynomial in the size of $\mathcal{A}$. As tableaux rules involve no branching, this implies the termination of the tableaux algorithm in polynomial time in the size of $\mathcal{A}$.

4 Soundness

In this section, we show the soundness of the tableaux algorithm for LE-$\mathcal{ALCR}$ ABox consistency. For any consistent ABox $\mathcal{A}$, we let its *completion* $\overline{\mathcal{A}}$ be its maximal expansion (which exists due to termination). If there is no clash (cf. beginning of Sect. 3.1) in $\overline{\mathcal{A}}$, we construct a relational structure $\mathbb{F} = (A, X, I, \mathcal{R}_\square, \mathcal{R}_\lozenge)$, where A and X are the sets of names of objects and features occurring in the expansion, and for any $a \in A$, $x \in X$, and any role names $R_\square \in \mathcal{R}_\square$, $R_\lozenge \in \mathcal{R}_\lozenge$, we have aIx, $aR_\square x$, $xR_\lozenge a$ iff such relational terms explicitly occur in $\overline{\mathcal{A}}$. We define an interpretation $\mathrm{I} = (\mathbb{F}, \cdot^\mathrm{I})$ on it as follows. For any object name a, and feature name x, we let $a^\mathrm{I} := a$ and $x^\mathrm{I} := x$. For any atomic concept name D, we define $D^\mathrm{I} = (x_D{}^\downarrow, a_D{}^\uparrow)$. The following lemmas show that I is a valid interpretation for LE-$\mathcal{ALCR}$. In what follows, we use object, feature, relation, and concept names to also denote their interpretation, when it is clear from the context.

Lemma 2. $x_D^{\downarrow\uparrow} = a_D^\uparrow$ and $a_D^{\uparrow\downarrow} = x_D^\downarrow$ for any $D \in \mathcal{C}$.

Proof. For any atomic concept name D, the creation and basic rules imply that $a_D I x_D \in \overline{\mathcal{A}}$. Therefore, $a_D^{\uparrow\downarrow} \subseteq x_D^\downarrow$. On the other hand, if bIy and $a_D I y$ are in $\overline{\mathcal{A}}$, then by the appending and basic rule, we get $bIx_D \in \overline{\mathcal{A}}$. Therefore, $x_D^\downarrow \subseteq a_D^{\uparrow\downarrow}$. This completes the proof of the second identity. The first one is proved analogously.

Lemma 3. *All* $R_\square \in \mathcal{R}_\square$ *and* $R_\lozenge \in \mathcal{R}_\lozenge$ *in* $\mathbb{F} = (\mathbb{P}, \mathcal{R}_\square, \mathcal{R}_\lozenge)$ *are I-compatible.*

Proof. We only prove that $R_\square$ is I-compatible, since the I-compatibility of $R_\lozenge$ is proved similarly. Firstly, we prove that $R_\square^{(0)}[y]$ is Galois-stable for any y. We reason by cases: if y is not of the form $\square z$ for any z, then, by rules $\square y$ and $R_\square$, we get $bR_\square y \in \overline{\mathcal{A}}$ iff $bI\square y \in \overline{\mathcal{A}}$. Therefore, $R_\square^{(0)}[y] = I^{(0)}[\square y]$ is Galois-stable. If y is of the form $\square z$ for some z, then $bR_\square y \in \overline{\mathcal{A}}$ implies, by $Ref_\square$, that $bIy \in \overline{\mathcal{A}}$. Conversely, if $bIy = bI\square z \in \overline{\mathcal{A}}$, then, $Trans_\square$ implies that $bR_\square \square z = bR_\square y \in \overline{\mathcal{A}}$. Hence, $R_\square^{(0)}[y] = I^{(0)}[y]$ is Galois-stable.

Let us show that $R_\square^{(1)}[b]$ is Galois-stable for any b. Reasoning by cases, if b is not of the form $\lozenge d$ for any d, then by rules $\lozenge b$ and $R_\square$, we get $bR_\square y \in \overline{\mathcal{A}}$ iff $\lozenge bIy \in \overline{\mathcal{A}}$. Therefore, $R_\square^{(1)}[b] = I^{(1)}[\lozenge b]$ is Galois-stable. If b is of the form $\lozenge d$ for some d, then $bR_\square y \in \overline{\mathcal{A}}$ implies, by the $Ref_\square$ rule, that $bIy \in \overline{\mathcal{A}}$. Conversely, if $bIy = \lozenge dIy \in \overline{\mathcal{A}}$, then $Trans_\lozenge$ implies that $yR_\lozenge \lozenge d \in \overline{\mathcal{A}}$. By $Sym_\square$, this implies that $\lozenge dR_\square y = bR_\square y \in \overline{\mathcal{A}}$. Therefore, $R_\square^{(1)}[b] = I^{(1)}[b]$ is Galois-stable.

Lemma 4. *Let* $\mathbb{F} = (A, X, I, \mathcal{R}_\square, \mathcal{R}_\lozenge)$ *be an enriched formal context constructed from a complete LE-$\mathcal{ALCR}$ ABox $\overline{\mathcal{A}}$ as discussed above. Then, $\mathbb{F}$ satisfies the following axioms: For any $R_\square \in \mathcal{R}_\square$, and $R_\lozenge \in \mathcal{R}_\lozenge$,*

1. **reflexivity:** $R_\square \subseteq I$, and $R_\blacksquare \subseteq I$,
2. **symmetry:** $R_\square \subseteq R_\blacksquare$, and $R_\blacksquare \subseteq R_\square$,
3. **transitivity:** $R_\square \subseteq R_\square ;_I R_\square$, and $R_\lozenge \subseteq R_\lozenge ;_I R_\lozenge$.

Proof. Let $b \in A$, $y \in X$. We only prove the first condition in each item, the proof of the second condition being dual.

1. If $bR_\square y$, then $bR_\square y \in \overline{\mathcal{A}}$. By $Ref_\square$, we get $bIy \in \overline{\mathcal{A}}$, implying bIy. This shows that $R_\square \subseteq I$.

2. If $bR_\Box y$, then $bR_\Box y \in \overline{\mathcal{A}}$. By $Sym_\Box$, we have $yR_\Diamond b \in \overline{\mathcal{A}}$ implying $yR_\Diamond b$. This shows that $R_\Box \subseteq R_\blacksquare$.
3. If $bR_\Box y$ and let $z \in I^{(1)}[R_\Box^{(0)}[y]]$, then, by construction, $bR_\Box y \in \overline{\mathcal{A}}$.

We reason by cases. If b is not of the form $\Diamond d$ for any d, then,
$$\Diamond bIy \in \overline{\mathcal{A}} \qquad \text{(by } R_\Box\text{)}$$
$$\implies yR_\Diamond \Diamond b \in \overline{\mathcal{A}} \qquad \text{(by } Trans_\Diamond\text{)}$$
$$\implies \Diamond bR_\Box y \in \overline{\mathcal{A}} \qquad \text{(by } Sym_\Diamond\text{)}$$
$$\implies \Diamond bR_\Box y \qquad \text{(by construction of } \mathbb{F}\text{)}$$
$$\implies \Diamond bIz \qquad \text{(as } z \in I^{(1)}[R_\Box^{(0)}[y]]\text{)}$$
$$\implies bR_\Box z \in \overline{\mathcal{A}} \qquad \text{(by } \Diamond b \text{ followed by } Sym_\Diamond\text{)}$$
$$\implies R_\Box^{(0)}[y] \subseteq R_\Box^{(0)}[I^{(1)}[R_\Box^{(0)}[y]]] \qquad \text{(by construction of } \mathbb{F}\text{)}$$
$$\implies R_\Box \subseteq R_\Box;_I R_\Box \qquad \text{(by def. of };_I\text{)}$$
If b is of the form $\Diamond d$ for some d, then
$$\Diamond dIz \in \overline{\mathcal{A}} \qquad \text{(as } z \in I^{(1)}[R_\Box^{(0)}[y]]\text{)}$$
$$\implies \Diamond dR_\Box z \in \overline{\mathcal{A}} \qquad \text{(by } Trans_\Diamond \text{ followed by } Sym_\Diamond\text{)}$$
$$\implies bR_\Box z \qquad \text{(by construction of } \mathbb{F} \text{ and } b = \Diamond d\text{)}$$
$$\implies R_\Box \subseteq R_\Box;_I R_\Box \qquad \text{(by def. of };_I\text{)}$$

From the lemmas above, it immediately follows that the tuple $M = (\mathbb{F}, \cdot^I)$ is an LE-$\mathcal{ALCR}$ model. The following lemma states that the interpretation of any concept C occurring in $\overline{\mathcal{A}}$ in the model M is completely determined by the terms of the form $bIx_{C'}$ and $a_{C'}Iy$ occurring in the tableau expansion, where $C' = RT(C)$.

Lemma 5. *Let $M = (\mathbb{F}, \cdot^I)$ be the model defined by the construction above. Then for any concept C and individuals b, y occurring in $\mathcal{A}$,*
(1) $b \in [\![C]\!]_M$ iff $bIx_{RT(C)} \in \overline{\mathcal{A}}$ (2) $y \in [\![C]\!]_M$ iff $a_{RT(C)}Iy \in \overline{\mathcal{A}}$.

Proof. By induction on the complexity of C. The base case (when C is atomic concept name) is immediate by the construction of the model. For $C = \top$, by definition $RT(\top) = \top$. For any $b \in A$, as $x_\top :: \top \in \overline{\mathcal{A}}$ by the creation rule, and $b : \top \in \overline{\mathcal{A}}$ by $\top$ rule, we have $bIx_\top \in \overline{\mathcal{A}}$ by basic rule. Therefore, $x_\top^\downarrow = A = [\![\top]\!]$. For item 2, for any $y \in Y$, if $a_\top Iy \in \overline{\mathcal{A}}$, then by the appending rule $y :: \top \in \overline{\mathcal{A}}$. Hence, by $\top$ and basic rules, $bIy \in \overline{\mathcal{A}}$ for all b. Thus, $[\![\top]\!] = A^\uparrow \subseteq a_\top^\uparrow$. Moreover, if $y \in [\![\top]\!]$, then $bIy \in \overline{\mathcal{A}}$ for any b by $\top$ and basic rules. In particular $a_\top Iy \in \overline{\mathcal{A}}$. Thus, $[\![\top]\!] = a_\top^\uparrow$. The proof for $C = \bot$ is analogous.

For the induction step, we proceed by cases. In the remainder, for any concept C, we let C' denote $RT(C)$.

If $C = C_1 \wedge C_2$, as to the first claim,
$$b \in [\![C_1 \wedge C_2]\!]$$
iff $b \in [\![C_1]\!], b \in [\![C_2]\!]$ (by def. of $\wedge$)
iff $bIx_{C_1'}, bIx_{C_2'} \in \overline{\mathcal{A}}$ (by induction)

Suppose $b \in [\![C_1 \wedge C_2]\!]$, then by above reasoning we have $bIx_{C_1'}, bIx_{C_2'} \in \overline{\mathcal{A}}$. By appending rules, we have $b : C_1', b : C_2' \in \overline{\mathcal{A}}$. As $C_1' \wedge C_2' = (C_1 \wedge C_2)'$, and $C_1 \wedge C_2$ occurs in $\mathcal{A}$, $C_1' \wedge C_2'$ occurs in $\mathcal{A}$. By creation rule, we have $X_{C_1' \wedge C_2'} :: C_1' \wedge C_2' \in \overline{\mathcal{A}}$. Then by rule $\wedge_X$, we get $bIx_{C_1' \wedge C_2'} = bIx_{(C_1 \wedge C_2)'} \in \overline{\mathcal{A}}$. Conversely, suppose

$bIx_{(C_1 \wedge C_2)'} \in \overline{\mathcal{A}}$. By appending rule, we have $b : (C_1 \wedge C_2)' = b : C_1' \wedge C_2' \in \overline{\mathcal{A}}$. By $\wedge_A$ rule, $b : C_1', b : C_2' \in \overline{\mathcal{A}}$. By creation and basic rule this implies $bIx_{(C_1')'}, bIx_{(C_2')'} \in \overline{\mathcal{A}}$, which is equivalent to $b \in [\![C_1 \wedge C_2]\!]$ as argued earlier.

For the second claim,
$$y \in ([\![C_1 \wedge C_2]\!])$$
$$\text{iff } \forall b(b \in [\![C_1 \wedge C_2]\!] \implies bIy) \quad \text{(by def. of } \wedge\text{)}$$
$$\text{iff } bIx_{(C_1 \wedge C_2)'} \in \overline{\mathcal{A}} \implies bIy \in \overline{\mathcal{A}} \quad \text{(by (1))}$$

If $y \in ([\![C_1 \wedge C_2]\!])$, by creation and basic rule for $(C_1 \wedge C_2)'$, we have $a_{(C_1 \wedge C_2)'} Ix_{(C_1 \wedge C_2)'} \in \overline{\mathcal{A}}$. By $bIx_{(C_1 \wedge C_2)'} \in \overline{\mathcal{A}} \implies bIy \in \overline{\mathcal{A}}$, we get $a_{(C_1 \wedge C_2)'} Iy \in \overline{\mathcal{A}}$. Conversely, suppose $a_{(C_1 \wedge C_2)'} Iy \in \overline{\mathcal{A}}$ and $bIx_{(C_1 \wedge C_2)'} \in \overline{\mathcal{A}}$. By appending and basic rule, we have $bIy \in \overline{\mathcal{A}}$. This completes the proof. The proof for $C = C_1 \vee C_2$ is similar to the previous one.

If $C = [R_\square]C_1$, then as to the first claim,
$$b \in [\![[R_\square]C_1]\!]$$
$$\text{iff } \forall y(y \in ([\![C_1]\!]) \Rightarrow bR_\square y) \quad \text{(def. of } [R_\square]\text{)}$$
$$\text{iff } \forall y(a_{C_1'} Iy \in \overline{\mathcal{A}} \Rightarrow bR_\square y \in \overline{\mathcal{A}}) \text{ (by induction)}$$

If $b \in [\![[R_\square]C_1]\!]$, by Lemma 1.2, $([R_\square]C_1)' = ([R_\square]C_1')'$. Hence $C_1' \in sub([R_\square]C_1)'$ which occurs in the tableaux after pre-processing. Hence $a_{C_1'} Ix_{C_1'} \in \overline{\mathcal{A}}$, by the creation and basic rules. Therefore, by $\forall y(a_{C_1'} Iy \in \overline{\mathcal{A}} \Rightarrow bR_\square y \in \overline{\mathcal{A}})$, we get $bR_\square x_{C_1'} \in \overline{\mathcal{A}}$. We reason by cases. If C_1' is of the form $[R_\square]C_2$ for some C_2, then
$$bR_\square x_{\square C_2} = bR_\square \square x_{C_2} \in \overline{\mathcal{A}}$$
implies $bI \square x_{C_2} = bIx_{\square C_2} \in \overline{\mathcal{A}}$ \hfill (by $Ref_\square$)
iff $bIx_{(\square C_1)'}$ \hfill ($([R_\square]C_1)' = [R_\square]C_2$ by Lemma 1.2)

If C_1' is not of the form $[R_\square]C_2$ for any C_2, then by Lemma 1.2, $C' = ([R_\square](C_1'))' = [R_\square]C_1'$. By the restricted $R_\square$ rule, $bI \square x_{C_1'} = bIx_{\square C_1'} \in \overline{\mathcal{A}}$. Hence, $bIx_{(\square C_1)'} \in \overline{\mathcal{A}}$.

Conversely, if $bIx_{(\square C_1)'} \in \overline{\mathcal{A}}$, and $a_{C_1'} Iy \in \overline{\mathcal{A}}$, then we reason by cases. If C_1 is of the form $[R_\square]C_2$, for some C_2, then, by Lemma 1.2, we get $([R_\square]C_1)' = [R_\square]C_2$. Hence, in this case, the terms reported in the following table can be added to the tableau.

Rule/rules	premises	added terms
$Trans_\square, R_\square$	$bIx_{\square C_2} = bI\square x_{C_2}$	$\Diamond bIx_{\square C_2}$
a_C, x_C, I	$a_{\square C_2} Iy, \Diamond bIx_{\square C_2}$	$\Diamond bIy$
$\Diamond b, Sym_\Diamond$	$\Diamond bIy$	$bR_\square y$

Indeed, we can apply the rule $R_\square$ in the first line in the above table to add a term with constant name $\Diamond b$ in it, since b does not contain any $\Diamond$ in it (as it appears in $\mathcal{A}$). (b) Suppose C_1 is not of the form $[R_\square]C_2$, for any C_2. Therefore, $([R_\square]C_1)' = [R_\square]C_1'$. By the appending rule, we have $b : ([R_\square]C_1)' = b : [R_\square]C_1' \in \overline{\mathcal{A}}$, and $y :: C_1' \in \overline{\mathcal{A}}$. Therefore, by $\square$ rule, we get $bR_\square y \in \overline{\mathcal{A}}$. Hence proved.

For the second claim,

$$y \in (\![R_\square]C_1]\!)$$
$$\text{iff } \forall b (b \in [\![R_\square]C_1]\!] \Rightarrow bIy) \quad \text{(def. of } [R_\square])$$
$$\text{iff } \forall b (bIx_{(\square C_1)'} \in \overline{\mathcal{A}} \Rightarrow bIy \in \overline{\mathcal{A}}) \quad \text{(by (1))}$$

Suppose $y \in (\![R_\square]C_1]\!]$. We consider two different cases. (a) Suppose C_1' is of the form $[R_\square]C_2$ for some C_2. Then, by item 2 of Lemma 1 $([R_\square]C_1)' = [R_\square]C_2$. Hence By creation rule for $[R_\square]C_2$ and basic rule, $a_{\square C_2}Ix_{\square C_2} \in \overline{\mathcal{A}}$. By $bIx_{(\square C_1)'} \in \overline{\mathcal{A}} \Rightarrow bIy \in \overline{\mathcal{A}}$, this implies $a_{\square C_2}Iy = a_{(\square C_1)'}Iy \in \overline{\mathcal{A}}$. (b) Suppose C_1' is not of the form $[R_\square]C_2$ for any C_2, then $([R_\square]C_1)' = [R_\square]C_1'$. Then, by creation and basic rule for $([R_\square]C_1)'$, we get $a_{(\square C_1)'}Ix_{(\square C_1)'} = a_{(\square C_1)'}Ix_{\square C_1'} \in \overline{\mathcal{A}}$. By $bIx_{(\square C_1)'} \in \overline{\mathcal{A}} \Rightarrow bIy \in \overline{\mathcal{A}}$, we get $a_{(\square C_1)'}Iy \in \overline{\mathcal{A}}$. Hence proved. Conversely, suppose $a_{(\square C_1)'}Iy \in \overline{\mathcal{A}}$, and $bIx_{(\square C_1)'} \in \overline{\mathcal{A}}$. Then by appending and basic rule, we get $bIy \in \overline{\mathcal{A}}$, as required. The proof for $C = \langle R_\diamond \rangle C_1$ is similar to the previous one.

Theorem 2 (Soundness). *The model $M = (\mathbb{F}, \cdot^I)$ defined above satisfies $\mathcal{A}$.*

Proof. We proceed by cases.

1. By construction, M satisfies all terms of the form $bR_\square y$, bIy, or $yR_\diamond b$ in $\mathcal{A}$.
2. By construction, any relational term is satisfied by M iff it explicitly occurs in $\overline{\mathcal{A}}$. Thus, either M satisfies all terms of the form $\neg(bR_\square y)$, $\neg(bIy)$, or $\neg(yR_\diamond b)$ occurring in $\mathcal{A}$, or some expansion of $\mathcal{A}$ contains a clash (cf. beginning of Sect. 3.1).
3. For the terms of the form $b : C$, $y :: C$, $\neg(b : C)$, and $\neg(y :: C)$, by Lemma 5, $b \in [\![C]\!]$ iff $bIx_{C'} \in \overline{\mathcal{A}}$, and $y \in (\![C]\!]$ iff $a_{C'}Iy \in \overline{\mathcal{A}}$, where $C' = RT(C)$. For any $b : C$, $y :: C$, $\neg(b : C)$, or $\neg(y :: C)$ occurring in $\mathcal{A}$, we have $b : C'$, $y :: C'$, $\neg(b : C')$, or $\neg(y :: C')$ in $\overline{\mathcal{A}}$ by preprocessing. Therefore, by creation and basic rules, we add $bIx_{C'}$, $a_{C'}Iy$, $\neg(bIx_{C'})$, or $\neg(a_{C'}Iy)$ to $\overline{\mathcal{A}}$. Thus, M satisfies $\mathcal{A}$.

The following corollary is an immediate consequence of the termination and soundness proofs for tableaux.

Corollary 1 (Finite Model Property). *For any consistent LE-$\mathcal{ALCR}$ ABox $\mathcal{A}$, there exists some model of $\mathcal{A}$ of size polynomial in $|\mathcal{A}|$.*

5 Completeness

In this section, we prove the completeness of the tableaux algorithm described in Sect. 3. The following lemma is crucial for this proof.

Lemma 6. *For any LE-$\mathcal{ALCR}$ ABox $\mathcal{A}$, any model $M = (\mathbb{F}, \cdot^I)$ of $\mathcal{A}$ can be extended to a model $M' = (\mathbb{F}', \cdot^{I'})$ such that $\mathbb{F}' = (A', X', I', \{R'_\square\}_{\square \in \mathcal{G}}, \{R'_\diamond\}_{\diamond \in \mathcal{F}})$, $A \subseteq A'$ and $X \subseteq X'$, and moreover for every $\square \in \mathcal{G}$ and $\diamond \in \mathcal{F}$:*

1. There exist $a_C \in A'$ and $x_C \in X'$ such that:

$$C^{I'} = (I'^{(0)}[x_C^{I'}], I'^{(1)}[a_C^{I'}]), \quad a_C^{I'} \in [\![C^{I'}]\!], \quad x_C^{I'} \in (\![C^{I'}]\!),$$

2. For every individual b in A, there exists $\Diamond b$ in A' such that:

$$I'^{(1)}[\Diamond b] = R'^{(1)}_{\square}[b^{I'}] \quad and \quad I'^{(1)}[\Diamond b] = R'^{(0)}_{\Diamond}[b^{I'}],$$

3. For every individual y in X, there exists $\square y$ in X' such that:

$$I'^{(0)}[\square y] = R'^{(1)}_{\Diamond}[y^{I'}] \quad and \quad I'^{(0)}[\square y] = R'^{(0)}_{\square}[y^{I'}],$$

4. For any C, $[\![C^I]\!] = [\![C^{I'}]\!] \cap A$ and $(\![C^I]\!) = (\![C^{I'}]\!) \cap X$.

Proof. Fix $\square \in \mathcal{G}$ and $\Diamond \in \mathcal{F}$. Let M' be defined as follows. For every concept C, we add new elements a_C and x_C to A and X (respectively) to obtain the sets A' and X'. For any $J \in \{I, R_\square\}$, any $a \in A'$ and $x \in X'$, we set $aJ'x$ iff one of the following holds:

1. $a \in A$, $x \in X$, and aJx;
2. $x \in X$, and $a = a_C$ for some concept C, and bJx for all $b \in [\![C^I]\!]$;
3. $a \in A$, and $x = x_C$ for some concept C, and aJy for all $y \in (\![C^I]\!)$;
4. $a = a_{C_1}$ and $x = x_{C_2}$ for some C_1, C_2, and bJy for all $b \in [\![C_1^I]\!]$, and $y \in (\![C_2^I]\!)$.

We set $xR'_\Diamond a$ iff one of the following holds:

1. $a \in A$, $x \in X$, and $xR_\Diamond a$;
2. $x \in X$, and $a = a_C$ for some concept C, and $xR_\Diamond b$ for all $b \in [\![C^I]\!]$;
3. $a \in A$, and $x = x_C$ for some concept C, and $yR_\Diamond a$ for all $y \in (\![C^I]\!)$;
4. $a = a_{C_1}$ and $x = x_{C_2}$ for some C_1, C_2, and $yR_\Diamond b$ for all $b \in [\![C_1^I]\!]$, $y \in (\![C_2^I]\!)$.

For any $b \in A$, $y \in X$, let $\Diamond b = a_{\Diamond \mathbf{b}}$, and $\square y = x_{\square \mathbf{y}}$, where $\mathbf{b}$, and $\mathbf{y}$ are concepts generated by b, and y, respectively. For any atomic concept name D, let $D^{I'} = (I'^{(0)}[x_D], I'^{(1)}[a_D])$. We claim that M' is an LE-$\mathcal{ALCR}$ model satisfying for any concept C, $[\![C^{I'}]\!] = [\![C^I]\!] \cup \{a_{C'} \mid C' \leq C\}$, and $(\![C^{I'}]\!) = (\![C^I]\!) \cup \{x_{C'} \mid C \leq C'\}$.

Firstly, we need to show that $D^{I'} = (I'^{(0)}[x_D], I'^{(1)}[a_D])$ is a valid concept. This is proved as follows. By construction item 4, we have $a_D I' x_D$. Therefore, $I'^{(0)}[I'^{(1)}[a_D]] \subseteq I'^{(0)}[x_D]$. By construction items 2 and 4, $I'^{(1)}[a_D] = (\![D^I]\!) \cup \{x_C \mid D^I \leq C^I\}$. For any $b \in I'^{(0)}[x_D]$, by construction items 3 and 4, $bI'y$ for all $y \in D^I$. Hence, by construction item 2, $bI'x_C$ for all C satisfying $D^I \leq C^I$. Therefore, $b \in I'^{(0)}[I'^{(1)}[a_D]]$. Therefore, $I'^{(0)}[I'^{(1)}[a_D]] = I'^{(0)}[x_D]$. We can similarly prove $I'^{(1)}[I'^{(0)}[x_D]] = I'^{(1)}[a_D]$.

Secondly, we need to show that relations $R'_\square$ and $R'_\Diamond$ are I-compatible. We give proof for $R'_\square$. The proof for $R'_\Diamond$ will be dual. By construction items 1 and 2, for any $y \in X$, $R'^{(0)}_{\square}[y] = R^{(0)}_{\square}[y] \cup \{a_C \mid [\![C^I]\!] \subseteq R^{(0)}_{\square}[y]\}$. By I-compatibility of $R_\square$ in M, $R^{(0)}_{\square}[y] = [\![\square y]\!]$. Hence, by construction items 2 and 4, $R'^{(0)}_{\square}[y] = I'^{(0)}[x_{\square y}]$ is Galois-stable. Similarly, by construction items 1 and 3, for any $b \in A$, $R'^{(1)}_{\square}[b] = R^{(1)}_{\square}[b] \cup \{x_C \mid (\![C^I]\!) \subseteq R^{(1)}_{\square}[b]\}$. By I-compatibility of $R_\square$ in M,

$R_\Box^{(1)}[b] = (\Box y)$. Hence, by construction items 3 and 4, $R_\Box'^{(1)}[b] = I'^{(1)}[a_{\Box y}]$ is Galois-stable in M'.

Thirdly, we need to show that M' satisfies the axioms reflexivity, symmetry, and transitivity. By construction of the model, it is clear that the complex algebra of M and M' are identical. Therefore, as M satisfies these axioms, so does M'.

We have proved that M' is a valid LE-$\mathcal{ALCR}$ model. We now show that for any concept C, $[\![C^{I'}]\!] = [\![C^I]\!] \cup \{a_{C'} \mid C'^I \leq C^I\}$, and $([\![C^{I'}]\!]) = ([\![C^I]\!]) \cup \{x_{C'} \mid C^I \leq C'^I\}$. The proof is by induction on length of C. The base case follows from the construction immediately. In inductive case, we give proofs for connectives $\vee$, and $\Diamond$. Proofs for $\wedge$ and $\Box$ are dual.

(1) Suppose $C = C_1 \vee C_2$. Then, $([\![C^{I'}]\!]) = ([\![C_1^{I'}]\!]) \cap ([\![C_2^{I'}]\!])$. By induction, $([\![C_1^I]\!]) \cap ([\![C_2^I]\!]) = (([\![C_1^I]\!]) \cup \{x_{C'} \mid C_1^I \leq C'^I\}) \cap (([\![C_2^I]\!]) \cup \{x_{C'} \mid C_2^I \leq C'^I\}) = (([\![C_1^I]\!]) \cap ([\![C_2^I]\!])) \cup \{x_{C'} \mid (C_1 \vee C_2)^I \leq C'^I\} = ([\![(C_1 \vee C_2)^I]\!]) \cup \{x_{C'} \mid (C_1 \vee C_2)^I \leq C'^I\}$. On the other hand, $[\![C^{I'}]\!] = I'^{(0)}[([\![C^{I'}]\!])]$. By previous part, $[\![C^{I'}]\!] = I'^{(0)}[([\![C^I]\!]) \cup \{x_{C'} \mid C^I \leq C'^I\}] = I'^{(0)}[([\![C^I]\!])] \cap I'^{(0)}[\{x_{C'} \mid C^I \leq C'^I\}]$. By construction items 1 and 3, $I'^{(0)}[([\![C^I]\!])] = I^{(0)}[([\![C^I]\!])] \cup \{a_{C'} \mid C'^I \leq C^I\} = [\![C^I]\!] \cup \{a_{C'} \mid C'^I \leq C^I\}$. By construction items 3 and 4, $I'^{(0)}[\{x_{C'} \mid C^I \leq C'^I\}] = [\![C^I]\!] \cup \{a_{C'} \mid C'^I \leq C^I\}$. Therefore, $I'^{(0)}[([\![C^I]\!])] \cap I'^{(0)}[\{x_{C'} \mid C^I \leq C'^I\}] = [\![C^I]\!] \cup \{a_{C'} \mid C'^I \leq C^I\}$. Hence proved.

(1) Suppose $C = \Diamond C_1$. Then, $([\![C^{I'}]\!]) = R_\Diamond'^{(0)}[[\![C_1^{I'}]\!]]$. By induction, $R_\Diamond'^{(0)}[[\![C_1^{I'}]\!]] = R_\Diamond'^{(0)}[[\![C_1^I]\!] \cup \{a_{C'} \mid C'^I \leq C_1^I\}] = R_\Diamond'^{(0)}[[\![C_1^I]\!]] \cap R_\Diamond'^{(0)}[\{a_{C'} \mid C'^I \leq C_1^I\}]$. By construction items 1 and 3, $R_\Diamond'^{(0)}[[\![C_1^I]\!]] = R_\Diamond^{(0)}[[\![C_1^I]\!]] \cup \{x_{C'} \mid C^I = (\Diamond C_1)^I \leq C'^I\}$. By construction items 2 and 4, $R_\Diamond'^{(0)}[\{a_{C'} \mid C'^I \leq C_1^I\}] = R_\Diamond^{(0)}[[\![C_1^I]\!]] \cup \{x_{C'} \mid C^I = (\Diamond C_1)^I \leq C'^I\}$. Therefore, $R_\Diamond'^{(0)}[[\![C_1^I]\!]] \cap R_\Diamond'^{(0)}[\{a_{C'} \mid C'^I \leq C_1^I\}] = R_\Diamond^{(0)}[[\![C_1^I]\!]] \cup \{x_{C'} \mid C^I = (\Diamond C_1)^I \leq C'^I\} = ([\![\Diamond C_1]\!]) \cup \{x_{C'} \mid C^I = (\Diamond C_1)^I \leq C'^I\}$. The proof for extension part is similar to the $\vee$ case. Hence proved.

From the claim it is easy to check that M' is a model of LE-$\mathcal{ALCR}$ satisfying the required properties.

We are now ready to prove the completeness result.

Theorem 3 (Completeness). *Let $\mathcal{A}$ be a consistent LE-$\mathcal{ALCR}$ ABox and $\mathcal{A}'$ be obtained via the application of any expansion rule applied to $\mathcal{A}$. Then $\mathcal{A}'$ is also consistent.*

Proof. If $\mathcal{A}$ is consistent, by Lemma 6, a model M' of $\mathcal{A}$ exists which satisfies items 1 to 4 stated in the lemma. We will show M' is also a model of $\mathcal{A}'$, where $\mathcal{A}'$ is obtained from $\mathcal{A}$ after the pre-processing and the application of some LE-$\mathcal{ALCR}$ tableaux expansion rule. The result for pre-processing step follows from the fact that models of LE-$\mathcal{ALCR}$ satisfy the axioms reflexivity and transitivity. The result for creation rules follows immediately from Lemma 6. For the basic, ⊤, ⊥, and rules for logical connectives follows straight-forwardly by semantics of these connectives. The result for adjunction rules follows by from I-compatibility and Lemma 6, along with the fact that symmetry axioms imply that the operators $\Box$ (resp. $\Diamond$) and ■ (resp. ♦) coincide. Thus, to complete the proof we need to show that the rough concept rules from Sect. 3 preserve the consistency of ABoxes. The result for reflexivity rules $Ref_\Box$, $Ref_\Diamond$, and symmetry rules $Sym_\Box$, $Sym_\Diamond$ follows immediately from the fact that models of LE-$\mathcal{ALCR}$ must satisfy the reflexivity and symmetry axioms. For the rule $Trans_\Diamond$, suppose $\Diamond bIy \in \mathcal{A}$. Then $M' \models \Diamond bIy$. We know that $I^{(1)}[\Diamond b] = I^{(1)}[\Diamond \mathbf{b}]$. Therefore, $y \in ([\Diamond \mathbf{b}])$. By symmetry, adjunction, and I-compatibility, we have $b \in [\![\Box \mathbf{y}]\!]$. Therefore, by transitivity, $b \in [\![\Box \Box \mathbf{y}]\!]$. Again by symmetry, adjunction and I-compatibility, we have $\Diamond \mathbf{b} \leq \Box \mathbf{y}$. Therefore, as $\Diamond b \in [\![\Diamond \mathbf{b}]\!]$ and $y \in ([\mathbf{y}])$, we must have $\Diamond bR_\Box y$. The proof for other transitivity axiom is similar.

6 TBox Consistency and Extensions of LE-$\mathcal{ALCR}$

In this section, we first extend the tableaux algorithm described in Sect. 3, to also include TBox axioms. We then define two expansions of LE-$\mathcal{ALCR}$: (1) with the concepts generated by a set of objects or features, and (2) with the feature-pair inconsistency terms, and also extend the tableaux algorithms to the aforementioned expansions.

6.1 Tableaux Algorithm for TBox Axioms

In this section, we expand the tableaux algorithm to TBox axioms. TBox in LE-$\mathcal{ALCR}$ consists of concept inclusion axioms of the form $C_1 \sqsubseteq C_2$ for some concepts C_1 and C_2. To extend the tableaux algorithm to the knowledge bases with TBoxes we add the following rules to the tableaux expansion rules.

$$\sqsubseteq_a \frac{b: C_1, C_1 \sqsubseteq C_2}{b: C_2} \qquad \frac{y :: C_2, C_1 \sqsubseteq C_2}{y :: C_1} \sqsubseteq_x$$

The soundness and completeness of the tableaux algorithm obtained is trivial. The termination proof remains similar.

Remark 1. Note that in LE-$\mathcal{ALC}$ adding the above rules naively can lead to infinite cycles for cyclic TBoxes. These cycles can stem from repeated application of a rule $\sqsubseteq_a$ or $\sqsubseteq_x$ along with some adjunction rule. For example, consider a knowledge base with a TBox axiom $C_1 \sqsubseteq \Box C_1$, and an ABox axiom $b: C_1$ in it. Repeated application of rules $\sqsubseteq_a$, $\Box$ and $R_\Box$ can lead to infinite cycles. However, in the case of LE-$\mathcal{ALCR}$, the restrictions on the application of adjunction rules ensure that such cycles can never occur (which is reflected in the fact that no constant name appearing in tableaux expansion contains more than one $\Box$ or $\Diamond$ operators in it).

6.2 Extending LE-$\mathcal{ALCR}$ with Generated Concepts

In many applications, we are interested in defining a concept to be the concept generated by some set of objects or features. To this end, we extend the logic LE-$\mathcal{ALCR}$ with the concepts generated by a given set of objects or features. We call this extended description logic LE-$\mathcal{ALCRO}$. For any set of object names (resp. feature names) B (resp. Y), the concept **B** (resp. **Y**) denotes the concept generated by it. For any generated concept **B** (resp. **Y**), its interpretation under and interpretation I is given by $((B^I)^{\uparrow\downarrow}, (B^I)^{\uparrow})$ (resp. $((Y^I)^{\downarrow}, (Y^I)^{\downarrow\uparrow})$, where $\uparrow$ and $\downarrow$ are interpreted in terms of I^I. From this definition, it is easy to check that a tableaux algorithm for LE-$\mathcal{ALCRO}$ can be obtained by adding the following rules to tableaux expansion rules for LE-$\mathcal{ALCR}$.

$$A \frac{b \in B}{b : \mathbf{B}} \qquad X \frac{y \in Y}{y :: \mathbf{Y}} \qquad \frac{bIy, \quad \forall b \in B}{a_B Iy} a_A \qquad \frac{bIy, \quad \forall y \in Y}{bIx_Y} x_X$$

6.3 Extending LE-$\mathcal{ALCR}$ with Feature-Pair Inconsistencies

In many applications, we may need to define two features that are inconsistent with each other, i.e. no element can have both the features together e.g. pairs of opposite features like even and odd, or positive and negative. In the classical case, such information is presented straightforwardly using negation. However, in the case of LE-logics, we do not have negation in the language. Therefore, we need to describe such information explicitly. To this end, we add a new term $N(y, z)$ to the LE-$\mathcal{ALCR}$ ABox, where y and z are feature names. For any interpretation $I = (\mathbb{F}, \cdot^I)$ of LE-$\mathcal{ALCR}$, with $\mathbb{F} = (A, X, I, \mathcal{R}_\square, \mathcal{R}_\diamond)$, I satisfies a term $N(y, z)$ iff for any $b \in A$, $bI^I y^I$ implies $\neg(bI^I z^I)$ and $bI^I z^I$ implies $\neg(bI^I y^I)$. That is, no $b \in A$ can have both y and z as features. We denote the expansion of LE-$\mathcal{ALCR}$ (resp. LE-$\mathcal{ALCRO}$) with the terms of the form $N(y, z)$ by LE-$\mathcal{ALCRN}$ (resp. LE-$\mathcal{ALCRON}$). The tableaux algorithm for LE-$\mathcal{ALCRN}$ (resp. LE-$\mathcal{ALCRON}$) is obtained by adding the following tableaux expansion rule to the expansion rules for LE-$\mathcal{ALCRN}$ (resp. LE-$\mathcal{ALCRON}$).

$$N_1 \frac{bIy, \quad N(y, z)}{\neg(bIz)} \qquad N_2 \frac{bIz, \quad N(y, z)}{\neg(bIy)}$$

Correctness and termination of the tableaux algorithm follow from the definition of satisfaction for terms of the form $N(y, z)$, and the fact that no new concept or constant names are added by the new rules N_i.

7 Example

We now give a toy example of an ontology described in the language LE-$\mathcal{ALCRON}$ and perform some reasoning in it. We describe a (partial) knowledge base for Whittaker's five kingdom classification of living things[6] in LE-$\mathcal{ALCRON}$. The following table lists atomic concept names in the knowledge base.

---

[6] This example is only given as a toy example for an application of the description logic LE-$\mathcal{ALCRON}$. We do not claim that this is a precise or detailed ontology for the five kingdom classification.

Description Logic for Rough Concepts   85

concept	symbol	concept	symbol
living things	C_1	animalia	C_2
plantae	C_3	fungi	C_4
protista	C_5	monera	C_6

This ontology is based on the following set of features:

feature	symbol	feature	symbol
nuclear membrane present	w_1	nuclear membrane absent	y_1
cellulose cell wall present	w_2	non-cellulose cell wall present	y_2
cell wall absent	z_2	eukaryotic	w_3
prokaryotic	y_3	autotrophic	w_4
heterotrophic	y_4	multi-cellular	w_5
uni-cellular	y_5	has DNA	w_6
has RNA	w_7	reproduction	w_8

Consider $Y_1 := \{w_1, y_2, w_3, w_5, y_4, w_6, w_7, w_8\}$, $Y_2 := \{w_1, z_2, w_3, w_5, y_4, w_6, w_7, w_8\}$. As discussed in the preliminaries, the operators $\vee$, and $\wedge$ denote the least upper bound and the the greatest upper bound (not necessarily set theoretic joins or meets) of the concepts, respectively. For any concept C, the concepts $\Box C$ and $\Diamond C$ represent the lower and the upper approximations of C, respectively. We assume that the lower and upper approximation operators satisfy all the axioms listed in table 1. These axioms are commonly used in rough set theory to characterize lower and upper approximations.

Let $\mathcal{K} = (\mathcal{A}, \mathcal{T})$ be a knowledge base such that $\mathcal{T} = \{C_1 \equiv C_2 \vee C_3 \vee C_4 \vee C_5 \vee C_6, C_1 \equiv C_2 \vee C_3, C_1 \equiv C_3 \vee C_4, C_1 \equiv C_2 \vee C_4, C_2 \wedge C_3 \sqsubseteq \bot, C_3 \wedge C_4 \sqsubseteq \bot, C_2 \wedge C_4 \sqsubseteq \bot, C_4 \equiv Y_1, C_2 \equiv Y_2\}$, and
$\mathcal{A} = \{$ dog:C_2, $\neg$(dog:C_3), $\neg$(dog:C_4), rose:C_3, $\neg$(rose:C_2), $\neg$(rose:C_4), mushroom:C_4 $\neg$(mushroom:C_2), $\neg$(mushroom:C_3), covid virus: $\Diamond C_1$, $\neg$ (covid virus I w_6), $\neg$ (Drosera: $\Box C_2$), $w_6 :: C_1$, $w_7 :: C_1$, $\neg$(stone:C_1), $w_8 :: \Diamond C_1$, $w_2 :: C_3$, $y_5 :: C_5$, $y_5 :: C_6$, $y_3 :: C_5$, $w_6 :: C_1$, $N(w_1, y_1)$, $N(w_2, y_2)$, $N(w_2, z_2)$, $N(y_2, z_2)$, $N(w_3, y_3)$, $N(w_4, y_4)$, $N(w_5, y_5)\}$.

From the axioms $C_1 \equiv C_2 \vee C_3$, $C_1 \equiv C_3 \vee C_4$, $C_1 \equiv C_2 \vee C_4$, $C_2 \wedge C_3 \sqsubseteq \bot$, $C_2 \wedge C_4 \sqsubseteq \bot$, $C_3 \wedge C_3 \sqsubseteq \bot$, any model for $\mathcal{K}$ contains the lattice M_3 as a sublattice, and hence must be non-distributive. Thus, LE-based (non-distributive) description logic is necessary to reason with this knowledge base.

Consider the following reasoning questions regarding the above ontology:

1. If we find a new living things does it have to belong to one of the five kingdoms?

2. Can a living thing be in both the kingdoms animalia and plantae?
3. Is there a feature which every living thing must have, but not every species/object in the upper approximation of living things (i.e. things which may possibly be considered to be living things) need to have?
4. If we find a new sample with features w_1, w_2, and w_3, what inference can we make about which kingdom the sample belongs to?

For the first question, we translate it into the problem of checking the consistency of the knowledge base $\mathcal{K}' = \{\mathcal{A}', \mathcal{T}\}$, where $\mathcal{A}' = \mathcal{A} \cup \{b : C_1\} \cup \{\neg(b : C_i) \mid 2 \leq i \leq 6\}$ where b is a new object name. If the knowledge base $\mathcal{K}$ implies that every living thing belongs to one of the five kingdoms, then the new knowledge base $\mathcal{K}'$ must be inconsistent. However, the tableaux algorithm shows that $\mathcal{K}'$ is consistent. Therefore, there can be living things that do not belong to any of the five kingdoms as per $\mathcal{K}$.[7]

For the second question, we can translate it into the problem of checking the consistency of the knowledge base $\mathcal{K}' = \{\mathcal{A}', \mathcal{T}\}$, where $\mathcal{A}'$ is obtained by adding the axioms $b : C_1$, $b : C_2$, and $b : C_3$ for some new object name b. The tableaux algorithm (partial expansion) applies to $\mathcal{K}'$ as follows:

Rule	Premises	Added terms
Creation		$x_{C_2 \wedge C_3} : C_2 \wedge C_3$
$\wedge_X$	$x_{C_2 \wedge C_3} : C_2 \wedge C_3, b : C_2, b : C_3$	$b : C_2 \wedge C_3$
I	$b : C_3, w_2 :: C_3$	bIw_2
$\sqsubseteq_a$	$b : C_2 \wedge C_3, C_2 \wedge C_3 \sqsubseteq \bot$	$b : \bot$
$\sqsubseteq_a$	$b : C_2, C_2 \sqsubseteq \mathbf{Y}_1$	$b : \mathbf{Y}_1$
X	$z_2 \in Y_1$	$z_2 :: \mathbf{Y}_1$
I	$b : \mathbf{Y}_1, z_2 :: \mathbf{Y}_1$	bIz_2
N_1	$bIw_2, N(w_2, z_2)$	$\neg(bIz_2)$

The term bIz_2 clashes with $\neg(bIz_2)$. Thus, $\mathcal{K}'$ is inconsistent. Therefore, according to $\mathcal{K}$ no element can be in both kingdoms animalia and plantae.

For the third question, if such a feature does not exists then we must have $(\!(C_1)\!) \subseteq (\!(\Diamond C_1)\!)$. Thus, this question can be translated into checking the consistency of the knowledge base $\mathcal{K}' = \{\mathcal{A}, \mathcal{T}'\}$, where $\mathcal{T}'$ is obtained by adding the axiom $\Diamond C_1 \sqsubseteq C_1$ to $\mathcal{T}$. Then, by applying rule $\sqsubseteq_a$ to covid virus:$\Diamond C_1$, covid virus:C_1 would be added to the tableaux expansion. Then, by rule I and $w_6 :: C_1$, term (covid virus)Iw_6 is added to the tableaux, giving a clash with the term $\neg$ (covid virus I w_6). Therefore, there is a feature that every living thing must have, but not every species/object in the upper approximation of living things (i.e. things that may be considered to be living things) needs to have.

---

[7] In fact, living things like lichens are not classified into any of the kingdoms in the five kingdom system.

For the fourth question, we can show that the sample species does not belong to the kingdom animalia. This is done by showing that if we add bIw_1, bIw_2, bIw_3, and $b : C_2$ to $\mathcal{A}$ for some new object name b, the resulting knowledge base $\mathcal{K}' = \{\mathcal{A}', \mathcal{T}\}$ is inconsistent. The tableaux algorithm (partial expansion) applies to $\mathcal{K}'$ as follows:

Rule	Premises	Added terms
$\sqsubseteq_a$	$b : C_2$, $C_2 \equiv \mathbf{Y}_2$	$b : \mathbf{Y}_2$
$\mathbf{X}$	$z_2 \in Y_2$	$z_2 :: \mathbf{Y}_2$
I	$b : \mathbf{Y}_2$, $z_2 :: \mathbf{Y}_2$	bIz_2
N_2	bIz_2, $N(w_2, z_2)$	$\neg(bIw_2)$

The term $\neg(bIw_2)$ clashes with bIw_2. Therefore, the new sample can not be in the kingdom animalia. By similar arguments, we can show that the new sample can not be a part of the kingdom fungi. However, it can be shown that the new sample can belong to the kingdom plantae. That is, there is a model which is consistent with axioms bIw_1, bIw_2, bIw_3, and $b : C_2$, where the object name b represents the new sample.

8 Conclusion and Future Work

In this paper, we extended the two-sorted non-distributive description logic LE-$\mathcal{ALC}$ to LE-$\mathcal{ALCR}$ by adding reflexivity, symmetry, and transitivity axioms to model rough concepts. Algebraically, this logic corresponds to complete lattices with modal operators satisfying these axioms. We provided tableaux a sound and complete polynomial time tableaux algorithm for LE-$\mathcal{ALCR}$. Additionally, we expanded LE-$\mathcal{ALCR}$ to include concepts generated from sets of objects/features and introduced feature-pair inconsistency terms, providing corresponding tableaux algorithms. We demonstrated knowledge representation and reasoning in LE-$\mathcal{ALCR}$ through an example involving a knowledge base for Whittaker's five kingdom classification.

This work can be extended in several interesting directions.

Description Logics for Weaker Logics for Rough Concepts. The description logic LE-$\mathcal{ALCR}$ corresponds to the non-distributive counterparts of S5 modal logic. In classical rough set theory, modal logics with approximation operators weaker than S5 have been explored [30,32]. Researching description logics corresponding to these weaker logics and their potential applications presents an intriguing avenue for future study.

Generalizing other Description Logics for Reasoning under Uncertainty to the Nondistributive Setting. In this work, we have provided a description logic to reason about approximate concepts within rough set theory. It would be interesting to study non-distributive description logics using other frameworks for reasoning with approximate or imprecise knowledge, such as fuzzy logics, probabilistic logics, and possibilistic logics. This could provide the possibility of establishing a closer and more formally explicit connection between approximate or imprecise reasoning methods in Formal Concept Analysis (FCA) and description logics.

References

1. Baader, F., Calvanese, D., McGuinness, D., Nardi, D., Patel-Schneider, P.: The Description Logic Handbook: Theory, Implementation and Applications. Cambridge University Press, Cambridge (2003)
2. Baader, F., Horrocks, I., Sattler, U.: Description logics as ontology languages for the semantic web. In: Hutter, D., Stephan, W. (eds.) Mechanizing Mathematical Reasoning. LNCS (LNAI), vol. 2605, pp. 228–248. Springer, Heidelberg (2005). https://doi.org/10.1007/978-3-540-32254-2_14
3. Benítez-Caballero, M.J., Medina, J., Ramírez-Poussa, E.: Characterizing one-sided formal concept analysis by multi-adjoint concept lattices. Mathematics **10**(7), 1020 (2022)
4. Berardi, D., Calvanese, D., De Giacomo, G.: Reasoning on UML class diagrams. Artif. Intell. **168**(1–2), 70–118 (2005)
5. van der Berg, I., Domenico, A.D., Greco, G., Manoorkar, K.B., Palmigiano, A., Panettiere, M.: Non-distributive description logic (2024)
6. Conradie, W., et al.: Modal reduction principles across relational semantics. Fuzzy Sets Syst. **481**, 108892 (2024). https://doi.org/10.1016/j.fss.2024.108892, https://www.sciencedirect.com/science/article/pii/S0165011424000381
7. Conradie, W., et al.: Rough concepts. Inf. Sci. **561**, 371–413 (2021)
8. Conradie, W., Frittella, S., Palmigiano, A., Piazzai, M., Tzimoulis, A., Wijnberg, N.M.: Toward an epistemic-logical theory of categorization. In: Electronic Proceedings in Theoretical Computer Science, EPTCS, vol. **251** (2017)
9. Conradie, W., Frittella, S., Palmigiano, A., Piazzai, M., Tzimoulis, A., Wijnberg, N.M.: Categories: how i learned to stop worrying and love two sorts. In: Väänänen, J., Hirvonen, Å., de Queiroz, R. (eds.) WoLLIC 2016. LNCS, vol. 9803, pp. 145–164. Springer, Heidelberg (2016). https://doi.org/10.1007/978-3-662-52921-8_10
10. Conradie, W., Palmigiano, A.: Algorithmic correspondence and canonicity for non-distributive logics. Ann. Pure Appl. Logic **170**(9), 923–974 (2019)
11. Conradie, W., Palmigiano, A., Robinson, C., Wijnberg, N.: Non-distributive logics: from semantics to meaning. In: Rezus, A. (ed.) Contemporary Logic and Computing, Landscapes in Logic, vol. 1, pp. 38–86. College Publications (2020)
12. Ganter, B., Wille, R.: Applied lattice theory: Formal concept analysis. In: In General Lattice Theory, G. Grätzer editor, Birkhäuser. Citeseer (1997)
13. Ganter, B., Wille, R.: Formal Concept Analysis: Mathematical Foundations. Springer, Cham (2012)
14. Greco, G., Jipsen, P., Liang, F., Palmigiano, A., Tzimoulis, A.: Algebraic proof theory for le-logics. ACM Trans. Comput. Log. **25**(1), 1–37 (2024)
15. Heinsohn, J.: Probabilistic description logics. In: de Mantaras, R.L., Poole, D. (eds.) Uncertainty in Artificial Intelligence, pp. 311–318. Morgan Kaufmann, San Francisco (CA) (1994). https://doi.org/10.1016/B978-1-55860-332-5.50044-4, https://www.sciencedirect.com/science/article/pii/B9781558603325500444
16. Hollunder, B.: An alternative proof method for possibilistic logic and its application to terminological logics. Int. J. Approximate Reasoning **12**(2), 85–109 (1995)

17. Horrocks, I., et al.: Daml+oil: a description logic for the semantic web. IEEE Data Eng. Bull. **25**(1), 4–9 (2002)
18. Jiang, Y., Tang, Y., Wang, J., Tang, S.: Reasoning within intuitionistic fuzzy rough description logics. Inf. Sci. **179**(14), 2362–2378 (2009). https://doi.org/10.1016/j.ins.2009.03.001, https://www.sciencedirect.com/science/article/pii/S0020025509001133, including Special Section - Linguistic Decision Making
19. Jiang, Y., Wang, J., Tang, S., Xiao, B.: Reasoning with rough description logics: an approximate concepts approach. Inf. Sci. **179**(5), 600–612 (2009). https://doi.org/10.1016/j.ins.2008.10.021, https://www.sciencedirect.com/science/article/pii/S0020025508004416, special Section - Quantum Structures: Theory and Applications
20. Liu, M., Shao, M., Zhang, W., Wu, C.: Reduction method for concept lattices based on rough set theory and its application. Comput. Math. Appl. **53**(9), 1390–1410 (2007)
21. Lukasiewicz, T.: Expressive probabilistic description logics. Artif. Intell. **172**(6), 852–883 (2008). https://doi.org/10.1016/j.artint.2007.10.017, https://www.sciencedirect.com/science/article/pii/S0004370207001877
22. Lukasiewicz, T., Straccia, U.: Managing uncertainty and vagueness in description logics for the semantic web. J. Web Semant. **6**(4), 291–308 (2008)
23. Orlowska, E.: A logic of indiscernibility relations. In: Skowron, A. (ed.) SCT 1984. LNCS, vol. 208, pp. 177–186. Springer, Heidelberg (1985). https://doi.org/10.1007/3-540-16066-3_17
24. Orłowska, E., Pawlak, Z.: Representation of nondeterministic information. Theoret. Comput. Sci. **29**(1–2), 27–39 (1984)
25. Pawlak, Z.: Rough sets. Int. J. Comput. Inf. Sci. **11**(5), 341–356 (1982)
26. Qi, G., Ji, Q., Pan, J.Z., Du, J.: Extending description logics with uncertainty reasoning in possibilistic logic. Int. J. Intell. Syst. **26**(4), 353–381 (2011)
27. Staab, S., Studer, R.: Handbook on ontologies. Springer Science & Business Media (2010)
28. Wei, L., Qi, J.J.: Relation between concept lattice reduction and rough set reduction. Knowl.-Based Syst. **23**(8), 934–938 (2010)
29. Yao, Y., Chen, Y.: Rough set approximations in formal concept analysis. In: Peters, J.F., Skowron, A. (eds.) Transactions on Rough Sets V, pp. 285–305. Springer, Berlin (2006)
30. Yao, Y.Y., Lin, T.Y.: Generalization of rough sets using modal logics. Intell. Autom. Soft Comput. **2**(2), 103–119 (1996)
31. Yen, J., et al.: Generalizing term subsumption languages to fuzzy logic. In: IJCAI vol. 91, pp. 472–477 (1991)
32. Zhu, W.: Generalized rough sets based on relations. Inf. Sci. **177**(22), 4997–5011 (2007)

Rule Induction and Machine Learning

Greedy Algorithm for Construction of Deterministic Decision Trees for Conventional Decision Tables from Closed Classes

Azimkhon Ostonov[(✉)] and Mikhail Moshkov

Computer, Electrical and Mathematical Sciences and Engineering Division and Computational Bioscience Research Center, King Abdullah University of Science and Technology (KAUST), Thuwal 23955-6900, Saudi Arabia
{azimkhon.ostonov,mikhail.moshkov}@kaust.edu.sa

Abstract. This paper examines specific classes of conventional decision tables (DTs) that are closed under operations of attribute (column) removal and decision modifications assigned to rows. For DTs belonging to any of these closed classes (CCs), we investigate a greedy algorithm that constructs a deterministic decision tree. We demonstrate that the number of steps performed by this algorithm is limited by twice the number of rows in the table. Furthermore, we compare the behavior of two functions. The first function describes the increase in the minimum complexity of a deterministic decision tree for a DT from the CC in the worst-case scenario, in relation to the complexity of the set of attributes associated with columns of the table. The second function characterizes the worst-case complexity growth of the deterministic decision tree constructed by the greedy algorithm for a DT from the CC, considering the growth of complexity of the set of attributes associated with columns of the table. We divide the entire collection of pairs consisting of a bounded complexity measure (BCM) and a CC into three subsets. For each subset, we establish lower and upper bounds for the second function based on the first function.

Keywords: closed class of decision tables · deterministic decision tree · greedy algorithm

1 Introduction

This paper focuses on the examination of a specific greedy algorithm used to construct deterministic decision trees [3]. The performance of this algorithm is analyzed in the context of conventional decision tables (DTs) that belong to classes closed under operations of removing columns (attributes) and changing the decisions assigned to rows. It is worth noting that the former of these two operations, mainly referred to as attribute reduction, has been extensively studied by the rough set community [2,9–11].

Decision tables, commonly employed in data analysis, consist of rows labeled with decisions. Decision trees, which are widely recognized as classification algorithms and a means of representing knowledge, have been extensively studied [1,7,8].

Closed classes (CCs) of decision tables (DTs) encompass various types of tables, with the most natural examples being those that relate to problems over finite and infinite information systems [6]. It is worth noting that the entire range of CCs of DTs, as mentioned in [5], extends beyond the CCs associated with information systems.

In the paper [5], the focus was on examining the behavior the function denoted as $\mathcal{F}_{\psi,A}$. This function explores the worst-case growth of the minimum complexity of a deterministic decision tree for a DT belonging to the CC A on the complexity of the set of attributes associated with the columns of the table. The complexity measure ψ in the definition of this function can be any bounded measure, such as depth or weighted depth.

This paper investigates the greedy algorithm $U_{R,\psi}$ introduced in [3]. The algorithm constructs a deterministic decision tree $U_{R,\psi}(T)$ for a given table T belonging to the class A. We demonstrate that the number of execution steps of the algorithm is no more than twice the number of rows in table the T. Additionally, we examine the function $\mathcal{U}_{\psi,A}$, which describes the worst-case complexity growth of the deterministic decision tree constructed by the $U_{R,\psi}$ algorithm for a DT from the class A with the growth of complexity of the set of attributes associated with the columns of the DT.

Using the results from [3,5], we prove that there are exactly three types of joint behavior of the functions $\mathcal{F}_{\psi,A}$ and $\mathcal{U}_{\psi,A}$:

(a) Both functions $\mathcal{F}_{\psi,A}$ and $\mathcal{U}_{\psi,A}$ are bounded from above by constants.
(b) Both functions $\mathcal{F}_{\psi,A}$ and $\mathcal{U}_{\psi,A}$ behave as logarithms.
(c) For any n and a positive constant c, $\mathcal{F}_{\psi,A}(n) \leq \mathcal{U}_{\psi,A}(n) \leq n$ and $\mathcal{U}_{\psi,A}(n) \leq c\mathcal{F}_{\psi,A}(n)^3$. For infinitely many n, $\mathcal{F}_{\psi,A}(n) = n$.

Type (a) is considered degenerate because the number of rows in decision tables from the closed class A is limited by a constant. Type (b) is the most significant type for our investigation. For this type, both the minimum complexity of deterministic decision trees and the complexity of decision trees constructed by the greedy algorithm in the worst-case exhibit behavior that can be described as logarithmic in the complexity of the set of attributes in the table. Type (c) is not particularly noteworthy as the minimum complexity of deterministic decision trees in the worst case is the same as the complexity of the set of attributes in the table.

The results presented in this paper are the essential generalization of the results obtained in [4] for the depth of decision trees.

Note that in the present paper, we repeat without any changes many of the definitions, notation and statements from [5] to give readers the opportunity to use both papers without additional difficulties.

The paper is organized into six sections. Section 2 covers the key definitions and notation. Section 3 presents the main results. In Sect. 4, we present auxiliary statements. Section 5 is dedicated to proving the main results. Finally, Sect. 6 provides brief conclusions.

2 Main Definitions and Notation

Let ω be the set $\{0, 1, 2, \ldots\}$. For any number $k \in \omega$ excluding 0 and 1, let $E_k = \{0, 1, \ldots, k-1\}$. We define P as the set of attributes, denoted as $\{f_i : i \in \omega\}$. In this context, two attributes f_l and f_p from P are considered distinct if $l \neq p$.

2.1 Decision Tables

First, we define the notion of a decision table [5].

Definition 1. *Let $k \in \omega \setminus \{0, 1\}$ and let $\mathcal{M}_k$ represent the set of rectangular tables where each cell is filled with a number taken from the set E_k. In these tables, every row is distinct, labeled with a number from ω, and each column is labeled with a distinct attribute from P. The rows can be interpreted as tuples of attribute values. It is worth noting that empty tables without any rows are included in the set $\mathcal{M}_k$. For convenience, the same symbol Λ is used to refer to these tables. The tables belonging to $\mathcal{M}_k$ are referred to as decision tables (DTs).*

We define $\mathcal{M}_k \mathcal{C}$ as the set of tables from $\mathcal{M}_k$ where each table has all its rows labeled with the same decision. Let $\Lambda \in \mathcal{M}_k \mathcal{C}$.

Consider a nonempty DT T from $\mathcal{M}_k$. We define $P(T)$ as the set of attributes associated with the columns of table T. Let $f_{i_1}, \ldots, f_{i_m} \in P(T)$ and $\delta_1, \ldots, \delta_m \in E_k$. We denote $T(f_{i_1}, \delta_1) \cdots (f_{i_m}, \delta_m)$ as the table obtained by removing all rows from T that do not satisfy the following condition: in the columns labeled with attributes $f_{i_1}, \ldots, f_{i_m}$, the rows have the corresponding numbers $\delta_1, \ldots, \delta_m$.

We will now introduce two operations that can be performed on decision tables: column removal and decision modification [5].

Definition 2. *Column removal. Consider a set D that is a subset of $P(T)$. We eliminate from table T all columns labeled with attributes from the set D. Within each group of rows that are identical in the remaining columns, we retain one row with the minimum decision. We denote the resulting table as $I(D, T)$. Specifically, $I(\emptyset, T)$ is equal to T, and $I(P(T), T)$ is equal to Λ. It is evident that $I(D, T)$ belongs to $\mathcal{M}_k$.*

Definition 3. *Decision modification. Consider the function $\nu : E_k^{|P(T)|} \to \omega$ (where E_k^0 is defined as the empty set). For each row $\bar{\delta}$ in table T, we replace the decision associated with that row with the value obtained by applying ν to $\bar{\delta}$. We represent the resulting table as $J(\nu, T)$. It is clear that $J(\nu, T)$ belongs to $\mathcal{M}_k$.*

Definition 4. *Denote $[T] = \{J(\nu, I(D, T)) : D \subseteq P(T), \nu : E_k^{|P(T) \setminus D|} \to \omega\}$. The set $[T]$ is the closure of the table T under the operations of column removal and decision modification [5].*

Definition 5. *Consider a set A that is a subset of $\mathcal{M}_k$, where A is not empty. Let $[A] = \bigcup_{T \in A} [T]$. The set $[A]$ represents the closure of the set A under the two operations being considered. A class or set of DTs, denoted by A, is referred to as a* closed class *(CC) [5] if $[A] = A$.*

Suppose A_1 and A_2 are CCs of DTs from $\mathcal{M}_k$. Then, the union of A_1 and A_2, denoted by $A_1 \cup A_2$, is also a CC of DTs from $\mathcal{M}_k$.

2.2 Deterministic Decision Trees

A *finite tree with a root* refers to a finite directed tree where only one specific node, known as the *root*, does not have any incoming edges. The nodes in the tree that do not have any outgoing edges are referred to as *leaf* nodes.

Definition 6. *A k-decision tree [5] is a finite tree with a root, consisting of a minimum of two nodes, wherein*

- *The root node and the edges originating from the root do not have any assigned labels.*
- *Every leaf node in the tree is marked with a decision taken from the set ω.*
- *Every node in the tree, excluding the root and leaf nodes, is assigned a label representing an attribute from the set P. Additionally, each outgoing edge from these non-root, non-leaf nodes is marked with a number from the set E_k.*

We utilize $P(\Gamma)$ to denote the set of attributes assigned to the nodes of Γ that are not the root or leaf nodes. A *complete path* in Γ is defined as a sequence $\rho = u_1, e_1, \ldots, u_l, e_l, u_{l+1}$, which consists of nodes and edges in Γ. In this sequence, u_1 is the root of Γ, u_{l+1} corresponds to a leaf node of Γ, and for $j = 1, \ldots, l$, the edge e_j connects the node u_j to the node u_{j+1}. Consider $T \in \mathcal{M}_k$. In case that $P(\Gamma)$ is a subset of the set $P(T)$, we can associate the table T and the complete path ρ with a table $T(\rho)$. If $l = 1$, then $T(\rho) = T$. In case that $l > 1$ and for $j = 2, \ldots, l$, the node u_j is assigned the attribute f_{i_j} and the edge e_j is assigned the number σ_j, then $T(\rho) = (f_{i_2}, \sigma_2) \cdots (f_{i_l}, \sigma_l)$.

Definition 7. *Consider T as an element of $\mathcal{M}_k \setminus \Lambda$. A deterministic decision tree [5] for the table T refers to a k-decision tree Γ that meets the following criterion:*

- *The root of Γ has only a single outgoing edge.*
- *For any node in Γ that is not the root or a leaf node, the outgoing edges from that node are labeled with pairwise different numbers.*
- *$P(\Gamma) \subseteq P(T)$.*
- *For every row in T, there is a complete path ρ in Γ such that the corresponding row is in the table $T(\rho)$.*
- *For every complete path ρ in Γ, one of the following conditions must hold: either $T(\rho) = \Lambda$ or the decision associated with the final node of ρ is identical with the decision attached to all rows of the table $T(\rho)$.*

2.3 Complexity Measures

Let B represent the set of all finite words over the alphabet P, where P is the set containing elements f_i for each i in the set w. The set B includes the empty word λ, and the operation of concatenation is defined in the set B.

Definition 8. *A complexity measure (CM)* [5] *is a function* $\psi : B \to w$. *This function* ψ *possesses the following properties for any words* ξ_1 *and* ξ_2 *in* B:

(p1) $\psi(\xi_1) = 0$ only if it is the case that $\xi_1 = \lambda$ – property of positivity.
(p2) $\psi(\xi_1) = \psi(\xi_1')$ where ξ_1' is any word which can be obtained by changing the order of letters of the word ξ_1 – property of commutativity.
(p3) $\psi(\xi_1) \leq \psi(\xi_1\xi_2)$ – property of nondecreasing.
(p4) $\psi(\xi_1\xi_2) \leq \psi(\xi_1) + \psi(\xi_2)$ – property of boundedness from above.

The following functions are CMs:

- Function h for which, for any word α in B, $h(\alpha)$ is equal to the length of the word α. In other words, $h(\alpha)$ returns the number of characters or symbols in the word α.
- An arbitrary function $\varphi : B \to w$ is defined such that $\varphi(\lambda) = 0$, where λ represents the empty word. For any element f_i in the alphabet P, the function φ assigns a value greater than zero, denoted as $\varphi(f_i) > 0$. Furthermore, for any nonempty word $f_{i_1} \cdots f_{i_m}$ in B,

$$\varphi(f_{i_1} \cdots f_{i_m}) = \sum_{j=1}^{m} \varphi(f_{i_j}). \tag{1}$$

- An arbitrary function $\rho : B \to w$ is defined with the following properties: $\rho(\lambda) = 0$, where λ represents the empty word. For any element f_i in the alphabet P, the function ρ assigns a value greater than zero, denoted as $\rho(f_i) > 0$. Additionally, for any nonempty word $f_{i_1} \cdots f_{i_m}$ in B, $\rho(f_{i_1} \cdots f_{i_m})$ is equal to the maximum value among $\rho(f_{i_j})$ for $j = 1, \ldots, m$.

Definition 9. *A CM* ψ *is called* bounded complexity measure (BCM) [5] *if it satisfies the property of* boundedness from below:

(p5) $\psi(\xi) \geq |\xi|$ for any word ξ in P^*.

Any CM that satisfies Eq. (1), including the function h, is a BCM. It can be demonstrated that if ψ_1 and ψ_2 are CMs, then ψ_3 and ψ_4 are also CMs. For any word α in B, $\psi_3(\alpha) = \psi_1(\alpha) + \psi_2(\alpha)$ and $\psi_4(\alpha) = \max(\psi_1(\alpha), \psi_2(\alpha))$. Furthermore, if ψ_1 is a BCM, then ψ_3 and ψ_4 are also BCMs.

Definition 10. *A CM* ψ *is called* 1-computable *if there exists an algorithm, which, for a given attribute* $f_i \in P$, *returns the value* $\psi(f_i)$.

The simplest example of a 1-computable BCM is the function h.

Definition 11. *Consider a CM ψ. We can extend this measure to apply to the set of all finite subsets of the set P. Let D be a finite subset of P. If D is an empty set, then we define $\psi(D)$ as 0. Now, suppose D is a non-empty set, written as $D = \{f_{i_1}, \ldots, f_{i_m}\}$ with $m \geq 1$. In this case, we define $\psi(D)$ as $\psi(f_{i_1} \cdots f_{i_m})$.*

Consider $\Gamma \in \mathcal{T}_k$ and $\tau = v_1, d_1, \ldots, v_m, d_m, v_{m+1}$, which represents a complete path in Γ. We can associate a word $F(\tau) \in B$ with this path τ. If $m = 1$, then $F(\tau) = \lambda$, representing the empty word. If $m > 1$, and for $j = 2, \ldots, m$, the node v_j is labeled with the attribute f_{i_j}, then $F(\tau) = f_{i_2} \cdots f_{i_m}$.

Definition 12. *Consider a CM ψ. We can extend this measure to apply to the set $\mathcal{T}_k$. Let $\Gamma \in \mathcal{T}_k$ be a decision tree. We define $\psi(\Gamma)$ as the maximum value obtained by applying ψ to the word representations $F(\tau)$ of all complete paths τ in Γ. In other words, $\psi(\Gamma) = \max\{\psi(F(\tau))\}$, where the maximum is taken over all complete paths in the decision tree Γ. The value $\psi(\Gamma)$ is referred to as the complexity of the decision tree Γ with respect to the CM ψ. Additionally, the value $h(\Gamma)$, obtained by applying the function h to Γ, is referred to as the depth of the decision tree Γ.*

2.4 Parameters of Decision Trees and Tables

Consider a CM ψ. We will now explain the functions ψ^d, W_ψ, S_ψ, $\hat{S}_\psi$, M_ψ, N [5], and R defined on the set $\mathcal{M}_k$ and taking values from the set ω. For each of these functions, their value for the DT Λ is defined as 0. Now, let $T \in \mathcal{M}_k \setminus \Lambda$.

- $\psi^d(T) = \min\{\psi(\Gamma)\}$, the value of $\psi^d(T)$ is defined as the minimum value obtained by evaluating $\psi(\Gamma)$, where the minimum is taken over all deterministic decision trees Γ for the table T.
- $W_\psi(T) = \psi(P(T))$. This refers to the complexity associated with the set of attributes that are attached to columns of the table T.
- Let $\bar{\sigma}$ represent a row from the table T. We define $S_\psi(T, \bar{\sigma})$ as the minimum value of $\psi(Q)$, where Q is a subset of the set $P(T)$ such that, in the columns of T labeled with attributes from Q, the row $\bar{\sigma}$ is distinct from all other rows in the table T. Consequently, $S_\psi(T)$ is defined as the maximum value among all $S_\psi(T, \bar{\sigma})$, where the maximum is taken over all rows $\bar{\sigma}$ in the table T.
- $\hat{S}_\psi(T) = \max\{S_\psi(T^*) : T^* \in [T]\}$.
- If T belongs to the set $\mathcal{M}_k\mathcal{C}$, then $M_\psi(T)$ is equal to 0. Now, let us consider the case where T does not belong to $\mathcal{M}_k\mathcal{C}$. Suppose the number of attributes in T is n, and the columns of T are labeled with the attributes $f_{t_1}, \ldots, f_{t_n}$. Let $\bar{\sigma} = (\sigma_1, \ldots, \sigma_n)$ be an element of E_k^n. We represent by $M_\psi(T, \bar{\sigma})$ the smallest number $q \in \omega$ such that there exist attributes $f_{t_{i_1}}, \ldots, f_{t_{i_l}} \in P(T)$ for which $T(f_{t_{i_1}}, \sigma_{i_1}) \cdots (f_{t_{i_l}}, \sigma_{i_l}) \in \mathcal{M}_k\mathcal{C}$ and $\psi(f_{t_{i_1}} \cdots f_{t_{i_l}}) = q$. Then, $M_\psi(T)$ is equal to the maximum value obtained from evaluating $M_\psi(T, \bar{\sigma})$ for every $\bar{\sigma} \in E_k^n$.
- $N(T)$ represents the count of rows present in the table T.
- $R(T)$ represents the count of unordered pairs of rows in the table T that have different decision labels.

For the complexity measure h, we denote $W(T) = W_h(T)$ and $S(T) = S_h(T)$. Note that $W(T)$ is the number of columns in the table T.

3 Main Results

Initially, we examine the outcomes presented in the study [5] concerning the function $\mathcal{F}_{\psi,A}$. These results will be utilized to establish a comparison between the functions $\mathcal{U}_{\psi,A}$ and $\mathcal{F}_{\psi,A}$.

Consider a BCM ψ and a nonempty CC A consisting of DTs from $\mathcal{M}_k$. We will now introduce a function $\mathcal{F}_{\psi,A} : \omega \to \omega$. Suppose $n \in \omega$. Then

$$\mathcal{F}_{\psi,A}(n) = \max\{\psi^d(T) : T \in A, W_\psi(T) \leq n\}.$$

The function $\mathcal{F}_{\psi,A}$ describes the worst-case increase in the minimum complexity of a deterministic decision tree for a DT belonging to the class A with the growth in the complexity of the set of attributes associated with the columns of the table.

Suppose $D = \{n_i : i \in \omega\}$ is an infinite subset of the set ω in which, for any i belonging to ω, n_i is less than n_{i+1}. Let H_D be a function mapping from ω to ω. Suppose n belongs to ω. In case n is less than n_0, then $H_D(n)$ is defined as 0. In case, for some i belonging to ω, $n_i \leq n < n_{i+1}$ holds, then $H_D(n)$ is equal to n_i.

Theorem 1. (Theorem 1 [5]) *Consider a BCM ψ and a nonempty CC A consisting of DTs from $\mathcal{M}_k$. Then the function $\mathcal{F}_{\psi,A}$ is a nondecreasing function defined for all values in ω. It satisfies the properties $\mathcal{F}_{\psi,A}(n) \leq n$ for any $n \in \omega$, and $\mathcal{F}_{\psi,A}(0) = 0$. Furthermore, for this function, one of the following statements is true:*

(a) *In case, the functions S_ψ and N have constant upper bounds on the class A, then there exists a constant q_0 belonging to $\omega \setminus \{0\}$, where $\mathcal{F}_{\psi,A}^\infty(n) \leq q_0$ holds for every n belonging to ω.*
(b) *In case, the function S_ψ has a constant upper bound on the class A and the function N has no constant upper bound on the class A, then there exist constants q_1, q_2, q_3, q_4 each belonging to $\omega \setminus \{0\}$, where $q_1 \log_2 n - q_2 \leq \mathcal{F}_{\psi,A}^\infty(n) \leq q_3 \log_2 n + q_4$ holds for every n belonging to $\omega \setminus \{0\}$.*
(c) *In case, the function S_ψ has no constant upper bound on the class A, then there exists an infinite subset D of the set ω, where $H_D(n)$ is less than or equal to $\mathcal{F}_{\psi,A}^\infty(n)$ holds for every n belonging to ω.*

Consider a 1-computable BCM ψ. We will now provide a description of an algorithm denoted as $U_{R,\psi}$, as outlined in [3]. This algorithm takes as input a decision table T from $\mathcal{M}_k \setminus \Lambda$ and constructs a deterministic decision tree $U_{R,\psi}(T)$ for the given table T.

Algorithm $U_{R,\psi}$

Step 1. Build a tree G that consists of two nodes, namely v_0 and v_1. Create an edge originating from node v_0 and connecting to node v_1. Assign the DT T as the label for node v_1.

Step 2. If the tree G does not have any nodes that are labeled with DTs, then the decision tree $U_{R,\psi}(T)$ is simply the tree G.

If there exists a node v in the tree G that is labeled with DT Q, we proceed in the following way: If all the rows in table Q are labeled with the same decision, we replace the label of node v with this decision instead of the table Q. We then move on to Step 2.

Otherwise, for each $f_i \in P(T)$, find the minimum number $\sigma_i \in E_k$ such that $R(Q(f_i, \sigma_i)) = \max\{R(Q(f_i, \sigma)) : \sigma \in E_k\}$. Set $K(Q) = \{f_i : f_i \in P(T), R(Q) > R(Q(f_i, \sigma_i))\}$. For each $f_i \in K(Q)$, set $d(f_i) = \max\{\psi(f_i), R(Q)/(R(Q) - R(Q(f_i, \sigma_i)))\}$. Let f_{i_0} be the attribute from $K(Q)$ with the minimum index i_0 for which $d(f_{i_0}) = \min\{d(f_i) : f_i \in K(Q)\}$. Label the node v with the attribute f_{i_0} instead of the table Q. For each $\sigma \in E_k$ such that $Q(f_{i_0}, \sigma) \neq \Lambda$, add to the tree G a node $v(\sigma)$ and an edge that leaves the node v and enters the node $v(\sigma)$. This edge is labeled with the number σ, and node $v(\sigma)$ is labeled with the table $Q(f_{i_0}, \sigma)$. We then move on to Step 2.

We will not consider a detailed analysis of the time complexity of this algorithm. Instead, we will focus on assessing the number of times Step 2 is repeated during the construction of the decision tree $U_{R,\psi}(T)$.

Theorem 2. *Assuming we have a 1-computable BCM ψ, the algorithm $U_{R,\psi}$ will repeat Step 2 no more than $2N(T)$ times while constructing the decision tree $U_{R,\psi}(T)$ for any DT T from $\mathcal{M}_k \setminus \Lambda$.*

Consider a 1-computable BCM ψ and a CC A consisting of DTs from $\mathcal{M}_k$. By definition, $\psi(U_{R,\psi}(\Lambda)) = 0$. Now, we will introduce a function $\mathcal{U}_{\psi,A} : \omega \to \omega$. For any $n \in \omega$, the function is defined as follows:

$$\mathcal{U}_{\psi,A}(n) = \max\{\psi(U_{R,\psi}(T)) : T \in A, W_\psi(T) \leq n\}.$$

The function $\mathcal{U}_{\psi,A}$ describes how the complexity of the deterministic decision tree, constructed by the algorithm $U_{R,\psi}$ for a DT from class A, grows in the worst-case scenario with the growth of the complexity of the set of attributes associated with the columns of the DT.

The following statement provides lower and upper bounds for the function $\mathcal{U}_{\psi,A}(n)$ based on the function $\mathcal{F}_{\psi,A}(n)$.

Theorem 3. *Consider a 1-computable BCM ψ and a nonempty CC A consisting of DTs from $\mathcal{M}_k$. The function $\mathcal{U}_{\psi,A}$ is a nondecreasing function defined for all values in ω. It satisfies the properties $\mathcal{F}_{\psi,A}(n) \leq \mathcal{U}_{\psi,A}(n) \leq n$ for any $n \in \omega$. Additionally, for this function, one of the following statements is true:*

(a) In case, the functions S_ψ and N have constant upper bounds on the class A, then there exists a positive constant c such that $\mathcal{U}_{\psi,A}(n) \leq \mathcal{F}_{\psi,A}(n) + c$ for any $n \in \omega$.

(b) In case, the function S_ψ has a constant upper bound on the class A and the function N has no constant upper bound on the class A, then there exist positive constants c and d such that $\mathcal{U}_{\psi,A}(n) \leq c\mathcal{F}_{\psi,A}(n) + d$ for any $n \in \omega \setminus \{0\}$

(c) In case, the function S_ψ has no constant upper bound on the class A, then there exists a positive constant c such that $\mathcal{U}_{\psi,A}(n) \leq c\mathcal{F}_{\psi,A}(n)^3$ for any $n \in \omega$.

4   Auxiliary Statements

The following statement is a consequence of Theorem 2.1 in the paper [3].

Lemma 1. *For an arbitrary CM ψ and an arbitrary DT T from $\mathcal{M}_k$,*
$$\psi^d(T) \geq M_\psi(T).$$

The following statement can be derived from Lemma 1.3 and Theorem 3.1 in the paper [3].

Lemma 2. *Let ψ be a 1-computable BCM. Then, for an arbitrary DT T from $\mathcal{M}_k \setminus \{\Lambda\}$,*
$$\psi(U_{R,\psi}(T)) \leq \begin{cases} M_\psi(T), & if\ M_\psi(T) \leq 1, \\ M_\psi(T)^2 \ln R(T) + M_\psi(T), & if\ M_\psi(T) \geq 2. \end{cases}$$

Lemma 3. *(Lemma 3 [5]) For an arbitrary CM ψ and an arbitrary DT T from $\mathcal{M}_k$,*
$$M_\psi(T) \leq 2\hat{S}_\psi(T).$$

Lemma 4. *(Lemma 4 [5]) For an arbitrary DT T from $\mathcal{M}_k \setminus \{\Lambda\}$,*
$$N(T) \leq (kW(T))^{S(T)}.$$

Lemma 5. *(Lemma 5 [5]) For an arbitrary DT T from $\mathcal{M}_k \setminus \{\Lambda\}$, there exists a mapping $\nu : E_k^{W(T)} \to \omega$ such that*
$$h^d(J(\nu, T)) \geq \log_k N(T).$$

Let Γ be a k-decision tree. We use $L(G)$ to represent the count of nodes in Γ, excluding the root node, and $L_0(\Gamma)$ to represent the count of terminal nodes in this tree.

Lemma 6. *Consider a k-decision tree Γ where each node, except for the root and terminal nodes, has at least two outgoing edges. In this case, the following statement holds:*
$$L(\Gamma) \leq 2L_0(\Gamma) - 1.$$

Proof. We prove this statement using induction based on the value of $L(\Gamma)$. When $L(\Gamma) = 1$, then the statement is obvious. Assume that $n \geq 2$, and suppose that the statement holds for every Γ with $L(\Gamma) < n$. Let $L(\Gamma) = n$. Choose in Γ a node v such that all edges $e_1, \ldots, e_t$ leaving v enter the terminal nodes $v_1, \ldots, v_t$, respectively. Let us remove from Γ edges $e_1, \ldots, e_t$ and nodes $v_1, \ldots, v_t$ and label the node v with the number 0. Let denote the obtained k-decision tree by Γ^*. Obviously, $L(\Gamma^*) = L(\Gamma) - t$ and $L_0(\Gamma^*) = L_0(\Gamma) - t + 1$. Since $L(\Gamma^*) < n$, by inductive hypothesis, $L(\Gamma^*) \leq 2L_0(\Gamma^*) - 1$. Thus, $L(\Gamma) - t \leq 2L_0(\Gamma) - 2t + 2 - 1$ and $L(\Gamma) \leq 2L_0(\Gamma) - (t-1)$. Considering the fact that $t \geq 2$, we have $L(\Gamma) \leq 2L_0(\Gamma) - 1$.

5 Proofs of Theorems 2 and 3

Proof (Proof of Theorem 2). One can show that the number of repetitions of Step 2 is equal to $L(U_{R,\psi}(T)) + 1$. Based on the description of the algorithm $U_{R,\psi}$, it can be concluded that for any complete path τ in the decision tree $U_{R,\psi}(T)$, the decision table $T(\tau)$ is not equal to Λ. Evidently, for any row of the table T, there exists exactly one complete path τ of the decision tree $U_{R,\psi}(T)$ such that the considered row belongs to the table $T(\tau)$. Hence, the total number of complete paths in $U_{R,\psi}(T)$ is at most $N(T)$, and the count of terminal nodes $L_0(U_{R,\psi}(T))$ is less than or equal to $N(T)$. According to the algorithm $U_{R,\psi}$, each node in $U_{R,\psi}(T)$, excluding the root and terminal nodes, has a minimum of two outgoing edges. By Lemma 6, $L(U_{R,\psi}(T)) \leq 2N(T) - 1$. Therefore, the algorithm $U_{R,\psi}$ during the construction of the decision tree $U_{R,\psi}(T)$ repeats Step 2 no more than $2N(T)$ times.

Proof (Proof of Theorem 3). Let $n \in \omega$. Since A is a CC, $\Lambda \in A$. By definition, $\psi^d_\psi(\Lambda) = \psi(U_{R,\psi}(\Lambda)) = W_\psi(\Lambda) = 0$. Therefore the set $A(n) = \{T : T \in A, W_\psi(T) \leq n\}$ is nonempty. Let $A(n) \setminus \{\Lambda\} \neq \emptyset$ and $T \in A(n) \setminus \{\Lambda\}$. It is clear that $\psi^d(T) \leq \psi(U_{R,\psi}(T))$. It is easy to see that, in any complete path of the decision tree $U_{R,\psi}(T)$, different nodes which are any nodes except the root and terminal nodes are labeled with different attributes. Since ψ is a BCM, we have $\psi(U_{R,\psi}(T)) \leq W_\psi(T)$. Now it is easy to show that the value $\mathcal{U}_{\psi,A}(n)$ is defined and $\mathcal{F}_{\psi,A}(n) \leq \mathcal{U}_{\psi,A}(n) \leq n$. Evidently, $\mathcal{U}_{\psi,A}$ is a nondecreasing function.

(a) Suppose the functions S_ψ and N be bounded from above on the class A by constants $a > 1$ and $b > 1$, respectively. Let $T \in A \setminus \{\Lambda\}$. Using Lemma 3, we obtain that $M_\psi(T) \leq 2a$. Evidently, $R(T) \leq b^2$. By Lemma 2, $\psi(U_{R,\psi}(T)) \leq c$, where $c = (2a)^2 \ln b^2 + 2a$. It is clear that $\psi(U_{R,\psi}(\Lambda)) \leq c$. Therefore $\mathcal{U}_{\psi,A}(n) \leq \mathcal{F}_{\psi,A}(n) + c$ for any $n \in \omega$.

(b) Suppose the function S_ψ be bounded from above on the class A by a constant $t \geq 1$ and the function N be not bounded from above on the class A. Using Theorem 1, we obtain that there exist positive constants a and b such that $\mathcal{F}_{\psi,A}(n) \geq a \log_2 n - b$ for any $n \in \omega \setminus \{0\}$. Let $T \in A \setminus \{\Lambda\}$. Using Lemma 3, we obtain that $M_\psi(T) \leq 2t$. Since ψ is a BCM, $S(T) \leq S_\psi(T)$. By Lemma 4, $N(T) \leq (kW(T))^t$ and $R(T) \leq (kW(T))^{2t}$. From here and from Lemma 2 it follows that $\psi(U_{R,\psi}(T)) \leq (2t)^2 2t \ln(W(T)k) + 2t$. Since ψ is a BCM, $W(T) \leq W_\psi(T)$. Using the equality $\psi(U_{R,\psi}(\Lambda)) = 0$, we obtain $\mathcal{U}_{\psi,A}(n) \leq \frac{8t^3 \log_2 n}{\log_2 e} + 8t^3 \ln k + 2t$ for any $n \in \omega \setminus \{0\}$. Taking into account that, for any $n \in \omega \setminus \{0\}$, $\log_2 n \leq \frac{\mathcal{F}_{\psi,A}(n)+b}{a}$, we obtain that, for any $n \in \omega \setminus \{0\}$, $\mathcal{U}_{\psi,A}(n) \leq \frac{8t^3}{a \log_2 e} \mathcal{F}_{\psi,A}(n) + \frac{8t^3 b}{a \log_2 e} + 8t^3 \ln k + 2t$. Thus, for any $n \in \omega \setminus \{0\}$, $\mathcal{U}_{\psi,A}(n) \leq c\mathcal{F}_{\psi,A}(n) + d$, where $c = \frac{8t^3}{a \log_2 e}$ and $d = \frac{8t^3 b}{a \log_2 e} + 8t^3 \ln k + 2t$.

(c) Suppose the function S_ψ be not bounded from above on the class A. Let $n \in \omega$, T be a table from $A \setminus \{\Lambda\}$ and $W_\psi(T) \leq n$. Using Lemma 1, we obtain $M_\psi(T) \leq \psi^d(T) \leq \mathcal{F}_{\psi,A}(n)$.

Let $M_\psi(T) \geq 2$. By Lemma 5, there exists a mapping $\nu : E_k^{W(T)} \to \omega$ such that $h^d(J(\nu, T)) \geq \log_k N(T)$. Denote $T^* = J(\nu, T)$. It is clear that $T^* \in A$

and $W_\psi(T^*) = W_\psi(T) \leq n$. Since ψ is a BCM, $h^d(T^*) \leq \psi^d(T^*)$. Therefore $\log_k N(T) \leq \mathcal{F}_{\psi,A}(n)$ and $\ln N(T) \leq \mathcal{F}_{\psi,A}(n)/\log_k e$. Since $R(T) \leq N(T)^2$, we have $\ln R(T) \leq 2\mathcal{F}_{\psi,A}(n)/\log_k e$. Using Lemma 2, we obtain $\psi(U_{R,\psi}(T)) \leq M_\psi(T)^2 \ln R(T) + M_\psi(T) \leq 2\mathcal{F}_{\psi,A}(n)^3/\log_k e + \mathcal{F}_{\psi,A}(n) \leq c\mathcal{F}_{\psi,A}(n)^3$, where $c = 2/\log_k e + 1$.

Let $M_\psi(T) \leq 1$. Then, by Lemma 2, $\psi(U_{R,\psi}(T)) \leq M_\psi(T)$. Therefore $\psi(U_{R,\psi}(T)) \leq \mathcal{F}_{\psi,A}(n) \leq c\mathcal{F}_{\psi,A}(n)^3$.

It is clear that $\psi(U_{R,\psi}(\Lambda)) \leq c\mathcal{F}_{\psi,A}(n)^3$. As a result, we obtain $\mathcal{U}_{\psi,A}(n) \leq c\mathcal{F}_{\psi,A}(n)^3$.

6 Conclusions

This paper focused on analyzing the complexity of deterministic decision trees created by a greedy algorithm for decision tables belonging to closed classes of conventional decision tables, where each row is associated with a single decision label. Subsequent research will be directed towards investigating closed classes of decision tables with many-valued decisions, where rows are labeled with finite sets of decisions.

Acknowledgements. Research reported in this publication was supported by King Abdullah University of Science and Technology (KAUST).

References

1. Breiman, L., Friedman, J.H., Olshen, R.A., Stone, C.J.: Classification and Regression Trees. Wadsworth and Brooks (1984)
2. Hedar, A.R., Omar, M.A., Sewisy, A.A.: Rough sets attribute reduction using an accelerated genetic algorithm. In: 2015 IEEE/ACIS 16th International Conference on Software Engineering, Artificial Intelligence, Networking and Parallel/Distributed Computing (SNPD), pp. 1–7 (2015). https://doi.org/10.1109/SNPD.2015.7176207
3. Moshkov, M.: Conditional tests. In: Yablonskii, S.V. (ed.) Problemy Kibernetiki (in Russian), vol. 40, pp. 131–170. Nauka Publishers, Moscow (1983)
4. Moshkov, M.: On depth of conditional tests for tables from closed classes. In: Markov, A.A. (ed.) Combinatorial-Algebraic and Probabilistic Methods of Discrete Analysis (in Russian), pp. 78–86. Gorky University Press, Gorky (1989)
5. Ostonov, A., Moshkov, M.: On complexity of deterministic and nondeterministic decision trees for conventional decision tables from closed classes. Entropy **25**(10), 1411 (2023)
6. Pawlak, Z.: Information systems theoretical foundations. Inf. Syst. **6**(3), 205–218 (1981)
7. Quinlan, J.R.: C4.5: Programs for Machine Learning. Morgan Kaufmann (1993)
8. Rokach, L., Maimon, O.: Data Mining with Decision Trees - Theory and Applications, Series in Machine Perception and Artificial Intelligence, vol. 69. World Scientific (2007)
9. Wang, C., Ou, F.: An attribute reduction algorithm in rough set theory based on information entropy. In: 2008 International Symposium on Computational Intelligence and Design, vol. 1, pp. 3–6. IEEE (2008)

10. Yuan, Z., Chen, H., Xie, P., Zhang, P., Liu, J., Li, T.: Attribute reduction methods in fuzzy rough set theory: an overview, comparative experiments, and new directions. Appl. Soft Comput. **107**, 107353 (2021)
11. Zhang, X., Yao, Y.: Tri-level attribute reduction in rough set theory. Expert Syst. Appl. **190**, 116187 (2022)

Study of Dependency Degree and Bayesian Networks for Conflict Scenarios

Małgorzata Przybyła-Kasperek[1] and Rafał Deja[2(✉)]

[1] Institute of Computer Science, University of Silesia, ul. Bedzinska 39, Sosnowiec, Poland
`malgorzata.przybyla-kasperek@us.edu.pl`
[2] Department of Computer Science, WSB University, ul. Cieplaka 1c, Dąbrowa Górnicza, Poland
`rdeja@wsb.edu.pl`

Abstract. The paper delves into analyzing conflict scenarios seen by attribute dependency, emphasizing the interaction between individual attributes. The paper presents two distinct methodologies: attributes dependency degree and Bayesian Networks for attributes. The study investigates the scope to which attributes influence one another within the context of conflict. Through empirical analysis of two real-world cases, the research uncovers significant attributes whose values offer crucial insights into the compatibility or divergence of numerous other attributes. Both methods illuminate attribute connections by employing legible graphical representations, facilitating a deeper understanding of the conflict situation. The findings highlight attributes that emerge as pivotal, providing valuable guidance for negotiation in complex conflict situations.

Keywords: Conflict analysis · Dependency degree · Bayesian Network · Statistical inference · Negotiations

1 Introduction

Understanding the dynamics of conflicts and their different factors is essential in various domains ranging from social sciences to decision-making processes. Within the realm of conflict analysis, the exploration of attribute connections plays a key role in comprehending the complicatedness of the conflict situation. Attributes representing distinct issues of a conflict situation often exhibit complex dependencies that can significantly influence the course of negotiations, decision-making, and conflict resolution strategies. This paper investigates the degree of dependency among attributes within conflict scenarios as well as statistical inference and Bayes Networks. Our research is motivated by the recognition that attributes within a conflict situation do not exist in isolation; rather, they interact with each other in different ways, shaping the overall dynamics of the conflict.

Two distinct methodologies are considered: the dependency degree approach and Bayesian Networks [7]. These methodologies offer complementary perspectives on attribute relations, providing valuable insights into the underlying structures of conflict scenarios. The dependency degree approach mainly focuses on the study of abstraction classes of indiscernibility relations and checking the degree of inclusion of these classes. Thus, we can say the measure verifies consistency of attribute values. In contrast, Bayesian Networks provide a probabilistic framework for modeling attribute relationships, leveraging statistical inference to uncover hidden dependencies in attributes values [3].

Through empirical analysis of real-world conflict cases, we aim to explore the efficacy of these methodologies in determining attribute relations and identifying pivotal attributes that can significantly modify conflict dynamics. By presenting our findings in legible graphical representations, we seek to enhance the interpretability of attribute connections and indicate negotiation strategies in conflict resolution efforts.

The main contributions of the paper are:

- Propose a method for recognizing attribute connections in a conflict situation using the degree of dependency.
- Proving some basic properties for the degree of dependency measure.
- Propose a method for identifying attribute relationships using statistical inference and Bayesian Networks.
- Illustrating the application of the introduced approaches on two real-world examples.

Conflict analysis methods considered in the literature include conflict-based real-time safety analysis using video sensors and computer vision techniques [2], Pawlak's model of conflict analysis [6], and an extended Pawlak conflict model with a trust recommendation mechanism [15] and other methods. The rough set (RS) model is applied to conflict analysis to analyze and resolve conflicts accurately. Pawlak's model of conflict analysis is generalized and extended using three-way decision theory [14,16] to examine trisections of agents, pairs of agents, and three levels of conflict. The developments in the Pawlak conflict analysis include the introduction of probabilistic models by Lang, Miao, and Cai, which generalize Pawlak's model by using pair thresholds instead of the threshold 0.5 [4]. Another model proposed by Lang, Miao, and Fujita utilizes Pythagorean fuzzy numbers and group decision-making to analyze conflicts [5]. Deja and Skowron extended the Pawlak model by considering variables that lead to conflicts, addressing issues such as reasons for conflicts and partitioning of agents and issues [1,11]. Sun et al. established an extended conflict model based on probabilistic rough set and three-way decision theory to find optimal feasible strategies [13]. Yao reformulated and extended the classical Pawlak conflict model, focusing on the static model of conflict analysis and finding optimal feasible strategies [16]. In [10] conflict analysis is used for classification based on imbalanced data from dispersed sources. Paper [9] reviews selected extensions of Pawlak's conflict analysis model. These developments provide new perspectives and mathematical tools for analyzing conflicts and reaching consensus in conflict situations.

In the subsequent sections of this paper, we delve into the theoretical support of the dependency degree approach and Bayesian networks, elucidating their respective methodologies and applications in conflict analysis. We then present the results of our empirical analysis, showcasing the insights obtained from applying these methodologies to real-world conflict scenarios. Finally, we discuss the implications of our findings and offer recommendations for future research directions in the field of conflict analysis and resolution.

2 The Dependency Degree of Attributes

Information system is a pair $IS = (U, I)$, where U is the universe – the set of agents, I is a set of issues, and the set of values of $a \in I$ is equal $V^a = \{-1, 0, +1\}$. Opinion of agent $u \in U$ about issue $a \in I$ is the value $a(u)$. The meaning of this value is as follows: $a(u) = +1$ means the agent u is in favor of the issue a; $a(u) = 0$ means the agent u is neutral to the issue a; $a(u) = -1$ means the agent u is against to the issue a.

One of the basic definition in the rough sets theory is an indiscernibility relation on a given attribute $a \in I$.

Definition 1. *For the information system $IS = (U, I)$ and attribute $a \in I$ an indiscernibility relation on a attribute a is $IND(\{a\}) = \{(u, u') \in U \times U : a(u) = a(u')\}$.*

This relation is reflexive, symmetric and transitive [12]. This means that this is an equivalence relation, and its abstraction classes form a division of the set of agents U. Let us denote the set of abstraction classes for relations $IND(\{a\})$ by $\tilde{a} = \{[u]_{IND(\{a\})} : u \in U\}$.

Attribute b is dependent on attribute a, we denote it as $a \longrightarrow b$, if $\tilde{a} \subseteq \tilde{b}$. Attribute b is independent from attribute a, we denote it as $a \not\longrightarrow b$, if $\tilde{a} \not\subseteq \tilde{b}$. Attributes a and b are equivalent $a \longleftrightarrow b$ if $\tilde{a} = \tilde{b}$. In conflict analysis, considering attribute dependencies during negotiations is extremely important. This can help us recognize coalitions by checking compliance on a given attribute and recognizing attributes on which this compliance is also achieved. The most significant attributes are those on which the largest number of attributes depend – because their abstraction classes determine the abstraction classes of the largest number of other attributes. In other words, consistency on these attributes determines consistency on the largest number of other attributes. However, in most situations the attributes dependencies according to the above definition is not fulfilled. Then agents can be persuaded to make certain concessions and change their opinions if this is adequately justified by, for example, coalition support on further important issues. Statistical inference methods can be used for such a purpose. One of the appropriate significance test for testing the dependence of attributes with categorical values is the chi-square independence test. This test assesses whether a relations exists between two variables, that is, whether the observed distribution of observations within one variable depends on the other variable. Another possibility is to use a measure defined based on the rough sets attribute dependence.

Definition 2. *For the information system $IS = (U, I)$, where $card\{U\} \geq 2$ and attributes $a, b \in I$ a dependency degree $(deg(a \longrightarrow b))$ of attribute b on attribute a is defined as follows*

$$deg(a \longrightarrow b) = \frac{\sum_{X \in \tilde{a}} \sum_{Y \in \tilde{b}: X \cap Y \neq \emptyset} card\{X \setminus Y\}}{card\{U\} \cdot (card\{U\} - 1)}$$

It should be noted that we assume here that the set of attribute's values V^a, $a \in I$ can be any finite set, not only $\{-1, 0, +1\}$.

It can be said that dependency degree measures the correspondence of the abstraction classes of indiscernibility relation for the attribute a against the attribute b. More specifically, how finely the abstraction classes of indiscernibility relation on the attribute a are shared among the abstraction classes of indiscernibility relation on the attribute b. We will give some properties of the dependency degree measure.

Proposition 1. *The value $deg(a \longrightarrow b)$ belongs to the interval $[0, 1]$ for any $IS = (U, I)$, where $card\{U\} \geq 2$ and attributes $a, b \in I$.*

It is obvious that the value $deg(a \longrightarrow b) \geq 0$. Furthermore, if $a \longrightarrow b$ then $deg(a \longrightarrow b) = 0$. Indeed, if $a \longrightarrow b$ then $\tilde{a} \subseteq \tilde{b}$. This and the fact that the indiscernibility relation is an equivalence relation, means that for each abstraction class $X \in \tilde{a}$ there is exactly one abstraction class $Y \in \tilde{b}$ for which $X \subseteq Y$ and it is the only abstraction class for which $X \cap Y \neq \emptyset$. Thus $card\{X \setminus Y\} = 0$ for each $X \cap Y \neq \emptyset$.

Now let us consider the situation in which we get the highest value of $deg(a \longrightarrow b)$. This is the case in which the difference of sets $X \in \tilde{a}$ and $Y \in \tilde{b}$ is as large as possible. The value $deg(a \longrightarrow b)$ is the largest, when we have $\tilde{a} = \{\{u : u \in U\}\}$ (there is only one abstraction class consisting of all objects) and $\tilde{b} = \{\{u\} : u \in U\}$ (there are one-element abstraction classes). In this situation, the abstraction class $X \in \tilde{a}$ ($X = U$) has a non-empty intersection with each of the abstraction classes $Y \in \tilde{b}$. Moreover $card\{X \setminus Y\} = card\{U \setminus \{u\}\} = card\{U\} - 1$ for each $Y \in \tilde{b}$ and

$$deg(a \longrightarrow b) = \frac{(card\{U\} - 1) \cdot card\{U\}}{card\{U\} \cdot (card\{U\} - 1)} = 1.$$

Proposition 2. *The dependency degree is not commutative.*

For example, if $U = \{u_1, u_2\}$, $I = \{a, b\}$, $\tilde{a} = \{\{u_1\}, \{u_2\}\}$ and $\tilde{b} = \{\{u_1, u_2\}\}$. Then, $deg(a \longrightarrow b) = 0$ and $deg(b \longrightarrow a) = 1$. Thus $deg(a \longrightarrow b) \neq deg(b \longrightarrow a)$.

In general, when the attributes a and b are dependent ($a \longrightarrow b$; $\tilde{a} \subseteq \tilde{b}$), the $deg(a \longrightarrow b)$ is equal to 0, while with increasing diversity of abstraction classes for both attributes, the value of dependency degree increases.

Example 1. Let us consider the Middle East conflict situation [6] described by the information system $IS = (U, I)$. The agents are the countries in the region i.e. $U = \{u_1, u_2, u_3, u_4, u_5, u_6\}$ where u_1: Israel, u_2: Egypt, u_3: Palestinians, u_4: Jordan, u_5: Syria, u_6: Saudi Arabia. The attributes are the disputed issues i.e. $I = \{a_1, a_2, a_3, a_4, a_5\}$ where
a_1: Israeli settlements in the West Bank,
a_2: Israeli military outpost along the Jordan River,
a_3: Israeli retains East Jerusalem,
a_4: Israeli military outposts on the Golan Heights,
a_5: Arab countries grant citizenship to Palestinians who choose to remain within their borders.

Table 1. Information system for the Middle East conflict.

	a_1	a_2	a_3	a_4	a_5
u_1	−1	+1	+1	+1	+1
u_2	+1	0	−1	−1	−1
u_3	+1	−1	−1	−1	0
u_4	0	−1	−1	0	−1
u_5	+1	−1	−1	−1	−1
u_6	0	+1	−1	0	+1

Based on the above, we can designate the abstraction classes of indiscernibility relations for individual attributes:

$$\tilde{a}_1 = \{\{u_1\}, \{u_2, u_3, u_5\}, \{u_4, u_6\}\}$$
$$\tilde{a}_2 = \{\{u_1, u_6\}, \{u_2\}, \{u_3, u_4, u_5\}\}$$
$$\tilde{a}_3 = \{\{u_1\}, \{u_2, u_3, u_4, u_5, u_6\}\}$$
$$\tilde{a}_4 = \{\{u_1\}, \{u_2, u_3, u_5\}, \{u_4, u_6\}\}$$
$$\tilde{a}_5 = \{\{u_1, u_6\}, \{u_2, u_4, u_5\}, \{u_3\}\}$$

It can easily be seen that attributes a_1 and a_4 are equivalent, as they generate the same indiscernibility classes, while attribute a_3 is dependent on both attribute a_1 and a_4, as $\tilde{a}_1 \subseteq \tilde{a}_3$ and $\tilde{a}_4 \subseteq \tilde{a}_3$. The dependency graph is presented on Fig. 1. The most significant attributes in this case are attributes a_1 and a_4 because consistency on these attributes determines consistency on the largest number of other attributes.

To determine the dependency degree $(deg(a_1 \longrightarrow a_2))$ of the attribute a_2 on the attribute a_1 we calculate

$$deg(a_1 \longrightarrow a_2) = \frac{*}{6 \cdot 5} = \frac{5}{30}$$

where $* = card\{\{u_1\} \backslash \{u_1, u_6\}\} + card\{\{u_2, u_3, u_5\} \backslash \{u_2\}\} + + card\{\{u_2, u_3, u_5\} \backslash \{u_3, u_4, u_5\}\} + card\{\{u_4, u_6\} \backslash \{u_1, u_6\}\} + + card\{\{u_4, u_6\} \backslash \{u_3, u_4, u_5\}\}$

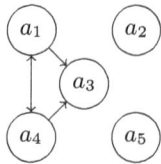

Fig. 1. Dependency graph for the Middle East conflict

Table 2. Values of dependency degree for the Middle East conflict.

	a_1	a_2	a_3	a_4	a_5
a_1	0	$\frac{5}{30}$	0	0	$\frac{5}{30}$
a_2	$\frac{5}{30}$	0	$\frac{2}{30}$	$\frac{5}{30}$	$\frac{3}{30}$
a_3	$\frac{5}{30}$	$\frac{11}{30}$	0	$\frac{5}{30}$	$\frac{10}{30}$
a_4	0	$\frac{6}{30}$	0	0	$\frac{5}{30}$
a_5	$\frac{5}{30}$	$\frac{3}{30}$	$\frac{2}{30}$	$\frac{5}{30}$	0

In Table 2, the values of dependency degree are given for each pair of attributes. The value $deg(a_i \longrightarrow a_j)$ is given in the a_i row and a_j column.

In the next step, if we consider that the values of dependency degree below $\frac{5}{30}$ are interesting for us the dependency graph is presented on Fig. 2. Based on the figure it can be concluded that the attributes a_1, a_4, a_2 and a_5 are equally important, since compliance on each of these attributes individually implies compliance on two other attributes. Of course, the values of dependency degree for the attributes $deg(a_2 \longrightarrow a_3)$, $deg(a_5 \longrightarrow a_3)$, $deg(a_3 \longrightarrow a_5)$ and $deg(a_5 \longrightarrow a_3)$ are greater than zero so forming coalitions equal to the abstraction classes on these attributes requires some negotiation and concessions. The abstraction classes of these attributes indicate the direction for negotiation – which agents should make concessions. The abstraction classes generated by the attributes a_3 and a_5 indicate that the key agents on which the formation of coalitions will depend are u_2 and u_3.

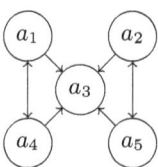

Fig. 2. Dependency graph based on dependency degree for the Middle East conflict

The section below will propose a method for investigating attribute dependencies in a conflict situation using a statistical approach.

3 Causal Discovery and Bayesian Networks

The dependencies between the issues in the conflict can be statistically verified using chi-square independence test. Though, the direction of this dependencies is unknown. To uncover causal relationships we use Bayesian Network (BN) [7,8]. A BN is represented as a directed acyclic graph (DAG) with variables as nodes of the graph and edges indicating conditional dependencies. A direction of an edge implies a causal relationship. If node A has an edge pointing to node B, then A influences B causally. Each node in BN has assigned Conditional Probability Tables (CPTs) that specifies the conditional probabilities given its parents. Based on evidence (observed variables), we can compute probabilities for other variables. Learning the BN structure from data involves identifying causal relationships. The most common techniques include constraint-based, score-based, and hybrid methods [3,18].

The constraint-based approach first identifies the independencies in the data and then searches for a graph that satisfies these independencies. The statistical tests help to determine whether two variables are independent given the presence of other variables. The algorithm seeks to construct the best DAG that captures causal relationships from observational data. The score-based approach first defines the measure of how well the BN fits the data, then searches over the space of possible DAGs for a graph with maximal score. The most common measures are Bayesian Information Criterion (BIC) and Akaike Information Criterion (AIC) for scoring the graph structure [19,20]. The number of possible graphs to be scored grows exponentially with the number of variables. Therefore the exhaustive search algorithm can be apply only for a small networks. In case of more nodes the heuristic search strategies can yield good results. The hybrid approach combines the constraint-based and score-based methods. The constraint-based method is used to identify the independencies and the score-based method is used to search for the best graph that satisfies these independencies.

Table 3. Independence test result for Middle East conflict situation.

source	target	p-value	chi-square	dof
a5	a3	0.003	12	2
a1	a3	3.1e−07	30	2
a1	a4	2.9e−12	60	4
a4	a3	3.1e−07	30	2
a2	a3	0.003	12	2
a2	a5	1.02e−06	33.33	4

As example the BN structure is inferred for Middle East conflict situation (described in Example 1). We apply constraint-based algorithm for inferring

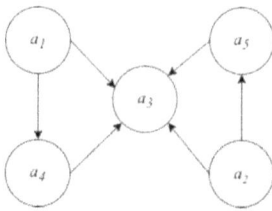

Fig. 3. Bayesian Network for Middle East conflict situation.

Bayesian Network. The independencies between the issues are verified using chi-square test. Since the chi-square test is sensitive to sample size we repeated the number of rows (number of agents) four times. This way we preserve the conflict structure while reaching the suitable number of values in each category; here -1, 0, +1. The strength of the edges computed with independence tests are presented in Table 3, the inferred BN in Picture 3 followed by the CPTs for each of the node in Tables 4, 5, 6, 7 and 8. The *dof* column contains the values of degrees of freedom in chi-square test. In CPTs we present the conditional probabilities between the input variables (parent nodes) and outcomes (child nodes).

Table 4. CPT of a_1.

$a_1\,(-1)$	0.33
$a_1\,(0)$	0.33
$a_1\,(+1)$	0.34

Table 5. CPT of a_2.

$a_2(-1)$	0.34
$a_2(0)$	0.33
$a_2(+1)$	0.33

Table 6. CPT of a_4.

a_1	$a_1(-1)$	$a_1(0)$	$a_1(1)$
$a_4(-1)$	0.33	0.32	0.36
$a_4(0)$	0.33	0.35	0.32
$a_4(+1)$	0.34	0.32	0.32

Table 7. CPT of a_5.

a_2	$a_2(-1)$	$a_2(0)$	$a_2(1)$
$a_5(-1)$	0.34	0.34	0.32
$a_5(0)$	0.33	0.33	0.32
$a_5(+1)$	0.32	0.33	0.35

It can be seen from Fig. 3 that the BN graph is similar to the graph created based on the dependency degree. The relationship between the attributes is straightforward in this example therefore using rough sets and statistical approach gives the same result. The BN is directed acyclic graph by the definition therefore only one direction between the attributes a_1, a_4 and a_2, a_5 is infer despite the equivalence visible for both pairs of attributes.

Table 8. CPT of a_3 (fragment).

a_1	$a_1(-1)$	$a_1(-1)$	$a_1(-1)$	...	$a_1(+1)$	$a_1(+1)$	$a_1(+1)$	$a_1(+1)$
a_2	$a_2(-1)$	$a_2(-1)$	$a_2(-1)$	...	$a_2(+1)$	$a_2(+1)$	$a_2(+1)$	$a_2(+1)$
a_4	$a_4(-1)$	$a_4(-1)$	$a_4(-1)$	...	$a_4(0)$	$a_4(+1)$	$a_4(+1)$	$a_4(+1)$
a_5	$a_5(-1)$	$a_5(0)$	$a_5(+1)$	...	$a_5(+1)$	$a_5(-1)$	$a_5(0)$	$a_5(+1)$
$a_3(-1)$	0.5	0.5	0.5	...	0.5	0.5	0.5	0.5
$a_3(+1)$	0.5	0.5	0.5	...	0.5	0.5	0.5	0.5

4 Parliamentary Elections in Poland in 2023

Conflicts and decision-making are ubiquitous in nearly every aspect of our lives. This paper explores a conflict scenario within the realm of politics, specifically drawing on a real-life case from the 2023 parliamentary elections in Poland. Consider the following example, named the "parliamentary elections". This example is derived from the Voting Lighthouse application, a product of the Center for Civic Education developed under Project No. POWR.03.01.00-00-T065/18, titled "Social and Civic Activation of Young People in the Development of Key Competencies"[17]. In this example, we have seven agents represented as $U = \{u_1, \ldots, u_7\}$ – these agents are the parties in parliamentary elections (ordered alphabetically): u_1 – Bezpartyjni Samorządowcy; u_2 – Koalicja Obywatelska; u_3 – Konfederacja; u_4 – Nowa Lewica; u_5 – Prawo i Sprawiedliwość; u_6 – Polska jest jedna; u_7 – Trzecia Droga; and twenty issues $A = \{a_1, \ldots, a_{20}\}$: a_1 – To decrease the influence of the European Union on Polish domestic policy; a_2 – To raise social transfers as a means of limiting the effects of inflation on citizens; a_3 – To finance private visits to specialists if the waiting time at a public facility exceeds three months; a_4 – To provide a free nursery place for every child; a_5 – To grant schools more freedom in choosing the content covered in the curriculum; a_6 – To build low-rent housing for rent by the government; a_7 – To increase taxes for top earners; a_8 – To introduce early retirement for those who have worked a certain number of years, regardless of their age; a_9 – To liberalize abortion laws; a_{10} – To make all entrepreneurs pay the same amount of health premium regardless of income; a_{11} – To make the result of any nationwide referendum binding regardless of the frequency of attendance; a_{12} – To adopt migrant relocation solutions endorsed by the European Union; a_{13} – To strengthen the independence of the legislature from parliament and government; a_{14} – To further increase the share of defense spending in Poland's GDP; a_{15} – To increase the powers of local governments at the expense of the central government; a_{16} – To quit coal mining in Poland no later than 2040; a_{17} – To block grain imports from Ukraine for Poland; a_{18} – To make Christian values the basis of state social policy; a_{19} – To reduce public media funding from the national budget; a_{20} – To limit the powers of secret services to track citizens' online activities.

The views of each agent to a specific issue is presented in Table 9, where, according to the Pawlak model of conflict analysis, +1 means agree, 0 have no opinion, −1 disagree.

Table 9. Information system for the 2023 parliamentary elections in Poland.

	a_1	a_2	a_3	a_4	a_5	a_6	a_7	a_8	a_9	a_{10}	a_{11}	a_{12}	a_{13}	a_{14}	a_{15}	a_{16}	a_{17}	a_{18}	a_{19}	a_{20}
u_1	+1	-1	+1	1	+1	1	-1	+1	1	-1	+1	-1	+1	1	+1	-1	-1	-1	+1	-1
u_2	-1	+1	1	+1	1	+1	-1	+1	1	+1	-1	-1	+1	1	+1	1	+1	-1	+1	+1
u_3	+1	-1	-1	-1	+1	-1	-1	-1	-1	+1	-1	-1	+1	1	+1	-1	+1	1	+1	+1
u_4	-1	+1	1	+1	1	+1	1	+1	1	-1	-1	+1	1	-1	+1	1	+1	-1	+1	+1
u_5	+1	1	0	+1	-1	+1	0	+1	-1	-1	0	-1	-1	+1	-1	-1	+1	1	-1	0
u_6	+1	-1	+1	1	-1	+1	-1	+1	-1	-1	+1	-1	+1	1	+1	-1	+1	1	+1	+1
u_7	+1	-1	+1	1	+1	1	-1	0	+1	-1	-1	-1	+1	1	+1	1	+1	1	+1	+1

Below, we will analyze the above example using the two previously proposed approaches. We start with using the concept of dependency degree.

Based on Table 9, we can designate the abstraction classes of indiscernibility relations for individual attributes:

$\widetilde{a_1} = \{\{u_1, u_3, u_5, u_6, u_7\}, \{u_2, u_4\}\}, \ \widetilde{a_2} = \{\{u_1, u_3, u_6, u_7\}, \{u_2, u_4, u_5\}\}$

$\widetilde{a_3} = \{\{u_1, u_2, u_4, u_6, u_7\}, \{u_3\}, \{u_5\}\}, \ \widetilde{a_4} = \{\{u_1, u_2, u_4, u_5, u_6, u_7\}, \{u_3\}\}$

$\widetilde{a_5} = \{\{u_1, u_2, u_3, u_4, u_7\}, \{u_5, u_6\}\}, \ \widetilde{a_6} = \{\{u_1, u_2, u_4, u_5, u_6, u_7\}, \{u_3\}\}$

$\widetilde{a_7} = \{\{u_1, u_2, u_3, u_6, u_7\}, \{u_4\}, \{u_5\}\}, \ \widetilde{a_8} = \{\{u_1, u_2, u_4, u_5, u_6\}, \{u_3\}, \{u_7\}\}$

$\widetilde{a_9} = \{\{u_1, u_2, u_4, u_7\}, \{u_3, u_5, u_6\}\}, \ \widetilde{a_{10}} = \{\{u_1, u_4, u_5, u_6, u_7\}, \{u_2, u_3\}\}$

$\widetilde{a_{11}} = \{\{u_1, u_6\}, \{u_2, u_3, u_4, u_7\}, \{u_5\}\}, \ \widetilde{a_{12}} = \{\{u_1, u_2, u_3, u_5, u_6, u_7\}, \{u_4\}\}$

$\widetilde{a_{13}} = \{\{u_1, u_2, u_3, u_4, u_6, u_7\}, \{u_5\}\}, \ \widetilde{a_{14}} = \{\{u_1, u_2, u_3, u_5, u_6, u_7\}, \{u_4\}\}$

$\widetilde{a_{15}} = \{\{u_1, u_2, u_3, u_4, u_6, u_7\}, \{u_5\}\}, \ \widetilde{a_{16}} = \{\{u_1, u_3, u_5, u_6\}, \{u_2, u_4, u_7\}\}$

$\widetilde{a_{17}} = \{\{u_1\}, \{u_2, u_3, u_4, u_5, u_6, u_7\}\}, \ \widetilde{a_{18}} = \{\{u_1, u_2, u_4\}, \{u_3, u_5, u_6, u_7\}\}$

$\widetilde{a_{19}} = \{\{u_1, u_2, u_3, u_4, u_6, u_7\}, \{u_5\}\}, \ \widetilde{a_{20}} = \{\{u_1\}, \{u_2, u_3, u_4, u_6, u_7\}, \{u_5\}\}$

We can see that attributes a_4 and a_6 are equivalent also a_{13}, a_{15} and a_{19} are equivalent and a_{12}, a_{14} are equivalent. There are also other attributes dependency, which are presented on Fig. 4. Only the attributes that are in some dependency relation are shown on the graph. It can be concluded that the most significant attributes in this case are attributes a_3 and a_7 because consistency on these attributes determines consistency on the largest number of other attributes.

In Table 10, the values of dependency degree are given for each pair of attributes. If we consider the values of dependency degree below 0.07 (values marked in blue in Table 10) the dependency graph is presented on Fig. 5. This analysis leads to the conclusion that attribute a_{11} is now as important as attributes a_3 and a_7. Convincing agents to change their views in the negotiation process can lead to the formation of strong coalitions equal to the abstraction classes with respect to the indiscernibility relation for this attribute.

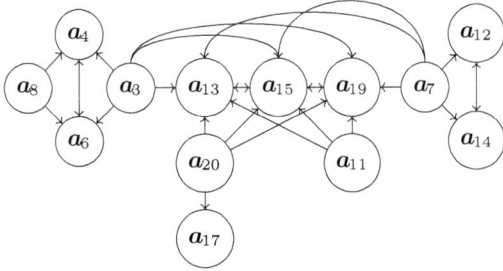

Fig. 4. Dependency graph for the Parliamentary elections in Poland in 2023

Table 10. Values of dependency degree for the Parliamentary elections in Poland in 2023.

	a_1	a_2	a_3	a_4	a_5	a_6	a_7	a_8	a_9	a_{10}	a_{11}	a_{12}	a_{13}	a_{14}	a_{15}	a_{16}	a_{17}	a_{18}	a_{19}	a_{20}
a_1	0	0.12	0.24	0.12	0.12	0.12	0.17	0.24	0.12	0.17	0.24	0.05	0.12	0.05	0.12	0.12	0.12	0.12	0.12	0.24
a_2	0.07	0	0.17	0.10	0.17	0.10	0.14	0.19	0.17	0.17	0.17	0.07	0.07	0.07	0.07	0.17	0.10	0.17	0.12	0.17
a_3	0.12	0.12	0	0	0.12	0	0.12	0.12	0.12	0.12	0.12	0.12	0	0.12	0	0.12	0.12	0.12	0	0.12
a_4	0.14	0.14	0.14	0	0.14	0	0.33	0.14	0.14	0.14	0.29	0.14	0.14	0.14	0.14	0.14	0.14	0.14	0.14	0.29
a_5	0.12	0.17	0.17	0.12	0	0.12	0.17	0.24	0.12	0.12	0.17	0.12	0.05	0.12	0.05	0.12	0.12	0.12	0.05	0.17
a_6	0.14	0.14	0.14	0	0.14	0	0.33	0.14	0.14	0.14	0.29	0.14	0.14	0.14	0.14	0.14	0.14	0.14	0.14	0.29
a_7	0.12	0.12	0.12	0.12	0.12	0.12	0	0.24	0.12	0.12	0.12	0	0	0	0	0.12	0.12	0.12	0	0.12
a_8	0.12	0.12	0.12	0	0.12	0	0.24	0	0.12	0.12	0.21	0.12	0.12	0.12	0.12	0.12	0.12	0.12	0.12	0.24
a_9	0.10	0.17	0.14	0.07	0.07	0.07	0.17	0.17	0	0.17	0.29	0.10	0.07	0.10	0.07	0.10	0.10	0.10	0.07	0.17
a_{10}	0.17	0.17	0.17	0.05	0.12	0.05	0.24	0.17	0.17	0	0.24	0.12	0.12	0.12	0.12	0.17	0.12	0.17	0.12	0.24
a_{11}	0.10	0.10	0.10	0.10	0.05	0.10	0.10	0.19	0.14	0.10	0	0.10	0	0.10	0	0.10	0.05	0.14	0	0.05
a_{12}	0.14	0.14	0.29	0.14	0.14	0.14	0.14	0.29	0.14	0.14	0.29	0	0.14	0	0.14	0.14	0.14	0.14	0.14	0.29
a_{13}	0.14	0.14	0.14	0.14	0.14	0.14	0.14	0.29	0.14	0.14	0.14	0.14	0	0.14	0	0.14	0.14	0.14	0	0.14
a_{14}	0.14	0.14	0.29	0.14	0.14	0.14	0.14	0.29	0.14	0.14	0.29	0	0.14	0	0.14	0.14	0.14	0.14	0.14	0.29
a_{15}	0.14	0.14	0.14	0.14	0.14	0.14	0.14	0.29	0.14	0.14	0.14	0.14	0	0.14	0	0.14	0.14	0.14	0	0.14
a_{16}	0.17	0.17	0.19	0.10	0.10	0.10	0.17	0.17	0.10	0.17	0.17	0.07	0.10	0.07	0.10	0	0.10	0.17	0.10	0.19
a_{17}	0.14	0.14	0.29	0.14	0.14	0.14	0.29	0.29	0.14	0.14	0.29	0.14	0.14	0.14	0.14	0.14	0	0.14	0.14	0.14
a_{18}	0.17	0.17	0.19	0.10	0.17	0.10	0.17	0.19	0.10	0.17	0.26	0.07	0.10	0.07	0.10	0.17	0.07	0	0.10	0.17
a_{19}	0.14	0.14	0.14	0.14	0.14	0.14	0.14	0.29	0.14	0.14	0.14	0.14	0	0.14	0	0.14	0.14	0.14	0	0.14
a_{20}	0.12	0.12	0.12	0.12	0.12	0.12	0.12	0.24	0.12	0.12	0.12	0.12	0	0.12	0	0.12	0	0.12	0	0

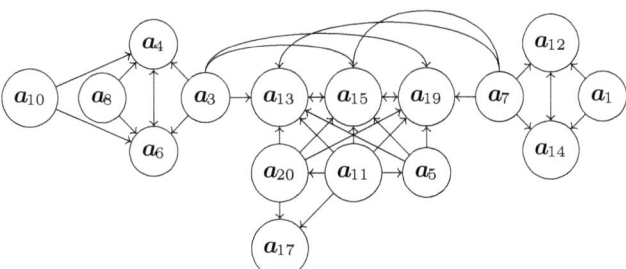

Fig. 5. Dependency graph based on dependency degree for the Parliamentary elections in Poland in 2023

Let us consider what exactly coalitions are defined by this attribute a_{11}. The abstraction classes of indiscernibility relations are equal to $\widetilde{a_{11}} = \{\{u_1, u_6\}, \{u_2, u_3, u_4, u_7\}, \{u_5\}\}$. Thus, we can see that political party u_5 – Prawo i Sprawiedliwość has views that are rather incompatible with the other parties. In addition, the parties u_1 – Bezpartyjni Samorządowcy and u_6 – Polska jest jedna are in one coalition, which is also rather in line with the overall assessment of the political scene in Poland. On the other hand, the occurrence of the coalition $\{u_2, u_3, u_4, u_7\}$ of parties Koalicja Obywatelska, Konfederacja, Nowa Lewica and Trzecia Droga is interesting. Currently in the Polish parliament, the coalition of Koalicja Obywatelska, Nowa Lewica and Trzecia Droga is in power, however, Konfederacja is not in this coalitions. However, the analysis shows that there are some attributes ($a_5, a_{11}, a_{13}, a_{15}, a_{17}, a_{19}, a_{20}$) that give grounds for joining Konfederacja to the coalition in power now. Of course, this would require concessions and negotiations with Konfederacja.

Table 11. Independence test result for the Parliamentary elections in Poland in 2023.

source	target	p-value	chi-square	dof
a_1	a_2	0.0006	12	2
a_8	a_6	8.3e−07	28	2
a_{15}	a_{20}	8.3e−07	28	2
a_{14}	a_7	8.3e−07	28	2
a_{14}	a_{12}	6.2e−06	20	1
a_4	a_3	8.3e−07	28	2
a_4	a_6	6.2e−06	20	1
a_4	a_8	8.3e−07	28	2
a_{17}	a_{20}	8.3e−07	28	2
a_{12}	a_7	8.3e−07	28	2
a_9	a_{18}	0.0003	12	1
a_{13}	a_{15}	6.2e−06	20	1
a_{13}	a_{19}	6.2e−06	20	1
a_{19}	a_{15}	6.2e−06	20	1

Now we analyze the example using statistical tools and Bayesian Networks. The BN presented in Fig. 6 is based on the independence test results presented in Table 11 and inferred using constraint-based algorithm. The CPTs for the BN are presented in Tables 12, 13, 14, 15, 16, 17, 18, 19, 20, 21, 22, 23, 24, 25, 26 and 27. The independence test was not statistically significant for any pair containing the attributes $a_5, a_{10}, a_{11}, a_{16}$, therefore they were not included in the BN. Also, due to lack of dependencies between some of the attributes the BN consists of several small graphs. Despite this we find the similarity between the BN and the dependency graph presented in Fig. 5. In fact, all BN graphs are the subgraphs of the dependency graph. Just the relations between the attributes a_2, a_1, and a_{18}, a_9 are missing in the dependency graphs. It can be said that the dependency degree is more sensitive measure than independence test – there are more dependencies revealed in the model.

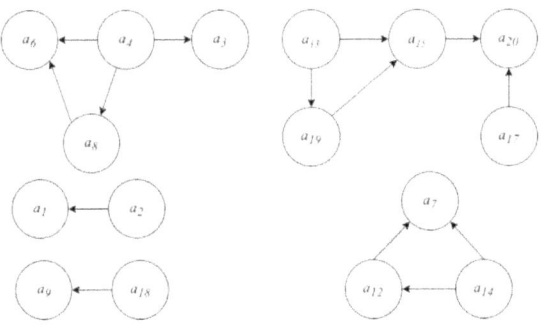

Fig. 6. Bayesian Network for for the Parliamentary elections in Poland in 2023.

Table 12. CPT of a_1.

a_1 (-1)	0.49
a_1 (+1)	0.51

Table 13. CPT of a_{14}.

$a_{14}(-1)$	0.49
$a_{14}(+1)$	0.51

Table 14. CPT of a_4.

a_4 (−1)	0.49
a_4 (+1)	0.51

Table 15. CPT of a_9.

$a_9(-1)$	0.49
$a_9(+1)$	0.51

Table 16. CPT of a_{13}.

a_{13} (−1)	0.49
a_{13} (+1)	0.51

Table 17. CPT of a_{17}.

$a_{17}(-1)$	0.49
$a_{17}(+1)$	0.51

Table 18. CPT of a_2.

a1	a1(−1)	a1(+1)
a2(−1)	0.5	0.5
a2(+1)	0.5	0.5

Table 19. CPT of a_{12}.

a14	$a_{14}(-1)$	$a_{14}(+1)$
$a_{12}(-1)$	0.5	0.5
$a_{12}(+1)$	0.5	0.5

Table 20. CPT of a_{19}.

a13	$a_{13}(-1)$	$a_{13}(+1)$
$a_{19}(-1)$	0.5	0.5
$a_{19}(+1)$	0.5	0.5

Comparing the two approaches proposed in this paper – Dependency Degree and Bayesian Networks – it can be concluded that the determination of attribute dependencies, in both cases, is based on the value of individual attributes and their correspondences. Thus, in the end, similar dependencies are discovered. However, it should be noted that theoretical background and graphical presentation differ for both approaches. However, the χ^2 test requires more data than when we use abstraction classes; the test statistic approximates a chi-squared distribution more closely as sample sizes increase.

Table 21. CPT of a_{18}.

a_9	$a_9(-1)$	$a_9(+1)$
$a_{18}(-1)$	0.5	0.5
$a_{18}(+1)$	0.5	0.5

Table 22. CPT of a_8.

a_4	$a_4(-1)$	$a_4(+1)$
$a_8(-1)$	0.34	0.32
$a_8(0)$	0.33	0.33
$a_8(+1)$	0.33	0.36

Table 23. CPT of a_3.

a_4	$a_4(-1)$	$a_4(+1)$
$a_3(-1)$	0.34	0.32
$a_3(0)$	0.33	0.33
$a_3(+1)$	0.33	0.36

Table 24. CPT of a_{20}.

a_{15}	$a_{15}(-1)$	...	$a_{15}(+1)$
a_{17}	$a_{17}(-1)$	...	$a_{17}(+1)$
$a_{20}(-1)$	0.33	...	0.31
$a_{20}(0)$	0.33	...	0.31
$a_{20}(+1)$	0.33	...	0.31

Table 25. CPT of a_7.

a_{12}	$a_{12}(-1)$	...	$a_{12}(+1)$
a_{14}	$a_{14}(-1)$	...	$a_{14}(+1)$
$a_7(-1)$	0.33	...	0.31
$a_7(0)$	0.33	...	0.31
$a_7(+1)$	0.33	...	0.31

Table 26. CPT of a_6.

a_4	$a_4(-1)$	$a_4(0)$	...	$a_4(+1)$	$a_4(+1)$
a_8	$a_8(-1)$	$a_8(0)$	...	$a_8(0)$	$a_8(+1)$
$a_6(-1)$	0.5	0.5	...	0.5	0.45
$a_6(+1)$	0.5	0.5	...	0.5	0.55

Table 27. CPT of a_{15}.

a_{13}	$a_{13}(-1)$	$a_{13}(-1)$	$a_{13}(+1)$	$a_{13}(+1)$
a_{19}	$a_{19}(-1)$	$a_{19}(+1)$	$a_{19}(-1)$	$a_{19}(+1)$
$a_{15}(-1)$	0.51	0.5	0.5	0.46
$a_{15}(+1)$	0.49	0.5	0.5	0.54

5 Conclusion

The paper analyzes a conflict situation stored in the form of an information system containing agents' opinions on given conflict issues. The main aspect considered in the paper is the relations between individual issues/attributes describing the conflict situation. Two methods for verifying such a relations have been proposed. The first one use the dependency degree, the second use the statistical inference and Bayesian networks. In both cases, attribute connections can be presented in the form of legible graphs and indicate the attributes that should be paid attention to during negotiations. The paper presents two real cases and analyzes using both approaches. It was noticed that certain attributes are particularly important (their values provide the important knowledge about the compatibility or distribution of values of large number of other attributes) and constitute important guidelines for creating coalitions in a conflict situation.

References

1. Deja, R.: Conflict analysis. Int. J. Intell. Syst. **17**(2), 235–253 (2002)
2. Fu, C., Sayed, T.: Bayesian dynamic extreme value modeling for conflict-based real-time safety analysis. Anal. Methods Accident Res. **34**, 100204 (2022)
3. Kitson, N.K., Constantinou, A.C., Guo, Z., Liu, Y., Chobtham, K.: A survey of Bayesian Network structure learning. Artif. Intell. Rev. **56**(8), 8721–8814 (2023)

4. Lang, G., Miao, D., Cai, M.: Three-way decision approaches to conflict analysis using decision-theoretic rough set theory. Inf. Sci. **406**, 185–207 (2017)
5. Lang, G., Miao, D., Fujita, H.: Three-way group conflict analysis based on Pythagorean fuzzy set theory. IEEE Trans. Fuzzy Syst. **28**(3), 447–461 (2019)
6. Pawlak, Z.: An inquiry into anatomy of conflicts. Inf. Sci. **109**(1–4), 65–78 (1998)
7. Pearl, J.: Probabilistic reasoning in intelligent systems: networks of plausible inference. Morgan Kaufmann (1988)
8. Pearl, J.: Causality. Cambridge University Press (2009)
9. Przybyła-Kasperek, M., Deja, R., Wakulicz-Deja, A.: Selected Approaches to Conflict Analysis Inspired by the Pawlak Model-Case Study. In: Campagner, A., et al. (eds.) International Joint Conference on Rough Sets, pp. 3–17. Springer, Cham (2023)
10. Przybyła-Kasperek, M.: Study of selected methods for balancing independent data sets in k-nearest neighbors classifiers with Pawlak conflict analysis. Appl. Soft Comput. **129**, 109612 (2022)
11. Skowron, A., Deja, R.: On some conflict models and conflict resolutions. Rom. J. Inform. Sci. Technol. **3**(1–2), 69–82 (2002)
12. Skowron, A., Rauszer, C.: The discernibility matrices and functions in information systems. In: Słowiński, R. (eds.) Intelligent Decision Support: Handbook of Applications and Advances of the Rough Sets Theory, pp. 331–362. Springer, Dordrecht (1992)
13. Sun, B., Chen, X., Zhang, L., Ma, W.: Three-way decision making approach to conflict analysis and resolution using probabilistic rough set over two universes. Inf. Sci. **507**, 809–822 (2020)
14. Tang, X., Zeng, T., Tan, Y., Ding, B.: Conflict analysis based on three-way decision theoretic fuzzy rough set over two universes. Ingenierie des Systemes d'Information **25**(1), 75 (2020)
15. Tong, S., Sun, B., Chu, X., Zhang, X., Wang, T., Jiang, C.: Trust recommendation mechanism-based consensus model for Pawlak conflict analysis decision making. Int. J. Approximate Reasoning **135**, 91–109 (2021)
16. Yao, Y.: Three-way conflict analysis: reformulations and extensions of the Pawlak model. Knowl. Based Syst. **180**, 26–37 (2019)
17. (CCE) The Center for Citizenship Education, Voting Lighthouse application. https://latarnikwyborczy.pl/. Accessed 15 Mar 2024
18. Waldmann, M.R., Martignon, L.: A Bayesian network model of causal learning. In Proceedings of the Twentieth Annual Conference of the Cognitive Science Society, pp. 1102–1107. Routledge (2022)
19. Watanabe, S.: A widely applicable Bayesian information criterion. J. Mach. Learn. Res. **14**(1), 867–897 (2013)
20. Vrieze, S.I.: Model selection and psychological theory: a discussion of the differences between the Akaike information criterion (AIC) and the Bayesian information criterion (BIC). Psychol. Methods **17**(2), 228 (2012)

Consideration of Detecting Data and Functional Dependency in Tabular Data with Missing Values by the Obtained Rules

Hiroshi Sakai[1,2(✉)], Michinori Nakata[3], Dominik Ślęzak[4], and Junzo Watada[1,5]

[1] Department of Data Science, Shimonoseki City University, 2-1-1 Daigakucho, Shimonoseki 751-8510, Japan
`sakai.scukit@gmail.com, sakai-hi@shimonoseki-cu.ac.jp`
[2] Kyushu Institute of Technology, Kitakyushu, Japan
[3] Tokyo University of Information Sciences, 4-1 Onaridai, Wakaba-ku, Chiba 265-8501, Japan
`nakatam@ieee.org`
[4] Institute of Informatics, University of Warsaw, Warsaw, Mazowieckie, Poland
`slezak@mimuw.edu.pl`
[5] Waseda University, Kitakyushu, Japan
`watada@waseda.jp`

Abstract. Functional dependency is an essential concept between attributes in relational algebra and tabular data analysis, and it is applied to the decomposition of tabular data. On the other hand, data dependency is a concept between attribute values. Functional dependency is usually given, and data dependency is often recognized after data analysis. We may recognize the hidden functional dependency as a particular case of data dependency. In this paper, we apply the rules obtained by the NIS-Apriori-based rule generator, which was implemented to handle rules from tabular and tabular data with missing values. We detect some candidates CONs of condition attributes that affect the decision attribute Dec using the obtained rules. We then apply the same rule generator specifying the detected one CON to determine the actual degree of dependency. This step eliminates the need to enumerate all CONs to understand dependencies and can handle extended dependencies for DIS and NIS. A running example using the implemented tools is also provided.

Keywords: tabular data · rule generation · missing value · detection of data dependency · the NIS-Apriori algorithm

1 Introduction and Background

We generate rules using the NIS-Apriori-based rule generator, which can handle tabular data with missing values. Due to the obtained rules, we can detect some candidates of CONs of condition attributes that affect the decision attribute Dec. Then, we apply the same rule generator specifying the detected one CON

to determine the actual degree of dependency. This step eliminates the need to enumerate all CONs to understand dependencies and can also handle extended dependencies for DIS and NIS. We may have a hidden functional dependency as a particular case of data dependency.

Functional dependency [5] will also apply to missing value estimation. In [2], a method to detect functional dependencies from a table with missing values is proposed. In [3], a similar process is proposed and applied to estimating missing values in the Restaurant dataset. The TANE system loosened the constraint of functional dependency to approximate dependency [7].

Our framework uses the rules that were obtained, differentiating it from the above research. To detect the data dependency of a decision attribute Dec, we determine a set CON of condition attributes using the obtained rules and calculate the degree of its dependency. We will also consider data and functional dependency in tabular data with missing values. Based on this new dependency, we will recognize a dependency from $\{P, R\}$ to Dec in Table 2.

Now, we confirm the basic framework in rule generation. We use the concept of equivalence classes defined by rough set theory [9]. Due to [8–10, 14], we employ the term DIS (*Deterministic Information System*) for tabular data and the term NIS (*Non-deterministic Information System*) for tabular data with missing values.

Table 1. An exemplary DIS ψ.

(INS)tance	P	Q	R	Dec
x1	b	s	x	2
x2	a	s	x	2
x3	c	s	y	1
x4	b	t	y	1
x5	a	t	z	1
x6	a	t	z	1
x7	b	t	x	2
x8	a	t	z	1

Table 2. An exemplary NIS Φ.

INS	P	Q	R	Dec
x1	b	s	x	2
x2	a	?	x	2
x3	c	s	?	1
x4	b	t	y	1
x5	?	t	z	1
x6	a	t	z	1
x7	b	?	x	2
x8	a	t	z	1

Let us consider Table 1, where we usually consider a set $\{P, Q, R\}$ of condition attributes and Dec of a decision attribute.

1. We term a pair $[attribute, value]$ a *descriptor*. An *implication* is a formula $\wedge_i T_i \Rightarrow [Dec, value]$ for descriptors T_i, like $[R, x] \Rightarrow [Dec, 2]$ and $[P, c] \wedge [Q, s] \Rightarrow [Dec, 1]$ in Table 1.
2. If we identify a descriptor with an item, we can see Table 1 expresses eight transactions [1]. The first itemset is $\{[P, b], [Q, s], [R, x], [Dec, 2]\}$, and the eighth itemset is $\{[P, a], [Q, t], [R, z], [Dec, 1]\}$. Thus, we can connect tabular data to Apriori-based rule generation [1].

3. We assign the *support*(τ) (the occurrence ratio) and *accuracy*(τ) (the consistency ratio) values to each implication τ. Let $eq([A, val_A])$ be a set of instances satisfying the descriptor $[A, val_A]$ and $|SET|$ be the number of the element in SET. For $\tau : \wedge_i T_i \Rightarrow [Dec, val]$,
    (i) $support(\tau) = |\cap_i eq(T_i) \cap eq([Dec, val])| / |INS|$ (INS: Instances),
    (ii) $accuracy(\tau) = |\cap_i eq(T_i) \cap eq([Dec, val])| / |\cap_i eq(T_i)|$.
    For $\tau : [Q, t] \Rightarrow [Dec, 1]$ in Table 1,
    (i) $support(\tau) = 1/2$ ($= 4/8$, τ occurs four times for eight instances),
    (ii) $accuracy(\tau) = 4/5$ (τ occurs four times for five instances of $[Q, t]$).
4. A *rule* from DIS is an implication τ satisfying below
    $support(\tau) \geq \alpha$ and $accuracy(\tau) \geq \beta$ for given two thresholds α and β.
    We may add the constraint that the condition part of τ is *minimal* [6]. The minimality of the condition part of rules is used to reduce the number of rules. We will have all redundant implications as rules if we remove the minimality constraint.

Now, we move to NIS like Table 2, where four missing values '?' are artificially added to Table 1 for explaining the framework. NIS was proposed for handling information incompleteness in DIS [6,8].

1. If we replace each '?' with its possible value respectively, we have one DIS. Such a DIS is termed a *derived* DIS, and a set $DD(\Phi)$ of all derived DISs is defined [10,13]. There are 36 ($=2^2 \times 3^2$) derived DISs from Φ in Table 2.
2. A *certain rule* τ is an implication satisfying below
    $support(\tau) \geq \alpha$ and $accuracy(\tau) \geq \beta$ in each $\psi \in DD(\Phi)$.
    A *possible rule* τ is an implication satisfying below
    $support(\tau) \geq \alpha$ and $accuracy(\tau) \geq \beta$ in at least one $\psi \in DD(\Phi)$.
3. The above definitions, based on possible world semantics, seem natural. However, the number of $DD(\Phi)$ increases exponentially. In the Mammographic data set [4], the number of elements in $DD(\Phi_{Mammo})$ exceeds 10^{100}. It is huge. The complexity problem occurs for realization.

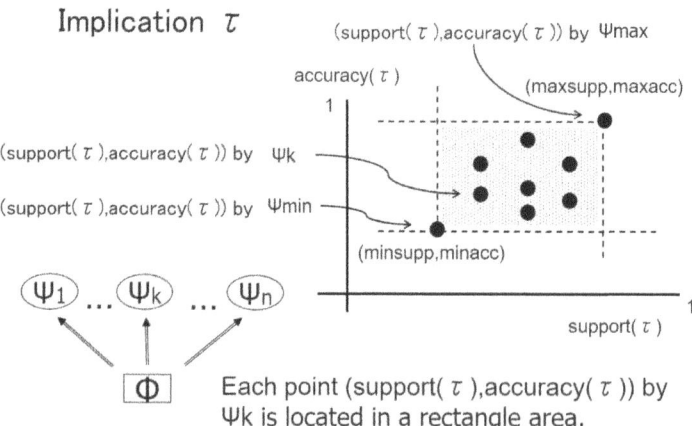

Fig. 1. The relation between ψ_{min}, ψ_{max}, and $\psi \in DD(\Phi)$ [10].

4. The complexity problem was solved due to the following results [10, 13].
   (Property 1) There exists the worst DIS $\psi_{min} \in DD(\Phi)$ for an implication τ, where both $support(\tau)$ and $accuracy(\tau)$ are the minimum. Similarly, there exists the best DIS $\psi_{max} \in DD(\Phi)$ for an implication τ, where both $support(\tau)$ and $accuracy(\tau)$ are the maximum. The image of this property is shown in Fig. 1.
   (Property 2) We can calculate two values $minsupp(\tau)$ ($=support(\tau)$ in ψ_{min}) and $minacc(\tau)$ ($=accuracy(\tau)$ in ψ_{min}). Here, $inf([A, val_A])$ is a set of instances with definite $[A, val_A]$, and $sup([A, val_A])$ is a set of instances with definite $[A, val_A]$ or $[A, ?]$.

$$minsupp(\tau) = \frac{|inf(\wedge_{A \in CON}[A, val_A]) \cap inf([Dec, val])|}{|INS|},$$

$$minacc(\tau) = \frac{|inf(\wedge_{A \in CON}[A, val_A]) \cap inf([Dec, val])|}{|inf(\wedge_{A \in CON}[A, val_A])| + |OUTACC|},$$

$$OUTACC = \{sup(\wedge_{A \in CON}[A, val_A])$$
$$\setminus inf(\wedge_{A \in CON}[A, val_A])\} \setminus inf([Dec, val]).$$

The above formulas do not depend on the number of $|DD(\Phi)|$. We can derive the similar formulas on $maxsupp(\tau)$ and $maxacc(\tau)$.
In NIS Φ, the following holds for each descriptor.
$inf([A, val]) \subset eq([A, val]) \subset sup([A, val])$
($eq([A, val])$ is an equivalence class in any $\psi \in DD(\Phi)$).
However, in DIS ψ,
$DD(\psi) = \{\psi\}$ and $inf([A, val]) = eq([A, val]) = sup([A, val])$.
Thus, $OUTACC = \emptyset$, $minsupp(\tau) = maxsupp(\tau)$ and $minacc(\tau) = maxacc(\tau)$ are derived.

Example 1. Let us consider Table 2 and an implication $\tau : [P, a] \wedge [Q, t] \Rightarrow [Dec, 1]$. We can know $minsupp(\tau)$ and $minacc(\tau)$ by checking 36 derived DISs sequentially. However, we can know them based on (Property 2).

1. $inf([P, a]) \cap inf([Q, t])$ (a set of definite instances)
   $=\{x2, x6, x8\} \cap \{x4, x5, x6, x8\} = \{x6, x8\}$,
2. $sup([P, a]) \cap sup([Q, t])$ (a set of possible instances)
   $=\{x2, x5, x6, x8\} \cap \{x2, x4, x5, x6, x7, x8\} = \{x2, x5, x6, x8\}$,
3. $inf([Dec, 1]) = sup([Dec, 1]) = \{x3, x4, x5, x6, x8\}$,
4. $OUTACC = \{x2, x5\} \setminus \{x3, x4, x5, x6, x8\} = \{x2\}$.
5. Due to the above equations,
   $minsupp(\tau) = |\{x6, x8\}|/8 = 2/8$,
   $minacc(\tau) = |\{x6, x8\}|/(|\{x6, x8\}| + |\{x2\}|) = 2/3$.
   To reduce the $accuracy(\tau)$, '?'=t is selected in $x2$ (to make the same condition and a different decision), and '?'= either b or c is selected in $x5$ (not to make the same condition and the same decision). One $\psi \in DD(\Phi)$ satisfying the above assignment is ψ_{min} for τ.

We added the above calculation to the Apriori algorithm and proposed the DIS-Apriori and the NIS-Apriori algorithms. We also implemented them on Windows PC (3.60GHz) in Python [11,12]. We made use of the following characteristics.

1. We identified a descriptor as an item and a tuple of an instance as an itemset. For example, the first tuple of Table 1 is identified as an itemset $\{[P,b],[Q,s],[R,x],[Dec,2]\}$.
2. Since the decision attribute Dec is usually decided, one $[Dec,\_]$ must be in every itemset. For example, an itemset $\{[P,b],[Q,s],[R,x]\}$ is meaningless because there is no $[Dec,\_]$.
3. Due to the condition of tabular data, at most, one descriptor for each attribute is in an itemset. For example, an itemset $\{[P,b],[Q,s],[Q,t],[Dec,2]\}$ is meaningless because both $[Q,s]$ and $[Q,t]$ exist.
4. In the Apriori algorithm, the set
   $C_{i+1}=\{CAN : itemset \mid |CAN|=i+1\}$
   of candidates is generated from
   $C_i=\{CAN : itemset \mid |CAN|=i,$ the constraint of $support$ is satisfied$\}$ [1].
   Since there is no concept of attributes, the set C_{i+1} of candidates is generated by combining all itemsets in C_i. However, we have attributes in tabular data and can reduce the number of itemsets based on the above characteristics.
5. As a result, one itemset corresponds to one implication. Thus, we can examine the constraint of $accuracy$ for each $CAN \in C_{i+1}$ after examining the constraint of $support$.

2 Detection of Data and Functional Dependency Using the Obtained Rules

This section applies the rules obtained for detecting data and functional dependency in DIS and NIS.

2.1 Strict Dependency (no Inconsistency) in DIS

Let's consider two implications $\tau_x : \wedge_i[A_i,val_i] \Rightarrow [Dec,val]$ from the instance x and $\tau_y : \wedge_i[A_i,val_i] \Rightarrow [Dec,val']$ from the instance y. If $val \neq val'$, we say x and y are $inconsistent$ for $CON=\cup_i\{A_i\}$ and Dec. If $val=val'$, we say x and y are $consistent$ for $CON=\cup_i\{A_i\}$ and Dec. In [9], the degree of dependency $\gamma(CON,Dec)$ from CON to Dec is defined by the ratio (the number of consistent instances)$/|INS|$.
We easily connect this degree with rules. Let $ATR(\tau_x)$ denote a set of condition attributes in τ_x, then
$\gamma(CON,Dec)=|\{x \in INS | ATR(\tau_x)=CON, accuracy(\tau_x)=1\}|/|INS|$.
If $\gamma(CON,Dec)=1$, we see there is a functional dependency from CON to Dec because each instance is consistent from CON to Dec.

Remark 1. We must find each CON for Dec with a high degree of $\gamma(CON, Dec)$ to detect data dependency. If we enumerate CONs for the set AT, we must consider $(2^{|AT|}-1)$ cases sequentially. However, we can apply the rules obtained for detecting degrees using Proposition 1. We can consider only condition attributes related to the obtained rules and avoid handling the $(2^{|AT|}-1)$ cases.

Proposition 1. *Let $RULE$ be the set of the obtained rules, whose constraints are support > 0, accuracy$=1$, and no minimality. Furthermore, let $RULE(CON)$ be $\{\wedge_{A \in CON}[A, \_] \Rightarrow [Dec, \_] \in RULE\}$. Then, $\gamma(CON, Dec) = \Sigma_{\tau \in RULE(CON)} support(\tau)$.*

Example 2. We have the following degrees using Table 1 and Fig. 2.

1. For $\gamma(\{P, Q\}, Dec)$, we summarize the *support* values of $[P, \_] \wedge [Q, \_] \Rightarrow [Dec\_]$ using Fig. 2 and have $\gamma(\{P, Q\}, Dec) = 3/4$ (lines 2, 4, 9, 12).
2. For $\gamma(\{P, R\}, Dec)$, we summarize the *support* values of $[P, \_] \wedge [R, \_] \Rightarrow [Dec\_]$ using Fig. 2 and have $\gamma(\{P, R\}, Dec) = 1$ (lines 1, 5, 6, 10, 13). We recognize a functional dependency from $\{P, R\}$ to Dec.

```
----------------------rule2---------------------
         condition           decision        sup      acc
 1:  ('P', 'b')('R', 'y')==>('Dec', '1')    0.125    1.0
 2:  ('Q', 's')('P', 'c')==>('Dec', '1')    0.125    1.0
 3:  ('Q', 's')('R', 'y')==>('Dec', '1')    0.125    1.0
 4:  ('P', 'a')('Q', 't')==>('Dec', '1')    0.375    1.0
 5:  ('P', 'a')('R', 'z')==>('Dec', '1')    0.375    1.0
 6:  ('P', 'c')('R', 'y')==>('Dec', '1')    0.125    1.0
 7:  ('R', 'y')('Q', 't')==>('Dec', '1')    0.125    1.0
 8:  ('Q', 't')('R', 'z')==>('Dec', '1')    0.375    1.0
 9:  ('P', 'b')('Q', 's')==>('Dec', '2')    0.125    1.0
10:  ('P', 'b')('R', 'x')==>('Dec', '2')    0.25     1.0
11:  ('Q', 's')('R', 'x')==>('Dec', '2')    0.25     1.0
12:  ('Q', 's')('P', 'a')==>('Dec', '2')    0.125    1.0
13:  ('R', 'x')('P', 'a')==>('Dec', '2')    0.125    1.0
14:  ('R', 'x')('Q', 't')==>('Dec', '2')    0.125    1.0
```

Fig. 2. The obtained all rules (with two conditions) from Table 1 under the constraints $support > 0$ and $accuracy = 1$. Here, the minimality constraint is not used to obtain all rules with two conditions. Since $accuracy(\tau)=1$ for $\tau : [P, c] \Rightarrow [Dec, 1]$, the 6th rule is redundant and usually is not generated.

2.2 Weakened Dependency (with Inconsistency) in DIS

Sometimes, the consistency condition is too strong, and we may not have any data dependency. We weaken the definition of consistency and define weakened dependency. In strict dependency, the consistency of an instance x is determined by $accuracy(\tau_x)=1$. We see an instance x is β-consistent, if there is τ_x satisfying $accuracy(\tau_x) \geq \beta$. Such β-consistency has already been proposed [2, 16].

Remark 2. We remark the following.

1. For $CON=\{P,Q\}$ in Table 1, we consider $\tau : [P,b] \Rightarrow [Dec,2]$ and its redundant implication $\tau' : [P,b] \wedge [Q, val_Q] \Rightarrow [Dec,2]$. The $accuracy(\tau)=2/3$. In case of val_Q=s, $accuracy(\tau')=1$ holds and $accuracy(\tau')$ increases. In case of val_Q=t, $accuracy(\tau')=1/2$ holds and $accuracy(\tau')$ decreases. Namely, the $accuracy$ value is non-monotonic for the redundant implication. We have a case that an instance x is not β-consistent for CON, but x is β-consistent for a subset of CON (**).
2. In strict dependency [9], $accuracy(\tau')=1$ holds if $accuracy(\tau)=1$ holds. Therefore, the above case does not occur, and we need to adjust the definition of consistency. Thus, we suppose that the case (**) is also β-consistent for CON and Dec. We need to consider subsets of CON so as not to miss the case of β-consistency.

By replacing consistency in $\gamma(CON, Dec)$ with β-consistency, we define below $\gamma_\beta(CON, Dec) = |\{x \in INS | ATR(\tau_x) \subset CON, accuracy(\tau_x) \geq \beta\}|/|INS|$. One simple solution is such that we examine $accuracy(\tau_x) \geq \beta$ for each τ_x ($x \in INS$). Then, we have the number of instances of β-consistency. However, we may have $\gamma_\beta(CON, Dec)$ like Example 3.

Example 3. Let's consider $\gamma_\beta(\{P,Q\}, Dec)$ in Table 1. We fix $\beta=0.75$.

1. Since Fig. 2 is the same for 0.75 consistency, we have the degree 3/4 based on Proposition 1. However, this calculation is wrong. Due to Remark 2, we need to examine $RULE(\{P\})= \{$2nd, 3rd in Fig. 3$\}$, $RULE(\{Q\})=\{$5th in Fig. 3$\}$. Instances $x4$ and $x7$ are inconsistent for $\{P,Q\}$ and Dec, but $x4$ is 0.75-consistent for $\{Q\}$ and Dec (due to the 5th rule in Fig. 3). Thus, $\gamma_{0.75}(\{P,Q\}, Dec)$ should be 7/8 based on (**).
2. If we consider the sets of rules without redundancy and summarize $support(\tau)$ for $\tau \in RULE(\{P\}) \cup RULE(\{Q\}) \cup RULE(\{P,Q\})$, the sum is more than (0.375+0.125) (for $RULE(\{P\})$ in Fig. 3) + 0.5 (for $RULE(\{Q\})$ in Fig. 3)=1.0. Thus, this calculation is also wrong.

We need to adjust each $support$ value for $\tau \in RULE(\{P\}) \cup RULE(\{Q\})$ and can obtain $\gamma_\beta(\{P,Q\}, Dec)$ in the following.

```
------------------rule1------------------
   condition     decision         sup     acc
1: ('R',  'x')==>('Dec',  '2')    0.375   1.0
2: ('P',  'a')==>('Dec',  '1')    0.375   0.75
3: ('P',  'c')==>('Dec',  '1')    0.125   1.0
4: ('R',  'y')==>('Dec',  '1')    0.25    1.0
5: ('Q',  't')==>('Dec',  '1')    0.5     0.8
6: ('R',  'z')==>('Dec',  '1')    0.375   1.0
```

Fig. 3. The obtained all rules (with one condition) from Table 1 under the constraints $support > 0$ and $accuracy \geq 0.75$.

3. For the 2nd rule τ_2 in Fig. 3, we adjust its *support* value using $\tau' : [P, a] \wedge [Q, \_] \Rightarrow [Dec, 1]$. We revise $support(\tau_2)$ to $support(\tau_2) - \Sigma_{\tau'} support(\tau')$. Here, the 4th rule in Fig. 1 hits, so its *support* value is revised to (0.375-0.375)=0.
4. For the 3rd rule τ_3 in Fig. 3, we consider $\tau' : [P, c] \wedge [Q, \_] \Rightarrow [Dec, 1]$. Here, the 2nd rule (*support*=0.125) in Fig. 1 hits, so the $support(\tau_3)$ value is revised to (0.125-0.125)=0. If we handle rules with the minimality constraint, τ' is not obtained. Therefore, we cannot revise the *support* value.
5. For the 5th rule τ_5 in Fig. 3, we adjust its *support* value using $\tau' : [P, \_] \wedge [Q, t] \Rightarrow [Dec, 1]$. We revise $support(\tau_5)$ to $support(\tau_5) - \Sigma_{\tau'} support(\tau')$. The 4th rule in Fig. 1 hits, so its *support* value is revised to (0.5-0.375)=0125.
6. Thus, we have
$\gamma_{0.75}(\{P, Q\}, Dec)$=0 (for $RULE(\{P\})$) + $(0.5 - 0.375)$ (for $RULE(\{Q\})$) + $(0.125 + 0.375 + 0.125 + 0.125)$ (for $RULE(\{P, Q\})$)=7/8.

Proposition 2. *For* CON=$\{P, Q\}$, *let* $RULE(\{P\})$, $RULE(\{Q\})$, $RULE(\{P, Q\})$ *be sets of the obtained rules without using the minimality constraint. Let* $RULE^*(\{P\})$ *and* $RULE^*(\{Q\})$ *be sets of rules, where support values are adjusted like Example 3. Then,*
$\gamma_\beta(\{P, Q\}, Dec)$=$\Sigma_{\tau \in RULE^*(\{P\}) \cup RULE^*(\{Q\}) \cup RULE(\{P,Q\})} support(\tau)$.

In a simple case, we can characterize $\gamma_\beta(CON, Dec)$ based on $RULE$. However, characterizing $\gamma_\beta(CON, Dec)$ will be complicated in a general case. The research on $\gamma_\beta(CON, Dec)$ using the set $RULE$ is in progress.

2.3 The Worst and the Best Strict Dependency in NIS

In NIS Φ, $\gamma(CON, Dec)$ depends on $\psi \in DD(\Phi)$. Thus, we may face with the degree $\gamma^{worst}(CON, Dec)$ of the worst strict dependency and the degree $\gamma^{best}(CON, Dec)$ of the best strict dependency from CON to Dec. Two dependencies are due to information incompleteness. Of course, we can apply the strict dependency in Sect. 2.1 to each $\psi \in DD(\Phi)$ sequentially and can obtain two degrees. We have the following from Table 2.
$\gamma^{worst}(\{P\}, Dec)$=1/8 ('?'=a or b), $\gamma^{best}(\{P\}, Dec)$=2/8 ('?'=c),
$\gamma^{worst}(\{P, R\}, Dec)$=$\gamma^{best}(\{P, R\}, Dec)$=1.
However, this method depends on the number of $DD(\Phi)$, which increases exponentially.

Remark 3. Here, we remind the characteristics of certain and possible rules. For simplicity, we consider Table 3, where the attribute values of P are c or d. Either $\tau_1 : [P, c] \Rightarrow [Dec, 1]$ or $\tau_2 : [P, d] \Rightarrow [Dec, 1]$ must occur twice.

1. In certain rule generation, the rule generator handles the case '?'=d for τ_1 and the case '?'=c for τ_2. Both τ_1 and τ_2 occur one time.
2. In possible rule generation, the rule generator handles the case '?'=c for τ_1 and the case '?'=d for τ_2. Both τ_1 and τ_2 occur twice.

Table 3. A Simple Table.

Instance	P	Q	R	Dec
$x3$	c	s	?	1
$x4$	d	t	y	1
$x5$	?	t	z	1

3. There is no problem in the above discussion. However, there is no $\psi \in DD(\Phi)$ causing τ_1 and τ_2 at the same time. Namely, the sum of the obtained $minsupp$ values may be smaller than the actual $support$ value. The sum of the obtained $maxsupp$ values may be larger than the actual $support$ value.

Example 4. Let's consider $\gamma^{worst}(\{P,R\}, Dec)$ in Table 2. Figure 4 shows all rules related to $\{P,R\}$ under the constraints $support > 0$ and $accuracy=1$. Namely, each certain rule is consistent in each of the 36 derived DISs, and each possible rule is consistent in at least one of the 36 derived DISs. If we apply the same formula in Proposition 1,
$\gamma^{worst}(\{P,R\}, Dec)=0.25+0.125+0.125+0.25=3/4$.
However, this is also a wrong result. For $\tau_{c1} : [P,c] \Rightarrow [Dec, 1]$ from $x3$ is a certain rule, but its redundant implication $\tau' : [P,c] \wedge [R, \_] \Rightarrow [Dec, 1]$ cannot be a certain rule because there is a missing value for the attribute R. We need to consider $CRULE(\{P\})$ and $CRULE(\{R\})$ for $CRULE(\{P,R\})$ again. Similar to Example 3, we adjust $support$ values.

1. For the c1 rule τ_{c1}, we adjust its $support$ value using $\tau' : [P,c] \wedge [R, \_] \Rightarrow [Dec, 1]$. We revise $support(\tau_{c1})$ to $support(\tau_{c1}) - 0 = 0.125$.
2. For the c2 rule τ_{c2}, we have $support(\tau')=1$ for $\tau' : [P, \_] \wedge [R, y] \Rightarrow [Dec, 1]$. The c6 rule hits here, so its $support$ value is revised to $(0.125-0.125)=0$.
3. For the c3 rule τ_{c3}, we adjust its $support$ value using $\tau' : [P, \_] \wedge [R, z] \Rightarrow [Dec, 1]$ The c7 rule hits here, so its $support$ value is revised to $(0.375-0.25)=0.125$.

```
------------------c_rules------------------              ------------------p_rules------------------
condition      decision       minsup     minacc          condition           decision         maxsup       maxacc
c1: ('P', 'c')==>('Dec', '1')   0.125   1.0              p1: ('R', 'x')==>('Dec', '2')          0.375    1.0
c2: ('R', 'y')==>('Dec', '1')   0.125   1.0              p2: ('P', 'c')==>('Dec', '1')          0.25     1.0
c3: ('R', 'z')==>('Dec', '1')   0.375   1.0              p3: ('R', 'y')==>('Dec', '1')          0.25     1.0
c4: ('P', 'b')('R', 'x')==>('Dec', '2')  0.25   1.0      p4: ('R', 'z')==>('Dec', '1')          0.5      1.0
c5: ('R', 'x')('P', 'a')==>('Dec', '2')  0.125  1.0      p5: ('P', 'b')('R', 'x')==>('Dec', '2') 0.25    1.0
c6: ('P', 'b')('R', 'y')==>('Dec', '1')  0.125  1.0      p6: ('R', 'x')('P', 'a')==>('Dec', '2') 0.125   1.0
c7: ('P', 'a')('R', 'z')==>('Dec', '1')  0.25   1.0      p7: ('P', 'b')('R', 'y')==>('Dec', '1') 0.125   1.0
                                                         p8: ('P', 'b')('R', 'z')==>('Dec', '1') 0.125   1.0
                                                         p9: ('R', 'x')('P', 'c')==>('Dec', '1') 0.125   1.0
                                                         p10: ('P', 'a')('R', 'z')==>('Dec', '1') 0.375  1.0
                                                         p11: ('P', 'c')('R', 'y')==>('Dec', '1') 0.125  1.0
                                                         p12: ('P', 'c')('R', 'z')==>('Dec', '1') 0.25   1.0
```

Fig. 4. The obtained all certain rules $CRULE$ (left side) and possible rules $PRULE$ (right side) related to $\{P,R\}$. Here, to obtain all rules, the minimality constraint is not used.

4. Due to Remark 3, we have
$\gamma^{worst}(\{P,R\}, Dec) \geq 0.125$ (for $RULE(\{P\})$) + 0.125 (for $RULE(\{R\})$) + (0.125 + 0.375 + 0.125 +0.125) (for $RULE(\{P,R\})$)=1. Namely, we have $\gamma^{worst}(\{P,R\}, Dec)$=1 and know that each implication from $\{P,R\}$ to Dec is consistent in each of the 36 derived DISs. We may say there is a functional dependency from $\{P,R\}$ to Dec in Table 2.

Proposition 3. *For $CON=\{P,R\}$, let $CRULE(\{P\})$, $CRULE(\{R\})$, $CRULE(\{P,R\})$ be sets of the obtained certain rules without using the minimality constraint. The constraints are support > 0 and accuracy$=1$. Let $CRULE^*(\{P\})$ and $CRULE^*(\{R\})$ be sets of rules, where support values are adjusted like Example 4. Then,*
$\gamma_{worst}(\{P,R\}, Dec) \geq \Sigma_{\tau \in CRULE^*(\{P\}) \cup CRULE^*(\{R\}) \cup CRULE(\{P,R\})} minsupp(\tau)$.

Example 5. Let's consider $\gamma^{best}(\{P,R\}, Dec)$ in Table 2. The rule p1 in Fig. 4 satisfies $accuracy(\tau)$=1 in $\psi_{max} \in DD(\Phi)$. Therefore, any redundant τ' also satisfies $accuracy(\tau')$=1 in ψ_{max} due to Remark 2. Thus, we do not consider $PRULE(\{P\})$ nor $PRULE(\{R\})$. It is enough to consider $PRULE(\{P,R\})$. Finally, we have
$\gamma^{best}(\{P,R\}, Dec) \leq \Sigma_{p5,p6,\cdots,p12} maxsupp(\tau)$=1.5.
Based on Example 4 and $1 \leq \gamma^{worst}(\{P,R\}, Dec) \leq \gamma^{best}(\{P,R\}, Dec)$=1.5, we have the following in this case.
$\gamma^{worst}(\{P,R\}, Dec)=\gamma^{best}(\{P,R\}, Dec)$=1.

Proposition 4. *For $CON=\{P,Q\}$, let $PRULE(\{P,Q\})$ be sets of the obtained possible rules without using the minimality constraint. The constraints are support > 0 and accuracy$=1$. Then,*
$\gamma^{best}(\{P,Q\}, Dec) \leq \Sigma_{\tau \in PRULE(\{P,Q\})} maxsupp(\tau)$.

3 Calculation of the Degree of Dependency

The main problem in detecting dependency is to find the set CON for Dec. In Sect. 2, we applied the obtained rules to detect the set CON. The presented propositions, examples, and remarks will help find CON.

We have the following two steps to have a high degree of dependency.
(i) We find the related set CON for Dec by the obtained rules.
(ii) We calculate the actual degree of dependency for the detected CON using the same rule generator.

Remark 4. We should remember Example 1. In the calculation of $support(\tau)$, $minsupp(\tau)$, and $maxsupp(\tau)$, we are handling $eq([A, val])$, $inf([A, val])$, and $sup([A, val])$, respectively. They are sets of instances. Therefore, we can know all instances supporting τ in rule generation. We mark each instance supporting τ ($ATR(\tau) \subset CON$), then $|\{marked\_instance\}|/|INS|$ becomes the degree of dependency. We can similarly apply this strategy to $\gamma(CON, Dec)$, $\gamma_\beta(CON, Dec)$ for DIS and $\gamma^{worst}(CON, Dec)$, $\gamma^{best}(CON, Dec)$ for NIS.

Input: Tabular data set DIS ψ, the obtained CON using rules, the decision attribute Dec, threshold values α, β.
Output: The degree $\gamma_\beta(CON, Dec)$.
1: Initialize an array $mark[x] \leftarrow 0$ for each instance $x \in INS$;
2: $step \leftarrow 1$;
3: **while** $step \leq |CON|$ **do**
4:     Generate rules $\{\tau\}$ ($support(\tau) \geq \alpha$, $accuracy(\tau) \geq \beta$),
5:     whose length of the condition part is $step$;
6:     **for all** τ **do**
7:         **if** $ATR(\tau) \subset CON$ **then** $mark[x] \leftarrow 1$ for each x supporting τ;
8:         (we can know each instance x using Remark 4)
9:         **end if**
10:    **end for**
11:    $step \leftarrow step + 1$;
12: **end while**
13: **return** $\gamma_\beta(CON, Dec) = |\{x \in INS \mid mark[x] = 1\}|/|INS|$

Fig. 5. The calculation of the actual degree of data dependency in DIS using rule generation.

Due to Remark 4, we initialize the array $mark[x]=0$ ($x \in INS$). Then, we execute the rule generation for CON and Dec. If τ is recognized as a rule and $ATR(\tau) \subset CON$, change $mark[x]=1$ for each x supporting τ in the rule generation process. After finishing rule generation, the degree is calculated by $|\{x \in INS \mid mark[x] = 1\}|/|INS|$. The pseudo-code execution flow is in Fig. 5.

4 Dependency for Another Decision Attribute

In the above discussion, we considered the fixed decision attribute Dec. However, we can assign another attribute to the decision attribute. This comes from the implementation of the DIS-Apriori and NIS-Apriori systems.

```
['1','P','b',1]['1','Q','s',1]['1','R','x',1]['1','Dec','2',1]
['2','P','a',1]['2','Q','s',2]['2','Q','t',2]['2','R','x',1]
['2','Dec','2',1]
['3','P','c',1]['3','Q','s',1]['3','R','x',2]['3','R','y',2]
['3','R','z',2]['3','Dec','1',1]
['4','P','b',1]['4','Q','t',1]['4','R','y',1]['4','Dec','1',1]
['5','P','a',2]['5','P','b',2]['5','P','c',2]['5','Q','t',1]
['5','R','z',1]['5','Dec','1',1]
['6','P','a',1]['6','Q','t',1]['6','R','z',1]['6','Dec','1',1]
['7','P','b',1]['7','Q','s',2]['7','Q','t',2]['7','R','x',1]
['7','Dec','2',1]
['8','P','a',1]['8','Q','t',1]['8','R','z',1]['8','Dec','1',1]
```

Fig. 6. The NRDF format from Table 2.

We first translate NIS to another format termed NRDF [15]. Figure 6 shows the translated NRDF format from Table 2. A tuple is translated to a list of [*instance*, *descriptor*, 1 or 2] (1: deterministic, 2: non-deterministic). There is less constraint in this format. Then, we specify the conditions of rules using a file termed *set*, whose contents are the specification of the decision attribute, the number of instances, α, and β. Namely, we can specify one decision attribute when executing rule generation. If we change the specification of the decision attribute, α, and β, we can easily have rules with different decision attributes. Figure 7 shows the obtained rules from Table 1, where *Dec*=*P*, the constraints are *support*>0 and *accuracy*=1.

```
-----------------------rule2---------------------
            condition              decision     sup      acc
1: ('Q', 's')('R', 'y')       ==>('P', 'c')    0.125    1.0
2: ('Q', 's')('Dec', '1')==>('P', 'c')         0.125    1.0
3: ('R', 'x')('Q', 't')       ==>('P', 'b')    0.125    1.0
4: ('Dec', '2')('Q', 't')==>('P', 'b')         0.125    1.0
5: ('R', 'y')('Q', 't')       ==>('P', 'b')    0.125    1.0
6: ('Dec', '1')('R', 'z')==>('P', 'a')         0.375    1.0
7: ('Q', 't')('R', 'z')       ==>('P', 'a')    0.375    1.0
```

Fig. 7. The obtained all rules with [P, _] as decision part.

From Fig. 7, we have the following.
$\gamma(\{Q, R\}, P)$=0.125 (1st) +0.125 (3rd) + 0.125 (5th) + 0.375 (7th) = 3/4,
$\gamma(\{Q, Dec\}, P)$=0.125 (2nd) +0.125 (4th) = 1/4,
$\gamma(\{R, Dec\}, P)$=0.375 (6th) = 3/8.
We know a data dependency from $\{Q, R\}$ to the decision attribute P may exist. We can easily have Table 4 by changing the decision attribute due to the above discussion.

Table 4. A table with the degree $\gamma(X, Y)$ of dependency from Table 1.

$X \Rightarrow Y$	P	Q	R	Dec
P	1 (P⇒P)	0.125 (P⇒Q)	0.125 (P⇒R)	0.125 (P⇒Dec)
Q	0 (Q⇒P)	1 (Q⇒Q)	0 (Q⇒R)	0 (Q⇒Dec)
R	0.375 (R⇒P)	0.375 (R⇒Q)	1 (R⇒R)	1 (R⇒Dec)
Dec	0 (Dec⇒P)	0 (Dec⇒Q)	0.375 (Dec⇒R)	1 (Dec⇒Dec)

If we handle tabular data with discrete values, having the correlation coefficient table in Statistics may be difficult. However, Table 4 will take a similar role for the correlation coefficient table.

5 Concluding Remarks

We applied the obtained rules to detecting data and functional dependency in tabular data with missing values. We will know the approximate dependencies

using the shown propositions and examples. Finally, we show one more actual example. For the Congressional Voting data set (US Congress data set, 435 instances, 16 condition attributes, decision attribute $a1$, decision attribute value is either $dem(ocrat)$ or $rep(ublic)$, 392 missing values, $|DD(\Phi_{Congress})| \geq 10^{100}$) [4], we had the following certain rules ($support \geq 0.1$ and $accuracy \geq 0.7$).
$\tau_{rep}: [a5, y] \Rightarrow [a1, republic]$ (minsupp=0.375, minacc=0.881),
$\tau_{dem}: [a5, n] \Rightarrow [a1, democrat]$ (minsupp=0.563, minacc=0.98).
Both rules are very strong because $accuracy(\tau_{rep}) \geq 0.881$ and $accuracy(\tau_{dem}) \geq 0.98$ for each $\psi \in DD(\Phi_{Congress})$. These rules will be reasonable for two political parties in the US. If we consider 0.88-consistency from $\{a5\}$ to $\{a1\}$, we have $\gamma_{0.88}^{worst}(\{a5\}, \{a1\}) \geq 0.375+0.563=0.938$. Probably, we soon detect a dependency from $\{a5\}$ to $\{a1\}$, even if there is imformation incompleteness. Some execution examples are in [12].

We will apply the results of this paper to the following issues,

(i) Missing value imputation by the obtained certain rules,
(ii) Machine learning from NIS to DIS.

Acknowledgments. The authors thank the reviewers for their helpful comments. Part of this work is supported by JSPS (Japan Society for the Promotion of Science) KAKENHI Grant Number JP20K11954.

References

1. Agrawal, R., Srikant, R.: Fast algorithms for mining association rules in large databases. In: Proceedings of the VLDB 1994, Morgan Kaufmann, pp. 487–499 (1994)
2. Berti-Equille, L., Harmouch, H., Naumann, F., Novelli, N., Thirumuruganathan, S.: Discovery of genuine functional dependencies from relational data with missing values. Proc. VLDB **2018**, 880–892 (2018)
3. Breve, B., Caruccio, L., Deufemia, V., Polese, G.: RENUVER: a missing value imputation algorithm based on relaxed functional dependencies. Proc. EDBT **2022**, 52–64 (2022)
4. Frank, A., Asuncion, A.: UCI machine learning repository, Irvine, CA: University of California, School of Information and Computer Science. http://mlearn.ics.uci.edu/MLRepository.html. Accessed 7 Jan 2022
5. Functional dependency, Wikipedia. https://en.wikipedia.org/wiki/Functional_dependency. Accessed 3 Apr 2022
6. Grzymała-Busse, J.W., Werbrouck, P.: On the best search method in the LEM1 and LEM2 algorithms. Incomplete Inf. Rough Set Anal. Stud. Fuzziness Soft Comput. **13**, 75–91 (1998)
7. Huhtala, Y., Karkkainen, J., Porkka, P., Toivonen, H.: TANE: an efficient algorithm for discovering functional and approximate dependencies. Comput. J. **42**(2), 100–111 (1999)
8. Orłowska, E., Pawlak, Z.: Representation of nondeterministic information. Theoret. Comput. Sci. **29**(1–2), 27–39 (1984)
9. Pawlak, Z.: Rough Sets. Kluwer Academic Publishers (1991)

10. Sakai, H., Nakata, M., Watada, J.: NIS-Apriori-based rule generation with three-way decisions and its application system in SQL. Inf. Sci. **507**, 755–771 (2020)
11. Jian, Z., Sakai, H., Ohwa, T., Shen, K.-Y., Nakata, M.: An adjusted apriori algorithm to itemsets defined by tables and an improved rule generator with three-way decisions. In: Bello, R., Miao, D., Falcon, R., Nakata, M., Rosete, A., Ciucci, D. (eds.) IJCRS 2020. LNCS (LNAI), vol. 12179, pp. 95–110. Springer, Cham (2020). https://doi.org/10.1007/978-3-030-52705-1_7
12. Sakai, H.: Execution logs by RNIA software tools. http://www.mns.kyutech.ac.jp/~sakai/RNIA. Accessed 3 Jan 2023
13. Sakai, H.: Studies on association rule-based table data analysis and its applications - new mathematics for data sciences -. J. Comb. Inf. Syst. Sci. **46**(1–4), 115–230 (2021)
14. Skowron, A., Rauszer, C.: The discernibility matrices and functions in information systems. In: Słowiński, R. (eds.) Intelligent Decision Support -. Handbook of Advances and Applications of the Rough Set Theory, Kluwer Academic Publishers, pp. 331–362. Springer, Dordrecht (1992). https://doi.org/10.1007/978-94-015-7975-9_21
15. Ślęzak, D., Sakai, H.: Automatic extraction of decision rules from non-deterministic data systems: theoretical foundations and SQL-based implementation. In: Ślęzak, D., Kim, T., Zhang, Y., Ma, J., Chung, K. (eds.) DTA 2009. CCIS, vol. 64, pp. 151–162. Springer, Heidelberg (2009). https://doi.org/10.1007/978-3-642-10583-8_18
16. Ziarko, W.: Variable precision rough set model. J. Comput. Syst. Sci. **46**(1), 39–59 (1993)

Distance-Based Fuzzy-Rough Sets and Their Application to the Classification Problem

Amrit Kumar[1(✉)] and Niladri Chatterjee[1,2]

[1] Department of Mathematics, Indian Institute of Technology Delhi, New Delhi, India
{maz228086,Niladri.Chatterjee}@maths.iitd.ac.in
[2] School of Information Technology, School of Artificial Intelligence,
Indian Institute of Technology Delhi, New Delhi, India

Abstract. We propose distance-based fuzzy-rough sets (DBFR) that rely on distance functions for the granulation of the underlying universe. The classification problem is investigated from a fuzzy perspective and cast as a concept approximation problem. DBFRs are employed to facilitate the approximation process. The geometrical nature of approximation emerging due to the use of distance functions is investigated. Naive classifiers based on DBFRs are proposed and experimentally evaluated on benchmark datasets.

Keywords: Fuzzy Rough Sets · Distance Functions · Classification

1 Introduction

Fuzzy sets [20] and rough sets [14] provide frameworks for dealing with imperfect data. While the former is concerned with vagueness (related to smooth boundaries of concepts), the latter deals with indistinguishability (related to granularity of knowledge). Fuzzy-rough sets (FRS) were proposed [5] as an amalgamation of the two approaches in an attempt to imbibe the best of both worlds so that both facets of imperfection can be dealt with at the same time. A two-step process is at the heart of most FRS models, viz., granulation of the universe by a suitable relation followed by approximation of the subsets of the universe using induced granules. Although much research is done vis-à-vis the approximation step [2,4,11,12,19], the granulation step has attracted comparatively little investigation, which is equally important, if not more. An important work in this direction is due to Hu et al. [6] that provided a systematic way to granulate the universe using kernel functions. Their main idea was that symmetric and reflexive kernels that are $[0,1]$-valued are fuzzy equivalence relations due to a theorem by Moser [13] and thus can be used to granulate the universe.

In this paper, we propose a distance-generator-based technique to granulate the universe and use it to define two types of "distance-based fuzzy-rough sets". The classification problem is formulated as that of approximation of concepts, followed by appropriately using the resulting approximations to build a classifier. While looking at classification as an approximation problem is not new (see, for instance, [7]), our work provides a conceptual framework to understand the classification problem from a fuzzy perspective rather than a probabilistic perspective, which is commonplace in machine learning. We detail the semantics of treating classification as an approximation problem and probe training data's role in the process, providing four different ways of interpreting the training data. Such a thorough investigation is lacking in the current literature. Furthermore, the use of distance function for granulation in the proposed FRS models equips them with a certain geometricality that is thoroughly investigated. Two naive classifiers are proposed and evaluated on benchmark datasets. The ability of proposed classifiers to unearth non-linear decision boundaries is shown.

The remainder of the paper is organized as follows. In Sect. 2, we recall the definition of distance functions and concepts from fuzzy and rough set theories that are relevant to this paper. Distance-based fuzzy-rough sets are proposed in Sect. 3. In Sect. 4, we investigate classification from a fuzzy perspective, formulate it as a problem of approximation of concepts, and use proposed DBFRs to realize the approximations. Section 5 details the geometrical aspects of the approximation process. Naive classifiers are proposed in Sect. 6, and results of their experimental evaluation are presented in Sect. 7. Finally, in Sect. 8, we conclude the paper and indicate future exploration avenues.

2 Preliminaries

2.1 Distances

Let X be a set. A function $d : X \times X \to [0, \infty]$ is called a distance if $d(x,x) = 0$ and $d(x,y) = d(y,x)$ for all $x, y \in X$. If a distance d satisfies the triangle inequality, i.e., $d(x,z) \leq d(x,y) + d(y,z)$ for all $x, y, z \in X$ then it is called a pseudo-metric, and the pair (X, d) is called a pseudo-metric space. For example, the L_2 pseudo-metric on $X = \mathbb{R}^n$ is defined by $L_2(x,y) = \sqrt{(x-y)^T(x-y)}$ for all $x, y \in X$.

2.2 Fuzzy Set Theory and Fuzzy Logic

Let X be the universe of discourse (called universe for short). A fuzzy subset A of X is a mapping $A : X \to [0,1]$. The collection of all fuzzy subsets of X is denoted by $\mathcal{F}(X)$. Given $A \in \mathcal{F}(X)$ and $x \in X$, $A(x) \in [0,1]$ is called the membership grade of x in A. A fuzzy subset $A \in \mathcal{F}(X)$ is called normal if there exists $x \in X$ such that $A(x) = 1$. Given fuzzy subset $A \in \mathcal{F}(X)$ and $Y \subseteq X$, the restriction of A on Y, denoted by $\underline{A}_Y$, is the fuzzy subset of Y defined by $\underline{A}_Y(y) = A(y)$ for all $y \in Y$. A fuzzy subset $A \in \mathcal{F}(X)$ is called non-zero if there exists $x \in X$ such that $A(x) > 0$. A family of fuzzy subsets $\mathcal{P} \subseteq \mathcal{F}(X)$ is called a fuzzy cover of X if for all $x \in X$ there exists $A \in \mathcal{P}$ such that $A(x) > 0$. Furthermore, it is called a normal fuzzy cover if all $A \in \mathcal{P}$ is normal.

A conjunctor $\mathcal{C}$ is a binary operation on $[0,1]$ that is increasing in both the arguments and satisfies $\mathcal{C}(0,0) = \mathcal{C}(0,1) = \mathcal{C}(1,0) = 0$ and $\mathcal{C}(1,1) = 1$. A t-norm/t-conorm $\mathcal{T}/\mathcal{S}$ is a binary operation on $[0,1]$ that is commutative, associative, increasing, and has $1/0$ as the neutral element. Both conjunctors and t-norms serve as fuzzy counterparts of the classical conjunction, albeit conjunctors form a larger class of operation and, therefore, have wider applicability. A negator $\mathcal{N}$ is an unary operation on $[0,1]$ that is decreasing and satisfies $\mathcal{N}(0) = 1$ and $\mathcal{N}(1) = 0$. Furthermore, a negator $\mathcal{N}$ is called involutive if $\mathcal{N} \circ \mathcal{N} = \text{id}$ and strict if it is continuous and strictly decreasing. The map $x \mapsto 1 - x$ for all $x \in [0,1]$ is called the standard negator and is denoted by $\mathcal{N}_S$. An implicator $\mathcal{I}$ is a binary operation on $[0,1]$ that is decreasing in the first argument, increasing in the second argument, and satisfies $\mathcal{I}(0,0) = 1 = \mathcal{I}(1,1)$ and $\mathcal{I}(1,0) = 0$. A coimplicator $\mathcal{I}^c$ is a binary operation on $[0,1]$ that is decreasing in the first argument, increasing in the second argument, and satisfies $\mathcal{I}^c(0,0) = 0 = \mathcal{I}^c(1,1)$ and $\mathcal{I}^c(0,1) = 1$. t-conorms, negators, and implicators are fuzzy counterparts of classical union, complement, and implication, respectively. Any t-norm $\mathcal{T}/$ t-conorm $\mathcal{S}$ induces an implicator/coimplicator called the R-implicator $\mathcal{I}_{\mathcal{T}}/$ R-coimplicator $\mathcal{I}_{\mathcal{S}}^c$ defined as $\mathcal{I}_{\mathcal{T}}(x,y) = \sup\{t \in [0,1] \,|\, \mathcal{T}(x,t) \leq y\}$ and $\mathcal{I}_{\mathcal{S}}^c(x,y) = \inf\{t \in [0,1] \,|\, \mathcal{S}(x,t) \geq y\}$ for all $x, y \in [0,1]$. Given conjunctor $\mathcal{C}$ and implicator $\mathcal{I}$, for any two fuzzy subsets $A, B \in \mathcal{F}(X)$, the degree of overlap between A and B, denoted by $\text{Overlap}(A,B)$, and degree of inclusion of A in B, denoted by $\text{Inc}(A,B)$, are defined as follows.

$$\text{Overlap}(A,B) = \sup_{x \in X} \mathcal{C}\left(A(x), B(x)\right) \; ; \; \text{Inc}(A,B) = \inf_{x \in X} \mathcal{I}\left(A(x), B(x)\right)$$

A generator [8] f is a $[0,1] \to [-\infty, \infty]$ mapping that is continuous, strictly decreasing, and satisfies $f(1) = 0$. Clearly, range of f is $[0, f(0)]$. The pseudo-inverse $f^{(-1)}$ of f is the $[-\infty, \infty] \to [0,1]$ mapping defined as $f^{(-1)}(x) = f^{-1}(\max(0, \min(f(0), x)))$ for all $x \in [-\infty, \infty]$. We call a generator f finite if $f(0) < \infty$ and infinite if $f(0) = \infty$. The following three generators are used in this paper.

$$f_{\cos}(t) = \cos^{-1}(t) \; ; \; f_{\times}(t) = -\ln(t) \; ; \; f_{\text{Ham}}(t) = \frac{1}{t} - 1$$

$f_{\cos}$ is a finite generator while $f_{\times}$ and f_{Ham} are infinite generators. Given a generator f, the functions $\mathcal{T}_f, \mathcal{I}_f : [0,1]^2 \to [0,1]$ defined as

$$\mathcal{T}_f(x,y) = f^{(-1)}\left(f(x) + f(y)\right) \; ; \; \mathcal{I}_f(x,y) = f^{(-1)}\left(f(y) - f(x)\right) \qquad (1)$$

for all $x, y \in [0,1]$ are a t-norm and implicator respectively. Moreover, if $f(0) < \infty$, then f also give rise to t-conorm $\mathcal{S}_f : [0,1]^2 \to [0,1]$, negator $\mathcal{N}_f : [0,1] \to [0,1]$ and co-implicator $\mathcal{I}_f^c : [0,1]^2 \to [0,1]$ defined as

$$\mathcal{S}_f(x,y) = f^{(-1)}(f(x) + f(y) - f(0)) \; ; \; \mathcal{N}_f(x) = f^{-1}(f(0) - f(x))$$
$$\mathcal{I}_f^c(x,y) = f^{(-1)}(f(0) - f(x) + f(y))$$

for all $x, y \in [0,1]$. In fact, $\mathcal{I}_f/\mathcal{I}_f^c$ is the R-implicator/R-coimplicator corresponding to $\mathcal{T}_f/\mathcal{S}_f$ and $\mathcal{N}_f$ is an involutive negator. Also, $\mathcal{T}_f$ & $\mathcal{S}_f$ and $\mathcal{I}_f$ & $\mathcal{I}_f^c$ are dual wrt $\mathcal{N}_f$[1]. Thus, given a finite generator f, a full set of fuzzy logic connectives can be extracted that are compatible with each other in that they satisfy duality identities. Denote by $\langle f \rangle$ the collection of induced connectives, i.e., $\langle f \rangle = \langle \mathcal{N}_f, \mathcal{T}_f, \mathcal{S}_f, \mathcal{I}_f, \mathcal{I}_f^c \rangle$. We define a similar collection of connectives for infinite generators using involutive negators. Given an infinite generator f and involutive negator $\mathcal{N}$, t-norm $\mathcal{T}_f$ and implicator $\mathcal{I}_f$ are defined as (1) and the t-conorm $\mathcal{S}_f$ and coimplicator $\mathcal{I}_f^c$ are defined as the $\mathcal{N}$-dual of $\mathcal{T}_f$ and $\mathcal{I}_f$ respectively. Denote by $\langle f, \mathcal{N} \rangle$ the collection of induced connectives, i.e., $\langle f, \mathcal{N} \rangle = \langle \mathcal{N}, \mathcal{T}_f, \mathcal{S}_f, \mathcal{I}_f, \mathcal{I}_f^c \rangle$.

A fuzzy (binary) relation R on X is a fuzzy subset of $X \times X$. For any $y \in X$, the R-foreset of y is the fuzzy set $Ry \in \mathcal{F}(X)$ defined by $Ry(x) = R(x, y)$ for all $x \in X$. R is called reflexive if $R(x,x) = 1$ for all $x \in X$ and symmetric if $R(x,y) = R(y,x)$ for all $x, y \in X$. R is called transitive with respect to t-norm $\mathcal{T}$ if $\mathcal{T}(R(x,y), R(y,z)) \leq R(x,z)$ for all $x, y, z \in X$. R is called a $\mathcal{T}$-equivalence for t-norm $\mathcal{T}$ or a fuzzy equivalence relation if it is reflexive, symmetric, and transitive wrt $\mathcal{T}$.

2.3 Fuzzy-Rough Set Theory

Let X be the universe and R be a crisp binary relation on X. Pawlak required R to model indistinguishability between the objects of the universe and, hence, chose it to be an equivalence relation. The 2-tuple (X, R) is called a crisp approximation space. The equivalence classes $\{Ry\}_{y \in X}$ can be seen as granules, each representing an atomic unit of knowledge in that they consist of elements that are indistinguishable from each other wrt relation R and, therefore, can not be further differentiated or divided. Thus, it is said that R granulates the universe X into granules $\{Ry\}_{y \in X}$. These granules are used to approximate subsets of X using lower approximation operator $R \downarrow$ and upper approximation operator $R \uparrow$ which are $\mathcal{P}(X) \to \mathcal{P}(X)$ mappings that send any given $A \subseteq X$ to its R-lower approximation $R \downarrow A$ and R-upper approximation $R \uparrow A$ defined as

$$R \downarrow A = \{y \in X \mid Ry \subseteq X\} \; ; \; R \uparrow A = \{y \in X \mid Ry \cap X \neq \emptyset\}$$

The pair $(R \downarrow A, R \uparrow A)$ is called a rough set. The inclusions $R \downarrow A \subseteq A \subseteq R \uparrow A$ hold, thereby justifying the usage of the words "lower" and "upper". Note, lower approximation and upper approximation can be interpreted as counterparts of necessity and possibility in modal logic [14] as $y \in R \downarrow A$ iff all elements indistinguishable from y, i.e., all the clones of y belong to A while $y \in R \uparrow A$ iff at least one element indistinguishable from y, i.e., at least one clone of y belong to A. In other words, y necessarily belongs to A iff $y \in R \downarrow A$ while y possibly belongs to A iff $y \in R \uparrow A$.

---

[1] Binary operations E_1 and E_2 on $[0,1]$ are called dual wrt negator $\mathcal{N}$ if the duality identities $\mathcal{N}(E_1(x,y)) = E_2(\mathcal{N}(x), \mathcal{N}(y))$ and $\mathcal{N}(E_2(x,y)) = E_1(\mathcal{N}(x), \mathcal{N}(y))$ hold for all $x, y \in [0,1]$. If $\mathcal{N}$ is involutive, then $E_1(x,y) := \mathcal{N}(E_2(\mathcal{N}(x), \mathcal{N}(y)))$ is called $\mathcal{N}$-dual of E_2 and $E_2(x,y) := \mathcal{N}(E_1(\mathcal{N}(x), \mathcal{N}(y)))$ and is called $\mathcal{N}$-dual of E_1.

Pawlak's rough sets are inadequate for handling real-world datasets as in almost all real-world datasets, objects are specified by features that take values from a continuous scale, thereby making indistinguishability or perfect equality between objects practically impossible. Fuzzy-rough sets (FRS) [5] come in handy here as they are based on similarity between objects captured via a fuzzy equivalence relation. Let R be a fuzzy equivalence relation on X. The pair (X, R) is called a fuzzy approximation space. The relation R granulates the universe into fuzzy or "soft" granules, viz., $\{Ry\}_{y \in X}$, and fuzzy subsets of X are approximated using these granules.

Many types of approximation operators are proposed in the literature, such as connectives-based operators [5,19], α-cut based operators [18], operators based on axiomatizations [12]. In this paper, we use two types of connectives-based approximation operators, viz. (θ, σ)-operators due to Mi & Zhang [11] and $(\mathcal{S}, \mathcal{T})$-operators due to Yeung et al. [19] to define distance-based fuzzy-rough sets. Both these types of operators give rise to implicator-conjunctor-based fuzzy rough sets [4], i.e., lower approximations are realized via an implicator while the upper approximation is realized via a conjunctor.

3 Distance-Based Fuzzy-Rough Sets

We present a new way of granulating the universe using distance functions, leveraging the following theorem due to Baets and Mesiar[2] [3].

Theorem 1. *Let (X, d) be a pseudo-metric space. For any t-norm $\mathcal{T}$ induced by generator f, the fuzzy relation $R := f^{(-1)} \circ d$ on X is a $\mathcal{T}$-equivalence relation.*

Hence, given a pseudo-metric d on X and a generator f, the relation $R_{d,f}$ defined as

$$R_{d,f}(x, y) = f^{(-1)}(d(x, y))$$

for all $x, y \in X$ is a fuzzy equivalence relation on X and thus can be used for granulation. We now introduce distance-based fuzzy-rough sets.

Definition 1. *Given a pseudo-metric space (X, d) and connectives set $\langle f \rangle$ or $\langle f, \mathcal{N} \rangle$, for any $A \in \mathcal{F}(X)$, the distance-based θ-lower and σ-upper approximation of A, denoted by $R_d \downarrow_\theta A$ and $R_d \uparrow_\sigma A$ respectively, are fuzzy subsets of X defined as*

$$(R_d \downarrow_\theta A)(y) = \inf_{x \in X} \theta\left(R_{d,f}(x, y), A(x)\right)$$
$$(R_d \uparrow_\sigma A)(y) = \sup_{x \in X} \sigma\left(\mathcal{N}(R_{d,f}(x, y)), A(x)\right)$$

for all $y \in X$. The pair $(R \downarrow_\theta A, R \uparrow_\sigma A)$ is called a distance-based (θ, σ) fuzzy-rough set.

---

[2] The actual theorem is much more general, but we only specify the part relevant to our work.

θ and σ in the above definition are precisely the implicator and co-implicator induced by f or f and $\mathcal{N}$ according as f is finite or infinite.

Definition 2. *Given a pseudo-metric space* (X, d) *and connectives set* $\langle f \rangle$ *or* $\langle f, \mathcal{N} \rangle$, *for any* $A \in \mathcal{F}(X)$, *the distance-based* $\mathcal{S}$-*lower and* $\mathcal{T}$-*upper approximation of* A, *denoted by* $R_d \downarrow_\mathcal{S} A$ *and* $R_d \uparrow_\mathcal{T} A$ *respectively, are fuzzy subsets of* X *defined as*

$$(R_d \downarrow_\mathcal{S} A)(y) = \inf_{x \in X} \mathcal{S}\left(\mathcal{N}(R_{d,f}(x,y)), A(x)\right)$$
$$(R_d \uparrow_\mathcal{T} A)(y) = \sup_{x \in X} \mathcal{T}\left(R_{d,f}(x,y), A(x)\right)$$

for all $y \in X$. *The pair* $(R \downarrow_\mathcal{S} A, R \uparrow_\mathcal{T} A)$ *is called a distance-based* $(\mathcal{S}, \mathcal{T})$ *fuzzy-rough set.*

$\mathcal{S}$ and $\mathcal{T}$ in the above definition are precisely the t-conorm and t-norm induced by f or f and $\mathcal{N}$ according as f is finite or infinite.

The inclusions $R_d \downarrow_\theta A \subseteq A \subseteq R_d \uparrow_\sigma A$ and $R_d \downarrow_\mathcal{S} A \subseteq A \subseteq R_d \uparrow_\mathcal{T} A$ hold. It can be verified that $(\alpha, \beta) \mapsto \sigma(\mathcal{N}(\alpha), \beta)$ is a conjunctor and $(\alpha, \beta) \mapsto \mathcal{S}(\mathcal{N}(\alpha), \beta)$ is an implicator. Thus, for any $y \in X$, $(R_d \downarrow_\theta A)(y)$ or $(R_d \downarrow_\mathcal{S} A)(y)$ is the degree of inclusion of Ry (a fuzzy granule) in the fuzzy set A while $(R_d \uparrow_\sigma A)(y)$ or $(R_d \uparrow_\mathcal{T} A)(y)$ is the degree of overlap of Ry and fuzzy set A. Hence, the semantics of lower and upper approximation as capturing necessity and possibility remain intact.

4 Classification Problem

4.1 Classification as Approximation of Concepts

The problem of classification is fundamental to machine learning. In a classification problem, we are given a labeled dataset $\mathcal{D} = \{(x_i, \ell_i)\}_{i=1}^m$ where for each i, ℓ_i is the label associated with object x_i, denoted in short by $\text{label}(x_i) = \ell_i$. For example, x_i's can be images of animals, and ℓ_i's can be the name of the animal in the image. The task is to learn a classifier λ which, for any input x, gives the label $\lambda(x)$.

We assume that $\mathcal{D}$ is a subset of $U \times L$, where U is the space from which objects come and is taken to be the universe, and L is the set of labels. Thus, the goal of the classification problem is to learn a classifier $\lambda : U \to L$ from $\mathcal{D} = \{(x_i, \ell_i)\}_{i=1}^m \subseteq U \times L$. $\mathcal{D}$ is commonly called the training data. Denote by T the set of all objects in the training data, i.e., $T = \{x_i\}_{i=1}^m \subseteq U$.

Objects in U are represented by a set of features. For illustration, paws, strips, skin color, horn, fur, etc., may be appropriate features for the classification of animal images. Denote the features as $\phi_1, \phi_2, \cdots, \phi_n$ where for each i, feature ϕ_i is seen as a $U \to \mathcal{V}_i$ mapping with $\mathcal{V}_i$ being the set of values ϕ_i may assume. Thus, any object $x \in U$ is identified as a vector of feature values $(\phi_i(x))_{i=1}^n$ residing in the feature space $\mathcal{V} := \prod_{i=1}^n \mathcal{V}_i$. We call $(\phi_i(x))_{i=1}^n$ the feature vector of x, and the map $\phi : U \to \mathcal{V}$ that sends objects in U to their feature vectors as the feature map. If all the features are real-valued, then $\mathcal{V} \subseteq \mathbb{R}^n$. The classifier λ is learnt using the feature vectors $\{\phi(x_i)\}_{i=1}^m \subseteq \mathcal{V}$. Hence, more appropriately, the goal is to learn the map $f : \mathcal{V} \to L$, and the classifier λ is given as $\lambda = f \circ \phi$.

For any label $\ell \in L$, the corresponding ℓ-th class D_ℓ is defined as the set of all objects in the training data with label ℓ, i.e., $D_\ell = \{x \in T \,|\, \text{label}(x) = \ell\}$. We interpret the classes D_ℓ's as containing specific examples of certain concepts. A "concept" is formalized as a fuzzy subset of the universe U. Consider the earlier example of x_i's being images of animals and ℓ_i's being the names of the animals in images. In this case, the universe U may be taken as the set of all animal images. Here, a class, such as $D_{\text{Capybara}} = \{x \in T \,|\, \text{label}(x) = \text{Capybara}\}$ contains specific examples of the concept "Capybara". We will distinguish a concept from the class of its specific examples via the hat symbol ($\hat{\ }$). Thus, the concept "Capybara" is denoted as $\hat{D}_{\text{Capybara}}$ and is understood as a fuzzy subset of U, i.e., $\hat{D}_{\text{Capybara}} \in \mathcal{F}(U)$. The class D_{Capybara} is interpreted as consisting of specific examples of the concept of Capybara in the training data. The notion of "specific examples" is formalized as follows.

$$\forall x \in D_{\text{Capybara}} \left(\hat{D}_{\text{Capybara}}(x) > 0 \right)$$

In other words, if the training data, which is the ground truth, indicates that the label of an object x is "Capybara", then we take it as sufficient evidence to conclude that x satisfies the concept of "Capybara" to at least a positive degree. Thus, $\hat{D}_{\text{Capybara}}$ is a non-zero fuzzy subset of U.

(a) x_1 ; label = Capybara   (b) x_2 ; label = Wombat   (c) x_3 ; label = Cat

Fig. 1. Objects x_1, x_2 and x_3 [10,16,17]

For the toy training data $\mathcal{D}_{toy} = \{(x_1, \text{Capybara}), (x_2, \text{Wombat}), (x_3, \text{Cat})\}$ (Fig. 1) values in Table 1 may be plausible.

Table 1. Membership degrees of x_i's to different concepts.

x	$\hat{D}_{\text{Capybara}}(x)$	$\hat{D}_{\text{Wombat}}(x)$	$\hat{D}_{\text{Cat}}(x)$
x_1	1	0.2	0
x_2	0.3	1	0
x_3	0	0	1

Formalizing a concept as a fuzzy subset of the universe enables us to encode vagueness that is often intrinsic to the training data. For example, a wombat-labeled image may consist of a wombat that looks similar to a capybara (Fig. 1 (b)) due to its position in the image. This kind of vagueness is unsuitable for probabilistic modeling, which is the common approach of encoding uncertainty in machine learning wherein one assumes the existence of a true albeit unknown conditional likelihood $p^*(\ell|x)$ over set L of labels or a joint likelihood $p^*(x,\ell)$ over object-label pairs, i.e., $U \times L$ with the classifier $\lambda : U \to L$ being such that for any $x \in U$, the expected error in approximating the random label obeying the distribution $p^*(.|x)$ by $\lambda(x)$ is the least possible or, equivalently, the classifier $\lambda : U \to L$ being such that the expected error between predicted label of random objects in U and random labels in L with object-label pairs obeying the joint distribution p^* is the least possible.

In general, labels ℓ_i's are names of different classes. For simplicity, we assume that $L = \{1, 2, \cdots, k\}$ so that for any object $x_j \in T$, the associated label $\ell_j \in \{1, \cdots, k\}$. Correspondingly, for each label $j \in L$, the j-th class $D_j := \{x \in T \mid \text{label}(x) = j\}$. Note, $T = \cup_{j=1}^{k} D_j$. The classes $D_1, \cdots, D_k$ contain specific examples of corresponding concepts $\hat{D}_1, \cdots, \hat{D}_k$. Formally, the following condition holds true.

$$\text{For each label } j \in L \text{ and for each object } x \in D_j, \ \hat{D}_j(x) > 0 \qquad (\text{I1})$$

Thus, $\{\hat{D}_j\}_{j \in L}$ is a collection of non-zero fuzzy subsets of U. Also, $\{\underline{\hat{D}_{j_T}}\}_{j \in L}$ forms a fuzzy cover of T.

As will be shown in Sect. 6, classifiers can be constructed by appropriately using the approximation of concepts that the training data gives rise to. The need for fuzzy-rough sets emerges naturally here as the framework of FRS enables the approximation of concepts (fuzzy subsets of the universe).

4.2 Crisp and Fuzzy Interpretations of Training Data

The interpretation of training data as consisting of specific examples of certain concepts, codified as condition (I1), can be refined to accommodate more precise ideas of what it may mean to be a "specific example". We propose the following three refinements, each being more strict than the previous one.

$$\text{For each label } j \in L \text{ and for each object } x \in D_j, \ \hat{D}_j(x) = 1 \qquad (\text{I2})$$

$$\begin{aligned}&\text{For each label } j \in L \text{ and for each object } x \in D_j, \ \hat{D}_j(x) = 1 \\ &\text{and for all label } i \in L \text{ with } i \neq j, \ \hat{D}_i(x) < 1\end{aligned} \qquad (\text{I3})$$

$$\begin{aligned}&\text{For each label } j \in L \text{ and for each object } x \in D_j, \ \hat{D}_j(x) = 1 \\ &\text{and for all label } i \in L \text{ with } i \neq j, \ \hat{D}_i(x) = 0^3\end{aligned} \qquad (\text{I4})$$

Condition (I2) essentially states that the objects in the training data are prototypical examples of their respective concepts, i.e., they fully satisfy their respective concepts. Note, $\{\hat{D}_{j_T}\}_{j \in L}$ is a normal fuzzy cover of T. This contrasts with the condition (I1) in that (I1) leads to $\{\hat{D}_{j_T}\}_{j \in L}$ being just a fuzzy cover of T. While condition (I3) amounts to saying that the objects in training data are prototypical examples of their and only their respective concepts. Observe Table 1 reflects this interpretation of training data $\mathcal{D}_{toy}$.

Condition (I4), on the other hand, means that the objects in the training data should be interpreted as being prototypical examples of their respective concepts while being completely unrepresentative of other concepts. This way of interpreting training data is more restrictive compared to previous conditions. This can also be seen from the fact that $\{\hat{D}_{j_T}\}_{j \in L} = \{D_j\}_{j \in L}$ is a crisp partition of T. We say that (I4) corresponds to a crisp interpretation of training data while (I1), (I2) and (I3) correspond to fuzzy interpretations of training data. Choosing a fuzzy interpretation entails the assignment of membership grade in $\hat{D}_j$'s to objects in T and, therefore, is more costly. Fuzzy interpretations may be useful for noisy data as noise-affected objects can be assigned lower membership grades to concepts, thereby minimizing their importance. In this paper, we will focus on the (I4) interpretation of the training data.

4.3 Approximating Concepts Using Distance-Based (θ, σ) and $(\mathcal{S}, \mathcal{T})$ Fuzzy-Rough Sets

Given a pseudo-metric d on the feature space $\mathcal{V}$ and a generator f, the relation $R_{d,f}$ on U defined as

$$R_{d,f}(x, y) = f^{(-1)}(d(\phi(x), \phi(y)))$$

for all $x, y \in U$ is a fuzzy equivalence relation due to theorem (1) and therefore granulates U into fuzzy granules $\{R_{d,f}\, y\}_{y \in U}$. These fuzzy granules are used to approximate the concepts $\{\hat{D}_j\}_{j \in L}$ via the proposed distance-based (θ, σ) and $(\mathcal{S}, \mathcal{T})$ fuzzy rough sets.

Given the connectives set $\langle f \rangle$ or $\langle f, \mathcal{N} \rangle$, for each concept $\hat{D}_j$ ($j \in L$), the distance-based (θ, σ)-approximations are

$$(R_d \downarrow_\theta \hat{D}_j)(y) = \inf_{x \in U} \theta \left(R_{d,f}(x, y), \hat{D}_j(x) \right)$$

$$(R_d \uparrow_\sigma \hat{D}_j)(y) = \sup_{x \in U} \sigma \left(\mathcal{N}(R_{d,f}(x, y)), \hat{D}_j(x) \right)$$

and the distance-based $(\mathcal{S}, \mathcal{T})$-approximations are

$$(R_d \downarrow_\mathcal{S} \hat{D}_j)(y) = \inf_{x \in U} \mathcal{S} \left(\mathcal{N}(R_{d,f}(x, y)), \hat{D}_j(x) \right)$$

$$(R_d \uparrow_\mathcal{T} \hat{D}_j)(y) = \sup_{x \in U} \mathcal{T} \left(R_{d,f}(x, y), \hat{D}_j(x) \right)$$

where $y \in U$. As the supremum (and infimum) is carried out over the entire universe U, thus to compute these approximations for a given $y \in U$, one needs to know $\hat{D}_j(x)$ for all $x \in U$ beforehand thereby defying the purpose of approximating $\hat{D}_j$ in the first place. Thus, we condition the approximations on the training data by carrying out infimum and supremum over T. The corresponding approximations are denoted by $(R_d \downarrow_\theta \hat{D}_j)|_T$, $(R_d \uparrow_\sigma \hat{D}_j)|_T$, $(R_d \downarrow_S \hat{D}_j)|_T$ and $(R_d \uparrow_T \hat{D}_j)|_T$ respectively. The following inequalities hold

$$(R_d \downarrow_\theta \hat{D}_j)|_T \geq R_d \uparrow_\sigma \hat{D}_j \text{ and } (R_d \uparrow_\sigma \hat{D}_j)|_T \leq R_d \uparrow_\sigma \hat{D}_j$$
$$(R_d \downarrow_S \hat{D}_j)|_T \geq R_d \downarrow_S \hat{D}_j \text{ and } (R_d \uparrow_T \hat{D}_j)|_T \leq R_d \uparrow_T \hat{D}_j$$

which shows that the conditioned lower approximation always overestimates the true lower approximation while the conditioned upper approximation always underestimates the true upper approximation. This indicates the inherent difficulty in computing the approximations of the concepts given only finitely many object-label pairs. For convenience, we will use the terms lower and upper approximations to refer to conditioned lower and conditioned upper approximations, respectively, from here on. Under the (I4) interpretation of the training data, we arrive at the following formulas.

$$(R_d \downarrow_\theta \hat{D}_j)|_T(y) = \min_{x \notin D_j} \theta\left(R_{d,f}(x,y), 0\right) \tag{2}$$

$$(R_d \uparrow_\sigma \hat{D}_j)|_T(y) = \max_{x \in D_j} \sigma\left(\mathcal{N}(R_{d,f}(x,y)), 1\right) \tag{3}$$

$$(R_d \downarrow_S \hat{D}_j)|_T(y) = \min_{x \notin D_j} \mathcal{N}(R_{d,f}(x,y)) \,;\, (R_d \uparrow_T \hat{D}_j)|_T(y) = \max_{x \in D_j} R_{d,f}(x,y) \tag{4}$$

For clarity, we emphasize that $x \notin D_j$ in Eqs. (2) and (4) means $x \in T \setminus D_j$. For illustration, consider the toy dataset depicted in Fig. 2 (a)[3] consisting of two classes D_1 and D_2. The corresponding concepts are $\hat{D}_1$ and $\hat{D}_2$. The feature space $\mathcal{V}$ is $\mathbb{R}^2$. Consider the connectives set $\langle f_{\cos} \rangle$ and pseudo-metric

$$d(a,b) = \sqrt{\left(\frac{a_x - b_x}{2\sigma_x}\right)^2 + \left(\frac{a_y - b_y}{2\sigma_y}\right)^2} \tag{5}$$

where $a = (a_x, a_y)$, $b = (b_x, b_y) \in \mathbb{R}^2$ and σ_x, σ_y are standard deviations of first and second feature's values respectively in the toy dataset. The relation $R_{d,f_{\cos}} = f_{\cos}^{(-1)} \circ d$ granulates the universe.

---

[3] The points in the Fig. 2 are actually the feature vectors of the objects in the toy dataset. For convenience, we won't explicate the distinction from here on.

Figure 2 (b) shows the distance-based $\mathcal{S}$-lower approximations of concepts $\hat{D}_1$ and $\hat{D}_2$ over the region $[-4,4]^2$ of feature space. The horizontal plane is the feature space $\mathbb{R}^2$ and the height corresponding to a point $P = \phi(y) \in \mathbb{R}^2$ for $y \in U$ is the value $(R_d \downarrow_{\mathcal{S}} \hat{D}_j)|_T(y)$ ($j = 1, 2$), i.e., the degree to which object y necessarily belongs to the concept $\hat{D}_j$ ($j = 1, 2$).

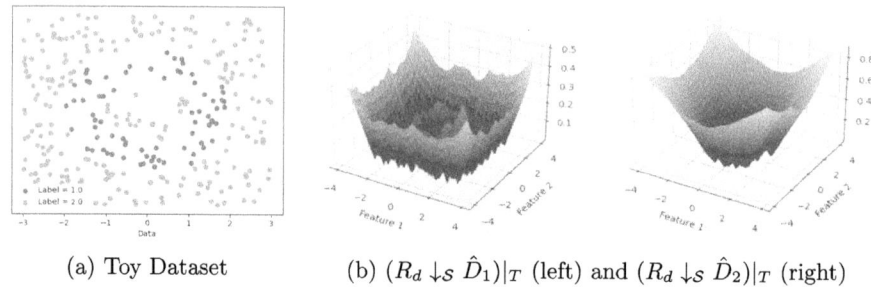

(a) Toy Dataset (b) $(R_d \downarrow_{\mathcal{S}} \hat{D}_1)|_T$ (left) and $(R_d \downarrow_{\mathcal{S}} \hat{D}_2)|_T$ (right)

Fig. 2. Toy Dataset and Approximation Fuzzy Sets

Figure 3 shows the contour plots of the approximation fuzzy sets $(R_d \downarrow_{\mathcal{S}} \hat{D}_1)|_T$ and $(R_d \downarrow_{\mathcal{S}} \hat{D}_2)|_T$ (shown in Fig. 2 (b)). A careful examination of these contour plots indicates that the value of $(R_d \downarrow_{\mathcal{S}} \hat{D}_1)|_T(y)$ for y near class D_1[4] is high (yellowish) while it is low (purplish) for y near class D_2. Similarly, the value of $(R_d \downarrow_{\mathcal{S}} \hat{D}_2)|_T(y)$ for y near class D_2 is high (yellowish) while it is low (purplish) for y near class D_1. This is consistent with our intuition as training data indicating that $y_0 \in D_1$, i.e., y_0 having label 1 hints that objects near y_0 may necessarily have label 1 to a greater degree than objects that are closer to a distinct class, viz., D_2.

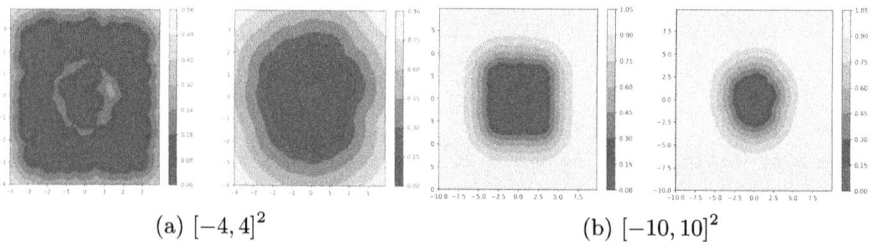

(a) $[-4, 4]^2$ (b) $[-10, 10]^2$

Fig. 3. Contour Plots of $(R_d \downarrow_{\mathcal{S}} \hat{D}_1)|_T$ and $(R_d \downarrow_{\mathcal{S}} \hat{D}_2)|_T$

---

[4] By "y near class D_1", it is meant that the feature vector of y, i.e., $\phi(y)$ is near feature vector of objects in class D_1, viz. blue points. As the feature space is equipped with a pseudo-metric, talking both "nearness" makes sense. For convenience, we won't emphasize this technicality from here on.

A closer look at the periphery of Fig. 3 (a) reveals that objects that are somewhat far away from the dataset belong to both $(R_d \downarrow_S \hat{D}_1)|_T$ and $(R_d \downarrow_S \hat{D}_2)|_T$ to a high degree. Indeed, plotting these lower approximations over a larger region ($[-10, 10]^2$) of the feature space as shown in Fig. 3 (b) confirms this. In fact, for $y \in U$ that is sufficiently far away from the dataset, experimentation indicates $(R_d \downarrow_S \hat{D}_j)|_T(y) = 1\ \forall\ j \in L$. This motivates us to define the "region of no information" (RONI).

Definition 3. *Given training data $\mathcal{D}$, pseudo-metric d on the feature space and set of connectives $\langle f \rangle$ or $\langle f, \mathcal{N} \rangle$, the region of no information, denoted by $RONI_\theta$ or $RONI_S$, is defined as*

$$RONI_{\theta/S} = \{y \in U \mid \text{For each label } j \in L,\ (R_d \downarrow_{\theta/S} \hat{D}_j)|_T(y) = 1\}$$

RONI can be seen as the region of the universe U where the given training data $\mathcal{D}$ fails to provide any discriminatory information about the different concepts. The following theorem gives insight as to what causes RONI.

Theorem 2. *Given training data $\mathcal{D}$, feature map ϕ, pseudo-metric d on the feature space and set of connectives $\langle f \rangle$ or $\langle f, \mathcal{N} \rangle$, for any $y \in U$ following holds under the (I4) interpretation of $\mathcal{D}$.*

$$y \in RONI_{\theta/S} \iff \text{For each label } j \in L \text{ and for each object } x \notin D_j,$$
$$d(\phi(x), \phi(y)) \geq f(0)$$

A corollary of the above theorem is that if $\text{dist}(y, T) \geq f(0)$[5], then $y \in RONI_{\theta/S}$. In other words, any object $y \in U$ that is sufficiently far away from the objects in the training data belongs to RONI; thus, nothing meaningful can be said about its label as it necessarily belongs to all the concepts to degree 1. Figure 4 (a) shows the RONI for the toy data set (Fig. 2).

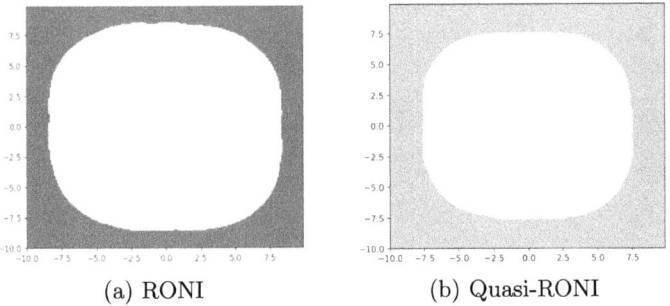

(a) RONI (b) Quasi-RONI

Fig. 4. RONI and Quasi-RONI of Toy Dataset

---

[5] $\text{dist}(y, T) := \min_{x \in T} d(\phi(x), \phi(y))$.

4.4 Thresholding Property and Region of No Information (RONI)

The generator f imparts "thresholding" property to the relation $R_{d,f}$, i.e., for any two objects $x, y \in U$,

$$d(\phi(x), \phi(y)) \geq f(0) \implies R_{d,f}(x,y) = 0 \qquad \text{(TP)}$$

In other words, any two sufficiently far apart objects are not related at all, i.e., related to degree 0. This means that given an object x in the training data, i.e., $x \in T$ and objects $y_1, y_2 \in U$ that are sufficiently far apart from x, viz., $d(\phi(x), \phi(y_i)) \geq f(0)$ $(i=1,2)$, $R_{d,f}(x, y_1) = 0 = R_{d,f}(x, y_2)$ even if $d(\phi(x), \phi(y_1)) > d(\phi(x), \phi(y_2))$ although it is desirable for a distinction to be made between y_1 and y_2 as y_1 is farther apart from x than y_2.

Infinite generators can be used to circumvent the thresholding problem as $f(0) = \infty$ means $d(\cdot, \cdot) \geq f(0)$ can never be satisfied[6] and thus $d(\phi(x), \phi(y_1)) > d(\phi(x), \phi(y_2)) \implies R_{d,f}(x, y_1) < R_{d,f}(x, y_2)$. Also, the use of infinite generators implies $RONI = \emptyset$, although for an object $y \in U$ that is far away from the objects in training data, $(R_d \downarrow_\mathcal{S} \hat{D}_j)|_T(y)$ is still close to 1 due to strictness of $\mathcal{N}(\cdot)$. This is evident from Fig. 4 (b) that shows a kind of quasi-RONI (region of the universe where $(R_d \downarrow_\mathcal{S} \hat{D}_j)|_T > 0.99$) of the toy data set (Fig. 2). The pseudo-metric used is the same as before (5), but the connectives set is induced by infinite generator $f_\times$ and the standard negator $\mathcal{N}_\mathcal{S}$.

5 Geometry of Approximation

Define $d_\theta, d_\mathcal{S} : U^2 \to [0,1]$ as

$$d_\theta(x,y) = \theta(R_{d,f}(x,y), 0) \quad ; \quad d_\mathcal{S}(x,y) = \mathcal{N}(R_{d,f}(x,y))$$

for all $x, y \in U$. Note d_θ and $d_\mathcal{S}$ are symmetric and $d_\theta(y,y) = 0 = d_\mathcal{S}(y,y)$ for all $y \in U$. Thus, d_θ and $d_\mathcal{S}$ are distance functions. For any $y \in U$ and label $j \in L$, define $NN_\theta^j(y), NN_\mathcal{S}^j(y)$ and $NN_d^j(y)$ to be the set of all nearest neighbor(s) of y in a distinct class (other than D_j) of the training data wrt distances $d_\theta, d_\mathcal{S}$ and d respectively. Formally,

$$NN_\theta^j(y) = \{x^* \in T \mid \forall \, x \in T \setminus D_j, \; d_\theta(x^*, y) \leq d_\theta(x, y)\}$$
$$NN_\mathcal{S}^j(y) = \{x^* \in T \mid \forall \, x \in T \setminus D_j, \; d_\mathcal{S}(x^*, y) \leq d_\mathcal{S}(x, y)\}$$
$$NN_d^j(y) = \{x^* \in T \mid \forall \, x \in T \setminus D_j, \; d(\phi(x^*), \phi(y)) \leq d(\phi(x), \phi(y))\}$$

Observe Eqs. (2) and (4) assume the following form

$$(R_d \downarrow_\theta \hat{D}_j)|_T(y) = d_\theta(x_\theta, y) \quad ; \quad (R_d \downarrow_\mathcal{S} \hat{D}_j)|_T(y) = d_\mathcal{S}(x_\mathcal{S}, y) \qquad (6)$$

where $x_\theta \in NN_\theta^j(y)$ and $x_\mathcal{S} \in NN_\mathcal{S}^j(y)$ for all $y \in U$. Equation (6) indicate that the degree to which an object $y \in U$ necessarily belongs to the concept $\hat{D}_j$ is

---

[6] Assuming that the pseudo-metric d is finite-valued which is typically true.

the d_θ or d_S distance between y and its nearest neighbor in a distinct class of the training data, i.e., in $T \setminus D_j$.

For $y \in U$ and $R > 0$, define R θ-ball $B_R^\theta(y)$, R S-ball $B_R^S(y)$ and R d-ball $B_R^d(y)$ as the set of objects in the training data that is within a distance of R wrt d_θ, d_S and d respectively. Formally,

$$B_R^\theta(y) = \{x \in T \mid d_\theta(x,y) < R\} \; ; \; B_R^S(y) = \{x \in T \mid d_S(x,y) < R\}$$
$$B_R^d(y) = \{x \in T \mid d(\phi(x), \phi(y)) < R\}$$

In the above definitions, y and R are called the center and radius of the corresponding balls.

5.1 Finite Generators

Theorem 3. *Given training data $\mathcal{D}$, feature map ϕ, pseudo-metric d on the feature space, finite generator f and connectives set $\langle f \rangle$, following holds under the (I4) interpretation of $\mathcal{D}$.*

(F1) For all $x, y \in U$, $d_{\theta/S}(x,y) = f^{-1}(\max\{0, f(0) - d(\phi(x), \phi(y))\})$

(F2) Given $y \in U$ and $j \in L$, if $x^* \in NN_d^j(y)$ and $d(\phi(x^*), \phi(y)) \geq f(0)$, then for all objects x in the training data from a distinct class, $d_{\theta/S}(x,y) = 1$. In particular, $d_{\theta/S}(x^*, y) = 1$.

(F3) For all $y \in U$, $B_R^\theta(y) = B_R^S(y) = \begin{cases} T & \text{if } R > 1 \\ B_{f(0)-f(R)}^d & \text{if } R \leq 1 \end{cases}$

(F1) indicates that under the use of a finite generator, both the d_θ and d_S distance functions become the same. (F2) indicates that if nearest neighbor of y in a distinct class is sufficiently far apart, then all objects in distinct class(es) of the training data are equidistant from y wrt distance $d_{\theta/S}$. (F3) establishes that geometry of θ/S-balls is same as the geometry of d-ball.

As $(R_d \downarrow_{\theta/S} \hat{D}_j)|_T(y)$ is the d_θ/d_S distance between y and its nearest neighbor in a distinct class of the training data, thus computing $(R_d \downarrow_{\theta/S} \hat{D}_j)|_T(y)$ is same as finding the nearest neighbor of y in a distinct class of the training data. This computation can be geometrically visualized as follows.

Consider a d_θ/d_S ball centered at y with a small radius. Start increasing the radius of the ball gradually and stop as soon as an object x^* of a distinct class in the training data is trapped. This is nearest neighbor of y in a distinct class. If the radius of this ball is less than or equal to 1, i.e., $x^* \in B_R^{\theta/S}(y)$ for $R \leq 1$, then $(R_d \downarrow_{\theta/S} \hat{D}_j)|_T(y) = d_{\theta/S}(x^*, y) < 1$ otherwise the nearest neighbor x^* lies in $B_R^{\theta/S}(y) \setminus B_1^{\theta/S}(y)$ where $R > 1$, i.e., $x^* \in T \setminus B_{f(0)}^d(y)$ and thus $(R_d \downarrow_{\theta/S} \hat{D}_j)|_T(y) = d_{\theta/S}(x^*, y) = 1$, i.e., y completely necessarily satisfy the concept $\hat{D}_j$. In fact, if nearest neighbor of y in a distinct class is not found within $B_1^{\theta/S}(y) = B_{f(0)}^d(y)$ then all objects in the training data outside $B_1^{\theta/S}(y)$ are nearest neighbors of y in a distinct class as any such object is at

$d_{\theta/S}$ distance of 1 from y and thus, in a sense, the whole of $T \setminus B_1^{\theta/S}(y)$ seems identical to y wrt distance $d_{\theta/S}$. In other words, the training data fails to provide any discriminatory information beyond $B_1^{\theta/S}(y)$ if a finite generator is used for granulating the universe. These observations are illustrated in Figs. 5 and 6.

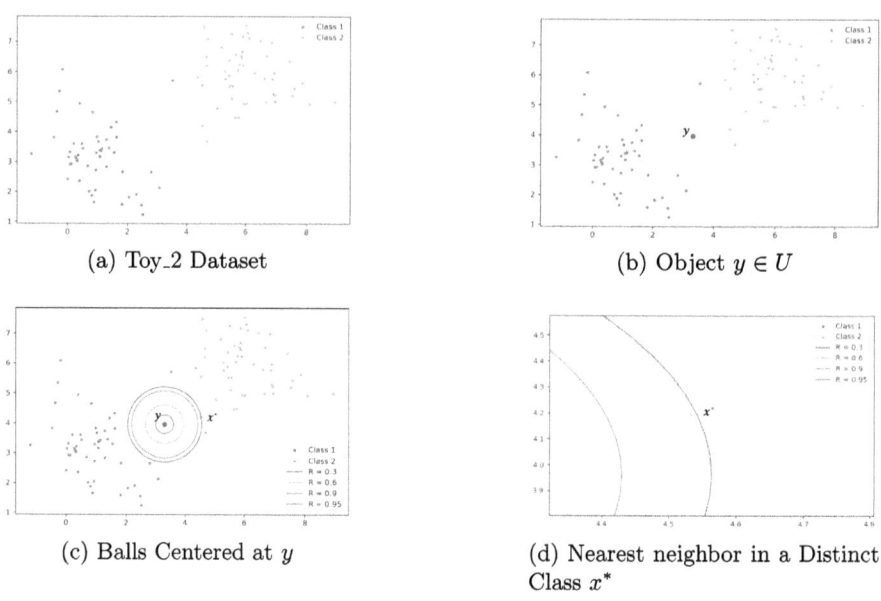

Fig. 5. Visualizing Computation of $(R_d \downarrow_{\theta/S} \hat{D}_1)|_T(y)$

Figures 5 (a) and 6 (a) show Toy_2 dataset and Toy_3 dataset respectively. Both datasets consist of two classes, D_1 and D_2. An object y is chosen from the universe, and computation of $(R_d \downarrow_{\theta/S} \hat{D}_1)|_T(y)$ as described in the previous paragraph is indicated. The generator $f_{\cos}$, connectives set $\langle f_{\cos} \rangle$ and $d = L_2$ is used. As shown in Fig. 5 (c), as radius R of the ball $B_R^{\theta/S}(y)$ is increased from 0.9 to 0.95, an object x^* from a distinct class is trapped. This is the nearest neighbor of y in a distinct class. In fact, $d_\theta(x^*, y) = 0.94706 = d_S(x^*, y)$ and thus $(R_d \downarrow_{\theta/S} \hat{D}_1)|_T(y) = 0.94706$. On the other hand, for the Toy_3 dataset, no object from a distinct class is trapped as the radius R of the ball $B_R^{\theta/S}(y)$ is gradually increased to 1 as indicated in Fig. 6 (c). Thus, all objects in the training data from a distinct class that is outside $B_1^{\theta/S}(y)$, viz. all the orange objects or the objects with label 2 are nearest neighbors of y in a distinct class as all of them are at a $d_{\theta/S}$ distance of 1 from y. Hence, $(R_d \downarrow_{\theta/S} \hat{D}_1)|_T(y) = 1$. Above considerations lead to the following theorem.

Theorem 4. *Given training data $\mathcal{D}$, feature map ϕ, pseudo-metric d on the feature space, finite generator f and connectives set $\langle f \rangle$, following holds under the (I4) interpretation of $\mathcal{D}$.*

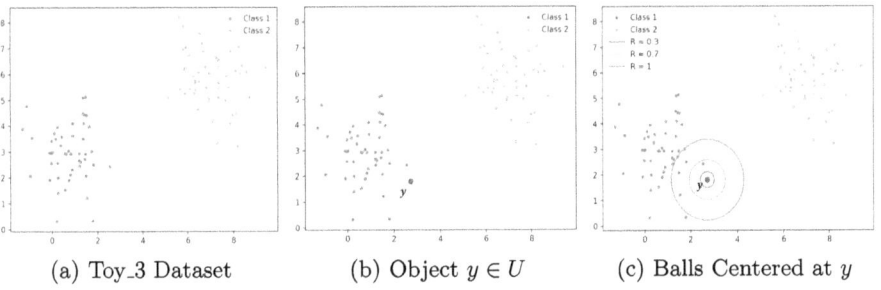

Fig. 6. Visualizing Computation of $(R_d \downarrow_{\theta/\mathcal{S}} \hat{D}_1)|_T(y)$

If $K \geq f(0)$, then for all label $j \in L$ and training objects $y \in D_j$, $(R_d \downarrow_{\theta/\mathcal{S}} \hat{D}_j)|_T(y) = 1$ and $(R_d \downarrow_{\theta/\mathcal{S}} \hat{D}_i)|_T(y) = 0$ for all labels $i \neq j$.
Here, K is the separation co-efficient of the training data $\mathcal{D}$ which is defined as $K = \min_{\substack{i,j \in L \\ i \neq j}} dist(D_i, D_j)$ where $dist(D_i, D_j) := \min\{d(\phi(x), \phi(y)) \mid x \in D_i, y \in D_j\}$ for all labels i and j.

Theorem 4 indicates that if the classes in the training data are sufficiently far apart, then $\theta/\mathcal{S}$-lower approximations of concepts restricted to the set of training objects, i.e., $\overline{(R_d \downarrow_{\theta/\mathcal{S}} \hat{D}_j)|_T}_T$ are essentially $\{D_j\}_{j \in L}$. For the Toy_3 dataset (Fig. 6 (a)), separation co-efficient $K > \frac{\pi}{2} = f_{\cos}(0)$ and thus $(R_d \downarrow_{\theta/\mathcal{S}} \hat{D}_1)|_T(y) = 1$ and $(R_d \downarrow_{\theta/\mathcal{S}} \hat{D}_2)|_T(y) = 0$ for all $y \in D_1$ while $(R_d \downarrow_{\theta/\mathcal{S}} \hat{D}_1)|_T(y) = 0$ and $(R_d \downarrow_{\theta/\mathcal{S}} \hat{D}_2)|_T(y) = 1$ for all $y \in D_2$.

5.2 Infinite Generators

Theorem 5. *Given training data $\mathcal{D}$, feature map ϕ, pseudo-metric d on the feature space, infinite generator f, involutive negator $\mathcal{N}$ and connectives set $\langle f, \mathcal{N} \rangle$, following holds under the (I4) interpretation of $\mathcal{D}$.*

(IF1) For all $x, y \in U$,

(a) $d_\theta(x, y) = \begin{cases} 1 & \text{if } d(\phi(x), \phi(y)) = \infty \\ 0 & \text{if } d(\phi(x), \phi(y)) < \infty \end{cases}$

(b) $d_\mathcal{S}(x, y) = \mathcal{N}\left(f^{-1}\left(d(\phi(x), \phi(y))\right)\right)$

(IF2) For all $y \in U$,

(a) $B_R^\theta(y) = \begin{cases} T & \text{if } R > 1 \\ T \setminus \{z \in T \mid d(\phi(z), \phi(y)) = \infty\} & \text{if } R \leq 1 \end{cases}$

(b) $B_R^\mathcal{S}(y) = \begin{cases} T & \text{if } R > 1 \\ T \setminus \{z \in T \mid d(\phi(z), \phi(y)) = \infty\} & \text{if } R = 1 \\ B_{f(\mathcal{N}(R))}^d & \text{if } R < 1 \end{cases}$

($IF1$) indicates that the d_θ distance is effectively useless when an infinite generator is used, as it can only differentiate between objects that are infinitely far apart and objects that are not infinitely far apart. So, no meaningful analysis of the distance d_θ is possible in the infinite generator case. ($IF2$) indicates that the geometry of $\mathcal{S}$-balls is same as the geometry of d-balls.

Similar to the case of finite generators, computation of $(R_d \downarrow_\mathcal{S} \hat{D}_j)|_T(y)$ where $y \in U$ can be geometrically visualized as starting with a $d_\mathcal{S}$ ball of small radius centered at y and then gradually increasing the radius of the ball until an object x^* from a distinct class in the training data is trapped. The object x^* is nearest neighbor of y in a distinct class. Due to the continuity and strict monotonicity of f and $\mathcal{N}$ we can get arbitrarily far in the universe U as $R \to 1^-$ and therefore x^* lies in $B_R^\mathcal{S}(y)$ for $R < 1$. Hence, $(R_d \downarrow_\mathcal{S} \hat{D}_j)|_T(y) = d_\mathcal{S}(x^*, y) < 1$. This is not true only in the highly unlikely case of all objects of distinct class(es) being infinitely far apart from y wrt d and thus at a $d_\mathcal{S}$ distance of 1 from y. In this case, all objects in distinct class(es) are nearest neighbors of y in a distinct class, and $(R_d \downarrow_\mathcal{S} \hat{D}_j)|_T(y) = 1$.

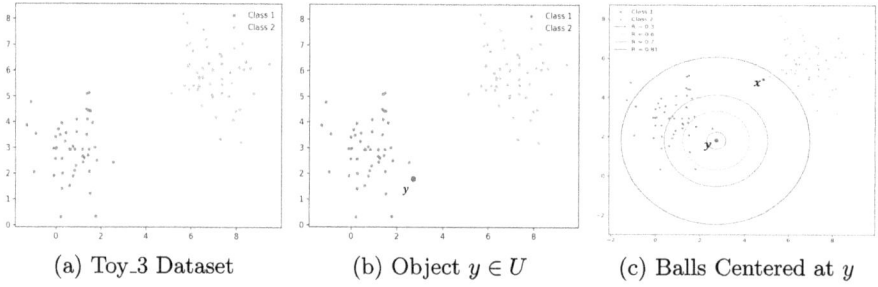

Fig. 7. Visualizing Computation of $(R_d \downarrow_{\theta/\mathcal{S}} \hat{D}_1)|_T(y)$

Figure 7 shows the Toy_3 dataset discussed earlier. An object y is chosen from the universe and computation of $(R_d \downarrow_{\theta/\mathcal{S}} \hat{D}_1)|_T(y)$ is indicated. The generator f_{Ham}, connectives set $\langle f_{\text{Ham}}, \mathcal{N}_s \rangle$ and $d = L_2$ is used. As shown in the figure, as the radius R of the ball $B_R^\mathcal{S}(y)$ is increased from 0.7 to 0.81, an object x^* from a distinct class is trapped. This is the nearest neighbor of y in a distinct class. In fact, $d_\mathcal{S}(x^*, y) = 0.8030$ and thus $(R_d \downarrow_\mathcal{S} \hat{D}_1)|_T(y) = 0.8030$.

6 Naive Classifiers

The approximation of the concepts $\hat{D}_j$'s can be used in different ways to predict the label of unseen objects, or in other words, to build a classifier $\lambda : U \to L$. One obvious albeit naive way to predict the label of an unseen object y (i.e., $y \in U \backslash T$) is to assign y that label such that y necessarily belongs to the associated concept to the maximal degree. Leveraging the fact that lower approximations capture the notion of "necessity", we formally define for any given $y \in U$,

$$\lambda_{\theta/\mathcal{S}}(y) = \arg\max_{j \in L}(R_d \downarrow_{\theta/\mathcal{S}} \hat{D}_j)|_T(y) \qquad (7)$$

In case arg max produces a non-singleton set, i.e., y belongs to more than one concept with the maximal degree, $\lambda(y)$ is arbitrarily chosen from the set of tied labels. We call λ_θ and $\lambda_\mathcal{S}$ the θ-naive and $\mathcal{S}$-naive classifier respectively. Under the (I4) interpretation of the training data, (7) assumes the following form.

$$\lambda_\theta(y) = \arg\max_{j \in L}\left(\min_{x \notin D_j} \theta\left(R_{d,f}(x,y), 0\right)\right) \; ; \; \lambda_\mathcal{S}(y) = \arg\max_{j \in L}\left(\min_{x \notin D_j} \mathcal{N}(R_{d,f}(x,y))\right)$$

The naive classifier (7) divides the universe U into $|L|$ regions, one for each label, with each such region consisting of those objects of the universe that are assigned a particular label by the naive classifier. Denote these regions by $\left\{\bar{D}_j^{\theta/\mathcal{S}}\right\}_{j \in L}$ and call them decision regions. Formally, for any $j \in L$,

$$\bar{D}_j^{\theta/\mathcal{S}} = \{y \in U \mid \lambda_{\theta/\mathcal{S}}(y) = j\}$$

The boundary separating the decision regions is referred to as the decision boundary. For the toy dataset (Fig. 2 (a)), the decision regions $\bar{D}_1^\mathcal{S}$ and $\bar{D}_2^\mathcal{S}$ are shown in Fig. 8. The plot is obtained by computing the label of objects with feature vectors in the region $[-10, 10]^2$, evenly spaced by a step-size of 0.1^7. The brown-colored area is the RONI. Figure 8 shows the power of the naive classifier. Non-linear boundaries are learned directly in the original feature space using just the standard L_2 pseudo-metric and a bunch of logic operations.

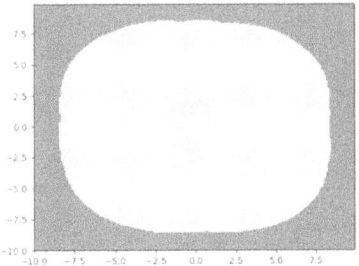

Fig. 8. Decision Regions $\bar{D}_1^\mathcal{S}$ (blue) and $\bar{D}_2^\mathcal{S}$ (orange) of Toy Dataset

6.1 Decision Regions and SVM

Figure 9 (a) shows another synthetic toy dataset (Toy_Linear). It consists of two classes D_1 and D_2 that are linearly separable. Infinite generator f_{Ham}, connectives set $\langle f_{\text{Ham}}, \mathcal{N}_\mathcal{S}\rangle$ and pseudo-metric $d = L_2$ is used. Figure 9 (b) shows the contours of the $\mathcal{S}$-lower approximations $(R_d \downarrow_\mathcal{S} \hat{D}_1)|_T$ and $(R_d \downarrow_\mathcal{S} \hat{D}_2)|_T$ over the

---

[7] The decision boundary can be more finely traced by computing the label of objects with feature vectors separated by a smaller step-size, say, 0.01.

region $[-7, 8] \times [0, 15]$ of the feature space $\mathbb{R}^2$. The decision regions induced by $\mathcal{S}$-naive classifier are indicated in Fig. 10 (a). The plot is obtained by computing the label of objects with feature vectors in the region $[-7, 8] \times [0, 15]$, evenly spaced by a step size of 0.01. The decision boundary appears roughly linear, which motivates the comparison with SVM. The SVM hyperplane is shown in Fig. 10 (b).

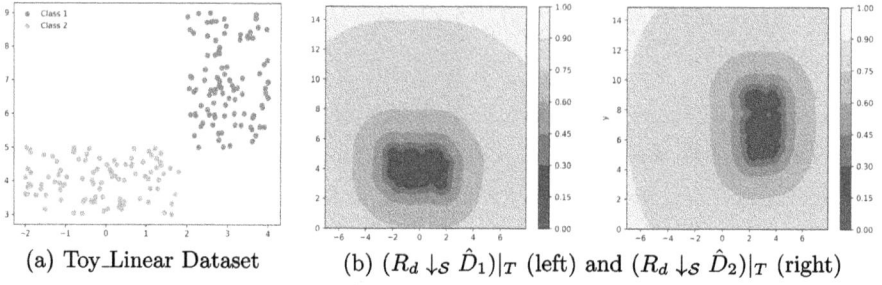

(a) Toy_Linear Dataset (b) $(R_d \downarrow_\mathcal{S} \hat{D}_1)|_T$ (left) and $(R_d \downarrow_\mathcal{S} \hat{D}_2)|_T$ (right)

Fig. 9. Toy_Linear Dataset and Contour Plots of Approximation Fuzzy Sets

The black dots correspond to the support vectors. Figure 10 (c) zooms in the region consisting of the support vectors. These support vectors effectively determine the SVM hyperplane. Figure 10 suggests that for a linearly separable dataset, the decision boundary obtained through $\mathcal{S}$-naive classifier seems to be optimal in a similar sense as SVM.

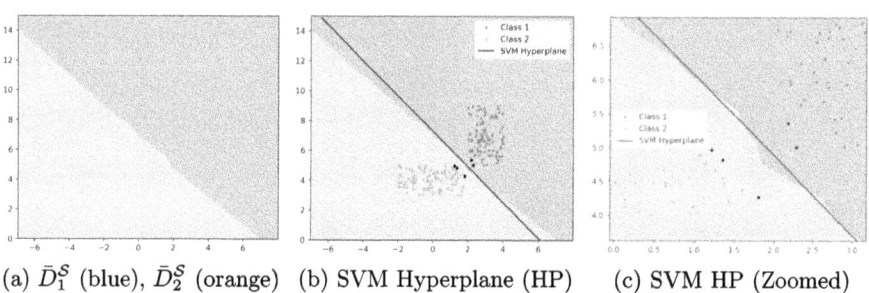

(a) $\bar{D}_1^\mathcal{S}$ (blue), $\bar{D}_2^\mathcal{S}$ (orange) (b) SVM Hyperplane (HP) (c) SVM HP (Zoomed)

Fig. 10. Decision Regions $\bar{D}_1^\mathcal{S}$ & $\bar{D}_2^\mathcal{S}$ and SVM Hyperplane

7 Experiments

In order to evaluate the performance of the naive classifiers, experiments were performed on 11 benchmark datasets. These datasets[8] are described in the

---

[8] All datasets except "Diabetes" are taken from [1] and "Diabetes" dataset, originally from the National Institute of Diabetes and Digestive and Kidney Diseases, is taken from UCI Machine Learning's Kaggle page.

Table 2. The performance of the naive classifiers is compared with four other widely used classifiers, viz., fuzzy-rough nearest neighbors (FRNN) [7], support vector machine (SVM), classification & regression tree (CART) and naive Bayes (NB). FRNN is implemented via fuzzy-rough-learn package [9] while SVM, CART, and NB are implemented via scikit-learn package [15].

Table 2. Datasets Description

Dataset	#Objects	#Features	#Labels
Toy (Fig. 2 (a))	315	2	2
Iris	150	4	3
Wine	178	13	3
Heart	270	13	2
Hepatitis	155	19	2
Glass	214	9	7
Cancer	569	30	2
Diabetes	768	8	2
Credit	690	14	2
Ecoli	336	7	8
Yeast	1484	8	10
Sonar	208	60	2

Tables 3 and 4 contain accuracy and F1-scores of the classifiers obtained via stratified 10-fold cross-validation.

Table 3. Performance of λ_θ and λ_S.

Dataset	λ_θ				λ_S			
	$\langle f_{\cos}\rangle$		$\langle f_\times, \mathcal{N}_s\rangle$		$\langle f_{\cos}\rangle$		$\langle f_\times, \mathcal{N}_s\rangle$	
	Acc	F1	Acc	F1	Acc	F1	Acc	F1
Toy	0.96	0.97	0.23	0.09	0.96	0.97	0.96	0.97
Iris	0.95	0.95	0.33	0.17	0.95	0.95	0.95	0.95
Wine	0.96	0.95	0.33	0.17	0.96	0.95	0.96	0.95
Heart	0.73	0.73	0.56	0.40	0.73	0.73	0.73	0.73
Hepatitis	0.81	0.84	0.21	0.07	0.81	0.84	0.81	0.85
Glass	0.70	0.72	0.33	0.16	0.70	0.72	0.70	0.72
Cancer	0.95	0.95	0.63	0.48	0.95	0.95	0.95	0.95
Diabetes	0.70	0.70	0.65	0.51	0.70	0.70	0.70	0.70
Credit	0.79	0.79	0.56	0.40	0.79	0.79	0.79	0.79
Ecoli	0.81	0.81	0.43	0.25	0.81	0.81	0.81	0.81
Yeast	0.52	0.54	0.31	0.15	0.52	0.54	0.52	0.54
Sonar	0.87	0.87	0.47	0.30	0.87	0.87	0.88	0.87

Furthermore, for FRNN, the number of nearest neighbors used is 5 except for Ecoli and Yeast datasets for which 1 and 4 nearest neighbors are used, respectively, as these are the sizes of the least populated class of respective datasets. These numbers are arrived at by evaluating FRNN with stratified 10-fold cross validation with t number of nearest neighbors where t varied from 1 to 10 or the size of the least populated class, whichever is the smallest. The results indicated that the FRNN performed best when the number of nearest neighbors used is 5. All the features were range normalized. L_2 pseudo-metric was used in all the experiments. The scores obtained when finite generator $f_{\cos}$ and connectives set $\langle f_{\cos} \rangle$ are used are indicated under $\langle f_{\cos} \rangle$ columns while the scores obtained when infinite generator $f_\times$ and the connectives set $\langle f_\times, \mathcal{N}_S \rangle$ are used are indicated under $\langle f_\times, \mathcal{N}_S \rangle$ columns.

Table 4. Performance of FRNN, SVM, CART and NB.

Dataset	FRNN		SVM		CART		NB	
	Acc	F1	Acc	F1	Acc	F1	Acc	F1
Toy	0.96	0.96	0.96	0.96	0.91	0.91	0.77	0.67
Iris	0.96	0.96	0.96	0.96	0.95	0.95	0.95	0.95
Wine	0.96	0.96	0.96	0.96	0.89	0.89	0.98	0.98
Heart	0.79	0.78	0.79	0.78	0.75	0.75	0.83	0.83
Hepatitis	0.84	0.85	0.84	0.85	0.77	0.78	0.63	0.66
Glass	0.67	0.65	0.67	0.65	0.67	0.70	0.44	0.57
Cancer	0.97	0.97	0.97	0.97	0.92	0.92	0.93	0.93
Diabetes	0.74	0.73	0.74	0.73	0.69	0.69	0.76	0.75
Credit	0.84	0.84	0.84	0.84	0.83	0.83	0.80	0.80
Ecoli	0.81	0.81	0.81	0.81	0.77	0.77	0.79	0.80
Yeast	0.55	0.56	0.55	0.56	0.52	0.55	0.14	0.16
Sonar	0.83	0.82	0.83	0.82	0.75	0.74	0.68	0.68

Note, θ-naive classifier and $\mathcal{S}$-naive classifier perform identically when finite generator $f_{\cos}$ is used. This is a direct consequence of the theorem (3) that states that d_θ and $d_\mathcal{S}$ distances become the same if the generator used is finite, thereby making both θ-naive classifier and $\mathcal{S}$-naive classifier essentially the same. Also, the θ-naive classifier performs very poorly when infinite generator $f_\times$ is used. This is because under the use of generator $f_\times$ and connectives set $\langle f_\times, \mathcal{N}_S \rangle$, $\theta(\alpha, 0) = 1$ if $\alpha = 0$ and $\theta(\alpha, 0) = 0$ for $\alpha > 0$, thus, $(R_{L_2} \downarrow_\theta \hat{D}_j)|_T(y) = \min_{x \notin D_j} \theta(R_{L_2,f}(x,y), 0) = 0$ for all $y \in U$ since for any object $y \in U$ and object x in training data, i.e., $x \in T$, $R_{L_2,f}(x,y) = f_\times^{(-1)}(L_2(\phi(x), \phi(y))) = e^{-L_2(\phi(x), \phi(y))} > 0$. This means the θ-naive classifier assigns labels arbitrarily in this case, thus the poor performance. Thus, for comparison with other classifiers, we consider $\lambda_{\theta/S}$ when finite generator $f_{\cos}$ is used and λ_S when infinite generator $f_\times$ is used.

Naive classifiers perform similarly to CART and NB on Iris while outperforming them on all the other datasets except on Wine and Heart. In comparison to FRNN and SVM, naive classifiers perform almost the same or better on Toy, Iris, Wine, Ecoli, and Sonar datasets. FRNN performs better on other datasets, which may be attributed to the fact that FRNN uses both lower and upper approximations while naive classifiers, naively enough, use only lower approximations. All in all, the proposed naive classifiers outperform CART and NB while performing comparably with FRNN and SVM, which is a fairly involved method.

8 Conclusion and Future Work

We proposed distance-based fuzzy-rough sets and applied them to the classification problem as understood from a fuzzy perspective of approximation of concepts. The geometry of the approximation process is explored in detail. Naive classifiers based on lower approximation of concepts are proposed and are evaluated on benchmark datasets. The ability of naive classifiers to unearth non-linear boundaries is illustrated.

In the future, we would like to investigate the conditions under which the decision boundary unearthed by naive classifiers is same as the SVM hyperplane. Methods to operationalize fuzzy interpretations of the training data and the performance of the naive classifiers when such interpretations are used will be studied and compared with other noise-tolerant fuzzy-rough models. The pseudo-metric requirement can be relaxed to allow for general distance functions to be used for granulation. Here, we will explore if metric learning techniques can come in handy.

Acknowledgments. Authors would like to thank Dr. Balasubramanian Jayaram for drawing our attention to De Baets & Mesiar's theorem.

References

1. Bache, K., Lichman, M.: UCI machine learning repository (2013). http://archive.ics.uci.edu/ml
2. Cornelis, C., Verbiest, N., Jensen, R.: Ordered weighted average based fuzzy rough sets. In: Yu, J., Greco, S., Lingras, P., Wang, G., Skowron, A. (eds.) RSKT 2010. LNCS (LNAI), vol. 6401, pp. 78–85. Springer, Heidelberg (2010). https://doi.org/10.1007/978-3-642-16248-0_16
3. Baets, D.B., Mesiar, R.: Pseudo-metrics and T-equivalences. J. Fuzzy Math. **5**(1997), 471–481 (1997)
4. D'eer, L., Verbiest, N., Cornelis, C., Godo, L.: A comprehensive study of implicator-conjunctor-based and noise-tolerant fuzzy rough sets: Definitions, properties and robustness analysis. Fuzzy Sets Syst. **275**, 1–38 (2015). https://doi.org/10.1016/j.fss.2014.11.018

5. Dubois, D., Prade, H.: Rough fuzzy sets and fuzzy rough sets*. Int. J. Gen Syst **17**(2–3), 191–209 (1990). https://doi.org/10.1080/03081079008935107
6. Hu, Q., Yu, D., Pedrycz, W., Chen, D.: Kernelized fuzzy rough sets and their applications. IEEE Trans. Knowl. Data Eng. **23**(11), 1649–1667 (2011). https://doi.org/10.1109/TKDE.2010.260
7. Jensen, R., Cornelis, C.: A new approach to fuzzy-rough nearest neighbour classification. In: Chan, C.-C., Grzymala-Busse, J.W., Ziarko, W.P. (eds.) RSCTC 2008. LNCS (LNAI), vol. 5306, pp. 310–319. Springer, Heidelberg (2008). https://doi.org/10.1007/978-3-540-88425-5_32
8. Klir, G.J., Yuan, B.: Fuzzy Sets and Fuzzy Logic: Theory and Applications. Prentice-Hall (2015)
9. Lenz, O., Cornelis, C., Peralta, D.: Fuzzy-rough-learn 0.2 : a Python library for fuzzy rough set algorithms and one-class classification. In: 2022 IEEE International Conference On Fuzzy Systems (FUZZ-IEEE), p. 8. IEEE (2022). https://doi.org/10.1109/fuzz-ieee55066.2022.9882778
10. Lorentey, K.: Capybara (2011). https://flic.kr/p/9wcG1g
11. Mi, J.S., Zhang, W.X.: An axiomatic characterization of a fuzzy generalization of rough sets. Inf. Sci. **160**(1), 235–249 (2004). https://doi.org/10.1016/j.ins.2003.08.017
12. Morsi, N.N., Yakout, M.: Axiomatics for fuzzy rough sets. Fuzzy Sets Syst. **100**(1), 327–342 (1998). https://doi.org/10.1016/S0165-0114(97)00104-8
13. Moser, B.: On the t-transitivity of kernels. Fuzzy Sets Syst. **157**(13), 1787–1796 (2006). https://doi.org/10.1016/j.fss.2006.01.007
14. Pawlak, Z.: Rough sets. Int. J. Comput. Inform. Sci. **11**, 341–356 (1982). https://doi.org/10.1007/BF01001956
15. Pedregosa, F., et al.: Scikit-learn: machine learning in Python. J. Mach. Learn. Res. **12**, 2825–2830 (2011)
16. Simmons, C.N.: Feral cat (2013). https://flic.kr/p/goCcNw
17. Trotman, K.: Wombat! (2006). https://flic.kr/p/khqre
18. Yao, Y.Y.: Combination of Rough and Fuzzy Sets Based on α-Level Sets, pp. 301–321. Springer, Boston (1997). https://doi.org/10.1007/978-1-4613-1461-5_15
19. Yeung, D., Chen, D., Tsang, E., Lee, J., Xizhao, W.: On the generalization of fuzzy rough sets. IEEE Trans. Fuzzy Syst. **13**(3), 343–361 (2005). https://doi.org/10.1109/TFUZZ.2004.841734
20. Zadeh, L.: Fuzzy sets. Inf. Control **8**(3), 338–353 (1965). https://doi.org/10.1016/S0019-9958(65)90241-X

Dealing with Missing Values Meaning Unknown in Probabilistic Approximations

Michinori Nakata[1(✉)], Hiroshi Sakai[2,3], and Takeshi Fujiwara[1]

[1] Faculty of Informatics, Tokyo University of Information Sciences, 4-1 Onaridai, Wakaba-ku, Chiba 265-8501, Japan
nakatam@ieee.org, fujiwara@rsch.tuis.ac.jp
[2] Department of Data Science, Shimonoseki City University, 2-1-1 Daigakucho, Shimonoseki City 751-8510, Japan
[3] Kyushu Institute of Technology, Kitakyushu, Japan

Abstract. We describe probabilistic approximations dealing with missing values. A relation used by Kryszkiewicz for indiscernibility of missing values is applied to deal with the missing values in previous work, but the relation considers only the indiscernibility of a missing value with another value. There are two possibilities for missing values. A missing value may be indiscernible from a value, or it may be discernible from that value. To deal with the two possibilities, we handle missing values in probabilistic approximations by using possible world semantics, as it is used by Lipski in incomplete databases. As a result, we obtain the lower and the upper bound of the actual probabilistic approximation. We can obtain these bounds without worrying about the number of possible tables.

Keywords: rough sets · probabilistic approximations · incomplete information · missing values · possible world semantics

1 Introduction

Rough sets, proposed by Pawlak [1] under complete information, are applied as an important tool to the fields of data mining and related topics. Approximations in the rough sets are represented by a pair of lower and upper approximations using the indiscernibility of objects. The indiscernibility is expressed by a binary relation showing a pair of objects are indiscernible if their properties, which are denoted by attributes, are equal.

Pawlak's rough sets are extended in various points. One of extensions is to introduce probabilistic concepts. The extension is proposed in several forms; for example, decision-theoretic rough sets, variable precision rough sets, Bayesian rough sets, generalized probabilistic approximations and so on [2–7]. In such extended rough sets, probabilistic approximations are proposed [7]. As the first attempt, this paper handles probabilistic approximations.

Pawlak's framework is constructed under data tables containing only complete information, but it is not always possible to obtain a data table that contains only complete information [8,9]. Therefore, Kryszkiewicz [10–12] proposed how to deal with missing values under Pawlak's rough sets. Kryszkiewicz embedded a missing value into the indiscernibility relation that is used by Pawlak under complete information without missing values. It is very significant to use her treatment of missing values because the indiscernibility of objects whose properties are missing can be handled by a binary relation. Most authors have applied Kryszkiewicz's method to data tables with missing values [12–22] and its revised versions have been used [14,21,23–25]. Missing values in probabilistic approximations also have been dealt with using Kryszkiewicz's method [7].

The indiscernibility relation of Kryszkiewicz as well as Pawlak is defined using value equality; concretely speaking, two objects are indiscernible when their properties are specified by the same value. If a value is missing, it is represented as a missing value using a symbol such as *. The missing value is not a normal value because it has two possibilities. One is the possibility of being equal to a value, and the other is the possibility of not being equal to that value. The extended relation by Kryszkiewicz deals with only the former possibility. This means that Kryszkiewicz's relation has the drawback that it neglects the latter possibility, and information loss occurs [26–28]. As a result, poor approximations are derived [21,26,28].

We check the previous work of probabilistic approximations and propose to deal with missing values using an approach based on possible world semantics in probabilistic approximations. This approach was developed by Lipski in incomplete databases [29]. There is no information loss in Lipski's approach because it treats all possibilities of missing values. He showed that there are limits to the information that can be obtained from incomplete data tables, that is, only the lower and the upper bound of the desired information can be extracted. This is true for rough sets that deal with data tables containing missing values, as is shown for fuzzy rough sets in [30–32]. Unfortunately, Lipski's work has not yet been reflected in research on probabilistic approximations. We first check the previous work of probabilistic approximations and we will derive the lower and the upper bound of probabilistic approximations by using a series of processes shown by Lipski.

The paper is organized as follows. In Sect. 2, the traditional approach of rough approximations and probabilistic approximations are briefly addressed under a complete data table. In Sect. 3, the previous work of probabilistic approximations is described under an incomplete data table. It is pointed out by an example that information loss occurs. In Sect. 4, we develop probabilistic approximations based on possible world semantics. We derive the lower and the upper bound of probabilistic approximations. In Sect. 5, we address conclusions.

2 Rough Approximations and Probabilistic Approximations

We suppose that the information of a set of objects is obtained as a two-dimensional data table with rows and columns. A column expresses a property called an attribute and a row the properties of an object. Pawlak dealt with a complete data table with only normal values that are single and definite values. Complete data table CT consists of $(U, AT, \cup_{a \in AT} V_a)$, where universe U is a non-empty finite set of objects, AT is a finite set of attributes, and $\cup_{a \in AT} V_a$ is a set of value domains of attributes from AT, where attributes are functions of the form $a : U \rightarrow V_a$, that is,

$$CT = \{o \in U \mid \forall a \in AT\ \exists v \in V_a\ a(o) = v\}, \quad (1)$$

where $a(o)$ is the value of attribute a for o. Binary relation I_A for indiscernibility of objects in U on subset $A \subseteq AT$ of attributes is defined as:

$$I_A = \{(o, o') \in U \times U \mid \forall a \in A\ a(o) = a(o')\}. \quad (2)$$

This relation is an equivalence relation, called indiscernibility relation. Equivalence class $E_A(o)$ containing object o is:

$$E_A(o) = \{o' \in U \mid (o, o') \in I_A\}.$$

Family FE_A of equivalence classes on A is:

$$FE_A = \cup_{o \in U} \{E_A(o)\}. \quad (3)$$

The objects are uniquely partitioned using the family. Lower approximation $\underline{apr}_A(T)$ and upper approximation $\overline{apr}_A(T)$ of target set T of objects by FE_A are defined as:

$$\underline{apr}_A(T) = \{o \in U \mid E_A(o) \subseteq T\}, \quad (4)$$
$$\overline{apr}_A(T) = \{o \in U \mid E_A(o) \cap T \neq \emptyset\}. \quad (5)$$

Example 1
Let CT be a complete data table obtained as follows:

$$CT$$

U	a_1	a_2	a_3
1	x	w	c
2	x	w	a
3	y	v	a
4	x	u	a
5	y	u	c
6	x	u	b
7	x	u	b
8	x	u	b

In information table CT, $U = \{o_1, o_2, o_3, o_4, o_5, o_6, o_7, o_8\}$. Domains V_{a_1}, V_{a_2}, and V_{a_3} of attributes a_1, a_2, and a_3 are $\{x, y\}$, $\{u, v, w\}$, and $\{a, b, c\}$, respectively. Indiscernibility relation $I_{\{a_1, a_2\}}$ on $\{a_1, a_2\}$ is:

$$I_{\{a_1,a_2\}} = \{(o_1, o_1), (o_1, o_2), (o_2, o_1), (o_2, o_2), (o_3, o_3), (o_4, o_4), (o_4, o_6), (o_4, o_7), (o_4, o_8),$$
$$(o_5, o_5), (o_6, o_4), (o_6, o_6), (o_6, o_7), (o_6, o_8), (o_7, o_4), (o_7, o_6), (o_7, o_7), (o_7, o_8),$$
$$(o_8, o_4), (o_8, o_6), (o_8, o_7), (o_8, o_8)\}.$$

Equivalence classes of individual objects on $\{a_1, a_2\}$ are:

$$E_{\{a_1,a_2\}}(o_1) = E_{\{a_1,a_2\}}(o_2) = \{o_1, o_2\},$$
$$E_{\{a_1,a_2\}}(o_3) = \{o_3\},$$
$$E_{\{a_1,a_2\}}(o_4) = E_{\{a_1,a_2\}}(o_6) = E_{\{a_1,a_2\}}(o_7) = E_{\{a_1,a_2\}}(o_8) = \{o_4, o_6, o_7, o_8\},$$
$$E_{\{a_1,a_2\}}(o_5) = \{o_5\}.$$

Family $FE_{\{a_1,a_2\}}$ of equivalence classes on $\{a_1, a_2\}$ is:

$$FE_{\{a_1,a_2\}} = \{\{o_1, o_2\}, \{o_3\}, \{o_4, o_6, o_7, o_8\}, \{o_5\}\}.$$

Let target set T be $\{o_2, o_3, o_4\}$ that is specified by constraint $a_3 = a$. Lower approximation $\underline{apr}(T)_{\{a_1,a_2\}}$ and upper approximation $\overline{apr}(T)_{\{a_1,a_2\}}$ by $FE_{\{a_1,a_2\}}$ are:

$$\underline{apr}_{\{a_1,a_2\}}(T) = \{o_3\},$$
$$\overline{apr}_{\{a_1,a_2\}}(T) = \{o_1, o_2, o_3, o_4, o_6, o_7, o_8\}.$$

Probabilistic approximation $Papr_A(T)^\alpha$ of T for A is defined as:

$$Papr_A(T)^\alpha = \{o \in U \mid Pr(T \mid E_A(o)) \geq \alpha\}, \quad (6)$$
$$Pr(T \mid E_A(o)) = |E_A(o) \cap T|/|E_A(o)|, \quad (7)$$

where $Pr(T \mid E_A(o))$ is the conditional probability of T under $E_A(o)$ and α is a threshold with $0 < \alpha \leq 1$. $Papr_A(T)^\alpha$ is equal to lower approximation $\underline{apr}_A(T)$ when α is equal to 1, and it becomes nearer to upper approximation $\overline{apr}_A(T)$ when α become smaller and is equal to $\overline{apr}_A(T)$ at a small value of α.

Example 2 (continuation of Example 1)
The respective probabilistic approximations for $\alpha = 1, 0.5$, and 0.25 are:

$$Papr_{\{a_1,a_2\}}(T)^1 = \{o_3\} = \underline{apr}_{\{a_1,a_2\}}(T),$$
$$Papr_{\{a_1,a_2\}}(T)^{0.5} = \{o_1, o_2, o_3\},$$
$$Papr_{\{a_1,a_2\}}(T)^{0.25} = \{o_1, o_2, o_3, o_4, o_6, o_7, o_8\} = \overline{apr}_{\{a_1,a_2\}}(T).$$

3 Probabilistic Approximations in Incomplete Data Tables

Probabilistic approximations were proposed using Kryszkiewicz's method [7]. Kryszkiewicz expresses binary relation KI_A of indiscernibility in data tables with missing values as follows:

$$KI_A = \{(o, o') \in U \times U \mid \forall a \in A \ a(o) = a(o') \vee a(o) = * \vee a(o') = *\}, \quad (8)$$

where $*$ is a missing value, meaning the value is present but unknown. This says that an object with missing values are indiscernible with an arbitrary object because a missing value may be equal to the value. However, a missing value is not a normal value with a deterministic value because there are two possibilities: it may be equal to a normal value, or it may not. Indiscernible class $KE_A(o)$ obtained from KI_A is defined as:

$$KE_A(o) = \{o' \in U \mid (o, o') \in KI_A\}. \quad (9)$$

Note that $KE_A(o)$ is not an equivalence class, so we call it an indiscernible class. Family FKE_A of indiscernible classes on A is derived from KI_A:

$$FKE_A = \cup_{o \in U} \{KE_A(o)\}. \quad (10)$$

Lower approximation $\underline{Kapr}_A(T)$ and upper approximation $\overline{Kapr}_A(T)$ of target set T of objects by FKE_A are:

$$\underline{Kapr}_A(T) = \{o \in U \mid KE_A(o) \subseteq T\}, \quad (11)$$
$$\overline{Kapr}_A(T) = \{o \in U \mid KE_A(o) \cap T \neq \emptyset\}. \quad (12)$$

Probabilistic approximation $PKapr_A(T)^\alpha$ of T for A is:

$$PKapr_A(T)^\alpha = \{o \in U \mid KPr(T \mid KE_A(o)) \geq \alpha\}, \quad (13)$$
$$KPr(T \mid KE_A(o)) = |KE_A(o) \cap T|/|KE_A(o)|. \quad (14)$$

Example 3
Let incomplete data table IT be obtained as follows:

	IT		
U	a_1	a_2	a_3
1	y	u	c
2	x	v	c
3	y	$*$	a
4	$*$	u	c
5	$*$	v	b
6	x	u	b
7	x	v	b

In table IT, universe U, and domains V_{a_1}, V_{a_2}, and V_{a_3} of attributes a_1, a_2, and a_3 are the same as CT in Example 1. By applying formula (8) to $\{a_1, a_2\}$ in table IT,

$$KI_{\{a_1,a_2\}} = \{(o_1, o_1), (o_1, o_3), (o_1, o_4), (o_2, o_2), (o_2, o_5), (o_2, o_7), (o_3, o_1),$$
$$(o_3, o_3), (o_3, o_4), (o_3, o_5), (o_4, o_1), (o_4, o_3), (o_4, o_4), (o_4, o_6), (o_5, o_2),$$
$$(o_5, o_3), (o_5, o_5), (o_5, o_7), (o_6, o_4), (o_6, o_6), (o_7, o_2), (o_7, o_5), (o_7, o_7)\}.$$

Class $KE_{\{a_1,a_2\}}(o_j)$ derived from $KI_{\{a_1,a_2\}}$ for o_j with $j = 1, \cdots, 7$ is:

$$KE_{\{a_1,a_2\}}(o_1) = \{o_1, o_3, o_4\},$$
$$KE_{\{a_1,a_2\}}(o_2) = \{o_2, o_5, o_7\},$$
$$KE_{\{a_1,a_2\}}(o_3) = \{o_1, o_3, o_4, o_5\},$$
$$KE_{\{a_1,a_2\}}(o_4) = \{o_1, o_3, o_4, o_6\},$$
$$KE_{\{a_1,a_2\}}(o_5) = \{o_2, o_3, o_5, o_7\},$$
$$KE_{\{a_1,a_2\}}(o_6) = \{o_4, o_6\},$$
$$KE_{\{a_1,a_2\}}(o_7) = \{o_2, o_5, o_7\}.$$

For family $FKE_{\{a_1,a_2\}}$,

$$FKE_{\{a_1,a_2\}} = \{\{o_1, o_3, o_4\}, \{o_2, o_5, o_7\}, \{o_1, o_3, o_4, o_5\}, \{o_1, o_3, o_4, o_6\}, \{o_2, o_3, o_5, o_7\}, \{o_4, o_6\}\}.$$

Let target set T be $\{o_1, o_2, o_4\}$ specified by constraint $a_3 = c$. The respective probabilistic approximations for $\alpha_{\{a_1,a_2\}} = 1, 0.66, 0.5, 0.33$, and 0.25 are:

$$PKapr_{\{a_1,a_2\}}(T)^1 = \emptyset = \underline{Kapr}_{\{a_1,a_2\}}(T),$$
$$PKapr_{\{a_1,a_2\}}(T)^{0.66} = \{o_1\},$$
$$PKapr_{\{a_1,a_2\}}(T)^{0.5} = \{o_1, o_3, o_4, o_6\},$$
$$PKapr_{\{a_1,a_2\}}(T)^{0.33} = \{o_1, o_2, o_3, o_4, o_6, o_7\},$$
$$PKapr_{\{a_1,a_2\}}(T)^{0.25} = \{o_1, o_2, o_3, o_4, o_5, o_6, o_7\} = \overline{Kapr}_{\{a_1,a_2\}}(T).$$

The probabilistic approximations in Example 3 are not correct. For example, o_1 has four possibilities: its value with respect to $\{a_1, a_2\}$ is either not equal to the value of any other object, equal to the value of o_3, equal to the value of o_4, or equal to the values of o_3 and o_4. In other words, the indiscernible class of o_1 is either $\{o_1\}$, $\{o_1, o_3\}$, $\{o_1, o_4\}$, or $\{o_1, o_3, o_4\}$. If the indiscernible class is $\{o_1\}$ or $\{o_1, o_4\}$, o_1 belongs to the probabilistic approximation with conditional probability 1. This possibility does not reflect in the probabilistic approximations of Example 3; namely, information loss occurs.

In the next section, to resolve the information loss, we develop an approach based on possible world semantics.

4 Probabilistic Approximations Based on Possible World Semantics

The footsteps that Lipski used [29] are applied to an incomplete information table with missing values as follows:

1. To derive the set of possible tables from an incomplete data table.
2. To apply the approach used in complete data tables to each possible table.
3. To aggregate the results obtained from the possible tables by using two operations: intersection and union ones.
4. To obtain lower and upper bounds of probabilistic approximations.

4.1 Possible Tables and Their Indiscernibility Relations

Possible table pt_A on set A of attributes is a table in which each missing value for every attribute $a \in A$ in incomplete data table IT is replaced with value $v \in V_a$.

$$pt_A = \{o \in U \mid (\forall a \in A \exists v \in V_a \; a_{pt}(o) = v) \wedge (\forall a \notin A \; a_{pt}(o) = a_{IT}(o))\}, \quad (15)$$

where $a_{pt}(o)$ and $a_{IT}(o)$ are the values of a for o in possible table pt and in incomplete data table IT, respectively. Note that every possible table is a complete data table. When missing values exist on set A of attributes in incomplete data table IT, set PT_A of possible tables on A is:

$$PT_A = \{pt_{A,1}, \cdots, pt_{A,n}\}, \quad (16)$$

where every possible table $pt_{A,i}$ has an equal possibility that it is actual and n is the number of possible tables.

Attribute $a \in A$ has a value in V_a in all possible tables on A. Indiscernibility relation $PI_{A,i}$ is obtained from possible table $pt_{A,i}$:

$$PI_{A,i} = \{(o, o') \in U \times U \mid \forall a \in A \; a(o)_i = a(o')_i\}, \quad (17)$$

where $a(o)_i$ is the value of attribute a for o in $pt_{A,i}$.

Example 4

We obtain twelve ($= 2 \times 2 \times 3$) possible tables $pt_{\{a_1,a_2\},1}, \cdots, pt_{\{a_1,a_2\},12}$ from IT on $\{a_1, a_2\}$ because missing value $*$ on attribute a_1 of objects o_4 and o_5 is replaced by one of domain elements x and y of attribute a_1 and missing value $*$ on attribute a_2 of object o_3 is replaced by one of domain elements u, v and w of attribute a_2.

$pt_{\{a_1,a_2\},1}$

U	a_1	a_2	a_3
1	y	u	c
2	x	v	c
3	y	u	a
4	y	u	c
5	y	v	b
6	x	u	b
7	x	v	b

$pt_{\{a_1,a_2\},2}$

U	a_1	a_2	a_3
1	y	u	c
2	x	v	c
3	y	v	a
4	y	u	c
5	y	v	b
6	x	u	b
7	x	v	b

$pt_{\{a_1,a_2\},3}$

U	a_1	a_2	a_3
1	y	u	c
2	x	v	c
3	y	w	a
4	y	u	c
5	y	v	b
6	x	u	b
7	x	v	b

$pt_{\{a_1,a_2\},4}$

U	a_1	a_2	a_3
1	y	u	c
2	x	v	c
3	y	u	a
4	y	u	c
5	x	v	b
6	x	u	b
7	x	v	b

$pt_{\{a_1,a_2\},5}$

U	a_1	a_2	a_3
1	y	u	c
2	x	v	c
3	y	v	a
4	y	u	c
5	x	v	b
6	x	u	b
7	x	v	b

$pt_{\{a_1,a_2\},6}$

U	a_1	a_2	a_3
1	y	u	c
2	x	v	c
3	y	w	a
4	y	u	c
5	x	v	b
6	x	u	b
7	x	v	b

$pt_{\{a_1,a_2\},7}$

U	a_1	a_2	a_3
1	y	u	c
2	x	v	c
3	y	u	a
4	x	u	c
5	y	v	b
6	x	u	b
7	x	v	b

$pt_{\{a_1,a_2\},8}$

U	a_1	a_2	a_3
1	y	u	c
2	x	v	c
3	y	v	a
4	x	u	c
5	y	v	b
6	x	u	b
7	x	v	b

$pt_{\{a_1,a_2\},9}$

U	a_1	a_2	a_3
1	y	u	c
2	x	v	c
3	y	w	a
4	x	u	c
5	y	v	b
6	x	u	b
7	x	v	b

$pt_{\{a_1,a_2\},10}$

U	a_1	a_2	a_3
1	y	u	c
2	x	v	c
3	y	u	a
4	x	u	c
5	x	v	b
6	x	u	b
7	x	v	b

$pt_{\{a_1,a_2\},11}$

U	a_1	a_2	a_3
1	y	u	c
2	x	v	c
3	y	v	a
4	x	u	c
5	x	v	b
6	x	u	b
7	x	v	b

$pt_{\{a_1,a_2\},12}$

U	a_1	a_2	a_3
1	y	u	c
2	x	v	c
3	y	w	a
4	x	u	c
5	x	v	b
6	x	u	b
7	x	v	b

By applying formula (17) to each possible table, indiscernibility relation $PI_{\{a_1,a_2\},i}$ on $\{a_1,a_2\}$ for $i = 1, \cdots, 12$ is:

$PI_{\{a_1,a_2\},1} = \{(o_1,o_1),(o_2,o_2),(o_3,o_3),(o_4,o_4),(o_5,o_5),(o_6,o_6),(o_7,o_7),(o_1,o_3),$
$(o_3,o_1),(o_1,o_4),(o_4,o_1),(o_2,o_7),(o_7,o_2),(o_3,o_4),(o_4,o_3)\},$

$PI_{\{a_1,a_2\},2} = \{(o_1,o_1),(o_2,o_2),(o_3,o_3),(o_4,o_4),(o_5,o_5),(o_6,o_6),(o_7,o_7),(o_1,o_4),$
$(o_4,o_1),(o_2,o_7),(o_7,o_2),(o_3,o_5),(o_5,o_3)\},$

$PI_{\{a_1,a_2\},3} = \{(o_1,o_1),(o_2,o_2),(o_3,o_3),(o_4,o_4),(o_5,o_5),(o_6,o_6),(o_7,o_7),(o_1,o_4),$
$(o_4,o_1),(o_2,o_7),(o_7,o_2)\},$

$PI_{\{a_1,a_2\},4} = \{(o_1,o_1),(o_2,o_2),(o_3,o_3),(o_4,o_4),(o_5,o_5),(o_6,o_6),(o_7,o_7),(o_1,o_3),$
$(o_3,o_1),(o_1,o_4),(o_4,o_1),(o_2,o_7),(o_7,o_2),(o_3,o_4),(o_4,o_3),(o_2,o_5),$
$(o_5,o_2),(o_5,o_7),(o_7,o_5)\},$

$$PI_{\{a_1,a_2\},5} = \{(o_1,o_1),(o_2,o_2),(o_3,o_3),(o_4,o_4),(o_5,o_5),(o_6,o_6),(o_7,o_7),(o_1,o_4),$$
$$(o_4,o_1),(o_2,o_5),(o_5,o_2),(o_5,o_7),(o_7,o_5),(o_2,o_7),(o_7,o_2)\},$$
$$PI_{\{a_1,a_2\},6} = \{(o_1,o_1),(o_2,o_2),(o_3,o_3),(o_4,o_4),(o_5,o_5),(o_6,o_6),(o_7,o_7),(o_1,o_4),$$
$$(o_4,o_1),(o_2,o_5),(o_5,o_2),(o_5,o_7),(o_7,o_5),(o_2,o_7),(o_7,o_2)\},$$
$$PI_{\{a_1,a_2\},7} = \{(o_1,o_1),(o_2,o_2),(o_3,o_3),(o_4,o_4),(o_5,o_5),(o_6,o_6),(o_7,o_7),(o_1,o_3),$$
$$(o_3,o_1),(o_2,o_7),(o_7,o_2),(o_4,o_6),(o_6,o_4)\},$$
$$PI_{\{a_1,a_2\},8} = \{(o_1,o_1),(o_2,o_2),(o_3,o_3),(o_4,o_4),(o_5,o_5),(o_6,o_6),(o_7,o_7),(o_2,o_7),$$
$$(o_7,o_2),(o_3,o_5),(o_5,o_3),(o_4,o_6),(o_6,o_4)\},$$
$$PI_{\{a_1,a_2\},9} = \{(o_1,o_1),(o_2,o_2),(o_3,o_3),(o_4,o_4),(o_5,o_5),(o_6,o_6),(o_7,o_7),(o_2,o_7),$$
$$(o_7,o_2),(o_4,o_6),(o_6,o_4)\},$$
$$PI_{\{a_1,a_2\},10} = \{(o_1,o_1),(o_2,o_2),(o_3,o_3),(o_4,o_4),(o_5,o_5),(o_6,o_6),(o_7,o_7),(o_1,o_3),$$
$$(o_3,o_1),(o_2,o_5),(o_5,o_2),(o_2,o_7),(o_7,o_2),(o_4,o_6),(o_6,o_4),(o_5,o_7),$$
$$(o_7,o_5)\},$$
$$PI_{\{a_1,a_2\},11} = \{(o_1,o_1),(o_2,o_2),(o_3,o_3),(o_4,o_4),(o_5,o_5),(o_6,o_6),(o_7,o_7),(o_2,o_5),$$
$$(o_5,o_2),(o_5,o_7),(o_7,o_5),(o_2,o_7),(o_7,o_2),(o_4,o_6),(o_6,o_4)\},$$
$$PI_{\{a_1,a_2\},12} = \{(o_1,o_1),(o_2,o_2),(o_3,o_3),(o_4,o_4),(o_5,o_5),(o_6,o_6),(o_7,o_7),(o_2,o_5),$$
$$(o_5,o_2),(o_5,o_7),(o_7,o_5),(o_2,o_7),(o_7,o_2),(o_4,o_6),(o_6,o_4)\}.$$

4.2 Equivalence Classes in Possible Tables

Equivalence class $PE_A(o)_i$ containing object o in $pt_{A,i}$ is:

$$PE_A(o)_i = \{o' \in U \mid (o,o') \in PI_{A,i}\}. \tag{18}$$

Family $FPE_A(o)$ of equivalence classes containing o consists of those containing o in the possible tables.

$$FPE_A(o) = \cup_i \{PE_A(o)_i\}. \tag{19}$$

The family has a lattice structure with the minimum and the maximum element [33]. Family $FPE_{A,i}$ of equivalence classes in possible table $pt_{A,i}$ is obtained from indiscernibility relation $PI_{A,i}$.

$$FPE_{A,i} = \cup_{o \in U} \{PE_A(o)_i\}. \tag{20}$$

4.3 Aggregation of Indiscernibility Relations in Possible Tables

We have two aggregations of indiscernibility relations: the minimum and the maximum indiscernibility relation. Minimum indiscernibility relation PI_A^{min} is defined as the intersection of $PI_{A,i}$:

$$PI_A^{min} = \cap_i PI_{A,i}. \tag{21}$$

Maximum indiscernibility relation PI_A^{max} is defined as the union of $PI_{A,i}$:

$$PI_A^{max} = \cup_i PI_{A,i}, \qquad (22)$$

where PI_A^{min} is an equivalence relation, but PI_A^{max} is a tolerance relation. Family$\{PI_A^{min}, PI_{A,1}, \cdots, PI_{A,n}, PI_A^{max}\}$ has a lattice structure with minimum element PI_A^{min} and maximum element PI_A^{max} [33]. Furthermore, PI_A^{min} is also expressed with the following formula [28]:

$$PI_A^{min} = \{(o,o') \in U \times U \mid (o = o') \vee$$
$$(\forall a \in A \; a(o) = a(o') \wedge a(o) \neq * \wedge a(o') \neq *)\}, (23)$$

whereas PI_A^{max} is equal to KI_A of formula (8) [28]. Therefore, PI_A^{min} and PI_A^{max} can be obtained without worrying about the number of possible tables.

We derive minimum indiscernible class $PE_A^{min}(o)$ of o and maximum indiscernible class $PE_A^{max}(o)$ from PI_A^{min} and PI_A^{max}:

$$PE_A^{min}(o) = \{o' \in U \mid (o',o) \in PI_A^{min}\}, \qquad (24)$$
$$PE_A^{max}(o) = \{o' \in U \mid (o',o) \in PI_A^{max}\}, \qquad (25)$$

where strictly speaking, $PE_A^{min}(o)$ is an equivalence class, but $PE_A^{max}(o)$ is a tolerance class.

Proposition 1

$$PE_A^{min}(o) = \cap_i PE_A(o)_i,$$
$$PE_A^{max}(o) = \cup_i PE_A(o)_i.$$

Proposition 2
For every $(o,o') \in PE_A^{min}(o)$, there exists $pt_{A,i} \in PT_A$ such that $(o,o') \in PI_{A,i}$, and for every $(o,o') \in PE_A^{max}(o)$, there exists $pt_{A,k} \in PT_A$ such that $(o,o') \in PI_{A,k}$.

The actual indiscernible class $PE_A(o)$ exists between $PE_A^{min}(o)$ and $PE_A^{max}(o)$; namely, $PE_A^{min}(o) \subseteq PE_A(o) \subseteq PE_A^{max}(o)$. Note that $PE_A^{min}(o)$ and $PE_A^{max}(o)$ are lower and upper bounds of $PE_A(o)$, but there may not always be possible tables from which they can be derived.

Families FPE_A^{min} of minimum indiscernible classes and FPE_A^{max} of maximum indiscernible classes are:

$$FPE_A^{min} = \cup_{o \in U} \{PE_A^{min}(o)\}, \qquad (26)$$
$$FPE_A^{max} = \cup_{o \in U} \{PE_A^{max}(o)\}. \qquad (27)$$

Example 5

By applying formulae (21) and (22) to indiscernibility relations of Example 4, minimum indiscernibility relation $PI^{min}_{\{a_1,a_2\}}$ and maximum indiscernibility relations $PI^{max}_{\{a_1,a_2\}}$ are:

$$PI^{min}_{\{a_1,a_2\}} = \{(o_1,o_1),(o_2,o_2),(o_3,o_3),(o_4,o_4),(o_5,o_5),(o_6,o_6),(o_2,o_7),$$
$$(o_7,o_2),(o_7,o_7)\},$$
$$PI^{max}_{\{a_1,a_2\}} = \{(o_1,o_1),(o_1,o_3),(o_1,o_4),(o_2,o_2),(o_2,o_5),(o_2,o_7),(o_3,o_1),$$
$$(o_3,o_3),(o_3,o_4),(o_3,o_5),(o_4,o_1),(o_4,o_3),(o_4,o_4),(o_4,o_6),(o_5,o_2),$$
$$(o_5,o_3),(o_5,o_5),(o_5,o_7),(o_6,o_4),(o_6,o_6),(o_7,o_2),(o_7,o_5),(o_7,o_7)\}.$$

By applying formulae (24) and (25) to $PI^{min}_{\{a_1,a_2\}}$ and $PI^{max}_{\{a_1,a_2\}}$, minimum indiscernible class $PE^{min}_{\{a_1,a_2\}}(o_j)$ and maximum indiscernible class $PE^{max}_{\{a_1,a_2\}}(o_j)$ containing object o_j on $\{a_1,a_2\}$ for $j=1,\cdots,7$ are:

$$PE^{min}_{\{a_1,a_2\}}(o_1) = \{o_1\}, \ PE^{max}_{\{a_1,a_2\}}(o_1) = \{o_1,o_3,o_4\},$$
$$PE^{min}_{\{a_1,a_2\}}(o_2) = \{o_2,o_7\}, \ PE^{max}_{\{a_1,a_2\}}(o_2) = \{o_2,o_5,o_7\},$$
$$PE^{min}_{\{a_1,a_2\}}(o_3) = \{o_3\}, \ PE^{max}_{\{a_1,a_2\}}(o_3) = \{o_1,o_3,o_4,o_5\},$$
$$PE^{min}_{\{a_1,a_2\}}(o_4) = \{o_4\}, \ PE^{max}_{\{a_1,a_2\}}(o_4) = \{o_1,o_3,o_4,o_6\},$$
$$PE^{min}_{\{a_1,a_2\}}(o_5) = \{o_5\}, \ PE^{max}_{\{a_1,a_2\}}(o_5) = \{o_2,o_3,o_5,o_7\},$$
$$PE^{min}_{\{a_1,a_2\}}(o_6) = \{o_6\}, \ PE^{max}_{\{a_1,a_2\}}(o_6) = \{o_4,o_6\},$$
$$PE^{min}_{\{a_1,a_2\}}(o_7) = \{o_2,o_7\}, \ PE^{max}_{\{a_1,a_2\}}(o_7) = \{o_2,o_5,o_7\},$$

For example, the actual indiscernible class $PE_{\{a_1,a_2\}}(o_1)$ exists between $\{o_1\}$ and $\{o_1,o_3,o_4\}$; namely, it is either $\{o_1\}$, $\{o_1,o_3\}$, $\{o_1,o_4\}$, or $\{o_1,o_3,o_4\}$. There exist possible tables from which these four indiscernible classes are obtained. On the other hand, the actual indiscernible class $PE_{\{a_1,a_2\}}(o_4)$ exists between $PE^{min}_{\{a_1,a_2\}}(o_4)(=\{o_4\})$ and $PE^{max}_{\{a_1,a_2\}}(o_4)(=\{o_1,o_3,o_4,o_6\})$, but there exist no possible tables from which $PE^{min}_{\{a_1,a_2\}}(o_4)$ and $PE^{max}_{\{a_1,a_2\}}(o_4)$ are obtained, which are lower and upper bounds. By using formulae (26) and (27),

$$FPE^{min}_{\{a_1,a_2\}} = \{\{o_1\},\{o_2,o_7\},\{o_3\},\{o_4\},\{o_5\},\{o_6\}\}.$$
$$FPE^{max}_{\{a_1,a_2\}} = \{\{o_1,o_3,o_4\},\{o_1,o_3,o_4,o_5\},\{o_1,o_3,o_4,o_6\},\{o_2,o_5,o_7\},\{o_2,o_3,o_5,o_7\},$$
$$\{o_4,o_6\}\}.$$

4.4 Lower and Upper Bounds of Probabilistic Approximations

An object has more than one indiscernible class, as is shown in the previous subsection. By using an indiscernible class of an object, its conditional probability is derived for the probabilistic approximation of a target set of objects. The conditional probability depends on the indiscernible class. An object may have more than one conditional probability in a probabilistic approximation. We cannot know which conditional probability is actual without additional information. We obtain the range of conditional probabilities that an object has.

Using the minimum and the maximum indiscernible classes, we obtain lower bound $Pr^{min}(T \mid PE_A(o))$ and upper bound $Pr^{max}(T \mid PE_A(o))$ of conditional probability of object o that belongs to a probabilistic approximation. Let X and XX be the minimum and the maximum value without considering part $(PE_A^{max}(o) \backslash PE_A^{min}(o)) \cap T$ and Y and YY be the minimum and the maximum value when considering part $(PE_A^{max}(o) \backslash PE_A^{min}(o)) \cap T$.

$$Pr^{min}(T \mid PE_A(o)) = \min(X, Y), \qquad (28)$$

$$X = \frac{|PE_A^{min}(o) \cap T|}{|PE_A^{min}(o) \cup (PE_A^{max}(o) \backslash PE_A^{min}(o) \backslash \cup Z_A(o))|}, \qquad (29)$$

$$Y = \min_{k,l} \frac{|(PE_A^{min}(o) \cap T) \cup (ZE_A^{k,l}(o) \cap T))|}{|PE_A^{min}(o) \cup ZE_A^{k,l}(o) \cup (PE_A^{max}(o) \backslash PE_A^{min}(o) \backslash T)|}, \qquad (30)$$

$$Pr^{max}(T \mid PE_A(o)) = \max(XX, YY), \qquad (31)$$

$$XX = \frac{|PE_A^{min}(o) \cap T|}{|PE_A^{min}(o)|}, \qquad (32)$$

$$YY = \max_{k,l} \frac{|(PE_A^{min}(o) \cap T) \cup (ZE_A^{k,l}(o) \cap T)|}{|PE_A^{min}(o) \cup ZE_A^{k,l}(o)|}, \qquad (33)$$

where family $Z_A(o)$ of indiscernible classes except $PE_A^{min}(o)$ have common elements with $PE_A^{max}(o) \backslash PE_A^{min}(o)$ and T is:

$$Z_A(o) = \cup_{o' \neq o} \{PE_A^{min}(o') \mid (PE_A^{min}(o') \cap T \neq \emptyset) \wedge \\ (PE_A^{min}(o') \cap (PE_A^{max}(o) \backslash PE_A^{min}(o)) \neq \emptyset)\}, \qquad (34)$$

Set $ZE_A^{k,l}(o)$ of elements in the case that k elements are selected from $Z_A(o)$ is:

$$ZE_A^{k,l}(o) = \cup_{i=1,k} \; e_i^l \in Z_A(o) \qquad (35)$$

where $1 \leq k \leq n_A(o)$, $n_A(o)$ is the number of elements in $Z_A(o)$, and the total number l of ways to choose k elements is $_{n_A(o)}C_k$. Note that $Pr^{min}(T \mid PE_A(o))$ and $Pr^{max}(T \mid PE_A(o))$ are derived by using both $PE_A^{min}(o)$ and $PE_A^{max}(o)$ and they can be calculated without worrying about the number of possible tables.

We have two probabilistic approximations. For probabilistic approximations $Papr_A^{min}(T)^\alpha$ and $Papr_A^{max}(T)^\alpha$ of T for A,

$$Papr_A^{min}(T)^\alpha = \{o \in U \mid Pr_{min}(T \mid PE_A(o)) \geq \alpha\}, \qquad (36)$$
$$Papr_A^{max}(T)^\alpha = \{o \in U \mid Pr_{max}(T \mid PE_A(o)) \geq \alpha\}. \qquad (37)$$

$Papr_A^{min}(T)^\alpha$ and $Papr_A^{max}(T)^\alpha$ is the lower and the upper bound of the actual probabilistic approximation, respectively.

Example 6

Applying formulae (28)–(35) to $PE_{\{a_1,a_2\}}^{min}(o_j)$ and $PE_{\{a_1,a_2\}}^{max}(o_j)$ of Example 5 under $T = \{o_1, o_2, o_4\}$, we derive $Pr^{min}(T \mid PE_{\{a_1,a_2\}}(o_j))$ and $Pr^{max}(T \mid PE_{\{a_1,a_2\}}(o_j))$ for o_j with $j = 1, \cdots, 7$. For example, by using $PE_{\{a_1,a_2\}}^{min}(o_1) = \{o_1\}$ and $PE_{\{a_1,a_2\}}^{max}(o_1) = \{o_1, o_3, o_4\}$ of Example 5,

$$X = \frac{|\{o_1\} \cap \{o_1, o_2, o_4\}|}{|\{o_1\} \cup (\{o_1, o_3, o_4\} \setminus \{o_1\} \setminus \{o_4\}|} = \frac{|\{o_1\}|}{|\{o_1, o_3\}|} = \frac{1}{2},$$

$$Y = \frac{|(\{o_1\} \cap \{o_1, o_2, o_4\}) \cup (\{o_4\} \cap \{o_1, o_2, o_4\})|}{|\{o_1\} \cup \{o_4\} \cup (\{o_1, o_3, o_4\} \setminus \{o_1\} \setminus \{o_1, o_2, o_4\})|} = \frac{|\{o_1, o_4\}|}{|\{o_1, o_3, o_4\}|} = \frac{2}{3},$$

$$XX = \frac{|\{o_1\} \cap \{o_1, o_2, o_4\}|}{|\{o_1\}|} = \frac{|\{o_1\}|}{|\{o_1\}|} = 1,$$

$$YY = \frac{|(\{o_1\} \cap \{o_1, o_2, o_4\}) \cup (\{o_4\} \cap \{o_1, o_2, o_4\})|}{|\{o_1\} \cup \{o_4\}|} = \frac{|\{o_1, o_4\}|}{|\{o_1, o_4\}|} = 1,$$

where $Z_{\{a_1,a_2\}}(o_1) = \{o_4\}$, $k = 1$, $l = 1$, and $ZE_{\{a_1,a_2\}}^{1,1}(o_1) = \{o_4\}$ by using formulae (34) and (35). Similarly, calculating for the other objects, we obtain as follows:

$$Pr^{min}(T \mid PE_{\{a_1,a_2\}}(o_1)) = 0.5, \ Pr^{max}(T \mid PE_{\{a_1,a_2\}}(o_1)) = 1,$$
$$Pr^{min}(T \mid PE_{\{a_1,a_2\}}(o_2)) = 0.33, \ Pr^{max}(T \mid PE_{\{a_1,a_2\}}(o_2)) = 0.5,$$
$$Pr^{min}(T \mid PE_{\{a_1,a_2\}}(o_3)) = 0, \ Pr^{max}(T \mid PE_{\{a_1,a_2\}}(o_3)) = 0.66,$$
$$Pr^{min}(T \mid PE_{\{a_1,a_2\}}(o_4)) = 0.33, \ Pr^{max}(T \mid PE_{\{a_1,a_2\}}(o_4)) = 1,$$
$$Pr^{min}(T \mid PE_{\{a_1,a_2\}}(o_5)) = 0, \ Pr^{max}(T \mid PE_{\{a_1,a_2\}}(o_5)) = 0.33,$$
$$Pr^{min}(T \mid PE_{\{a_1,a_2\}}(o_6)) = 0, \ Pr^{max}(T \mid PE_{\{a_1,a_2\}}(o_6)) = 0.5,$$
$$Pr^{min}(T \mid PE_{\{a_1,a_2\}}(o_7)) = 0.33, \ Pr^{max}(T \mid PE_{\{a_1,a_2\}}(o_7)) = 0.5.$$

Note that 0.33 of $Pr^{min}(T \mid PE_{\{a_1,a_2\}}(o_4))$ is the lower bound and the minimum value that is obtained from possible tables is 0.5. When $\alpha = 1, 0.66, 0.5, 0.3$, by applying formulae (36) and (37) to $Pr^{min}(T \mid PE_{\{a_1,a_2\}}(o_j))$ and $Pr^{max}(T \mid PE_{\{a_1,a_2\}}(o_j))$, the lower and the upper bound of probabilistic approximation are:

$$Papr_{\{a_1,a_2\}}^{min}(T)^1 = \emptyset,$$
$$Papr_{\{a_1,a_2\}}^{max}(T)^1 = \{o_1, o_4\},$$
$$Papr_{\{a_1,a_2\}}^{min}(T)^{0.66} = \emptyset,$$
$$Papr_{\{a_1,a_2\}}^{max}(T)^{0.66} = \{o_1, o_3, o_4\},$$

$$Papr^{min}_{\{a_1,a_2\}}(T)^{0.5} = \{o_1\},$$
$$Papr^{max}_{\{a_1,a_2\}}(T)^{0.5} = \{o_1, o_2, o_3, o_4, o_6, o_7\},$$
$$Papr^{min}_{\{a_1,a_2\}}(T)^{0.33} = \{o_1, o_2, o_4, o_7\},$$
$$Papr^{max}_{\{a_1,a_2\}}(T)^{0.33} = \{o_1, o_2, o_3, o_4, o_5, o_6, o_7\},$$

where $Papr^{min}_{\{a_1,a_2\}}(T)^0 = Papr^{max}_{\{a_1,a_2\}}(T)^0 = U$.

5 Conclusions

We have described probabilistic approximations in incomplete data tables where missing values appear. The previous approaches use the relation of object indiscernibility that Kryszkiewicz proposed. The relation is accompanied with information loss because it deals with only one of possibilities that a missing value has. To resolve the information loss, we have constructed probabilistic approximations under possible world semantics. Finally, we have obtained the lower and the upper bound of probabilistic approximations. We can calculate these bounds without worrying about the number of possible tables because only the lower and upper bounds of indiscernible classes are used.

The future work will be to examine how rule induction is performed from probabilistic rough sets with lower and upper bounds.

Acknowledgment. The authors wish to thank the anonymous reviewers for their valuable comments.

References

1. Pawlak, Z.: Rough Sets: Theoretical Aspects of Reasoning about Data. Kluwer Academic Publishers, Dordrecht (1991). https://doi.org/10.1007/978-94-011-3534-4
2. Pawlak, Z., Wong, S.K.M., Ziarko, W.: Rough sets: probabilistic versus deterministic approach. Int. J. Man Mach. Stud. **29**, 81–95 (1988)
3. Yao, Y.Y., Wong, S.K.M.: A decision theoretic framework for approximate concepts. Int. J. Man Mach. Stud. **37**, 793–809 (1992)
4. Ziarko, W.: Variable precision rough set model. J. Comput. Syst. Sci. **46**, 39–59 (1993)
5. Ślęzak, D., Ziarko, W.: The investigation of the Bayesian rough set model. Int. J. Approximate Reasoning **40**, 81–91 (2005)
6. Yao, Y.: Probabilistic rough set approximations. Inf. Sci. **49**, 255–271 (2008)
7. Grzymala-Busse, J. W.: Generalized probabilistic approximations. Trans. Rough Sets **XVI**, 1–16 (2013)
8. Parsons, S.: Current approaches to handling imperfect information in data and knowledge bases. IEEE Trans. Knowl. Data Eng. **8**, 353–372 (1996)
9. Parsons, S.: Addendum to "current approaches to handling imperfect information in data and knowledge bases". IEEE Trans. Knowl. Data Eng. **10**, 862 (1998)
10. Kryszkiewicz, M.: Rough set approach to incomplete information systems. Inf. Sci. **112**, 39–49 (1998)

11. Kryszkiewicz, M.: Properties of incomplete information systems in the framework of rough sets. Rough Sets Knowl. Discov. **1**, 422–450 (1998)
12. Kryszkiewicz, M.: Rules in incomplete information systems. Inf. Sci. **113**, 271–292 (1999)
13. Greco, S., Matarazzo, B., Słowinski, R.: Handling missing values in rough set analysis of multi-attribute and multi-criteria decision problems. In: Zhong, N., Skowron, A., Ohsuga, S. (eds.) RSFDGrC 1999. LNCS (LNAI), vol. 1711, pp. 146–157. Springer, Heidelberg (1999). https://doi.org/10.1007/978-3-540-48061-7_19
14. Grzymala-Busse, J.W.: Data with missing attribute values: generalization of indiscernibility relation and rule induction. Trans. Rough Sets **I**, 78–95 (2004)
15. Grzymala-Busse, J.W.: Characteristic relations for incomplete data: a generalization of the indiscernibility relation. Trans. Rough Sets **IV**, 58–68 (2005)
16. Latkowski, R.: Flexible indiscernibility relations for missing values. Fund. Inform. **67**, 131–147 (2005)
17. Leung, Y., Li, D.: Maximum consistent techniques for rule acquisition in incomplete information systems. Inf. Sci. **153**, 85–106 (2003)
18. Nakata, M., Sakai, H.: Rough sets handling missing values probabilistically interpreted. In: Ślezak, D., Wang, G., Szczuka, M., Düntsch, I., Yao, Y. (eds.) RSFDGrC 2005. LNCS (LNAI), vol. 3641, pp. 325–334. Springer, Heidelberg (2005). https://doi.org/10.1007/11548669_34
19. Nakata, M., Sakai, H.: Twofold rough approximations under incomplete information. Int. J. Gen. Syst. **42**, 546–571 (2013). https://doi.org/10.1080/17451000.2013.798898
20. Sakai, H.: Effective procedures for handling possible equivalence relation in nondeterministic information systems. Fund. Inform. **48**, 343–362 (2001)
21. Stefanowski, J., Tsoukiàs, A.: Incomplete information tables and rough classification. Comput. Intell. **17**, 545–566 (2001)
22. Sun, L., Wanga, L., Ding, W., Qian, Y., Xu, J.: Neighborhood multi-granulation rough sets-based attribute reduction using Lebesgue and entropy measures in incomplete neighborhood decision systems. Knowl.-Based Syst. **192**, 105373 (2020)
23. Wang, G.: Extension of rough set under incomplete information systems. In: 2002 IEEE World Congress on Computational Intelligence. IEEE International Conference on Fuzzy Systems. FUZZ-IEEE 2002, pp. 1098–1103 (2002). https://doi.org/10.1109/FUZZ.2002.1006657
24. Nguyen, D.V., Yamada, K., Unehara, M.: Extended tolerance relation to define a new rough set model in incomplete information systems. Adv. Fuzzy Syst. **2013**, 10 (2013). https://doi.org/10.1155/2013/372091
25. Rady, E.A., Abd El-Monsef, M.M.E., Adb El-Latif, W.A.: A modified rough sets approach to incomplete information systems. J. Appl. Math. Decisi. Sci. **2007**, 13 (2007)
26. Nakata, M., Sakai, H.: Applying rough sets to information tables containing missing values. In: Proceedings of 39th International Symposium on Multiple-Valued Logic, pp. 286–291. IEEE Press (2009). https://doi.org/10.1109/ISMVL.2009.1
27. Yang, T., Li, Q., Zhou, B.: Related family: a new method for attribute reduction of covering information systems. Inf. Sci. **228**, 175–191 (2013)
28. Nakata, M., Saito, N., Sakai, H., Fujiwara, T.: Kryszkiewicz's relation for indiscernibility of objects in data tables containing missing values. In: Campagner, A., Urs Lenz, O., Xia, S., Ślezak, D., Was, J., Yao, J. (eds) Proceedings of IJCRS 2023, Lecture Notes in Compuer Science, vol. 14481, pp. 170–184. Springer, Cham (2023). https://doi.org/10.1007/978-3-031-50959-9_12

29. Lipski, W.: On semantics issues connected with incomplete information databases. ACM Trans. Database Syst. **4**, 262–296 (1979)
30. Jensen, R., Shen, Q.: Interval-valued fuzzy-rough feature selection in datasets with missing values. In: Proceedings of FUZZ-IEEE 2009, pp. 610–615. IEEE Press (2009). https://doi.org/10.1109/FUZZY.2009.5277289
31. Couso, I., Dubois, D.: Rough sets, coverings and incomplete information. Fund. Inform. **108**(3–4), 223–347 (2011)
32. Lenz, O.U., Peralta, D., Cornelis, C.: Adapting fuzzy rough sets for classification with missing values. In: Ramanna, S., Cornelis, C., Ciucci, D. (eds.) IJCRS 2021. LNCS (LNAI), vol. 12872, pp. 192–200. Springer, Cham (2021). https://doi.org/10.1007/978-3-030-87334-9_16
33. Nakata, M., Saito, N., Sakai, H., Fujiwara, T.: Structures derived from possible tables in an incomplete information table. In: Proceedings of SCIS 2022, p. 6. IEEE Press (2022). https://doi.org/10.1109/SCISISIS55246.2022.10001919

On Complexity of Deterministic and Nondeterministic Decision Trees for Decision Tables with Many-Valued Decisions from Closed Classes

Azimkhon Ostonov(✉) and Mikhail Moshkov

Computer, Electrical and Mathematical Sciences and Engineering Division and Computational Bioscience Research Center, King Abdullah University of Science and Technology (KAUST), Thuwal 23955-6900, Saudi Arabia
{azimkhon.ostonov,mikhail.moshkov}@kaust.edu.sa

Abstract. Decision trees (DTRs) and decision rules are extensively examined and applied in various domains of computer science. The theory of DTRs and rules highlights several crucial inquiries, such as how the complexity of deterministic decision trees (DDTRs) and decision rule systems depends on the complexity of the set of attributes associated with the columns of the decision table (DT). In this research paper, we focus on the analysis of nondeterministic decision trees (NDTRs) as a substitute for decision rule systems. NDTRs can be seen as representations of decision rule systems. We examine classes of DTs featuring multi-valued decisions (known as multi-label DTs) that maintain closure under attribute removal (columns) and modifications to the sets of assigned decisions for rows. We examine the behavior of functions that describe the worst-case dependence of the minimum complexity of DDTRs and NDTRs on the complexity of the set of attributes associated with the columns of tables belonging to a closed class (CC) of DTs closed under the above two operations. We enumerate all possible types of behavior exhibited by these functions.

Keywords: Decision tables with many-valued decisions · Closed classes of decision tables · Deterministic decision trees · Nondeterministic decision trees

1 Introduction

In several branches of computer science, decision trees (DTRs) and decision rules are extensively researched and applied. It appears that the following queries are crucial to the theory of DTRs and rules: the relationship between the complexity of the set of attributes associated with the columns of decision table (DT) and the complexity of DTRs and decision rule systems.

This research paper focuses on the analysis of nondeterministic decision trees (NDTRs) as representations of decision rule systems instead of studying decision rule systems directly. We investigate classes of DTs with many-valued decisions (known as multi-label DTs) that are closed under attribute removal (columns) and modifications in the assigned set of decisions for rows. Within an arbitrary closed class (CC) of DTs, we examine functions that describe the worst-case relationship between the minimum complexity of deterministic decision trees (DDTRs) and NDTRs and the complexity of the set of attributes associated with the columns. Our analysis involves enumerating all possible types of behavior exhibited by these functions.

A DT with many-valued decisions is a rectangular table where columns are labeled with attributes, rows are distinct from each other, and each row is associated with a nonempty, finite set of decisions. These rows represent tuples of attribute values, and the task is to determine the appropriate decision for a given row from its associated set of decisions. To achieve this, we can utilize queries that involve selecting an attribute and inquiring about its value for the specific row. This research focuses on two types of algorithms that rely on these queries: DDTRs and NDTRs. NDTRs can be interpreted as a means of representing any system of true decision rules for the table that cover all rows. The study focuses on bounded complexity measures (BCMs) that assess the time complexity of DTRs, such as the depth of the DTRs.

Many-valued DTs, referred to as multi-label DTs, are frequently encountered in data analysis [4,26,27]. These tables also find application in various fields like combinatorial optimization, computational geometry, and fault diagnosis, where they serve as a means to depict and investigate problems [1,12].

DTRs [1,21,23] and decision rule systems have extensive applications as classifiers, tools for knowledge representation, and algorithms for solving a range of problems including combinatorial optimization and fault diagnosis [3,6,11,19,24]. These models are highly interpretable and are widely employed in data analysis [8].

We investigate classes of DTs that have many-valued decisions and are closed under the removal of columns (attributes) and modifications in the decisions (specifically, sets of decisions). A common example of such classes is the CCs of DTs derived from information systems [18]. An information system comprises a set of objects (known as the universe) and a set of attributes (functions) defined on the universe, with values taken from a finite set. A problem related to an information system is defined by a finite number of attributes that partition the universe into nonempty domains, each having fixed attribute values. Each domain is associated with a nonempty finite set of decisions. The objective is to determine the decision from the set attached to the domain containing a given object from the universe.

A DT with many-valued decisions can be directly associated with this problem in a natural manner. The columns of the table represent the attributes under consideration, while the rows correspond to the domains and are labeled with the sets of decisions attached to domains. The collection of DTs that correspond to problems over an information system forms a CC that is derived from the system itself. It is important to note that the family of all CCs is significantly broader than the family of CCs generated specifically by information systems. Specifically, the combination of two CCs derived from two different information systems results in another CC. However, in general, there may not exist an information system that can generate this combined class.

Numerous classes of objects that exhibit closure under different operations have been extensively investigated. Notably, there are classes of Boolean functions that are closed under the operation of superposition [20], as well as minor-closed classes of graphs [22]. DTs serve as intriguing mathematical objects that warrant further mathematical exploration, particularly in the context of studying the CCs of DTs.

This research paper builds upon previous work on CCs of DTs, which was initiated in the study [9] and continued in subsequent works [13,15,16].

Consider a class of DTs with many-valued decisions, denoted as A, which is closed under the removal of columns and modifications in the assigned decisions. Let ψ represent a BCM. The focus of this research paper is on the study of two functions: $\mathcal{F}^{\infty}_{\psi,A}(n)$ and $\mathcal{G}^{\infty}_{\psi,A}(n)$.

The function $\mathcal{F}^{\infty}_{\psi,A}(n)$ captures the worst-case increase in the minimum complexity of a DDTR for a DT belonging to the class A, as the complexity of the set of attributes attached to the table's columns grows. In this paper, we provide proofs that demonstrate the function $\mathcal{F}^{\infty}_{\psi,A}(n)$ falls into one of three categories: it is either bounded by a constant from above, grows logarithmically with n, or exhibits almost linear growth depending on n (where it is bounded from above by n and equals n for infinitely many values of n).

The function $\mathcal{G}^{\infty}_{\psi,A}(n)$ represents the worst-case increase in the minimum complexity of a NDTR for a DT belonging to the class A, as the complexity of the set of attributes attached to the table's columns grows. Through our proofs, we establish that the function $\mathcal{G}^{\infty}_{\psi,A}(n)$ falls into one of two categories: it is either bounded by a constant from above or exhibits almost linear growth depending on n (where it is bounded from above by n and equals n for infinitely many values of n).

The findings presented in this paper regarding the functions $\mathcal{F}^{\infty}_{\psi,A}(n)$ and $\mathcal{G}^{\infty}_{\psi,A}(n)$ and their application to DTs with many-valued decisions closely align with the results obtained in [16] for conventional DTs. However, in this paper, we have chosen to provide rigorous proofs for these results. This decision is motivated by the fact that DTs with many-valued decisions generally exhibit significant differences compared to conventional DTs.

To demonstrate this point, let's consider an example. A complete DT consists of n columns filled with zeros and ones, and it is considered complete if it contains 2^n pairwise distinct rows. For a conventional complete DT, it can be shown that the minimum depth of a DDTR is at most the square of the minimum depth of a NDTR. This result is a straightforward generalization of the findings obtained for Boolean functions in [2,7,25] (also see [5]). However, we can construct an infinite sequence of complete DTs with many-valued decisions where the minimum depth of DDTRs increases while the minimum depth of NDTRs remains at most 3.

Note that the obtained results were mentioned without proof in our conference paper [14]. It should also be noted that this paper significantly differs from our concurrent work [17] where we present a greedy approach to constructing deterministic decision trees for conventional decision tables from closed classes.

The paper is organized into five sections. Section 2 covers the main definitions and notation used throughout the paper. In Sect. 3, we present the main results of our research. The proofs of these results are provided in Sect. 4. Finally, Sect. 5 offers brief conclusions based on the findings.

2 Main Definitions and Notation

Let ω be the set $\{0, 1, 2, \ldots\}$, and let $\mathcal{P}(\omega)$ represent the set of finite subsets of ω that are not empty. Additionally, for any number $k \in \omega$ excluding 0 and 1, let $E_k = \{0, 1, \ldots, k-1\}$. We define P as the set of attributes, denoted as $\{f_i : i \in \omega\}$. In this context, two attributes f_l and f_p from P are considered distinct if $l \neq p$.

2.1 Decision Tables

We now consider the definition of a decision table (DT) [14].

Definition 1. *Let $k \in \omega \setminus \{0, 1\}$ and let $\mathcal{M}_k^\infty$ represent the set of rectangular tables where each cell is filled with a number taken from the set E_k. In these tables, every row is distinct, labeled with a set from $\mathcal{P}(\omega)$ (set of decisions), and each column is labeled with a distinct attribute from P. The rows can be interpreted as tuples of attribute values. It is worth noting that empty tables without any rows are included in the set $\mathcal{M}_k^\infty$. For convenience, the same symbol Λ is used to refer to these tables. The tables belonging to $\mathcal{M}_k^\infty$ are referred to as decision tables with many-valued decisions (decision tables, DTs).*

Consider a table T belonging to the set $\mathcal{M}_k^\infty$. We define $\Delta(T)$ as the collection of rows in table T, and $\Pi(T)$ represents the intersection of the decision sets associated with each row in T. The decisions contained in $\Pi(T)$ are referred to as the *common decisions* for the table T.

Example 1. Figure 1 illustrates a DT from $\mathcal{M}_2^\infty$.

Let $\mathcal{M}_k^{\infty c}$ represent the set of tables from $\mathcal{M}_k^\infty$ containing at least one common decision. Let Λ be an element of $\mathcal{M}_k^{\infty c}$.

Consider a nonempty table T belonging to the set $\mathcal{M}_k^\infty$. Let $P(T)$ represent the set of attributes associated with the columns of table T. Let denote $\Omega_k(T)$ as the collection of finite sequences (words) composed of symbols from the alphabet $\{(f_i, \sigma) : f_i \in P(T), \sigma \in E_k\}$, which includes the empty sequence λ. A subtable of the table T corresponding to the word $\xi \in \Omega_k(T)$ is denoted by $T\xi$. If $\xi = \lambda$, then $T\xi = T$. If $\xi \neq \lambda$ and can be expressed as $\xi = (f_{i_1}, \sigma_1) \ldots (f_{i_m}, \sigma_m)$, then $T\xi$ is derived from T by eliminating all rows that do not meet the following

f_1 f_2 f_3	
1 0 0	{1}
0 1 0	{2}
0 0 1	{3}
1 1 0	{1,2}
1 0 1	{1,3}
0 1 1	{2,3}
1 1 1	{1,2,3}

Fig. 1. DT from $\mathcal{M}_2^\infty$

f_2 f_3	
1 1	{1}
0 1	{0,1}
1 0	{0,1}
0 0	{0}

Fig. 2. DT derived from the DT illustrated in Fig. 1 by applying operations of column removal and decision modification

criteria: in the columns labeled with attributes $f_{i_1}, \ldots, f_{i_m}$, the row contains values $\sigma_1, \ldots, \sigma_m$ respectively.

We will now introduce two operations that can be performed on DTs: column removal and decision modification [14]. Consider a table T from the set $\mathcal{M}_k^\infty$.

Definition 2. Column removal. *Consider a set Q which is a subset of the set of attributes $P(T)$. We proceed to remove from the table T all columns that are assigned attributes from the set Q. Within each group of rows that share identical values in the remaining columns, we retain only the initial row. The resulting table is denoted as $I(Q,T)$. Specifically, when Q is an empty set, we have $I(\emptyset, T) = T$, and when Q consists of all the attributes of T, we have $I(P(T), T) = \Lambda$. Evidently, the table $I(Q,T)$ belongs to the set $\mathcal{M}_k^\infty$.*

Definition 3. Decision modification. *Consider a function $\mu : E_k^{|P(T)|} \to \mathcal{P}(\omega)$, where $E_k^0 = \emptyset$ as per the definition. For every row $\bar{\sigma}$ in table T, we substitute the set of decisions associated with that row with $\mu(\bar{\sigma})$. The resulting table is denoted as $J(\mu, T)$. Evidently, the table $J(\mu, T)$ belongs to the set $\mathcal{M}_k^\infty$.*

Definition 4. *Denote $[T] = \{J(\mu, I(Q,T)) : Q \subseteq P(T), \mu : E_k^{|P(T)\setminus Q|} \to \mathcal{P}(\omega)\}$. The set $[T]$ represents the closure of the table T achieved by applying operations of column removal and decision modification [14].*

Example 2. Figure 2 illustrates the table $J(\mu, I(Q, T_0))$, where T_0 is the table illustrated in Fig. 1, $Q = \{f_1\}$ and $\mu(x_1, x_2) = \{x_1 \vee x_2, x_1 \wedge x_2\}$.

Definition 5. *Consider a nonempty set A that is a subset of $\mathcal{M}_k^\infty$. Let denote $[A]$ as the union of the closures of all tables T in A, i.e., $[A] = \bigcup_{T \in A} [T]$. The set*

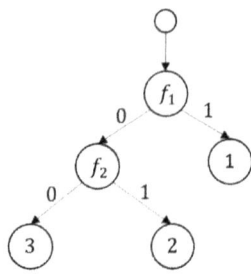

Fig. 3. A DDTR for the DT illustrated in Fig. 1

$[A]$ represents the closure of the set A under the specified operations of column removal and decision modification. If $[A] = A$, then the class (set) of DTs A is referred to as a closed class (CC) [14].

A CC will be referred to as *nontrivial* (NCC) if it includes DTs that are nonempty.

Suppose A_1 and A_2 are CCs belonging to $\mathcal{M}_k^\infty$. In that case, the union of A_1 and A_2, denoted as $A_1 \cup A_2$, forms a CC which also belongs to $\mathcal{M}_k^\infty$.

2.2 Deterministic and Nondeterministic Decision Trees

A *finite tree with a root* [14] refers to a finite directed tree where only one specific node, known as the *root*, does not have any incoming edges. The nodes in the tree that do not have any outgoing edges are referred to as *leaf* nodes.

Definition 6. *A k-decision tree (k-DTR) [14] is a finite tree with a root, consisting of a minimum of two nodes, wherein*

- *The root node and the edges originating from the root do not have any assigned labels.*
- *Every leaf node in the tree is marked with a decision taken from the set ω.*
- *Every node in the tree, excluding the root and leaf nodes, is assigned a label representing an attribute from the set P. Additionally, each outgoing edge from these non-root, non-leaf nodes is marked with a number from the set E_k.*

Example 3. Figures 3, 4 depict 2-DTRs.

We define $\mathcal{T}_k$ as the set of all k-DTRs. Let Γ belong to $\mathcal{T}_k$. We utilize $P(\Gamma)$ to denote the set of attributes assigned to the nodes of Γ that are not the root or leaf nodes. A *complete path* in Γ is defined as a sequence $\rho = u_1, e_1, \ldots, u_l, e_l, u_{l+1}$, which consists of nodes and edges in Γ. In this sequence, u_1 is the root of Γ, u_{l+1} corresponds to a leaf node of Γ, and for $j = 1, \ldots, l$, the edge e_j connects the node u_j to the node u_{j+1}. Consider $T \in \mathcal{M}_k^\infty$. In case that $P(\Gamma)$ is a subset of the set $P(T)$, we can associate the table T and the complete path ρ with a word $\pi(\rho) \in \Omega_k(T)$. If $l = 1$, then $\pi(\rho) = \lambda$. In case that $l > 1$ and for $j = 2, \ldots, l$, the node u_j is assigned the attribute f_{i_j} and the edge e_j is assigned the number σ_j, then $\pi(\rho) = (f_{i_2}, \sigma_2) \cdots (f_{i_l}, \sigma_l)$. We define $T(\rho) = T\pi(\rho)$.

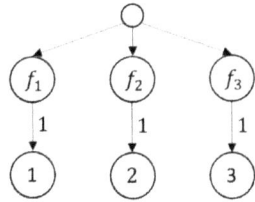

Fig. 4. A NDTR for the DT depicted in Fig. 1

Definition 7. *Consider T as an element of $\mathcal{M}_k^\infty \setminus \Lambda$. A deterministic decision tree (DDTR) [14] for the table T refers to a k-DTR Γ that meets the following criterion:*

- *The root of Γ has only a single outgoing edge.*
- *For any node in Γ that is not the root or a leaf node, the outgoing edges from that node are labeled with pairwise different numbers.*
- $P(\Gamma) \subseteq P(T)$.
- *For every row in T, there is a complete path ρ in Γ such that the corresponding row is part of the table $T(\rho)$.*
- *For every complete path ρ in Γ, one of the following conditions must hold: either $T(\rho) = \Lambda$ or the decision associated with the final node of ρ is a common decision for the table $T(\rho)$.*

Example 4. The 2-DTR depicted in Fig. 3 is a DDTR for the DT depicted in Fig. 1.

Definition 8. *A nondeterministic decision tree (NDTR) [14] for the table T, where $T \in \mathcal{M}_k^\infty \setminus \Lambda$, is a k-DTR Γ that fulfills the following requirements:*

- $P(\Gamma) \subseteq P(T)$.
- *For every row in T, there is a complete path ρ in Γ such that the corresponding row is part of the table $T(\rho)$.*
- *For every complete path ρ in Γ, one of two conditions must hold: either $T(\rho) = \Lambda$, or the decision associated with the final node of ρ is a common decision for the table $T(\rho)$.*

Example 5. The 2-DTR depicted in Fig. 4 is a NDTR for the DT depicted in Fig. 1.

2.3 Complexity Measures

Let P be an alphabet consisting of symbols $f_i, i \in \omega$. The set P^* refers to the collection of all finite sequences of symbols (words) from P, along with the empty sequence denoted as λ.

Definition 9. A partially bounded complexity measure *(partially BCM)* is any function denoted as $\psi : P^* \to \omega$ which possesses the following characteristics: for any two words ξ_1 and ξ_2 from P^*,

(p1) $\psi(\xi_1) = 0$ only if it is the case that $\xi_1 = \lambda$ - property of positivity.
(p2) $\psi(\xi_1) = \psi(\xi_1')$ where ξ_1' is any word which can be obtained by changing the order of letters of the word ξ_1 - property of commutativity.
(p3) $\psi(\xi_1) \leq \psi(\xi_1\xi_2)$ - property of nondecreasing.
(p4) $\psi(\xi_1\xi_2) \leq \psi(\xi_1) + \psi(\xi_2)$ - property of boundedness from above.

The subsequent functions qualify as partially BCMs:

– The function h is defined such that, for any word ξ belonging to P^*, $h(\xi)$ is equal to the length of ξ, denoted as $|\xi|$. This particular function is referred to as the *depth*.
– Any mapping $\varphi : P^* \to \omega$ such that $\varphi(\lambda) = 0$. Furthermore, for every symbol f_i in P, the function φ must yield a value greater than 0 (i.e., $\varphi(f_i) > 0$). Additionally, for any nonempty word $f_{i_1} \cdots f_{i_l} \in P^*$, the function φ should satisfy the following condition,

$$\varphi(f_{i_1} \cdots f_{i_l}) = \sum_{j=1}^{l} \varphi(f_{i_j}). \tag{1}$$

This mapping is referred to as the *weighted depth* function.
– Any function $\tau : P^* \to \omega$ that fulfills the following conditions: $\tau(\lambda) = 0$ for the empty word λ, $\tau(f_i) > 0$ for any symbol f_i in P, and for any nonempty word $f_{i_1} \cdots f_{i_l} \in P^*$, $\tau(f_{i_1} \cdots f_{i_l})$ is equal to the maximum value among $\tau(f_{i_j})$ for $j = 1, \ldots, l$.

Definition 10. A complexity measure ψ is called bounded complexity measure (BCM) *[14]* if it is a partially BCM and it satisfies the property of boundedness from below:

(p5) $\psi(\xi) \geq |\xi|$ for any word ξ in P^*.

Any partially BCM that satisfies the equality (1), including the function h, can be classified as a BCM. It can be demonstrated that if ψ_1 and ψ_2 are partially BCMs, then the functions ψ_3 and ψ_4 are also partially BCMs. Here, for any word $\xi \in P^*$, $\psi_3(\xi)$ is defined as the sum of $\psi_1(\xi)$ and $\psi_2(\xi)$, while $\psi_4(\xi)$ is defined as the maximum value between $\psi_1(\xi)$ and $\psi_2(\xi)$. If the function ψ_1 is a BCM, then both ψ_3 and ψ_4 are also BCMs.

Definition 11. Consider ψ as a partially BCM. We can broaden its application to include all finite subsets of the set P. Suppose Q is a finite subset of P. If Q is an empty set, then $\psi(Q)$ is defined as 0. However, if Q is a nonempty set with elements $f_{i_1}, \ldots, f_{i_l}$, where $l \geq 1$, then $\psi(Q)$ is equal to $\psi(f_{i_1} \cdots f_{i_l})$.

Definition 12. Assume ψ is a partially BCM. We expand its scope to encompass the set of finite words Ω formed from the alphabet $\{(f_i, \sigma) : f_i \in P, \sigma \in \omega\}$, which includes the empty word λ. Consider ξ as an element of the set Ω. If ξ is equal to λ, then $\psi(\xi)$ is defined as 0. Alternatively, if ξ takes the form $(f_{i_1}, \sigma_1) \cdots (f_{i_l}, \sigma_l)$, where $l \geq 1$, then $\psi(\xi)$ is equal to $\psi(f_{i_1} \cdots f_{i_l})$.

2.4 Parameters of Decision Trees and Tables

Definition 13. *Consider ψ as a partially BCM. We expand the application of the function ψ to include the set $\mathcal{T}_k$. Let Γ be an element of $\mathcal{T}_k$. Then $\psi(\Gamma)$ is defined as the maximum value obtained from evaluating $\psi(\pi(\rho))$, where the maximum value is determined by considering all complete paths ρ in the DTR Γ. When referring to a given partially BCM ψ, the value $\psi(\Gamma)$ will be referred to as the* complexity *of the DTR Γ. Additionally, the value $h(\Gamma)$ will be known as the* depth *of the DTR Γ.*

Assume ψ is a partially BCM. We will now explain the functions ψ^d, ψ^a, W_ψ, S_ψ, $\hat{S}_\psi$, N, Sep, and M_ψ [14,16], which are defined on the set $\mathcal{M}_k^\infty$. According to their definitions, each of these functions evaluates to 0 when applied to the input Λ. Consider T as an element of $\mathcal{M}_k^\infty \setminus \{\Lambda\}$, and let n belong to the set ω.

- $\psi^d(T) = \min\{\psi(\Gamma)\}$, the value of $\psi^d(T)$ is defined as the minimum value obtained by evaluating $\psi(\Gamma)$, where the minimum is taken over all DDTRs Γ for the table T.
- $\psi^a(T) = \min\{\psi(\Gamma)\}$, the value of $\psi^a(T)$ is defined as the minimum value obtained by evaluating $\psi(\Gamma)$, where the minimum is taken over all NDTRs Γ for the table T.
- $W_\psi(T) = \psi(P(T))$. This refers to the complexity associated with the set of attributes that are attached to the columns of the table T.
- $N(T)$ represents the count of rows present in the table T.
- Let $\bar{\sigma}$ represent a row from the table T. We define $S_\psi(T, \bar{\sigma})$ as the minimum value of $\psi(Q)$, where Q is a subset of the set $P(T)$ such that, in the columns of T labeled with attributes from Q, the row $\bar{\sigma}$ is distinct from all other rows in the table T. Consequently, $S_\psi(T)$ is defined as the maximum value among all $S_\psi(T, \bar{\sigma})$, where the maximum is taken over all rows $\bar{\sigma}$ in the table T.
- $\hat{S}_\psi(T) = \max\{S_\psi(T^*) : T^* \in [T]\}$.
- A set Q that is a subset of $P(T)$ is referred to as a *separating set* for the table T if, in the set of columns labeled with attributes from Q, the rows of table T are all distinct from each other. The term $Sep(T)$ represents the minimum number of elements required in a separating set for the table T.
- If T belongs to the set $\mathcal{M}_k^{\infty c}$, then $M_\psi(T)$ is equal to 0. Now, let us consider the case where T does not belong to $\mathcal{M}_k^{\infty c}$. Suppose the number of attributes in T is n, and the columns of T are labeled with the attributes $f_{t_1}, \ldots, f_{t_n}$. Let $\bar{\sigma} = (\sigma_1, \ldots, \sigma_n)$ be an element of E_k^n. We represent by $M_\psi(T, \bar{\sigma})$ the smallest number $q \in \omega$ such that there exist attributes $f_{t_{i_1}}, \ldots, f_{t_{i_l}} \in P(T)$ for which $T(f_{t_{i_1}}, \sigma_{i_1}) \cdots (f_{t_{i_l}}, \sigma_{i_l}) \in \mathcal{M}_k^{\infty c}$ and $\psi(f_{t_{i_1}} \cdots f_{t_{i_l}}) = q$. Then, $M_\psi(T)$ is equal to the maximum value obtained from evaluating $M_\psi(T, \bar{\sigma})$ for every $\bar{\sigma} \in E_k^n$.

For the complexity measure h which is bounded, we use the notation $W(T)$ to represent $W_h(T)$, $M(T)$ to represent $M_h(T)$, and $M(T, \bar{\sigma})$ to represent $M_h(T, \bar{\sigma})$. It should be noted that $W(T)$ specifies the count of columns present in the table T.

Example 6. Let T_0 be the DT depicted in Fig. 1. It can be demonstrated that $h^d(T_0) = 2$, $h^a(T_0) = 1$, $W(T_0) = 3$, $N(T_0) = 7$, $S_\psi(T_0) = 3$, $\hat{S}_\psi(T_0) = 3$, $Sep(T_0) = 3$, $M(T_0) = 2$.

3 Main Results

This section focuses on the outcomes obtained for the functions $\mathcal{F}_{\psi,A}^\infty$ and $\mathcal{G}_{\psi,A}^\infty$.

3.1 Function $\mathcal{F}_{\psi,A}^\infty$

Suppose ψ is a BCM and A is a NCC of DTs from $\mathcal{M}_k^\infty$. Let us define a function $\mathcal{F}_{\psi,A}^\infty$ [14] mapping from ω to ω. Suppose n is a number from ω. Then

$$\mathcal{F}_{\psi,A}^\infty(n) = \max\{\psi^d(T) : T \in A, W_\psi(T) \leq n\}.$$

The function $\mathcal{F}_{\psi,A}^\infty$ describes the growth in the worst case of the minimum complexity of DDTRs for DTs from A with the growth of the complexity of the sets of attributes attached to the columns of these DTs.

Suppose $D = \{n_i : i \in \omega\}$ is an infinite subset of the set ω in which, for any i belonging to ω, n_i is less than n_{i+1}. Let H_D be a function mapping from ω to ω. Suppose n belongs to ω. In case n is less than n_0, then $H_D(n)$ is defined as 0. In case, for some i belonging to ω, $n_i \leq n < n_{i+1}$ holds, then $H_D(n)$ is equal to n_i.

Theorem 1. *Suppose ψ is a BCM and A is a NCC of DTs from $\mathcal{M}_k^\infty$. Then $\mathcal{F}_{\psi,A}^\infty$ is an everywhere defined nondecreasing function, where $\mathcal{F}_{\psi,A}^\infty(n) \leq n$ holds for every n belonging to ω and $\mathcal{F}_{\psi,A}^\infty(0) = 0$. One of the following claims holds for this function:*

(a) *In case, the functions S_ψ and N have constant upper bounds on the class A, then there exists a constant q_0 belonging to $\omega \setminus \{0\}$, where $\mathcal{F}_{\psi,A}^\infty(n) \leq q_0$ holds for every n belonging to ω.*
(b) *In case, the function S_ψ has a constant upper bound on the class A and the function N has no constant upper bound on the class A, then there exist constants q_1, q_2, q_3, q_4 each belonging to $\omega \setminus \{0\}$, where $q_1 \log_2 n - q_2 \leq \mathcal{F}_{\psi,A}^\infty(n) \leq q_3 \log_2 n + q_4$ holds for every n belonging to $\omega \setminus \{0\}$.*
(c) *In case, the function S_ψ has no constant upper bound on the class A, then there exists an infinite subset D of the set ω, where $H_D(n)$ is less than or equal to $\mathcal{F}_{\psi,A}^\infty(n)$ holds for every n belonging to ω.*

3.2 Function $\mathcal{G}_{\psi,A}^\infty$

Suppose ψ is a BCM and A is a NCC of DTs from $\mathcal{M}_k^\infty$. Let us define a function $\mathcal{G}_{\psi,A}^\infty$ [14]. Suppose n is a number from ω. Then

$$\mathcal{G}_{\psi,A}^\infty(n) = \max\{\psi^a(T) : T \in A, W_\psi(T) \leq n\}.$$

The function $\mathcal{G}_{\psi,A}^\infty$ describes the growth in the worst case of the minimum complexity of NDTRs for DTs from A with the growth of the complexity of the sets of attributes attached to the columns of these tables.

Theorem 2. *Suppose ψ is a BCM and A be a NCC of DTs from $\mathcal{M}_k^\infty$. Then $\mathcal{G}_{\psi,A}^\infty$ is an everywhere defined nondecreasing function, where $\mathcal{G}_{\psi,A}^\infty(n)$ is less than or equal to n holds for every n belonging to ω and $\mathcal{G}_{\psi,A}^\infty(0) = 0$. One of the following claims holds for this function:*

(a) *In case, the function S_ψ has a constant upper bound on the class A, then there exists a constant q belonging to $\omega \setminus \{0\}$, where $\mathcal{G}_{\psi,A}^\infty(n)$ is less than or equal to q holds for every n belonging to ω.*
(b) *In case, the function S_ψ has no constant upper bound on the class A, then there exists an infinite subset D of the set ω, where $H_D(n)$ is less than or equal to $\mathcal{G}_{\psi,A}^\infty(n)$ holds for every n belonging to ω.*

4 Proofs of Theorems 1 and 2

Initially, we consider a number of auxiliary lemmas.

The subsequent upper bound on the minimum complexity of DDTR for a DT can be easily proven.

Lemma 1. *For an arbitrary complexity measure ψ and an arbitrary DT T from $\mathcal{M}_k^\infty$,*
$$\psi^d(T) \leq W_\psi(T).$$

The subsequent upper bound on the minimum complexity of DDTRs for a DT can be derived from Corollary 5.2 from [10].

Lemma 2. *For an arbitrary complexity measure ψ and an arbitrary DT T from $\mathcal{M}_k^\infty$,*
$$\psi^d(T) \leq \begin{cases} 0, & T \in \mathcal{M}_k^\infty \mathcal{C}, \\ M_\psi(T) \log_2 N(T), & T \notin \mathcal{M}_k^\infty \mathcal{C}. \end{cases}$$

The following four statements are straightforward extensions of similar results that were proven in [16] for conventional DTs.

Lemma 3. *For an arbitrary complexity measure ψ and an arbitrary DT T belonging to $\mathcal{M}_k^\infty$,*
$$M_\psi(T) \leq 2\hat{S}_\psi(T).$$

Lemma 4. *For an arbitrary DT T belonging to $\mathcal{M}_k^\infty \setminus \{\Lambda\}$,*
$$N(T) \leq (kW(T))^{S(T)}.$$

Lemma 5. *For an arbitrary DT T belonging to $\mathcal{M}_k^\infty \setminus \{\Lambda\}$, there exists a mapping $\mu: E_k^{W(T)} \to \mathcal{S}(\omega)$ such that*
$$h^d(J(\mu, T)) \geq \log_k N(T).$$

Lemma 6. *For an arbitrary complexity measure ψ and an arbitrary DT T belonging to $\mathcal{M}_k^\infty$, such that $|\Delta(T)| \geq 2$ holds, there exists a DT $T^* \in [T]$ where the following statement is valid*

$$\psi^a(T^*) = \psi^d(T^*) = W_\psi(T^*) = S_\psi(T^*) = S_\psi(T).$$

Establishing an upper bound on the minimum cardinality of a separating set for a DT can be accomplished through induction on the number of rows in the table. The proof of this upper bound is not difficult.

Lemma 7. *For an arbitrary DT T belonging to $\mathcal{M}_k^\infty \setminus \{\Lambda\}$,*

$$Sep(T) \leq N(T) - 1.$$

Proving the following upper bounds on the minimum complexity of NDTRs for a DT is not a difficult task.

Lemma 8. *For an arbitrary complexity measure ψ and an arbitrary DT T belonging to $\mathcal{M}_k^\infty$,*

$$\psi^a(T) \leq S_\psi(T) \leq W_\psi(T).$$

We now prove Theorems 1 and 2.

Proof (of Theorem 1). Considering the fact that A is a CC, $\Lambda \in A$. By definition of the function W_ψ we have $W_\psi(\Lambda) = 0$. Taking into account this fact and Lemma 1, we can conclude that $\mathcal{F}_{\psi,A}^\infty$ is an everywhere defined function and $\mathcal{F}_{\psi,A}^\infty(n)$ is less than or equal to n for any number n belonging to ω. Obviously, $\mathcal{F}_{\psi,A}^\infty$ is a nondecreasing function. Suppose T belongs to A and $W_\psi(T)$ is less than or equal to 0. From property (p1) of the function ψ, we derive that T is equal to Λ. Thus, $\mathcal{F}_{\psi,A}^\infty(0)$ is equal to 0.

(a) Suppose the functions S_ψ and N are bounded from above on the class A. Then there exist constants $v \geq 1$ and $w \geq 2$ where $S_\psi(T)$ is less than or equal to v and $N(T)$ is less than or equal to w hold for any DT $T \in A$. Suppose $T \in A$. Using the fact that A is a CC, we derive $\hat{S}_\psi(T) \leq v$. By Lemma 3, $M_\psi(T) \leq 2v$. From this inequality, inequality $N(T) \leq w$ and from Lemma 2 it follows that $\psi^d(T) \leq 2v \log_2 w$. Denote $q_0 = 2v \log_2 w$. Since T is an arbitrary DT from the class A, we derive that $\mathcal{F}_{\psi,A}^\infty(n) \leq q_0$ for any number n belonging to ω.

(b) Suppose the function S_ψ is bounded from above on the class A and the function N is not bounded from above on A. Then there exists a constant $v \geq 2$ where, for any DT $Q \in A$,

$$S_\psi(Q) \leq v. \tag{2}$$

Suppose $n \in \omega \setminus \{0\}$ and T is an arbitrary DT from A where $W_\psi(T) \leq n$. If T belongs to $\mathcal{M}_k^\infty \mathcal{C}$, then, evidently, $\psi^d(T) = 0$. Suppose $T \notin \mathcal{M}_k^\infty \mathcal{C}$. Using property (p5) of the function ψ, we derive $W(T)$ is less than or equal to n and $S(T)$ is less than or equal to v. From these inequalities and Lemma 4 we can conclude that $N(T) \leq (kn)^v$. From (2) and Lemma 3 it follows that $M_\psi(T) \leq 2v$. Taking into account the last two inequalities and Lemma 2, we derive

$$\psi^d(T) \leq 2v^2 \log_2 n + 2v^2 \log_2 k. \tag{3}$$

Denote $q_3 = 2v^2$ and $q_4 = 2v^2 \log_2 k$. Then $\psi^d(T) \leq q_3 \log_2 n + q_4$. From the fact that n is an arbitrary number belonging to $\omega \setminus \{0\}$ and T is an arbitrary DT belonging to A such that $W_\psi(T) \leq n$, we can conclude that, for any $n \in \omega \setminus \{0\}$,

$$\mathcal{F}^\infty_{\psi,A}(n) \leq q_3 \log_2 n + q_4. \tag{4}$$

Suppose $n \in \omega \setminus \{0\}$. Denote $q_1 = 1/\log_2 k$ and $q_2 = \log_k v$. Let us show that

$$\mathcal{F}^\infty_{\psi,A}(n) \geq q_1 \log_2 n - q_2. \tag{5}$$

Denote $m = \lfloor n/v \rfloor$. Since the function N has no constant upper bound on the set A, we derive that there exists a DT T belonging to A where $N(T) \geq k^m$ holds. Suppose C be a separating set for the DT T with the minimum cardinality where $\psi(f_i)$ is less than or equal to v holds for every f_i belonging to C. The existence of such a set can be deduced from the inequality (2) and properties (p2) and (p3) of the function ψ. Obviously, the cardinality of the set C is greater than or equal to m. Suppose D be a subset of the set C with m elements. Denote $T^0 = I(P(T) \setminus D, T)$. It can be demonstrated that, for any attribute of T^0, there are two rows of the DT T^0 that differ only in the column labeled with this attribute. Thus, D is a separating set for the DT T^0 with the smallest possible cardinality. By Lemma 7, $N(T^0) \geq m + 1 \geq n/v$.

By Lemma 5, we derive that there exists a mapping $\mu : E_k^m \to \omega$ where

$$h^d(J(\mu, T^0)) \geq \log_k(n/v). \tag{6}$$

Denote $T^* = J(\mu, T^0)$. Taking into account that ψ is a BCM and inequality (6), we obtain $\psi^d(T^*) \geq \log_k(n/v)$. Using the property (p4) of the function ψ, we derive $W_\psi(T^*) \leq \lfloor n/v \rfloor v \leq n$. Therefore the inequality (5) is valid. From (4) and (5) we can conclude that $q_1 \log_2 n - q_2 \leq \mathcal{F}^\infty_{\psi,A}(n) \leq q_3 \log_2 n + q_4$ for every number n belonging to $\omega \setminus \{0\}$.

(c) Suppose the function S_ψ has no constant upper bound on the class A. Taking into account Lemma 6, we derive that the set $D = \{W_\psi(T) : T \in A, \psi^d(T) = W_\psi(T)\}$ is infinite. From the fact that the A is CC, $\Lambda \in A$ and, since $\psi^d(\Lambda) = W_\psi(\Lambda) = 0$, $0 \in D$. Obviously, for every $n \in D$, $\mathcal{F}^\infty_{\psi,A}(n)$ is greater than or equal to n. Since $\mathcal{F}_{\psi,A}$ is a nondecreasing function, we derive that $\mathcal{F}^\infty_{\psi,A}(n)$ is greater than or equal to $H_D(n)$ for any n belonging to $\omega \setminus \{0\}$. □

Proof (of Theorem 2). Taking into account that A is a CC, we have $\Lambda \in A$. From the definition of the function W_ψ, we obtain $W_\psi(\Lambda)$ is equal to 0. From this fact and Lemma 8, we derive that $\mathcal{G}^\infty_{\psi,A}$ is an everywhere defined function and $\mathcal{G}^\infty_{\psi,A}(n) \leq n$ for every n belonging to ω. Obviously, $\mathcal{G}^\infty_{\psi,A}$ is a nondecreasing function. Suppose T belongs to A and $W_\psi(T)$ is less than or equal to 0. From property (p1) of the function ψ, we derive $T = \Lambda$. Thus, $\mathcal{G}^\infty_{\psi,A}(0)$ is equal to 0.

(a) Suppose the function S_ψ is bounded from above on the class A. Then there is a constant $q > 0$ where $S_\psi(T) \leq q$ holds for every T belonging to A. By Lemma 8, $\psi^a(T) \leq S_\psi(T) \leq q$ for every $T \in A$. Thus, for any n belonging to ω, $\mathcal{G}_{\psi,A}^\infty(n) \leq q$.

(b) Suppose the function S_ψ has no constant upper bound on the class A. By Lemma 6, we derive that the set $D = \{W_\psi(T) : T \in A, \psi^a(T) = W_\psi(T)\}$ is infinite. Taking into account that A is CC, $\Lambda \in A$ and, since $\psi^a(\Lambda) = W_\psi(\Lambda) = 0$, 0 belongs to D. Obviously, for every n belonging to D, $\mathcal{G}_{\psi,A}^\infty(n)$ is greater than or equal to n. Since $\mathcal{G}_{\psi,A}^\infty$ is a nondecreasing function, we derive that $\mathcal{G}_{\psi,A}^\infty(n)$ is greater than or equal to $H_D(n)$ for every n belonging to $\omega \setminus \{0\}$. □

5 Conclusions

This paper examines the connections between three parameters of tables derived from CCs of DTs with many-valued decisions: the minimum complexity of a DDTR, the minimum complexity of a NDTR, and the complexity of the set of attributes associated with the columns. Further investigation will be focused on exploring the relationships between the time and space complexity of DDTRs and NDTRs for DTs belonging to closed classes.

Acknowledgements. Research reported in this publication was supported by King Abdullah University of Science and Technology (KAUST).

References

1. Alsolami, F., Azad, M., Chikalov, I., Moshkov, M.: Decision and Inhibitory Trees and Rules for Decision Tables with Many-valued Decisions. ISRL, vol. 156. Springer, Cham (2020). https://doi.org/10.1007/978-3-030-12854-8
2. Blum, M., Impagliazzo, R.: Generic oracles and oracle classes (extended abstract). In: 28th Annual Symposium on Foundations of Computer Science, Los Angeles, California, USA, 27–29 October 1987, pp. 118–126. IEEE Computer Society (1987)
3. Boros, E., Hammer, P.L., Ibaraki, T., Kogan, A., Mayoraz, E., Muchnik, I.B.: An implementation of logical analysis of data. IEEE Trans. Knowl. Data Eng. **12**(2), 292–306 (2000)
4. Boutell, M.R., Luo, J., Shen, X., Brown, C.M.: Learning multi-label scene classification. Pattern Recogn. **37**(9), 1757–1771 (2004)
5. Buhrman, H., de Wolf, R.: Complexity measures and decision tree complexity: a survey. Theor. Comput. Sci. **288**(1), 21–43 (2002)
6. Fürnkranz, J., Gamberger, D., Lavrac, N.: Foundations of Rule Learning. Cognitive Technologies. Springer, Cham (2012). https://doi.org/10.1007/978-3-540-75197-7
7. Hartmanis, J., Hemachandra, L.A.: One-way functions, robustness, and the non-isomorphism of NP-complete sets. In: Proceedings of the Second Annual Conference on Structure in Complexity Theory, Cornell University, Ithaca, New York, USA, 16–19 June 1987. IEEE Computer Society (1987)
8. Molnar, C.: Interpretable Machine Learning. A Guide for Making Black Box Models Explainable, 2nd edn. (2022). https://christophm.github.io/interpretable-ml-book/

9. Moshkov, M.: On depth of conditional tests for tables from closed classes. In: Markov, A.A. (ed.) Combinatorial-Algebraic and Probabilistic Methods of Discrete Analysis, pp. 78–86. Gorky University Press, Gorky (1989). (in Russian)
10. Moshkov, M.: Comparative Analysis of Deterministic and Nondeterministic Decision Trees. ISRL, vol. 179. Springer, Cham (2020). https://doi.org/10.1007/978-3-030-41728-4
11. Moshkov, M., Piliszczuk, M., Zielosko, B.: Partial Covers, Reducts and Decision Rules in Rough Sets - Theory and Applications. Studies in Computational Intelligence, vol. 145. Springer, Cham (2008). https://doi.org/10.1007/978-3-540-69029-0
12. Moshkov, M., Zielosko, B.: Combinatorial Machine Learning - A Rough Set Approach. Studies in Computational Intelligence, vol. 360. Springer, Cham (2011). https://doi.org/10.1007/978-3-642-20995-6
13. Ostonov, A., Moshkov, M.: Comparative analysis of deterministic and nondeterministic decision trees for decision tables from closed classes. arXiv:2304.10594 [cs.CC] (2023)
14. Ostonov, A., Moshkov, M.: Deterministic and nondeterministic decision trees for decision tables with many-valued decisions from closed classes. In: Campagner, A., Urs Lenz, O., Xia, S., Slezak, D., Was, J., Yao, J. (eds.) IJCRS 2023. Lecture Notes in Computer Science, vol. 14481, pp. 89–104. Springer, Cham (2023). https://doi.org/10.1007/978-3-031-50959-9_7
15. Ostonov, A., Moshkov, M.: Deterministic and strongly nondeterministic decision trees for decision tables from closed classes. arXiv:2305.06093 [cs.CC] (2023)
16. Ostonov, A., Moshkov, M.: On complexity of deterministic and nondeterministic decision trees for conventional decision tables from closed classes. Entropy **25**(10), 1411 (2023)
17. Ostonov, A., Moshkov, M.: Greedy algorithm for construction of deterministic decision trees for conventional decision tables from closed classes. In: Proceedings of the International Joint Conference on Rough Sets (2024, to appear)
18. Pawlak, Z.: Information systems theoretical foundations. Inf. Syst. **6**(3), 205–218 (1981)
19. Pawlak, Z., Skowron, A.: Rudiments of rough sets. Inf. Sci. **177**(1), 3–27 (2007)
20. Post, E.: Two-Valued Iterative Systems of Mathematical Logic. Annals of Mathematics Studies, vol. 5. Princeton University Press, London (1941)
21. Quinlan, J.R.: C4.5: Programs for Machine Learning. Morgan Kaufmann, Burlington (1993)
22. Robertson, N., Seymour, P.D.: Graph minors. XX. Wagner's conjecture. J. Comb. Theory, Ser. B **92**(2), 325–357 (2004)
23. Rokach, L., Maimon, O.: Data Mining with Decision Trees - Theory and Applications. Series in Machine Perception and Artificial Intelligence, vol. 69. World Scientific (2007)
24. Skowron, A., Rauszer, C.: The discernibility matrices and functions in information systems. In: Slowinski, R. (ed.) Intelligent Decision Support. TDLD, vol. 11, pp. 331–362. Springer, Cham (1992). https://doi.org/10.1007/978-94-015-7975-9_21
25. Tardos, G.: Query complexity, or why is it difficult to separate $NP^A \cap coNP^A$ from P^A by random oracles A? Combinatorica **9**(4), 385–392 (1989)
26. Vens, C., Struyf, J., Schietgat, L., Dzeroski, S., Blockeel, H.: Decision trees for hierarchical multi-label classification. Mach. Learn. **73**(2), 185–214 (2008)
27. Zhou, Z., Zhang, M., Huang, S., Li, Y.: Multi-instance multi-label learning. Artif. Intell. **176**(1), 2291–2320 (2012)

Simulating Functioning of Decision Trees for Tasks on Decision Rule Systems

Kerven Durdymyradov[✉] and Mikhail Moshkov

Computer, Electrical and Mathematical Sciences and Engineering Division and Computational Bioscience Research Center, King Abdullah University of Science and Technology (KAUST), Thuwal 23955-6900, Saudi Arabia
{kerven.durdymyradov,mikhail.moshkov}@kaust.edu.sa

Abstract. DRSs (Decision Rule Systems) and DTs (Decision Trees) are well known as classification tools, knowledge representation methods, and algorithms. Their clarity and ease of interpretation in data analysis are widely recognized. The study of the relationship between DTs and DRSs is an important problem in computer science. There are established methods for converting DTs to DRSs. In this work, we explore the inverse transformation problem, which is challenging. Rather than constructing a full DT that answers the tasks on DRSs, our research provides a greedy algorithm that simulates the functioning of a DT for an input array of feature values.

Keywords: Decision trees · Decision rule systems · Greedy algorithm

1 Introduction

DRSs (Decision Rule Systems) [6,7,11,15,34–37] and DTs (Decision Trees) [3,4,8,26,35,41] are critical in their roles as classification tools, methods for representing knowledge, and as algorithms. Their clarity and ease of interpretation in data analysis are well-recognized [10,16,24,42].

Exploring the interconnections of these models is a critical task in computer science. Established techniques exist for transforming DTs into DRSs [38–40]. However, this work concentrates on exploring the task of transforming DRSs into DTs, which presents its own set of complexities.

The study associated with this task has a variety of focus areas:

- The dual-stage development of DTs is a method that first, establishes DRs (Decision Rules) from the given data, and then develops DTs by the established DRs. The advantages of this dual-stage development are detailed in [1,2,18–23,43].
- Connections between the depths of deterministic and nondeterministic DTs in the computation of Boolean functions [5,9,17,27,44]. Here, it's possible to represent nondeterministic DTs as DRSs.

- Connections between the depths of nondeterministic DTs (representation of DRSs) and deterministic DTs for the tasks within the information systems [25,28,30,32,33]. An information system is composed of the universe of objects and the set of features defined on this universe [36].

Our work expands on the syntactic approach introduced in earlier studies [29,31]. It operates under the assumption that a DRS is available, but there is an absence of information about the input data, and the goal is to convert this DRS into a DT.

Suppose S is a DRS, which consists of DRs in the form:

$$(f_{i_1} = b_1) \wedge \cdots \wedge (f_{i_m} = b_m) \to t,$$

where $f_{i_1}, \ldots, f_{i_m}$ represent features, $b_1, \ldots, b_m$ represent the respective values of the features, while t denotes the decision. There are two tasks related to this DRS:

- It is necessary to identify at least one DR that is feasible for a given input - values of features from S (having a true left part) or show that there are no such DRs.
- It is necessary to identify all the right parts of DRs that are feasible for a given input - values of features from S (we should return feasible rules with these right parts), or show that there are no such DRs.

For every task, we evaluate two versions. The initial version assumes that for each feature in the input, only the values present for this feature in the DRS S are permissible. In the alternate version, it is assumed that every feature in the input can take on any value, and we will say that this is the extended version of the tasks above.

In fact, there is another task that is associated with the system:

- For given values of features from S, it is necessary to identify all the feasible rules.

However, we have already considered this task in our previous work [13], so we will not consider it in here.

This study aims to reduce the quantity of queries needed to evaluate feature values. For that purpose, DTs are investigated as algorithmic solutions for addressing the considered four tasks.

In our prior research [13], we explored the minimum depth of DTs for these tasks and determined the bounds of the depth. The depth's bounds are subject to these characteristics of DRS: the length of the longest DR in the DRS, the total count of unique features, and the number of values for the feature with the most different values. We have shown that for certain DRSs, the minimum depth of the DTs can be considerably less than the total number of features in the DRS. This discovery shows that DTs are an effective option for such DRSs.

In our other paper [12], we analyzed the complexity involved in building DTs and the acyclic decision graphs that represent DTs. It has been observed that often, the minimum number of vertices in DT can increase superpolynomially relative to the dimensions of the DRSs. To handle this challenge, we proposed two variants of acyclic decision graphs as alternatives to DTs. Nevertheless, the concurrent reduction of the depth of these graphs and the vertex count of them presents a complex dual-criterion optimization task.

This task has been reserved for subsequent studies, and leading us to adopt a different method: rather than building the complete DT, we created a polynomial time algorithm that simulates the DT's functioning for an input - array of feature values. This algorithm relies on a supporting algorithm designed to find a vertex cover of a hypergraph that aligns with a DRS, where the hypergraph's vertices represent features and its hyperedges represent DRs from the DRS. The supporting algorithm, in each step appends into the developing cover, all features associated with a DR that hasn't been covered yet. Note that this is not a greedy algorithm.

In this study, we introduce an innovative algorithm that shows the functioning of a DT that answers the mentioned tasks for an input - array of feature values. For it, we use a supporting algorithm from [14] that solves the vertex cover problem greedily. It leads us to categorize the whole algorithm that shows the functioning of DT as greedy. We evaluated the accuracy of this algorithm by comparing the depth of the DT described with the minimum depth of a DT. The resulting bounds are not as good as for the algorithm discussed in [12]. Nonetheless, we believe that these algorithms complement each other: the earlier algorithm is more effective for DRSs with shorter DRs, while the new one shows promise for DRSs with longer DRs. For future, computer experiments are planned to investigate this hypothesis further.

The study reiterates the key definitions originally presented in [13] and provides some of the lemmas from [13], omitting their proofs.

This study contains five sections. In Sect. 2, we introduce key definitions. Section 3 presents supporting statements. Section 4 explains the greedy algorithm, showing the functioning of the DT, and the final Sect. 5 provides a brief conclusion and plans for the future.

2 Definitions

This section focuses on the definitions associated with DRSs and DTs. Actually, we are restating the relevant definitions from [13].

2.1 DRSs – Decision Rule Systems

Suppose $\mathbb{N}_0 = \{0, 1, 2, \ldots\}$, and $F = \{f_i : i \in \mathbb{N}_0\}$ is the set elements of which are called *features*.

Definition 1. *The expression below:*

$$(f_{i_1} = b_1) \wedge \cdots \wedge (f_{i_m} = b_m) \to t,$$

is a DR (Decision Rule). *Here $m \in \mathbb{N}_0$, $f_{i_1}, \ldots, f_{i_m}$ are distinct features from F and $b_1, \ldots, b_m, t \in \mathbb{N}_0$.*

This DR is represented as r. Here, t is the *right part* and $(f_{i_1} = b_1) \wedge \cdots \wedge (f_{i_m} = b_m)$ is the *left part* in the DR r. The *length* of the DR r is m. Let, $F(r) = \{f_{i_1}, \ldots, f_{i_m}\}$ and $E(r) = \{f_{i_1} = b_1, \ldots, f_{i_m} = b_m\}$. In the case of $m = 0$, $F(r)$ and $E(r)$ are empty sets. We assume that each DR has an *identifier* and identifiers are linearly ordered.

Definition 2. *The DRs r_1 and r_2 are equal when $E(r_1) = E(r_2)$ and also the right parts of the DRs r_1 and r_2 are equal.*

Definition 3. *A DRS (Decision Rule System) S is a nonempty and finite set of DRs with pairwise different identifiers. The DRS S can have equal DRs with different identifiers.*

Suppose $F(S) = \bigcup_{r \in S} F(r)$, $n(S) = |F(S)|$, $D(S)$ the set of the right parts of DRs from S, and $l(S)$ the length of a DR with the maximum length from the DRS S. Suppose $n(S) > 0$. For $f_i \in F(S)$, let $V_S(f_i) = \{b : f_i = b \in \bigcup_{r \in S} E(r)\}$ and $EV_S(f_i) = V_S(f_i) \cup \{*\}$, here $*$ is interpreted as a number from $\mathbb{N}_0$ that is not included in the set $V_S(f_i)$. Let $k(S) = \max |\{b : f_i = b \in \bigcup_{r \in S} E(r)\}|$, where the maximum is taken over all features $f_i \in F(S)$. In the case of $n(S) = 0$, we will have $k(S) = 0$. Let us denote the set of DRSs by Σ.

Suppose $S \in \Sigma$, $n(S) > 0$, and $F(S) = \{f_{j_1}, \ldots, f_{j_n}\}$, where $j_1 < \cdots < j_n$. Let $V(S) = V_S(f_{j_1}) \times \cdots \times V_S(f_{j_n})$ and $EV(S) = EV_S(f_{j_1}) \times \cdots \times EV_S(f_{j_n})$. For $\bar{b} = (b_1, \ldots, b_n) \in EV(S)$, let us denote $E(S, \bar{b}) = \{f_{j_1} = b_1, \ldots, f_{j_n} = b_n\}$.

Definition 4. *It will be stated that a DR r from S is feasible for an array $\bar{b} \in EV(S)$ if $E(r) \subseteq E(S, \bar{b})$.*

It is obvious that the DRs with an empty left part will be feasible for the array $\bar{b}$.

Let's introduce the tasks for the DRS S and $V \in \{V(S), EV(S)\}$.

Definition 5. *For (S, V), task All Decisions: for an array $\bar{b} \in V$, the objective is to identify a set Z of DRs from S meeting the subsequent criteria:*

– *Any DR from Z is feasible for the array $\bar{b}$.*
– *For any $t \in D(S) \setminus D(Z)$, any DR from S with the right part equal to t is not feasible for the array $\bar{b}$.*

Definition 6. *For (S, V), task Some Rules: for an array $\bar{b} \in V$, the objective is to identify a set Z of DRs from S meeting the subsequent criteria:*

– *Any DR from Z is feasible for the array $\bar{b}$.*
– *When $Z = \emptyset$, that means all DRs from S is not feasible for the array $\bar{b}$.*

For $(S, V(S))$ and $(S, EV(S))$ the tasks All Decisions will be denoted as $AD(S)$ and $EAD(S)$, respectively, and the tasks Some Rules will be denoted as $SR(S)$ and $ESR(S)$, respectively.

If $n(S) = 0$, any DR from S have an empty left part. Then, we can say (i) any subset Z of the set S with $D(Z) = D(S)$ is a solution of the tasks $AD(S)$ and $EAD(S)$, and (ii) any nonempty subset Z of the set S is a solution of the tasks $SR(S)$ and $ESR(S)$.

2.2 DTs - Decision Trees

A FDT (Finite Directed Tree) that has only one vertex with no entering edges will be called *FDT with root*. This vertex is the *root* of FDT. The *leaf* vertices of FDT are the vertices with no leaving edges. The *active* vertices are the vertices that are not leaf. In an FDT with root, a *full path* is a path from the root to a leaf vertex of the form $\mathcal{P} = v_1, e_1, \ldots, v_m, e_m, v_{m+1}$ where v_1 stands for the root, v_{m+1} for a leaf vertex and, for $i = 1, \ldots, m$, the edge e_i connects v_i and v_{i+1}.

There are two types of DTs: o-DTs (ordinary DTs) and e-DTs (extended DTs).

Definition 7. *A DT based on DRS S is a marked FDT with root $\mathcal{G}$, which fulfills the criteria below:*

- *Every active vertex in the DT $\mathcal{G}$ is marked with a feature from $F(S)$.*
- *Suppose an active vertex v in the DT $\mathcal{G}$ is marked with a feature f_i. If $\mathcal{G}$ is an o-DT, then the number of leaving edges of the vertex v is exactly $|V_S(f_i)|$ and they are marked with distinct elements from $V_S(f_i)$. If $\mathcal{G}$ is an e-DT, then the number of leaving edges of the vertex v is exactly $|EV_S(f_i)|$ and they are marked with distinct elements from $EV_S(f_i)$.*
- *Every leaf vertex in the DT $\mathcal{G}$ is marked with a subset of S.*

Suppose $\mathcal{G}$ is a DT based on DRS S. The set of full paths within the DT $\mathcal{G}$ is represented as $FP(\mathcal{G})$. Suppose $\mathcal{P} = v_1, e_1, \ldots, v_m, e_m, v_{m+1}$ is a full path in $\mathcal{G}$. That path is associated with a set of features $F(\mathcal{P})$ and the SE (System of Equations) $E(\mathcal{P})$. When $m = 0$ and $\mathcal{P} = v_1$, sets $F(\mathcal{P})$ and $E(\mathcal{P})$ will be empty sets. Suppose $m > 0$ and, for $j = 1, \ldots, m$, the vertex v_j is marked with the feature f_{i_j} and the edge e_j is marked with $b_j \in \mathbb{N}_0 \cup \{*\}$. So, $F(\mathcal{P}) = \{f_{i_1}, \ldots, f_{i_m}\}$ and $E(\mathcal{P}) = \{f_{i_1} = b_1, \ldots, f_{i_m} = b_m\}$. The set of DRs attached to the leaf vertex v_{m+1} is denoted by $\tau(\mathcal{P})$.

Definition 8. *A SE $\{f_{i_1} = b_1, \ldots, f_{i_m} = b_m\}$, with $f_{i_1}, \ldots, f_{i_m} \in F$ and $b_1, \ldots, b_m \in \mathbb{N}_0 \cup \{*\}$, will be called* incompatible *SE when we have $l, k \in \{1, \ldots, m\}$ where $l \neq k$, $i_l = i_k$, and $b_l \neq b_k$. In case it is not incompatible, it will be called* compatible.

Suppose S is a DRS and $\mathcal{G}$ be a DT based on S. Now let's define the tasks $AD(S)$ - that requires to identify all the right parts of DRs that are feasible for a given input or show that there are no such DRs, and $SR(S)$ - that requires to identify at least one DR that is feasible for a given input or show that there are no such DRs.

Definition 9. *It will be stated that $\mathcal{G}$ answering to the task $AD(S)$ (the task $EAD(S)$, respectively) if $\mathcal{G}$ is an o-DT (an e-DT, respectively) and any path $\mathcal{P} \in FP(\mathcal{G})$ with compatible SE $E(\mathcal{P})$ fulfills the criteria below:*

- $E(r) \subseteq E(\mathcal{P})$, for all $r \in \tau(\mathcal{P})$.
- The SE $E(r) \cup E(\mathcal{P})$ is incompatible for all $r \in S \setminus \tau(\mathcal{P})$ such that the right part of r does not belong to the set $D(\tau(\mathcal{P}))$.

Definition 10. It will be stated that $\mathcal{G}$ answering to the task $SR(S)$ (the task $ESR(S)$, respectively) if $\mathcal{G}$ is an o-DT (an e-DT, respectively) and any path $\mathcal{P} \in FP(\mathcal{G})$ with compatible SE $E(\mathcal{P})$ fulfills the criteria below:

- $E(r) \subseteq E(\mathcal{P})$, for all $r \in \tau(\mathcal{P})$.
- If $\tau(\mathcal{P}) = \emptyset$, then the SE $E(r) \cup E(\mathcal{P})$ is incompatible for all $r \in S$

For every full path $\mathcal{P} \in FP(\mathcal{G})$, the number of active vertices in $\mathcal{P}$ is indicated as $h(\mathcal{P})$. The *depth* of the DT $\mathcal{G}$ is the number $h(\mathcal{G}) = \max\{h(\mathcal{P}) : \mathcal{P} \in FP(\mathcal{G})\}$.

Suppose S is a DRS and $C \in \{AD, EAD, SR, ESR\}$. Then $h_C(S)$ represents the minimum depth of a DT based on DRS S that answers to the task $C(S)$.

Let $n(S) = 0$. If $C \in \{AD, EAD\}$, then the set of DTs answering to the task $C(S)$ corresponds to the set of trees containing a single vertex, which is marked with a subset Z of the set S with $D(Z) = D(S)$. If $C \in \{SR, ESR\}$, then the set of DTs answering to the task $C(S)$ corresponds to the set of trees containing a single vertex, which is marked with a nonempty subset Z of the set S. Thus, for $C \in \{AD, EAD, SR, ESR\}$, $h_C(S) = 0$ when $n(S) = 0$.

3 Supporting Statements

Initially, we present several observations from [13], and after we demonstrate novel ones.

Suppose S be a DRS and $\alpha = \{f_{i_1} = b_1, \ldots, f_{i_m} = b_m\}$ be a compatible SE such that $f_{i_1}, \ldots, f_{i_m} \in F$ and $b_1, \ldots, b_m \in \mathbb{N}_0 \cup \{*\}$. Let's introduce a DRS S_α. Consider r as a DR where the SE $E(r) \cup \alpha$ is compatible. The notation r_α refers to the DR derived from r by eliminating all equations of the left part of r that are included in α. The rules r and r_α have the same identifiers. We will say that the rule r_α *corresponds* to the rule r. Subsequently, S_α is the set of DRs r_α for which $r \in S$ and the SE $E(r) \cup \alpha$ is compatible.

Lemma 1. *(derived from Lemma 6 [13]) Suppose S is a DRS where $n(S) > 0$, $C \in \{AD, EAD, SR, ESR\}$, $\alpha = \{f_{i_1} = b_1, \ldots, f_{i_m} = b_m\}$ is a compatible SE where $f_{i_1}, \ldots, f_{i_m} \in F(S)$ and, for $j = 1, \ldots, m$, $b_j \in EV_S(f_{i_j})$ if $C \in \{EAD, ESR\}$ and $b_j \in V_S(f_{i_j})$ if $C \in \{AD, SR\}$. Then $h_C(S) \geq h_C(S_\alpha)$.*

A hypergraph $G(S)$ corresponding to a DRS S is a hypergraph with the set of vertices $F(S)$ and the set of hyperedges $\{F(r) : r \in S\}$. The vertex cover for that hypergraph is defined as a subset B of the vertex set $F(S)$, ensuring that $F(r) \cap B \neq \emptyset$ holds for any $r \in S$ with $F(r) \neq \emptyset$. In the case of $F(S) = \emptyset$, the empty set will be the only vertex cover of $G(S)$. The notation $\beta(S)$ is used to represent the minimum cardinality of the vertex cover of $G(S)$.

Let $I_{SR}(S)$ be a subsystem of the DRS S as follows: if S does not contain DRs with length 0, then $I_{SR}(S) = S$. Otherwise, $I_{SR}(S)$ consists of all DRs from S of the length 0.

Let $I_{AD}(S)$ be a subsystem of the DRS S as follows: let, $D_0(S)$ the set of the right parts of DRs from S, which length is equal to 0, then $I_{AD}(S)$ contains all DRs from S of the length 0 and all DRs from S for which the right parts do not belong to $D_0(S)$.

Lemma 2. *(derived from Lemma 7 [13])*

$$h_{EC}(S) \geq \beta(I_C(S))$$

for any DRS S, where $C \in \{AD, SR\}$.

Let S be a DRS, and let's define $R_{SR}(S)$ a subsystem of the DRS S, it contains all DRs $r \in S$ which fulfills the criteria: there is no DR $r' \in S$ such that $E(r') \subset E(r)$. The DRS S will be called *SR-reduced* if $R_{SR}(S) = S$.

Let's define $R_{AD}(S)$ a subsystem of the DRS S, it contains all DRs $r \in S$ which fulfills the criteria: there is no DR $r' \in S$ such that $E(r') \subset E(r)$ and the right parts of the DRs r and r' are equal. The DRS S will be called *AD-reduced* if $R_{AD}(S) = S$.

Lemma 3. *(derived from Lemma 8 [13]) For a DRS S:*

(a) $h_{ESR}(S) \geq l(S)$, if S is an SR-reduced DRS.
(b) $h_{EAD}(S) \geq l(S)$, if S is an AD-reduced DRS.

Lemma 4. *(Lemma 9 [13]) For a DRS S:*

$$h_{ESR}(S) = h_{ESR}(R_{SR}(S))$$

and

$$h_{EAD}(S) = h_{EAD}(R_{AD}(S)).$$

Let's define S^+ as the set of all DRs of the length $l(S)$ from the DRS S. DRs r_1 and r_2 from S^+ are called *equivalent* if $E(r_1) = E(r_2)$. This equivalence relation divides the set S^+ into distinct equivalence classes. The set $S^{\max}$ is defined as a set of DRs with exactly one representative from every equivalence class while excluding any additional DRs. It is clear that a set of features is a vertex cover for the hypergraph $G(S^+)$ if and only if this set is a vertex cover for the hypergraph $G(S^{\max})$.

Lemma 5. *For an SR-reduced DRS S with $n(S) > 0$:*

$$h_{ESR}(S) \geq \ln |S^{\max}| / \ln(k(S) + 1).$$

Proof. Let $r \in S^{\max}$ and the DR r be equal to $(f_{i_1} = b_1) \wedge \cdots \wedge (f_{i_m} = b_m) \to t$. We now define an array $\bar{b}(r) \in EV(S)$. For $j = 1, \ldots, m$, the array $\bar{b}(r)$ in the position corresponding to the feature f_{i_j} contains the number b_j. All other positions of the array $\bar{b}(r)$ are filled with the symbol $*$. We denote by $S(\bar{b}(r))$ the set of DRs from S that are feasible for the array $\bar{b}(r)$. Using the fact that the DRS S is SR-reduced, one can show that the set $S(\bar{b}(r))$ consists of the DR r and all DRs from S^+, which are equivalent to r. From here it follows that any DT addressing the task $ESR(S)$ has at least $|S^{\max}|$ leaf vertices.

Let $\mathcal{G}$ be a DT, which answers the task $ESR(S)$ and for which $h(\mathcal{G}) = h_{ESR}(S)$. It is easy to show that the number of leaf vertices in $\mathcal{G}$ is at most $(k(S)+1)^{h(\mathcal{G})}$. Therefore $(k(S)+1)^{h(\mathcal{G})} \geq |S^{\max}|$ and $h(\mathcal{G}) \geq \ln|S^{\max}|/\ln(k(S)+1)$. Thus, $h_{ESR}(S) \geq \ln|S^{\max}|/\ln(k(S)+1)$.

Lemma 6. *For an AD-reduced DRS S with $n(S) > 0$:*

$$h_{EAD}(S) \geq \ln|S^{\max}|/\ln(k(S)+1).$$

Proof. Let $r \in S^{\max}$ and the right part of r be equal to t. We define the array $\bar{b}(r) \in EV(S)$ just as in the proof of Lemma 5. We denote by $S(\bar{b}(r))$ the set of DRs from S that are feasible for the array $\bar{b}(r)$. The set $S^{\max}(\bar{b}(r))$ is defined in a similar way. Using the fact that the DRS S is AD-reduced, one can show that in $S(\bar{b}(r))$ only DR r and DRs equal to r have right parts equal to t and r is the only DR in $S^{\max}(\bar{b}(r))$. From here it follows that any DT addressing the task $EAD(S)$ has at least $|S^{\max}|$ leaf vertices. Further, just as in the proof of Lemma 5, we show that $h_{EAD}(S) \geq \ln|S^{\max}|/\ln(k(S)+1)$.

4 Algorithms

This section explores a supporting algorithm $\mathcal{A}_{greedy}$ that finds a vertex cover for the hypergraph associated with a DRS, and an algorithm $\mathcal{A}^C$, where $C \in \{AD, EAD, SR, ESR\}$, that shows the functioning of a DT answering to the task $C(S)$ for a DRS S where $n(S) > 0$.

4.1 Supporting Algorithm $\mathcal{A}_{greedy}$

In this section, we consider an algorithm for construction of vertex covers. In fact, we repeat the algorithm from [14].

Suppose S is a DRS where $n(S) > 0$. We now explain a polynomial time algorithm $\mathcal{A}_{greedy}$ designed for creating a vertex cover B for a given hypergraph $G(S^+)$ satisfying the inequality $|B| \leq \beta(S^+)\ln|S^{\max}| + 1$.

Algorithm $\mathcal{A}_{greedy}$

At each step, the algorithm selects a feature $f_i \in F(S^{\max})$ having the lowest index i that covers the maximum number of previously uncovered DRs from $S^{\max}$ and includes it in the set B (a feature f_i is considered to cover a DR $r \in S^{\max}$ if $f_i \in F(r)$). $\mathcal{A}_{greedy}$ completes its work once all DRs from $S^{\max}$ have been covered.

The algorithm in discussion fundamentally mirrors the widely recognized greedy algorithm for the set cover problem, as detailed in Sect. 4.1.1 of the book [35].

Lemma 7. *(derived from Theorem 4.1 [35]) Suppose S is a DRS where $n(S) > 0$. The algorithm $\mathcal{A}_{greedy}$ finds a vertex cover B for the hypergraph $G(S^+)$ satisfying the inequality $|B| \leq \beta(S^{\max}) \ln |S^{\max}| + 1 = \beta(S^+) \ln |S^{\max}| + 1$.*

4.2 Greedy Algorithm $\mathcal{A}^C$, $C \in \{SR, ESR, AD, EAD\}$

Let S be a DRS with $n(S) > 0$, $C \in \{SR, ESR, AD, EAD\}$. Let $V_C = V(S)$ if $C \in \{SR, AD\}$, and $V_C = EV(S)$ if $C \in \{ESR, EAD\}$. Let $R_C = R_{SR}$ if $C \in \{SR, ESR\}$ and $R_C = R_{AD}$ if $C \in \{AD, EAD\}$. We now describe a polynomial time algorithm $\mathcal{A}^C$ that, for a given array of feature values $\bar{b}$ from the set V_C, describes the functioning on this array of a DT $\mathcal{G}$, which answers the task $C(S)$.

Algorithm $\mathcal{A}^C$

Set $Q := S$.

Step 1. Set $P := R_C(Q)$. If $P = \emptyset$ or all DRs from P have an empty left part, then the tree $\mathcal{G}$ finishes its functioning. The result of this work is the set of DRs from S that correspond to the DRs with an empty left part from the set P. Otherwise, we move on to Step 2.

Step 2. Using the algorithm $\mathcal{A}_{greedy}$, we construct a node cover B of the hypergraph $G(P^+)$. The decision tree $\mathcal{G}$ sequentially computes values of the features from the set B ordered by ascending feature indices. As a result, we obtain a system α consisting of $|B|$ equations of the form $f_{i_j} = b_j$, where $f_{i_j} \in B$ and b_j is the computed value of the feature f_{i_j} (value of this feature from the array $\bar{b}$). Set $Q := P_\alpha$ and move on to Step 1.

We now consider bounds on accuracy for the algorithm $\mathcal{A}^C$, where $C \in \{SR, ESR, AD, EAD\}$.

Let S be a DRS with $n(S) > 0$. We denote by $E(S)$ the set of compatible SE $\{f_{i_1} = b_1, \ldots, f_{i_m} = b_m\}$ such that $f_{i_j} \in A(S)$ and $b_j \in V_S(f_{i_j})$ for $j = 1, \ldots, m$. Denote $\Phi(S) = \max\{\beta(S_\alpha) : \alpha \in E(S)\}$.

Theorem 1. *Let S be a DRS with $n(S) > 0$.*

(a) *If $C \in \{ESR, EAD\}$, then the algorithm $\mathcal{A}^C$ describes the functioning of a DT $\mathcal{G}$, which answers the task $C(S)$ and for which $h(\mathcal{G}) \leq h_C(S)^3 \ln(k(S)+1) + h_C(S)$.*

(b) *If $C \in \{SR, AD\}$, then the algorithm $\mathcal{A}^C$ describes the functioning of a DT $\mathcal{G}$, which answers the task $C(S)$ and for which $h(\mathcal{G}) \leq l(S)\Phi(S) \ln |S| + l(S)$.*

Proof. (a) Let us consider Step 2 of the algorithm. Note that $P := R_C(Q)$. Using Lemmas 1 and 4, we obtain $h_C(Q) = h_C(P) \geq h_C(P_\alpha)$. If we apply the same relations to the previous repetitions of Step 2, we obtain $h_C(S) \geq h_C(Q) = h_C(P)$.

It is clear that $l(R_C(S)) \geq l(P) > l(P_\alpha)$. From these inequalities, it follows that the number of repetitions of Step 2 is at most $l(R_C(S))$. Using Lemmas

3 and 4, we obtain $h_C(S) = h_C(R_C(S)) \geq l(R_C(S))$. Thus, the number of repetitions of Step 2 is at most $h_C(S)$.

From Lemma 7 it follows that the number of features values of which are computed by $\mathcal{G}$ during Step 2 is at most $\beta(P^{\max}) \ln |P^{\max}| + 1$. Evidently, $P^{\max} \subseteq P = R_C(Q) \subseteq I_C(Q)$. Therefore $\beta(P^{\max}) \leq \beta(I_C(Q))$. By Lemma 2, $\beta(I_C(Q)) \leq h_C(Q)$. We know that $h_C(Q) \leq h_C(S)$. Thus, $\beta(P^{\max}) \leq h_C(S)$. Evidently, $k(P) \leq k(S)$. By Lemmas 5 and 6, $\ln |P^{\max}| \leq h_C(P) \ln(k(P) + 1) \leq h_C(S) \ln(k(S) + 1)$. Therefore the number of features values of which are computed by $\mathcal{G}$ during Step 2 is at most $h_C(S)^2 \ln(k(S) + 1) + 1$. The number of repetitions of Step 2 is at most $h_C(S)$. Thus, the depth of the DT $\mathcal{G}$ is at most $h_C(S)^3 \ln(k(S) + 1) + h_C(S)$.

(b) Let us consider Step 2 of the algorithm. Note that $P := R_C(Q)$. It is clear that $l(S) \geq l(P) > l(P_\alpha)$. From these inequalities, it follows that the number of repetitions of Step 2 is at most $l(S)$. From Lemma 7 it follows that the number of features values of which are computed by $\mathcal{G}$ during Step 2 is at most $\beta(P^+) \ln |P^{\max}| + 1$. One can show that $P = S'_\gamma$ for some subsystem S' of the DRS S and some subsystem γ of the SE $E(S, \bar{b})$. It is clear that $P^+ \subseteq P = S'_\gamma \subseteq S_\gamma$. Therefore $\beta(P') \leq \beta(S_\gamma) \leq \Phi(S)$. It is clear that $|P^{\max}| \leq |S|$. As a result, the depth of the DT $\mathcal{G}$ is at most $l(S)\Phi(S) \ln |S| + l(S)$.

5 Conclusion and Plans for Future

In this study, we explored a novel algorithm $\mathcal{A}^C$, for $C \in \{SR, ESR, AD, EAD\}$, designed to simulate the functioning of a DT that answers to the task $C(S)$ for an input array of feature values. We evaluated the accuracy of this algorithm by comparing the depth of the DT described with the minimum depth of a DT. The resulting bound is not as good as for the algorithm discussed in [12]. Nonetheless, we believe that these algorithms complement each other: the earlier algorithm is more effective for DRSs with shorter DRs, while the new one shows promise for DRSs with longer DRs. Future plans include conducting computer experiments to test this assumption and to create a dynamic programming algorithm to minimize the depth of DTs and to experimentally compare the depths of DTs generated by both algorithms with the minimum depth.

Acknowledgements. Research reported in this publication was supported by King Abdullah University of Science and Technology (KAUST).

References

1. Abdelhalim, A., Traoré, I., Nakkabi, Y.: Creating decision trees from rules using RBDT-1. Comput. Intell. **32**(2), 216–239 (2016)
2. Abdelhalim, A., Traore, I., Sayed, B.: RBDT-1: a new rule-based decision tree generation technique. In: Governatori, G., Hall, J., Paschke, A. (eds.) RuleML 2009. LNCS, vol. 5858, pp. 108–121. Springer, Heidelberg (2009). https://doi.org/10.1007/978-3-642-04985-9_12

3. AbouEisha, H., Amin, T., Chikalov, I., Hussain, S., Moshkov, M.: Extensions of Dynamic Programming for Combinatorial Optimization and Data Mining. ISRL, vol. 146. Springer, Cham (2019). https://doi.org/10.1007/978-3-319-91839-6
4. Alsolami, F., Azad, M., Chikalov, I., Moshkov, M.: Decision and Inhibitory Trees and Rules for Decision Tables with Many-valued Decisions. ISRL, vol. 156. Springer, Cham (2020). https://doi.org/10.1007/978-3-030-12854-8
5. Blum, M., Impagliazzo, R.: Generic oracles and oracle classes (extended abstract). In: 28th Annual Symposium on Foundations of Computer Science, Los Angeles, California, USA, 27–29 October 1987, pp. 118–126. IEEE Computer Society (1987)
6. Boros, E., Hammer, P.L., Ibaraki, T., Kogan, A.: Logical analysis of numerical data. Math. Program. **79**, 163–190 (1997)
7. Boros, E., Hammer, P.L., Ibaraki, T., Kogan, A., Mayoraz, E., Muchnik, I.B.: An implementation of logical analysis of data. IEEE Trans. Knowl. Data Eng. **12**(2), 292–306 (2000)
8. Breiman, L., Friedman, J.H., Olshen, R.A., Stone, C.J.: Classification and Regression Trees. Wadsworth and Brooks (1984)
9. Buhrman, H., de Wolf, R.: Complexity measures and decision tree complexity: a survey. Theor. Comput. Sci. **288**(1), 21–43 (2002)
10. Cao, H.E.C., Sarlin, R., Jung, A.: Learning explainable decision rules via maximum satisfiability. IEEE Access **8**, 218180–218185 (2020)
11. Chikalov, I., Lozin, V.V., Lozina, I., Moshkov, M., Nguyen, H.S., Skowron, A., Zielosko, B.: Three Approaches to Data Analysis - Test Theory, Rough Sets and Logical Analysis of Data. Intelligent Systems Reference Library, vol. 41. Springer, Cham (2013). https://doi.org/10.1007/978-3-642-28667-4
12. Durdymyradov, K., Moshkov, M.: Construction of decision trees and acyclic decision graphs from decision rule systems. arXiv:2305.01721 [cs.AI] (2023)
13. Durdymyradov, K., Moshkov, M.: Bounds on depth of decision trees derived from decision rule systems with discrete attributes. Ann. Math. Artif. Intell. (2024). https://doi.org/10.1007/s10472-024-09933-x
14. Durdymyradov, K., Moshkov, M.: Greedy algorithm for inference of decision trees from decision rule systems. arXiv:2401.06793 [cs.AI] (2024)
15. Fürnkranz, J., Gamberger, D., Lavrac, N.: Foundations of Rule Learning. Cognitive Technologies, Springer, Cham (2012). https://doi.org/10.1007/978-3-540-75197-7
16. Gilmore, E., Estivill-Castro, V., Hexel, R.: More interpretable decision trees. In: Sanjurjo González, H., Pastor López, I., García Bringas, P., Quintián, H., Corchado, E. (eds.) HAIS 2021. LNCS (LNAI), vol. 12886, pp. 280–292. Springer, Cham (2021). https://doi.org/10.1007/978-3-030-86271-8_24
17. Hartmanis, J., Hemachandra, L.A.: One-way functions, robustness, and the non-isomorphism of NP-complete sets. In: Proceedings of the Second Annual Conference on Structure in Complexity Theory, Cornell University, Ithaca, New York, USA, 16–19 June 1987. IEEE Computer Society (1987)
18. Imam, I.F., Michalski, R.S.: Learning decision trees from decision rules: a method and initial results from a comparative study. J. Intell. Inf. Syst. **2**(3), 279–304 (1993)
19. Imam, I.F., Michalski, R.S.: Should decision trees be learned from examples or from decision rules? In: Komorowski, J., Raś, Z.W. (eds.) ISMIS 1993. LNCS, vol. 689, pp. 395–404. Springer, Heidelberg (1993). https://doi.org/10.1007/3-540-56804-2_37
20. Imam, I.F., Michalski, R.S.: Learning for decision making: the FRD approach and a comparative study. In: Raś, Z.W., Michalewicz, M. (eds.) ISMIS 1996. LNCS,

vol. 1079, pp. 428–437. Springer, Heidelberg (1996). https://doi.org/10.1007/3-540-61286-6_167
21. Kaufman, K.A., Michalski, R.S., Pietrzykowski, J., Wojtusiak, J.: An integrated multi-task inductive database VINLEN: initial implementation and early results. In: Džeroski, S., Struyf, J. (eds.) KDID 2006. LNCS, vol. 4747, pp. 116–133. Springer, Heidelberg (2007). https://doi.org/10.1007/978-3-540-75549-4_8
22. Michalski, R.S., Imam, I.F.: Learning problem-oriented decision structures from decision rules: the AQDT-2 system. In: Raś, Z.W., Zemankova, M. (eds.) ISMIS 1994. LNCS, vol. 869, pp. 416–426. Springer, Heidelberg (1994). https://doi.org/10.1007/3-540-58495-1_42
23. Michalski, R.S., Imam, I.F.: On learning decision structures. Fundam. Informaticae **31**(1), 49–64 (1997)
24. Molnar, C.: Interpretable Machine Learning. A Guide for Making Black Box Models Explainable, 2nd edn. (2022). https://christophm.github.io/interpretable-ml-book/
25. Moshkov, M.J.: Comparative analysis of deterministic and nondeterministic decision tree complexity local approach. In: Peters, J.F., Skowron, A. (eds.) Transactions on Rough Sets IV. LNCS, vol. 3700, pp. 125–143. Springer, Heidelberg (2005). https://doi.org/10.1007/11574798_7
26. Moshkov, M.J.: Time complexity of decision trees. In: Peters, J.F., Skowron, A. (eds.) Transactions on Rough Sets III. LNCS, vol. 3400, pp. 244–459. Springer, Heidelberg (2005). https://doi.org/10.1007/11427834_12
27. Moshkov, M.: About the depth of decision trees computing Boolean functions. Fundam. Informaticae **22**(3), 203–215 (1995)
28. Moshkov, M.: Comparative analysis of deterministic and nondeterministic decision tree complexity. Global approach. Fundam. Informaticae **25**(2), 201–214 (1996)
29. Moshkov, M.: Some relationships between decision trees and decision rule systems. In: Polkowski, L., Skowron, A. (eds.) RSCTC 1998. LNCS (LNAI), vol. 1424, pp. 499–505. Springer, Heidelberg (1998). https://doi.org/10.1007/3-540-69115-4_68
30. Moshkov, M.: Deterministic and nondeterministic decision trees for rough computing. Fundam. Informaticae **41**(3), 301–311 (2000)
31. Moshkov, M.: On transformation of decision rule systems into decision trees. In: Proceedings of the Seventh International Workshop Discrete Mathematics and its Applications, Moscow, Russia, 29 January–2 February 2001, Part 1, pp. 21–26. Center for Applied Investigations of Faculty of Mathematics and Mechanics, Moscow State University (2001). (in Russian)
32. Moshkov, M.: Classification of infinite information systems depending on complexity of decision trees and decision rule systems. Fundam. Informaticae **54**(4), 345–368 (2003)
33. Moshkov, M.: Comparative Analysis of Deterministic and Nondeterministic Decision Trees. ISRL, vol. 179. Springer, Cham (2020). https://doi.org/10.1007/978-3-030-41728-4
34. Moshkov, M., Piliszczuk, M., Zielosko, B.: Partial Covers, Reducts and Decision Rules in Rough Sets - Theory and Applications. Studies in Computational Intelligence, vol. 145. Springer, Cham (2008). https://doi.org/10.1007/978-3-540-69029-0
35. Moshkov, M., Zielosko, B.: Combinatorial Machine Learning - A Rough Set Approach. Studies in Computational Intelligence, vol. 360. Springer, Cham (2011). https://doi.org/10.1007/978-3-642-20995-6
36. Pawlak, Z.: Rough Sets - Theoretical Aspects of Reasoning about Data, Theory and Decision Library: Series D, vol. 9. Kluwer (1991)

37. Pawlak, Z., Skowron, A.: Rudiments of rough sets. Inf. Sci. **177**(1), 3–27 (2007)
38. Quinlan, J.R.: Generating production rules from decision trees. In: McDermott, J.P. (ed.) Proceedings of the 10th International Joint Conference on Artificial Intelligence. Milan, Italy, 23–28 August 1987, pp. 304–307. Morgan Kaufmann (1987)
39. Quinlan, J.R.: C4.5: Programs for Machine Learning. Morgan Kaufmann (1993)
40. Quinlan, J.R.: Simplifying decision trees. Int. J. Hum Comput Stud. **51**(2), 497–510 (1999)
41. Rokach, L., Maimon, O.: Data Mining with Decision Trees - Theory and Applications, Series in Machine Perception and Artificial Intelligence, vol. 69. World Scientific (2007)
42. Silva, A., Gombolay, M.C., Killian, T.W., Jimenez, I.D.J., Son, S.: Optimization methods for interpretable differentiable decision trees applied to reinforcement learning. In: Chiappa, S., Calandra, R. (eds.) The 23rd International Conference on Artificial Intelligence and Statistics, AISTATS 2020, Palermo, Sicily, Italy, 26–28 August 2020. Proceedings of Machine Learning Research, vol. 108, pp. 1855–1865. PMLR, Online (2020)
43. Szydlo, T., Sniezynski, B., Michalski, R.S.: A rules-to-trees conversion in the inductive database system VINLEN. In: Klopotek, M.A., Wierzchon, S.T., Trojanowski, K. (eds.) Intelligent Information Processing and Web Mining Advances in Soft Computing, vol. 31, pp. 496–500. Springer, Cham (2005). https://doi.org/10.1007/3-540-32392-9_60
44. Tardos, G.: Query complexity, or why is it difficult to separate $NP^A \cap coNP^A$ from P^A by random oracles A? Combinatorica **9**(4), 385–392 (1989)

RIONIDA: A Novel Algorithm for Imbalanced Data Combining Instance-Based Learning and Rule Induction

Grzegorz Góra[1]([✉]) and Andrzej Skowron[2]

[1] University of Warsaw, Stefana Banacha 2, 02-097 Warsaw, Poland
ggora@mimuw.edu.pl
[2] Systems Research Institute PAS, Newelska 6, 01-447 Warsaw, Poland
skowron@mimuw.edu.pl

Abstract. The article presents the RIONIDA learning algorithm based on combination of two widely-used empirical approaches: rule induction and instance-based learning for imbalanced data classification. The algorithm is a substantial extension of the well-known RIONA algorithm developed for balanced data.

RIONIDA is relatively fast and significantly outperforms the state-of-the-art algorithms analysed in the paper.

Keywords: Imbalanced Learning · Classification · Supervised Learning · Instance-based Learning · k Nearest Neighbours · Rule Induction

1 Introduction

Learning algorithms induce from training sets so-called classifiers that provide decisions for test objects. Any learning algorithm can be applied to a wide range of data sets and generates a classifier based on a given training set. Numerous learning algorithms have been developed so far yet new ones are still being proposed [33]. We focus on the development of the new learning algorithm for imbalanced data called RIONIDA, based on combination of rule induction and instance-based learning.

Recently, much scientific effort has been put into supervised learning that concerns learning from so-called imbalanced data [11,23,24], In classification tasks for imbalanced data the correct classification of objects into one specific decision class is much more important than into others. For the classification task with a binary decision, which we focus on, there is just one class of special importance. Usually, this class includes a much smaller number of objects than the other one. Therefore it is referred to as the minority class and the other one as the majority class. There are several reasons explaining why the standard classifiers (i.e. classifiers induced by learning algorithms designed for balanced data) do not work well with imbalanced data [23, pp. 24–26].

Methods for learning from imbalanced data can be divided into two groups: data-level solutions and algorithmic-level solutions. RIONIDA belongs to the algorithmic-level approach as it is a modification of RIONA [15,18]. We are following the suggestion of Vladimir N. Vapnik:

If you possess a restricted amount of information for solving some problem, try to solve the problem directly and never solve a more general problem as an intermediate step. It is possible that the available information is sufficient for a direct solution but is insufficient for solving a more general intermediate problem. [37, p. 12]

RIONIDA uses rules more general than in RIONA and realises additional concepts in comparison to RIONA, namely optimisation of (i) any explicitly given performance measure defined over the confusion matrix, (ii) weights for two classes, (iii) parameter related to the idea of scaled rules.

RIONIDA, with decisions explainable for the user, is relatively fast and significantly outperforms the state-of-the-art algorithms analysed in the paper, in particular on difficult regions regarding analysis of minority class [27]. The paper reports several results from the PhD thesis [16] that have not yet been published.

The approach described in this paper is closely connected to rough sets. We propose a generalisation of approximation spaces [30]. The existing approaches for concept approximation based on these spaces are based on reasoning about the (partial) inclusion of some specified granules constructed around test cases into decision classes corresponding to approximated concepts. The proposed generalisation requires detailed analysis of training cases within the granules. This kind of reasoning was successfully applied to imbalanced data. In this case, roughly speaking, decision-making for a given test case follows a certain process. Initially, a granule of relevant training cases is identified as a neighbourhood of a given test case using the k-nearest neighbours (k-nn) method, utilising a specialised distance measure and optimised value of k. Subsequently, a specific rule is created for each training case within the neighbourhood of the test case. The granules defined by the left hand sides of these rules are 'narrowed down' and it is checked for each training case from the neighbourhood if obtained in this way a sub-granule is formed with training cases labelled with the same decision as the considered training case. If confirmed, this strengthens the argument for assigning the same decision label to the test case (as it has the training case used for construction of the rule).

Finally, for obtaining the final decision for the considered test case, all arguments obtained for training cases within the neighbourhood are analysed; considering that arguments from minority class cases hold more weight than those from the majority class.

It is important to note that our proposed generalisation of approximation spaces emphasises the crucial role of rough set approach in AI for concept approximation.

The paper is structured as follows. Section 2 discusses related works. Section 3 presents the basic concepts and notation used in the subsequent sections. Section 4 describes the RIONA algorithm designed for balanced data. Section 5 introduces RIONIDA, a modification of RIONA, designed for classification of imbalanced data. Section 6 shows how RIONIDA internal parameters can be learned very efficiently. It also presents time and space complexity of RIONIDA. Section 7 describes the results of experiments in which RIONIDA was compared with some state-of-the-art algorithms for imbalanced data on benchmarks and real-life data sets. Finally, Sect. 8 concludes the paper.

The code of both RIONIDA and RIONA are publicly available [1] (they can be also used in WEKA platform).

2 Related Work

In the past, there have been some attempts to combine instance- and rule-based approaches, however only for balanced data (see e.g. [10,25,35,38]).

The approach used in developing RIONIDA is different from the ones presented in the literature. To our knowledge the only algorithm designed for imbalanced data analysis that combines the instance- and rule-based approaches and at the same time belongs to the algorithmic level approach (which modify algorithms for balanced data) is BRACID (see e.g. [27]). BRACID is a modification of the RISE algorithm to make it applicable for imbalanced data. There are some substantial differences between BRACID and RIONIDA: (i) BRACID calculates rules in the learning phase (in advance), while RIONIDA does it in the testing phase (i.e. according to the lazy approach); (ii) BRACID starts from rules equivalent to instances and induces quasi-optimal rules for the given data set; RIONIDA adopts a different strategy and takes into account a large space of parametrised rules formulated in a specific language; different parametrisations correspond to different approaches, including (a) pure instance-based approach, (b) pure rule-based approach, and (c) approaches that combine them; for the given data set, RIONIDA selects the optimal parameter settings of rules, and does it very efficiently; (iii) BRACID optimises rules for F-measure, while RIONIDA can optimise any performance measure specified by a user (defined on the basis of *confusion matrix*), and does it effectively (in comparison to direct searching of RIONIDA enormous parameter space).

RIONIDA is an extension of RIONA [15,18]. RIONA is using lazy learning approach. It was reported in the literature as one of the most accurate classification methods in many experimental comparisons done by various researchers, to name a few (the authors use the most common RIONA implementation from WEKA platform named RseslibKnn): Facebook content recognition [9, Chapter 1] (RIONA was the best one of 21 tested algorithms), environmental sound recognition [19] (best of 9 algorithms), metabolic pathway prediction of plant enzymes [29] (2nd of 47 algorithms), acoustic-based environment monitoring [34] (2nd of 8 algorithms), context awareness of a service robot [20] (2nd of 8 algorithms), student performance prediction [4] (5th of 47 algorithms).

One can also consider the NGE algorithm and its modifications [35,38] as examples of algorithms based on combination of instance based and rule based approaches (they are classifying new data points by computing their distance to the nearest "generalised exemplar", i.e. either a point or an axis-parallel rectangle). However, the reported experimental results [38] are showing that NGE and its modifications are inferior in comparison with kNN on the most of tested data sets while experiments with RIONIDA (extending RIONA) are showing that in the most of the tested data sets RIONIDA outperforms kNN (with proper filters).

3 Basic Notions

The domain of learning is a space of objects $\mathbf{X}$. Each object $x \in \mathbf{X}$ is described by a finite set of pairs $(a, a(x))$, where a is a conditional attribute from a given set A of (conditional) attributes, i.e. $a : \mathbf{X} \to V_a$ for $a \in A$, where the codomain V_a of a is the set of values of a and $a(x)$ is the value of a on the object $x \in \mathbf{X}$. We consider two types of attributes: numerical and symbolic. We denote the sets of symbolic and numerical attributes by A_{sym} and A_{num}, respectively.

Definition 1. *Any tuple* $(\mathbf{X}, A, d, \{\varrho_a\}_{a \in A})$, *where* $\mathbf{X}$ *is the space of objects, A is a set of attributes, $d : \mathbf{X} \to V_d$ is a decision attribute, and for any attribute $a \in A$ is given metric ϱ_a[1] on the respective value set V_a (i.e. for any $a \in A$, (V_a, ϱ_a) is a metric space) is called* metric decision system.

We use $V_d = \{d_{maj}, d_{min}\}$, where d_{min}, indicates *minority class* and d_{maj} indicates the *majority class*. By default, we assume that the aggregated metric ρ is simply defined as the sum of individual metrics, denoted as $Agr(\{\varrho_a\}_{a \in A})$.

Definition 2. *Let* $\mathbb{D} = (\mathbf{X}, A, d, \{\varrho_a\}_{a \in A})$ *be a metric decision system. A decision* rule *over the decision system $\mathbb{D}$ is an expression of the form $if\ t_1 \wedge t_2 \wedge \ldots \wedge t_m\ then\ d = v$, where $v \in V_d$, m is the number of attributes, t_i is a condition of the form $a_i \in V$ (where $V \subseteq V_{a_i}$ is defined using ϱ_{a_i}) for an attribute a_i ($i = 1, 2 \ldots, m$).*

For training set $trnSet \subset \mathbf{X}$ and test example $tst \in \mathbf{X} \setminus trnSet$ we define $N(tst, trnSet, k, \varrho)$ (when used in paper with smaller arguments, other are fixed) as the set of k training examples that are most similar to tst according to distance function ϱ.

4 The RIONA Algorithm

RIONA computes rules in a lazy manner (see e.g. [2,35,38]), that is it induces a very limited set of decision rules relevant only for the test example. The classification performed by RIONA on a given test object is based on rules induced only from a neighbourhood of the given test example. The empirical study showed

---
[1] ρ_a is used to define for a given $v \in V_a$ a neighbourhood of similar values for v. RIONIDA as RIONA learns (as default setting) SVDM metrics [10] for symbolic attributes; and for numerical attributes uses Euclidean metric on $\mathbb{R}$.

that it is enough to consider a small neighbourhood to achieve classification accuracy comparable to the algorithm induced from the whole learning set. Moreover, it uses rules with conditions of the form: *attribute belongs to a set of values*. These sets of values are specified by grouping both numerical and symbolic values of attributes. In particular, RIONA does not require discretisation (or value grouping). In voting for the decision by rules *covering* the example being classified, the aggregation of the support sets of such rules is used. RIONA constructs object neighbourhoods of the optimal size. The learning of the optimal neighbourhood is based on the idea of dynamic programming (see e.g. [7]), which makes the computational time complexity of this step low.

RIONA algorithm is presented in Algorithm 1. It uses function $isConsistent$ $(r, verifySet)$, which checks whether rule r is consistent with $verifySet$ i.e. if all the examples belonging to $verifySet$ satisfying the left-hand side of r are labelled by the same decision as the decision of r. This algorithm predicts the most common class among the training examples that are covered by the rules satisfied by a given test example and are in the specified neighbourhood. It uses g-rule, which informally is built of conditions chosen in such a way, that both the training and the test example satisfy the rule and the conditions are maximally specific. Its definition is generalised for RIONIDA. Thus, formally g-rule can be defined as sg-rule (see Definition 3) with fixed $s = 1$. For numerical attributes, the interval's endpoints are determined by the attribute values of the examples tst and trn used to form the rule. For symbolic attributes, the condition represents the specific group of values defined by a metric ball centered in attribute value of test example with radius such that the ball contains also training example.

Algorithm 1: RIONA-classify$(tst, trnSet, k, \{\varrho_a\}_{a \in A})$

Input: test example tst, training set $trnSet$, positive integer k, family of metrics for attributes $\{\varrho_a\}_{a \in A}$
Output: predicted decision for tst

1 begin
2 $\quad \varrho = Agr(\{\varrho_a\}_{a \in A})$
3 $\quad neighbourSet = N(tst, trnSet, k, \varrho)$
4 $\quad$ foreach $decision\ v \in V_d$ do
5 $\quad\quad | \quad supportSet(v) = \emptyset$
6 $\quad$ end
7 $\quad$ foreach $trn \in neighbourSet$ do
8 $\quad\quad v = d(trn)$
9 $\quad\quad$ if $isConsistent\left(g\text{-}rule\left(tst, trn, \{\varrho_a\}_{a \in A_{sym}}\right), neighbourSet\right)$ then
10 $\quad\quad\quad | \quad supportSet(v) = supportSet(v) \cup \{trn\}$
11 $\quad\quad$ end
12 $\quad$ end
13 $\quad$ return $\arg\max_{v \in V_d} |supportSet(v)|$
14 end

For every decision class, the RIONA algorithm computes the support set restricted to the neighbourhood $N(tst, k)$. For every training object trn from the neighbourhood $N(tst, k)$ the algorithm constructs the rule

$$g\text{-}rule\left(tst, trn, \{\varrho_a\}_{a \in A_{sym}}\right)$$

based on the considered example trn and the test example tst. Then, it checks whether this g-rule is consistent with the remaining training examples from the neighbourhood $N(tst, k)$. If the local decision rule is consistent, then the training example trn used to construct the rule is added to the support set of the appropriate decision. Finally, the algorithm selects the decision with the support set of the highest cardinality.

Below we describe the algorithm for estimation of the optimal value k for the neighbourhood $N(tst, k)$. This can be done in an analogous way to searching for the optimal value k for the kNN method. The leave-one-out method is used on a training set to estimate the Accuracy of the classifier for different values of k ($1 \leq k \leq k_{max}$); then the value of k with the highest estimated Accuracy is selected. Applying it directly would require repeating leave-one-out estimation k_{max} times. However, using dynamic programming technique, we emulate this process in time comparable to the single leave-one-out test for k equal to the maximal possible value $k = k_{max}$. Below we present the algorithm implementing this idea.

Algorithm 2: getClassificationVector(trn, $trnSet$, k_{max}, $\{\varrho_a\}_{a \in A}$)

Input: currently considered example $trn \in trnSet$, training set $trnSet$, number k_{max}, family of metrics for attributes $\{\varrho_a\}_{a \in A}$
Output: vector A of leave-one-out classification for trn for different values of parameter $k = 1, 2, \ldots, k_{max}$

1 **begin**
2     $\varrho = Agr(\{\varrho_a\}_{a \in A})$
3     $N = N(trn, trnSet \setminus \{trn\}, k_{max}, \varrho)$
4   vector $nn_1, \ldots, nn_{|N|} = N$ sorted according to the distance $\varrho(trn, \cdot)$
5   **foreach** $decision\ v \in V_d$ **do**
6 | $decStrength[v] = 0$
7   **end**
8     $currentDec = $ the most frequent decision in $trnSet$
9   **for** $k = 1$ **to** $|N|$ **do**
10   **if** $isConsistent(g\text{-}rule\left(trn, nn_k, \{\varrho_a\}_{a \in A_{sym}}\right), N)$ **then**
11           $v = d(nn_k)$
12           $decStrength[v] = decStrength[v] + 1$
13   **if** $decStrength[v] > decStrength[currentDec]$ **then**
14 | $currentDec = v$
15   **end**
16   **end**
17       $A[k] = currentDec$
18   **end**
19   **return** A
20 **end**

Algorithm 3: findOptimalK($trnSet$, k_{max}, $\{\varrho_a\}_{a \in A}$)

Input: training set $trnSet$, number k_{max}, family of metrics for attributes $\{\varrho_a\}_{a \in A}$
Output: optimal k

1 **begin**
2   **foreach** $trn \in trnSet$ **do**
3 | $A_{trn} = getClassificationVector(trn, trnSet, k_{max}, \{\varrho_a\}_{a \in A})$
4   **end**
5   **for** $k = 1$ **to** k_{max} **do**
6 | $estimatedAccuracy[k] = |\{trn \in trnSet : d(trn) = A_{trn}[k]\}|$
7   **end**
8   **return** $\arg\max_{1 \leq k \leq k_{max}} estimatedAccuracy[k]$
9 **end**

For a training example trn, the function $getClassificationVector(\ldots)$ (see Algorithm 2) finds k_{max} examples from $trnSet \setminus \{trn\}$ nearest to the example trn and sorts them according to the distance $\varrho(trn, \cdot)$ (i.e. we consider the metric ϱ with the first argument fixed). It should be pointed out that as in the testing phase, it is necessary to consider the neighbourhood $N(tst, k_{max})$, which, in general, may contain more than k_{max} objects. Next, it returns the vector of decisions that the RIONA classifier would return for successive values of k. Algorithm 3 calls this routine for every training object, and then it selects the value k for which the global estimation of Accuracy is maximal.

Moreover, the ONN algorithm was proposed, which instead of rules uses kNN method and learns the optimal neighbourhood in a similar way as RIONA does.

5 The RIONIDA Algorithm

In the construction of RIONIDA, some significant changes have been made in comparison with RIONA aiming to develop classifiers for imbalanced data with the highest possible quality.

Scaled generalised local decision rules are used in RIONIDA which are more general than rules used for RIONA. The idea is to select between rule-based method (as RIONA does) or kNN method (as the ONN algorithm does), and, on the other hand, to allow a smooth transition between these approaches.

Definition 3. *For any test object tst and any training object trn, we define the* scaled generalised local decision rule *(for short, the* sg-rule*), denoted by* sg-rule $(tst, trn, \{\varrho_a\}_{a \in A_{sym}}, s)$ *or simply sg-rule* (tst, trn, s) *(whenever parameters* $\{\varrho_a\}_{a \in A_{sym}}$ *are clear from the context or irrelevant due to generality), the decision rule with the decision $d(trn)$ and the following conditions t_a for each* attribute a:

$$t_a = \begin{cases} a \in [a(tst), a(tst) + l \cdot s] & \text{when } s \geq 0, cond_1 \\ a \in [a(tst) - l \cdot s, a(tst)] & \text{when } s \geq 0, cond_2 \\ a \in B\left(a(tst), r_a \cdot s\right) & \text{when } s \geq 0, cond_3 \\ a \in \emptyset & \text{when } s < 0, \end{cases}$$

where $cond_1 \equiv a$ is numerical, $a(tst) \leq a(trn)$, $cond_2 \equiv a$ is numerical, $a(tst) > a(trn)$, $cond_3 \equiv a$ is symbolic, $l = |a(tst) - a(trn)|$, $r_a = \varrho_a(a(tst), a(trn))$, and $B(c, R)$ is the closed metric ball of radius R centred at point c for metric ϱ_a, $s \in [-1, 1]$ is the scaling parameter of the rule.

Any sg-rule covers the test example, and the interval or ball corresponding to each attribute is scaled by the given parameter s in comparison to the rules used in RIONA[2]

RIONIDA, compared to RIONA additionally: (i) adds the possibility of choosing of the performance measure to be optimised; in fact, those more relevant for imbalanced data are taken into account (e.g. F-measure or G-mean), (ii) sets sensitivity constraint (for the minority class) to a higher level; furthermore, this sensitivity is adjusted to the currently analysed data, (iii) provides not only a possibility to learn when to use ONN (kNN like method) and when rule-based method, but also a combination of both types of algorithms (by tuning levels of rules inconsistency provided in Definition 3 a smooth transition between both types of algorithms is incorporated), (iv) automatically induces features not only those embedded in RIONA (optimal neighbourhood size and optimal metric), but also others.

In RIONIDA, after choosing the performance measure (which is relevant to the user needs), the learning phase is performed relative to this chosen performance measure. In consequence, the same chosen performance measure is used both in training and testing. The internal parameters (size of the neighbourhood – parameter k, sensitivity to the minority class – parameter p, the degree to which the rules are used – parameter s) are learned during the learning phase. It is important to note that we present an efficient in time methods of learning all of these parameters by the dynamic programming technique.

For each parameter, there is a set of values that we take into account. The sets of admissible values for the parameters k, p, s, we denote by K, P and S, respectively. Thus, the set of possible classes of classifiers that we search for are of the cardinality $|K| \cdot |P| \cdot |S|$. The space of all possible classifiers is determined not only by these parameters but also by the training set (analogously to kNN method when a distinct classifier is defined for each training set).

---

[2] For $s = 1$, this definition is equivalent to conditions used in RIONA. For $s = 0$ we have the rule covering only the test example and the training examples identical with the test example for all numerical attributes and distanced by 0 for all symbolic attributes. For $s < 0$, the premise of this rule is always false (formally speaking, not satisfied by any example) what relates to elimination of consistency checking and in consequence to working as the kNN algorithm. The parameter s such that $0 < s < 1$ defines the scaling of the satisfiability area of the rule.

Parameter p is responsible for making the minority class more important than the majority one. This importance is expressed by the degree of importance of the minority class expressed by p. One can relate this to cost sensitive learning. On the other hand, the parameter s is responsible for what kind of data we have: whether the data is more suitable for the rule-based classification, or maybe more for the kNN-type classification. This parameter allows us to fine-tune the type: completely kNN data, completely rule-based data, slightly rule-based data (e.g. 30%) and more kNN (e.g. 70%). It is the latter that corresponds to the smooth transition from the rule-based type of classifier to the kNN type classifier. Of course, all these parameters are learned from the training sample. This type of motivation resulted from previous experiments. In [16] is presented the analysis showing why the introduced parameters in RIONIDA and their proper selection are important factors for obtaining high performance of RIONIDA.

The learning process of these optimal parameters is discussed in Sect. 6. Algorithm 4 presents the RIONIDA algorithm for the testing phase. In the input of the algorithm all metrics are given (used for computation of the final metric), but in the sg-rule only metrics for symbolic attributes are used.

Algorithm 4: RIONIDA-classify

Input: test example tst, training set $trnSet$, positive integer k, number $p \in [0, 1]$, number $s \in [-1, 1]$, family of metrics for attributes $\{\varrho_a\}_{a \in A}$

$\varrho = Agr(\{\varrho_a\}_{a \in A})$
$nS = N(tst, trnSet, k, \varrho)$
$supportSet(d_{min}) = \emptyset$
$supportSet(d_{maj}) = \emptyset$
for all $trn \in nS$ **do**
    $v = d(trn)$
    if $isConsistent\left(sg\text{-}rule\left(tst, trn, \{\varrho_a\}_{a \in A_{sym}}, s\right), nS\right)$ **then**
        $supportSet(v) = supportSet(v) \cup \{trn\}$
    end if
    $p_{current} = \frac{|supportSet(d_{min})|}{|nS|}$
    if $p_{current} \geq p$ **then**
        return d_{min}
    else
        return d_{maj}
    end if
end for

At the step of classification, a performance measure dedicated to imbalanced data does not appear in RIONIDA. Here, it is assumed that these parameters have been optimised for this chosen (by a user) measure.

RIONA (ONN) is obtained from RIONIDA after setting the threshold p at 0.5, the parameter s at 1.0 (-1.0), and the optimisation measure to Accuracy. Thus, RIONIDA is an extension of RIONA and ONN as well.

6 Estimating the Optimal Values of Parameters for RIONIDA

Our preliminary analysis showed that the performance of RIONIDA can significantly depend on the chosen values of the parameters k, p, s (see [16]). The optimal values of these parameters depend on the analysed data set and the selected optimisation measure. Therefore, it is essential to find the optimal values of these parameters relative to the optimisation measure specified by a user. It should be noted that the domains of the parameters k, p, s (maximal admissible sets for these parameters) are as follows: $K_{max} = \{1, 2, \ldots, |trnSet|\}$, $P_{max} = [0, 1]$, $S_{max} = \{-0.1\} \cup [0, 1]$. We would like to search for the optimal triple values in the Cartesian product of these sets. From the algorithmic point of view, one should restrict the search to some finite subsets of these sets.

Analogously as in the case of RIONA, to construct an efficient algorithm one should take into account the following question: For given finite sets K, P, S, how to learn the optimal triple values efficiently from $K \times P \times S$?

By default we use sets $K = \{1, 2, \ldots, 100\}$, $P = \{0.01, 0.02, \ldots 0.5\}$, $S = \{-0.1, 0.0, 0.1, \ldots, 1.0\}$ with a size of 100, 50 and 12 respectively.

In [17] the estimation of the parameter p was done for some specific situations.

6.1 Efficient Learning of the Optimal Values of Parameters for RIONIDA

Here, we discuss the algorithm for estimation of the optimal values of the parameters k, p, s for RIONIDA. This can be done in an analogous way to searching for the optimal value of k in the case of RIONA. The leave-one-out method is used on the given training set to estimate the value of the performance measure (chosen by a user) for different values of $(k, p, s) \in K \times P \times S$ and the triple values of k, p, s for which the estimation of the measure value is the greatest is selected. The direct calculations require repeating leave-one-out estimation $|K| \cdot |P| \cdot |S|$ times. However, using the dynamic programming technique, we emulate this process in time comparable to the single leave-one-out test for k equal to the maximal possible value $k = k_{max} = |K|$. Below we present Algorithm 6 implementing this idea.

For a training example trn the function $getClassificationMatrix$ (see Algorithm 5) finds k_{max} examples from $trnSet \setminus \{trn\}$ nearest to the example trn and sorts them according to the distance $\varrho(trn, \cdot)$ from the trn object.

Next, for any example nn_k from the selected neighbourhood and any $s \in S$, the sg-rule is built on trn (treated as a testing object) and nn_k (treated as a training object), i.e. the rule $sg\text{-}rule\left(trn, nn_k, \{\varrho_a\}_{a \in A_{sym}}, s\right)$. The algorithm checks consistency of this sg-rule with the objects from the neighbourhood for different levels of $s \in S$ and stores this information in the entry corresponding to s of the array assigned to the object nn_k.

Next, it calculates the matrix of decisions that the RIONIDA classifier would return for different triple values $(k, p, s) \in K \times P \times S$ and this matrix is returned as a result.

Algorithm 5: getClassificationMatrix($trn, trnSet, K, P, S, \{\varrho_a\}_{a \in A}$)

1: **Input:** currently considered example $trn \in trnSet$, training set $trnSet$,
   K, P, S – sets of admissible values for parameters k, p, s, respectively,
   family of metrics for attributes $\{\varrho_a\}_{a \in A}$
2: **Output:** 3 dimensional matrix (for different triple values $(k, p, s) \in K \times P \times S$) of
   leave-one-out classification for trn
3: $k_{max} = |K|$ (we assume that K is the set of consequent natural numbers)
4: $\varrho = Agr(\{\varrho_a\}_{a \in A})$
5: $N = N(trn, trnSet \setminus \{trn\}, k_{max}, \varrho)$
6: vector $nn_1, \ldots, nn_{|N|} = N$ sorted according to the distance $\varrho(trn, \cdot)$
7: **for** $k = 1$ **to** $|N|$ **do**
8:   **for all** $s \in S$ **do**
9:     $nn_k.isConsistentOnLevel[s] = isConsistent\left(sg\text{-}rule\left(trn, nn_k, \{\varrho_a\}_{a \in A_{sym}}, s\right), N\right)$
10:   **end for**
11: **end for**
12: **for all** $s \in S$ **do**
13:   $decStrength[d_{min}] = 0$
14:   $decStrength[d_{maj}] = 0$
15:   **for** $k = 1$ **to** $|N|$ **do**
16:     **if** $nn_k.isConsistentOnLevel[s]$ **then**
17:       $v = d(nn_k)$
18:       $decStrength[v] = decStrength[v] + 1$
19:     **end if**
20:   $p = \frac{decStrength[d_{min}]}{decStrength[d_{min}]+decStrength[d_{maj}]}$
21:   **for all** $p_{current} \in P$ **do**
22:     $currentDec = d_{min}$
23:     **if** $p_{current} > p$ **then**
24:       $currentDec = d_{maj}$
25:     **end if**
26:     $M[k, p, s] = currentDec$
27:   **end for**
28:   **end for**
29: **end for**
30: **return** M

Algorithm 6: findOptimalParams3D($trnSet, K, P, S, optMeasure, \{\varrho_a\}_{a \in A}$)

1: **Input:** training set $trnSet$,
   K, P, S – sets of admissible values for parameters k, p, s, respectively,
   optimisation measure $optMeasure$ from {F-measure, G-mean, Accuracy},
   family of metrics for attributes $\{\varrho_a\}_{a \in A}$
2: **Output:** triple of the optimal values of parameters k, p, s
3: **for all** $trn \in trnSet$ **do**
4:   $M_{trn} = getClassificationMatrix(trn, trnSet, K, P, S, \{\varrho_a\}_{a \in A})$
5: **end for**
6: fill $estimatedConfusionMatrix$ with values 0
7: **for all** $(k, p, s) \in K \times P \times S$ **do**
8:   **for all** $trn \in trnSet$ **do**
9:     $realDec = d(trn)$
10:     $classifierDec = M_{trn}[k, p, s]$
11:     $estimatedConfusionMatrix[k, p, s][realDec, classifierDec] + +$
12:   **end for**
13:   count $estimatedMeasure[k, p, s]$ from $estimatedConfusionMatrix[k, p, s]$ based on $optMeasure$
14: **end for**
15: **return** $\arg\max\limits_{(k,p,s) \in K \times P \times S} estimatedMeasure[k, p, s]$

Algorithm $findOptimalParams3D$ (see Algorithm 6) calls the function $getClassificationMatrix(\ldots)$ for every training object. Next, it creates a matrix with the confusion matrix as its entry for each triple $(k, p, s) \in K \times P \times S$. The entry of this matrix corresponding to the index defined by the values of the parameters k, p, s consists of the confusion matrix consisting of information for leave-one-out classification for these values of the parameters k, p, s over all training examples (excluding the considered one). Any confusion matrix (in the matrix of confusion matrices $estimatedConfusionMatrix$) is transformed into one value calculated using the selected optimisation measure $optMeasure$ (and stored in the matrix $estimatedMeasure$). Finally, it selects the triple of the optimal values of the parameters k, p, s for which the global estimation of the chosen optimisation measure is maximal.

This algorithm is analogous to the algorithm learning of the optimal parameter k in RIONA [18]. In this algorithm, the triple of the optimal values of the parameters k, p, s, rather than only one value of the parameter k is returned. Moreover, the optimal parameters relative to the given optimisation measure instead of the Accuracy measure are returned.

The algorithm $findOptimalParams3D$ has arguments K, P, S specifying the sets of admissible values for the parameters k, p, s, respectively. We assume that $K = \{1, 2, \ldots, k_{max}\}$, i.e. the admissible values of the parameter k are consecutive natural numbers.

Another argument of the algorithm is the optimisation measure $optMeasure$. In the current implementation F-measure, G-mean or Accuracy can be substituted here as the value of this argument. However, from the description of the algorithm, it is clear that any optimisation measure, which is the function of the confusion matrix, could also be used.

Assume that $|N| = |N(trn, k_{max})| \leq c \cdot k_{max}$ for all $trn \in trnSet$, where c is a constant very close to 1, $k_{max} = |K|$ is the parameter used to define the maximal size of the neighbourhood to be analysed. The time complexity of RIONIDA in the testing phase is $O(m(n + k^2))$ for a single test object, where $n = |trnSet|$, $m = |A|$, k is the parameter used to define the size of neighbourhood (usually learned during the learning phase). The time complexity of the learning phase of RIONIDA is $O(mn^2 + n|S| \cdot k_{max} \cdot (mk_{max} + |P|))$, where P, S are sets of admissible values of the parameters p, s, respectively. The space complexity of the learning phase for RIONIDA is $O(n \cdot |K| \cdot |P| \cdot |S|)$. For details see [16].

7 Experiments and Main Results

We compare our RIONIDA algorithm with other state-of-the-art algorithms on several benchmarks. The experiments were designed to be reproducible.

To evaluate learning algorithms (and also generated by them classifiers), we use two performance measures: F-measure and G-mean.[3]

---

[3] We have not used AUC measure because of (i) the criticism about it (see e.g. [22,32]); (ii) BRACID, one of the important learning algorithms that we wanted to

For estimation of the mentioned performance measure, we use 10 times repeated 10-fold stratified cross-validation. Partial results of each 10-fold stratified cross-validation are micro-averaged. As the final estimation of the desired measure, the average of ten repetitions of this procedure is used. For all compared learning algorithms, the same splits in the cross-validation process are used. It can be thought that the estimation is done in parallel for all learning algorithms. In practice, we simply use the same random seed (used for random partitions of data sets) for all learning algorithms in the process of estimating the chosen measure. This guarantees the reproducibility of experiments.

Data sets used in experiments are based mainly on UCI Machine Learning Repository [26].[4] Data sets containing originally more than two classes were transformed into the binary classification task by choosing one small class or joining several small classes into one (minority) class; other classes were joined into another (majority) class. Detailed information on used data sets is provided in [16]. These data sets were selected to create a relevant base for experiments since we have selected 20 fairly diverse imbalanced data sets considering many aspects related to difficulty of imbalanced data classification (see [28]; for details see [16]).

We use the Friedman statistical test (see [13]; see also [8]) for comparing multiple learning algorithms on multiple data sets. If this test passes, then we use the post-hoc tests Nemenyi or Finner. The first one enables us to compare all learning algorithms against each other, and the second one is used to compare the RIONIDA learning algorithm with other algorithms used in the comparative experiments. For all tests, we use the significance level $\alpha = 5\%$. The statistical analysis was done using the R Project for Statistical Computing, commonly known as the R (see https://www.R-project.org).

Generally, one of the aims of performed experiments was to compare the new RIONIDA algorithm with some other state-of-the-art learning algorithms. Some of them are specially designed for the classification of imbalanced data, and some are not. One can also use state-of-the-art learning algorithms developed for balanced data and apply them to the results of sampling methods (filters) dedicated to imbalanced data. We use two types of well-known filters (and additionally one trivial `Null-filter` for cases when no filter is used). Below we describe all learning algorithms and filters used in the experiments.

Algorithms developed for the analysis of imbalanced data together with their short descriptions are as follows.

BRACID [27] analogously to RISE uses an integrated representation of rules and single instances. It comprehensively addresses the issues associated with imbalanced data. It uses the strategy of bottom-up induction of rules from single examples with the specific generalisation strategy. A conflict resolution is based on the supports of the nearest rules to the test example.

---

compare with, does not return probabilities for the two decision classes (only the deterministic decision is returned).

[4] Only *mammography* data set is not publicly available and was supported by Nitesh Chawla [6].

MODLEM-C [21] is an extension of MODLEM (see below) with the possibility to strengthen Sensitivity. The rule strength is multiplied for all rules describing the minority class by the same real number called the *strength multiplier* given as a parameter.

Remaining algorithms (generally dedicated to balanced data) used in the experiments are: kNN [3] (it can select the relevant value of k based on cross-validation); MODLEM [36]; J48 (decision tree learning algorithm) [31]; PART [12]; RIPPER [14]; RISE [10]; RIONA.

In our experiments, we use the configuration of filters SMOTE+ENN, i.e. first SMOTE is used, and then ENN is used. The motivation for selecting SMOTE+ENN comes from [5], where it was shown that this combination of filters provides in practice very good performance in comparison to other combination of filters for data sets with a small number of positive examples. We use this configuration of filters for algorithms which are not dedicated to imbalanced data.[5]

7.1 Comparison of RIONIDA with the Selected State-of-the-Art Algorithms for G-Mean

In this subsubsection, we assume that the performance measure we are interested in is G-mean. Thus, the particular parameter of RIONIDA is set to optimise G-mean. The algorithm with this setting is called RIONIDA_G.

For each learning algorithm, the representative scores of G-mean for all used data sets were computed. Then the algorithms were ranked for each data set.

We do not present here, the representative scores of G-mean and obtained ranks for each learning algorithm. Such details can be found in [16]. However, we present the description of these results.

For considered 20 data sets, 15 times RIONIDA wins with all other algorithms (achieves the best score; rank equal to 1), and once (for *new-thyroid*) its score is equal to the other algorithm (namely, RIONA) with the best score (in this case RIONIDA has rank equal to 1.5).

The average of all ranks for RIONIDA gives the result 1.375, which is the best outcome. The difference between the average rank of RIONIDA and the second-lowest average rank (4.6 for BRACID) is relatively high (3.225).

The Friedman statistic was computed, and the obtained p-value is much smaller than $\alpha = 0.05$. Thus, we can perform post-hoc test.

Finner statistical test was used to compare all learning algorithms with the RIONIDA_G algorithm set as the control one.

For all learning algorithms used in comparisons with RIONIDA_G, the corresponding p-value is (much) smaller than $\alpha = 0.05$ (for details see [16]). Thus, we can (confidently) reject all the null hypotheses corresponding to the algorithms used in the comparison (at 0.05 level of significance). In other words, RIONIDA_G is significantly better than any other learning algorithm (with default algorithm and filter settings) relative to the G-mean performance measure.

---

[5] In fact, we describe in detail only the simplified version of experiments (see also Subsect. 7.3).

Additionally, we also present in Fig. 1 the critical difference plot for the Nemenyi statistical test in this case (with the level of significance at 0.05). This test, although conservative, shows that RIONIDA$_G$ is significantly better (with the level of significance at 0.05) than all other algorithms used in the comparison. For more detailed description see [16].

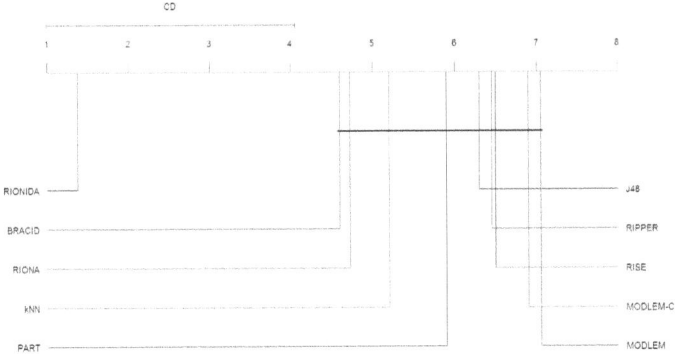

Fig. 1. Comparison of G-mean for all algorithms used in the comparison against each other with the Nemenyi statistical test. Groups of algorithms that are not significantly different (with the level of significance at 0.05) are connected.

7.2 Comparison of RIONIDA with the Selected State-of-the-Art Algorithms for F-Measure

This subsection is analogous to Subsect. 7.1 using F-measure instead of G-mean. Thus, RIONIDA is set to optimise F-measure (such algorithm is called RIONIDA$_F$). For half of the 20 data sets, RIONIDA wins with all other algorithms (has rank equal to 1). RIONIDA again achieved the best average rank (2.15). It is smaller by 2.05 from the second-lowest average rank (4.2 for BRACID). The Friedman statistical test returns again (much) less p-value than 0.05. All the adjusted p-values of the Finner procedure are less than 0.05. Thus, we can claim that RIONIDA is significantly better than any other algorithm used in the comparison. However, the highest p-value (around 0.03 for RIONIDA) is close to the threshold of 0.05. It shows that RIONIDA outperforms BRACID not as evidently as in the case for G-mean. Probably this is because BRACID was implemented to optimise F-measure. The second-highest p-value is for RIONA (with its optimal parameter settings and filter) and is smaller than 10^{-2}. All other p-values (related to other algorithms) are smaller than 10^{-3}. More details can be found in [16].

7.3 Conclusions for Experiments

To sum up, we performed experiments to compare our new proposed RIONIDA algorithm with the selected nine state-of-the-art algorithms with possible use of state-of-the-art configuration of filters. Both for G-mean and F-measure it was shown that RIONIDA significantly outperforms any algorithm used in the comparison for selected imbalanced data sets.

Let us note, that the parameter optimisation is built into the RIONIDA algorithm, and one can ask to perform a comparison with mentioned algorithms using a similar extent of parameter optimisation. Thus, additionally, we performed experiments with other two levels of increasing challenge for RIONIDA in relation to the other algorithms used in comparison. First, we performed experiments with different than default settings of the algorithms and different configurations of filters (SMOTE+ENN or SMOTE). For such new experimental setup the conclusions were the same as presented previously (for G-mean and F-measure). Second, we emulated learning of optimal algorithms settings and optimal filter configuration. For each algorithm (besides of RIONIDA and BRACID), we simply selected the maximal score of performance measure for many its configurations and used such maximal value (favouring other algorithms used in our comparison!) for the final statistical comparison in Friedman statistical test. For such new experimental setup the conclusions were generally the same as presented previously (for G-mean and F-measure).[6] For details see [16].

The obtained experimental results seem to be exceptionally good.

We also analysed the real-time of running of RIONIDA (training and testing). We found that it is comparable to other algorithms. We observed that the realtime of training and testing for RIONIDA is significantly less than given by the theoretical bounds (due to using indexing trees). For details see [16].

8 Conclusions

RIONIDA is an extension of the well-known RIONA. RIONIDA in comparison to the other state-of-the-art algorithms for imbalanced data obtains exceptionally good results, and has relatively low time complexity. It uses only the objects from the neighbourhood of testing object to generate rules and uses specific, complex kind of reasoning for concept approximation. RIONIDA with its excellent results illustrates very successful implementation of the mentioned Vapnik idea.

References

1. Rseslib 3: Rough set and machine learning open source in Java. http://rseslib.mimuw.edu.pl
2. Aha, D.W. (ed.): Lazy Learning, 1st edn. Springer, Dordrecht (1997). https://doi.org/10.1007/978-94-017-2053-3

---

[6] The only exception was for F-measure and kNN: RIONIDA achieved better average rank (2.85) than kNN (4.45) but the difference between RIONIDA and kNN is not statistically significant for scores computed in such a way (adjusted p-value in Finner statistical test for kNN was 0.09). However, one should also bear in mind that real meta-learning algorithm using kNN would obtain worse results than we took into comparison (as it was mentioned, we took maximal values of possible scores).

3. Aha, D.W., Kibler, D., Albert, M.K.: Instance-based learning algorithms. Mach. Learn. **6**(1), 37–66 (1991). https://doi.org/10.1023/A:1022689900470
4. Almasri, A., Celebi, E., Alkhawaldeh, R.S.: EMT: ensemble meta-based tree model for predicting student performance. Sci. Program. **2019**, 1–13 (2019). https://doi.org/10.1155/2019/3610248. Article No. 3610248
5. Batista, G.E.A.P.A., Prati, R.C., Monard, M.C.: A study of the behavior of several methods for balancing machine learning training data. ACM SIGKDD Explor. Newsl. **6**(1), 20–29 (2004). https://doi.org/10.1145/1007730.1007735
6. Chawla, N.V., Bowyer, K.W., Hall, L.O., Kegelmeyer, W.P.: SMOTE: synthetic minority over-sampling technique. J. Artif. Intell. Res. **16**, 321–357 (2002). https://doi.org/10.1613/jair.953
7. Cormen, T.H., Leiserson, C.E., Rivest, R.L., Stein, C.: Introduction to Algorithms, 2nd edn. The MIT Press and McGraw-Hill Book Company, Cambridge (2001)
8. Demšar, J.: Statistical comparisons of classifiers over multiple data sets. J. Mach. Learn. Res. **7**, 1–30 (2006)
9. Dey, N., Borah, S., Babo, R., Ashour, A.S. (eds.): Social Network Analytics: Computational Research Methods and Techniques, 1st edn. Academic Press, London (2019). https://doi.org/10.1016/C2017-0-02844-6
10. Domingos, P.: Unifying instance-based and rule-based induction. Mach. Learn. **24**(2), 141–168 (1996). https://doi.org/10.1007/BF00058656
11. Fernández, A., García, S., Galar, M., Prati, R.C., Krawczyk, B., Herrera, F.: Learning from Imbalanced Data Sets, 1st edn. Springer, Cham (2018). https://doi.org/10.1007/978-3-319-98074-4
12. Frank, E., Witten, I.H.: Generating accurate rule sets without global optimization. In: Proceedings of the 15th International Conference on Machine Learning (ICML 1998), pp. 144–151. Morgan Kaufmann, San Francisco (1998)
13. Friedman, M.: The use of ranks to avoid the assumption of normality implicit in the analysis of variance. J. Am. Stat. Assoc. **32**(200), 675–701 (1937). https://doi.org/10.1080/01621459.1937.10503522
14. Fürnkranz, J., Widmer, G.: Incremental reduced error pruning. In: Proceedings of the 11th International Conference on Machine Learning (ICML 1994), pp. 70–77. Morgan Kaufmann, San Francisco (1994). https://doi.org/10.1016/B978-1-55860-335-6.50017-9
15. Góra, G., Skowron, A., Wojna, A.: Explainability in RIONA algorithm combining rule induction and instance-based learning. In: Ganzha, M., Maciaszek, L.A., Paprzycki, M., Ślęzak, D. (eds.) Proceedings of the 18th Conference on Computer Science and Intelligence Systems, FedCSIS 2023, Warsaw, Poland, 17–20 September 2023. Annals of Computer Science and Information Systems, vol. 31, pp. 485–496. IEEE (2023). https://annals-csis.org/proceedings/2023/
16. Góra, G.: Combining instance-based learning and rule-based methods for imbalanced data. Ph.D. thesis, University of Warsaw, Warsaw (2022). https://www.mimuw.edu.pl/sites/default/files/gora_grzegorz_rozprawa_doktorska.pdf
17. Góra, G., Skowron, A.: On kNN class weights for optimising G-mean and F1-score. In: Campagner, A., Urs Lenz, O., Xia, S., Ślęzak, D., Wąs, J., Yao, J. (eds.) IJCRS 2023. LNCS, vol. 14481, pp. 414–430. Springer, Cham (2023). https://doi.org/10.1007/978-3-031-50959-9_29
18. Góra, G., Wojna, A.: RIONA: a classifier combining rule induction and K-NN method with automated selection of optimal neighbourhood. In: Elomaa, T., Mannila, H., Toivonen, H. (eds.) ECML 2002. LNCS, vol. 2430, pp. 111–123. Springer, Heidelberg (2002). https://doi.org/10.1007/3-540-36755-1_10

19. Grama, L., Rusu, C.: Choosing an accurate number of mel frequency cepstral coefficients for audio classification purpose. In: Proceedings of the 10th International Symposium on Image and Signal Processing and Analysis (ISPA 2017), pp. 225–230 (2017). https://doi.org/10.1109/ISPA.2017.8073600
20. Grama, L., Rusu, C.: Adding audio capabilities to TIAGo service robot. In: 2018 International Symposium on Electronics and Telecommunications (ISETC), pp. 1–4 (2018). https://doi.org/10.1109/ISETC.2018.8583897
21. Grzymala-Busse, J.W., Goodwin, L.K., Grzymala-Busse, W.J., Zheng, X.: An approach to imbalanced data sets based on changing rule strength. In: Pal, S.K., Polkowski, L., Skowron, A. (eds.) Rough-Neural Computing: Techniques for Computing with Words. Cognitive Technologies, pp. 543–553. Springer, Heidelberg (2004)
22. Hand, D.J.: Measuring classifier performance: a coherent alternative to the area under the ROC curve. Mach. Learn. **77**(1), 103–123 (2009). https://doi.org/10.1007/s10994-009-5119-5
23. He, H., Ma, Y.: Imbalanced Learning: Foundations, Algorithms, and Applications, 1st edn. Wiley-IEEE Press, Piscataway (2013)
24. Kaur, H., Pannu, H.S., Malhi, A.K.: A systematic review on imbalanced data challenges in machine learning: applications and solutions. ACM Comput. Surv. **52**(4), 1–36 (2019). https://doi.org/10.1145/3343440
25. Li, J., Dong, G., Ramamohanarao, K., Wong, L.: DeEPs: a new instance-based lazy discovery and classification system. Mach. Learn. **54**(2), 99–124 (2004). https://doi.org/10.1023/B:MACH.0000011804.08528.7d
26. Lichman, M.: UCI machine learning repository (2013). http://archive.ics.uci.edu/ml
27. Napierała, K.: Improving rule classifiers for imbalanced data. Ph.D. thesis, Poznań University of Technology, Poznań (2012)
28. Napierała, K., Stefanowski, J.: Types of minority class examples and their influence on learning classifiers from imbalanced data. J. Intell. Inf. Syst. **46**(3), 563–597 (2016). https://doi.org/10.1007/s10844-015-0368-1
29. de Oliveira Almeida, R., Valente, G.T.: Predicting metabolic pathways of plant enzymes without using sequence similarity: Models from machine learning. The Plant Genome **13**(3), e20043 (2020). https://doi.org/10.1002/tpg2.20043
30. Pawlak, Z., Skowron, A.: Rough sets: some extensions. Inf. Sci. **177**(28–40), 1 (2007). https://doi.org/10.1016/j.ins.2006.06.006
31. Quinlan, J.R.: C4.5: Programs for Machine Learning. Morgan Kaufmann, San Mateo (1993)
32. Raeder, T., Forman, G., Chawla, N.V.: Learning from imbalanced data: evaluation matters. In: Holmes, D.E., Jain, L.C. (eds.) Data Mining: Foundations and Intelligent Paradigms. Intelligent Systems Reference Library, vol. 23, pp. 315–331. Springer, Heidelberg (2012). https://doi.org/10.1007/978-3-642-23166-7_12
33. Russell, S.J., Norvig, P.: Artificial Intelligence: A Modern Approach, 4th edn. Pearson Education, Hoboken (2021)
34. Rusu, C., Grama, L.: Recent developments in acoustical signal classification for monitoring. In: 2017 5th International Symposium on Electrical and Electronics Engineering (ISEEE), pp. 1–10 (2017). https://doi.org/10.1109/ISEEE.2017.8170705
35. Salzberg, S.: A nearest hyperrectangle learning method. Mach. Learn. **6**, 251–276 (1991)

36. Stefanowski, J.: Rough set based rule induction techniques for classification problems. In: Proceedings of 6th European Congress on Intelligent Techniques & Soft Computing (EUFIT 1998),D vol. 1, pp. 109–113. Verlag Mainz, Aachen (1998)
37. Vapnik, V.N.: Statistical Learning Theory, 1st edn. Wiley-Interscience, New York (1998)
38. Wettschereck, D., Dietterich, T.G.: An experimental comparison of the nearest-neighbor and nearest-hyperrectangle algorithms. Mach. Learn. **19**, 5–27 (1995)

Granular Computing

Information System in the Light of Interactive Granular Computing

Soma Dutta[1]([✉]) and Andrzej Skowron[2]

[1] University of Warmia and Mazury in Olsztyn, Słoneczna 54, 10-710 Olsztyn, Poland
soma.dutta@matman.uwm.edu.pl
[2] Systems Research Institute, Polish Academy of Sciences, Newelska 6, 01-447 Warsaw, Poland
skowron@mimuw.edu.pl

Abstract. This paper is an attempt to develop a notion of interactive information system, which contrary to the classical notion of information system allows real physical interactions to play a role in the process of gathering information about the objects or situations. The proposed notion of interactive information system is grounded on the basic building blocks of Interactive Granular Computing (IGrC), known as complex granules (c-granules). The main point of departure of IGrC from other existing mathematical tools for modeling complex real physical phenomenon is to include a possibility of introducing a process of learning and perceiving through interactions initiated in the real physical world. Thus, the model is not completely restricted in a pure mathematical manifold. It has both the abstract module and implementational module connected by a notion of control.

Keywords: Interactions · (Interactive) Granular Computing · Perception · Complex granule · Informational granule · Control of c-granule · Information system

1 Introduction

Classical notion of rough set is defined based on a notion of information system and an equivalence relation partitioning different objects or situations based on the information available about them. Usually, an information system is presented by a tuple of the form (U, A) where U is a set of objects, A is a set of attributes, and for each $a \in A$ and $u \in U$, $a(u)$ represents a value. Thus, a row corresponding to an object is basically a vector of values over the set of attributes, and is often called the *information signature* of an object [12–14,19].

Usually such information signatures relative to a set of attributes A aim to model some real physical fragment or phenomenon of the world. The attributes of A create a window specification through which the real physical objects, denoted by U, are perceived, and this perceived information is stored as the information signature presented in some abstract language. Once the data is represented in the form of a table usually we work with the data and do not bother about the constraints (e.g. processes, space-time etc.) of collecting the data. Thus, the data, we work with, is just a snapshot of a real physical environment at a time point and often does not reflect the dynamical nature of the concerned environment. There are several papers [2,3,20] dedicated to dynamic information systems. However, in the existing approaches dynamics of these systems is defined by an apriori given function defining changes of states (or perceived situations). However, according to many researchers [1] mathematical models of dynamic systems dealing with complex phenomena cannot be obtained using classical mathematical modeling.

To model complex concepts about the real world we cannot deny the fact that with time the context, the relevant features of the same physical environment may change. The standard way of modelling based on data does not incorporate the information, e.g., how the process of perceiving the values of attributes is realised, where and how to access the concerned objects in the physical space, why a specific set of attributes are selected etc. Though there might be some human led pre-processing phase of data, such knowledge is not incorporated in the model itself. Hence, clearly the perception and action are out of the scope of such practices of modelling [4]. However, this is crucial for many tasks addressing complex phenomena [8] in the real world. The existing developments of rough sets of course work nicely when we are interested in a simplified model of the physical reality. This fails when we would like to apply the theory to such problems or tasks which require modelling beyond a mathematical manifold [1,4].

Let us imagine a human being reasoning about a vague phenomenon over a real physical fragment of the world. It neither does a mere calculation by replacing values on a specific formula, nor bases on a priori given formula describing the phenomenon. It rather *perceives and learns* the properties of the physical environment and their interrelations, *form a language* for describing them, and then manipulate by some *reasoning mechanisms* to determine whether or not to attach certain properties to a new situation. It is also flexible to update all these above three components based on the changes of the context and time [11]. Now while developing an intelligent agent simulating behaviour of a human being, we need a new tool, replacing the classical notion of information system, so that it can incorporate the above mentioned features.

So, what are the aspects of a classical information system that need to be modified? Firstly, while abstracting information about the objects the model should be able to perceive the real physical interactions bringing the objects into co-existence. Secondly, in order to be *intelligent* the model should also be aware of how the information about the real physical fragment, of which U is a denotation, is obtained. Thirdly, the model should also be able to represent the process of assigning values to the elements of U with respect to a concerned set of attributes. In this regard, we propose the model of *Interactive Granular Computing* (IGrC) [5–7,10,15,16,18][1] as a foundation of information systems. An initial motivation and idea behind modifying the existing mathematical tools

---

[1] See also papers related to IGrC at https://dblp.org/pid/s/AndrzejSkowron.html.

such as rough sets, fuzzy sets, are discussed in [17]. However, developing a formal notion, having one end grounded to a mathematical modeling and another end grounded to the real physical world, does not appear an easy task. Here, we attempt to, as much as possible, put forward the idea of interactive information system based on IGrC formally.

The paper is distributed as follows. Section 2 presents the basics of IGrC. Section 3 presents a definition for *interactive information system*. Section 4 illustrates the idea of interactive information system through a prototypical example.

2 A Brief Idea About IGrC

IGrC deals with both the abstract objects and the physical objects, which are not constructs of the abstract space (i.e., not mathematical in nature!). We assume that some portion of matter (*hunks* according to [9]) can be perceived in the physical space with the help of complex granules (c-granules, in short), the basic constructs of IGrC. As the physical objects are out of the *mathematical manifold*, following Edmund Husserl [4], one cannot talk about sets of such objects. Hence, we will use the term collection (or configuration) of physical objects.

The c-granules are constructed based on a collection $\mathcal{P}$ of physical objects and a set $\mathcal{I}$ of informational objects containing relevant information about the objects from $\mathcal{P}$.

c-granule: A c-granule g can be such that the control of it is associated to itself or is associated to another c-granule; the former kind is considered as a c-granule with control and the latter as a c-granule without control. A c-granule without control is constituted based on three collections of objects from $\mathcal{P}$, known as soft_suit, link_suit and hard_suit. Together these collections of objects determine the scope of the c-granule.

Soft_suit: The soft_suit refers to those objects which are directly accessible or about which some information is already encoded by the control of the c-granule g at some earlier time point.
Hard_suit: The hard_suit refers to those objects which are not yet accessed by the control of the c-granule g.
Link_suit: The link_suit refers to the (chain of) objects which create a communication channel between the soft_suit and the hard_suit.

ic-granule: A c-granule with control usually has an additional information layer attached to it, and such a c-granule is called informational granule. This information layer contains different relevant information about the objects lying in its scope. Let us note that any ic-granule may have several sub ic-granules, which are themselves ic-granules having their scopes contained in the scope of the parent ic-granule. Activities of the sub ic-granules are controlled by the control of the parent ic-granule (c-granule), and so, we can for the sake of simplicity only focus on the control of the parent ic-granule.

Control: The control of a c-granule (or an ic-granule) g is endowed with the information layers of all its sub ic-granules and with a mechanism of reasoning and implementing, which allows it to link between the abstract specification of the information and the respective real physical semantics.

Network of c-granules: A network of c-granules itself can be considered as an ic-granule where all the c-granules from the network are considered as the sub c-granules of the whole ic-granule. Let us note that, as the network of c-granules is itself regarded as an ic-granule, it is endowed with the abilities of controlling (sequentially or concurrently) the behaviour of its components.

As mentioned before, at any point of time t, the control of a c-granule (ic-granule) g has access to the information layers of all the ic-granules lying in the scope of g. Thus using some relevant information and information related to the perceived situation of the existing configuration of the ic-granules, it can dynamically change (by aggregating or decomposing) existing configuration of ic-granules to a different configuration. Let us denote the collection of all such configurations of ic-granules, starting from the initial time point t_0 to the current time point, as N. That is, by N_t ($\in N$) we mean the configuration of the ic-granules that is perceived/described by the control of g at time point t. In other words, N_t represents the state of the control at the time point t. $Inf(N_t)$ is the information associated to the ic-granules from N_t; formally $Inf(N_t)$ can be considered as a set of pairs consisting of names of the ic-granules and the perception related information associated to them. Later, we will describe how the control of g conducts the transition from one configuration N_t to another.

The reasoning mechanism of the control of a c-granule is responsible for aggregating, deleting, or generating new information from the existing information, generating action plans, converting the plan of actions to an implementational level language, and updating new information that is perceived after implementation of actions. Here to be noted that in order to initiate the real physical actions the control translates the specification of the abstract plan of actions in the language of the actuators; in IGrC, it is considered that going beyond the mathematical specifications of actions the real physical actions will be embedded through the real physical actuators. The specifications of the abstract plan of actions and the translation of that plan in the language of actuators are in general described in different languages. For a certain control, associated to a particular network of ic-granules, such languages are fixed; however, in order to transform an abstract plan of actions to the level of the actuators a finite number of translations from one language to a comparatively lower level language, which is more close to the language of the actuators, may be required. The real physical actions, which are supposed to be embedded on some real physical objects following the specifications of the actions in the language of the actuators, are out of the realm of the abstract module of the IGrC model. Basically, IGrC allows a time lapse between the phases where the prescribed plan is transferred from the abstract module to the implememtational module, and the real physical interactions take place through embedding the specification of the plan into the physical space. After this time interval, a new cycle of the control starts by perceiving the changes occurred in the environment after initiation of actions.

The role of the implementational module of the control of a c-granule is as follows. The control sends the implementational module a formal specification of the task to be performed in the physical space. This specification contains a description of the task as well as information, e.g., *when*, *how* and *where* this task should be realised. The implementational module first verifies whether this task can be realised; if it is possible, this module is responsible for preparation of the relevant configuration of the physical objects matching to the given specification. While preparing the required configuration, it encodes the information (following the specification) related to initiation of interactions on the objects belonging to the soft-suits of the relevant ic-granules. After initiation of the interactions, in a specified time interval, it encodes back the properties of the relevant objects from the soft-suits of the ic-granules of the concerned configuration and their interactions into the respective information layers of the ic-granules.

The current state of the control, at a given time t, is characterised by the current configuration of ic-granules. The current state of the control is changed by the control following the specification of information, available in the information layers of the ic-granules. So, clearly the state of the control of a network of ic-granules varies over time. For example, after initiation of real physical interactions the configurations of the previously considered ic-granules may change, and accordingly the part of the ic-granules, which was not possible to access before, may become perceivable. So, the respective information layers of the ic-granules also change. Thus, while defining the notion of transformation realised by the control of a c-granule g we specify two aspects; one is the time point t, and the other is Nt_t, the network of ic-granules at t describing the state of the control of g. The other components[2] of the control of g also may vary over time with the change of the network of ic-granules; however as the components of the control implicitly depend on the network of ic-granules at some time point, for simplicity of presentation, they are not indexed with the time point.

Below we present the definition of control of a given c-granule. It is a slightly modified version of definition proposed in [5].

Definition 1. *The control of a c-granule g, denoted as $Cont(g)$, is represented by a tuple*

$$(IG, INF, Lan_{upp}, Lan_{low}, \mathcal{R}_{inf}, \mathcal{R}_{imp}, Mod_{imp})$$

where the components are defined as follows.

(i) *IG is a network of (collections of) ic-granules whose real physical configuration can be potentially generated by the implementational module Mod_{imp} using information from the information layers of those ic-granules.*

---

[2] The components e.g., languages for abstract specifications and implementational level specifications as well as two relations connecting information regarding the ic-granules and descriptions of the plans of actions also may vary over time.

(ii) INF is a set of functions

$$inf : Nam \to \bigcup_{nam \in Nam} V_{nam},$$

where Nam is a finite subset of the set of names (e.g., addresses, spatio-temporal windows) and for any $nam \in Nam$, $inf(nam) = v \in V_{nam}$ where v represents the information labelled by nam.[3] We assume that the content of any information layer of the ic-granules from IG at a given moment of time t is represented by an element $inf \in INF$ and the set Nam is the set of names of the ic-granules from the configuration of ic-granules at time t.

(iii) Lan_{upp} is an abstract level language for representing the specifications of the plans of actions.

(iv) Lan_{low} is a lower level language representing each specification from Lan_{upp} into a language of actuators, or language of implementation.

(v) $\mathcal{R}_{inf} \subseteq INF \times Lan_{upp}$; in particular an expression of the form '$inf\ \mathcal{R}_{inf}\ \alpha$' represents that $inf \in INF$ contains information characterising the currently perceived situation[4] and α is the specification of the task to be performed[5]; α specifies a plan, expressed in the high level language Lan_{upp}, that to be realised in the currently perceived situation.

(vi) $\mathcal{R}_{imp} \subseteq Lan_{upp} \times Lan_{low}$ represents a relation connecting the abstract specification of an action into the language of the implementation (actuators).

(vii) Mod_{imp} represents a module which takes the instructions generated by $\mathcal{R}_{imp}$ and then implements those instructions by embedding them into the physical world pertaining to the relevant configuration of real physical objects.

In the definition of control $\mathcal{R}_{imp}$ is presented as a relation which transforms a plan of actions from an abstract level language to a lower level language. However, in the process of translating the actions to the language of the actuators, a finite number of translations from an upper level language to a lower level language may be required. Though Mod_{imp} is a component of the control of a c-granule, the real physical actions, performed through the functionalities of Mod_{imp}, are outside the realm of mathematical formalization. More specifically, Mod_{imp} may be considered as a technically equipped module through which the real physical actions of the actuators are run in the real physical world. The control of a c-granule g only can specify the description of actions for Mod_{imp}.

The results of the implementation obtained by initiation of Mod_{imp} may be different from the expected outcome specified by the control. The control is

---

[3] Elements of V_{nam} may be compound; inparticular, they may represent states of an information system in different moments of time.

[4] In particular inf can represent information about the perceived objects till the concerned time point, fragments of domain knowledge extracted from domain knowledge bases and/or general knowledge regarding physical laws.

[5] In general α may describe *what*, *where*, *how*, *when* the specified task should be performed; this also include information about the expected results after implementation of the plan of actions by the control as well as the specification of the networks of c-granules to be implemented or activated on the way of realising the given task.

expected to take into account the difference by adapting the required behaviour to the newly perceived changes.

It is worthwhile to see the role of control as a dynamical system. One can consider some analogy with a transition system. Usually a transition system consists of a set of states S and a transition relation R, specifying transition of a state s_1 to another state s_2. Here, the role of the control is analogous in the sense that it specifies how a computation process over a c-granule moves/changes from one configuration of ic-granules to another. By changes of the configurations of ic-granules we mean the changes in the information layers of the ic-granules and/or changes in the real physical fragments from the scope of the ic-granules. So, in general by control of a c-granule we mean a pair of the form $(S, tran)$, where S can be identified with IG of Definition 1 and $tran \subseteq IG \times IG$; however understanding either the states of S of a control or the transition from one state to another is not that simple as in the case of standard transition relation. Here $tran$ is much complex and is realised with the help of the other components of $Cont(g)$. A state of the control of a c-granule is defined by a configuration of ic-granules from IG. The control can only perceive some properties of this configuration which are available in the information layers of those ic-granules.

Let us explain how by the control of a c-granule g the currently perceived situation, characterised by the information inf related to the current configuration of ic-granules, is transformed into a new one, say inf' corresponding to a new configuration of ic-granules. Given any nam respective to a configuration of ic-granules at time t, say $Nt_t \in IG$, inf contains the information encoded in the information layers of those ic-granules; Among different information encoded in inf, there are information related to the perceived objects from the ic-granules of N_t. Let us denote the perception related information of N_t by $Inf(N_t)$. Let that in the process of realising a given task, the control of g aim to transform N_t into a new configuration of ic-granules. So, given $Inf(N_t)$ and some general information from inf, the control generates a plan α in the language Lan_{upp}, and consequently it is translated to a plan β in the language Lan_{low}. The plan β is then submitted to Mod_{imp}, which is responsible for embedding the specification of β in the physical world. This real physical interaction generates a new configuration of ic-granules $N_{t'}$ at a time point t' ahead of t. After a period of time, specified in β, Mod_{imp} starts interactions with $N_{t'}$ to perceive properties of the newly generated configuration of ic-granules. The specification of such actions is also encoded in α. Then the perceived properties of the objects from $N_{t'}$ and their interactions are encoded in the information layers of ic-granules of $N_{t'}$. Thus, inf is updated to inf' with the information of the new network $N_{t'}$.

3 Information Systems from the Perspective of IGrC

An attempt to extend the notion of information system in the light of IGrC requires changes in the existing definition of information systems. One of the aspects, where we need to depart, is to change the way of viewing attributes as apriori given mathematical functions. In this regard, Fig. 1 and Fig. 2 may give

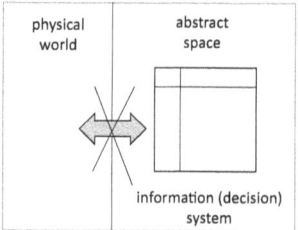

Fig. 1. In the existing approach to rough sets interactions with the physical world are eliminated. Attributes are mathematical functions.

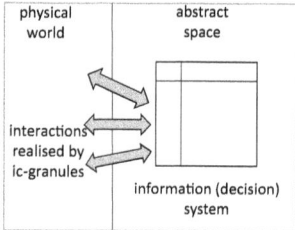

Fig. 2. Rough sets in IGrC: (i) perceiving values of attributes are based on interactions with the physical world, (ii) attributes are not pure mathematical functions; they are realised by ic-granules.

an intuitive idea about the proposed approach [17]. The existing notion of an information system can be visualized as a table or database where the entries corresponding to a particular row and a particular column indicates the value that is assigned to an object u by a given attribute a. However, questions like, why the attribute a is chosen, how the value of u with respect to a is measured, what is the spatio-temporal location of the object u etc., remain outside the realm of the model of the information system. In the context of IGrC, an information system in the existing sense is only an one time approximation or snapshot of the physical reality. Thus, based on such an image it is not possible to model a complex concept (phenomenon) of the real physical world. So, usually the rule based systems, developed base on static entities like a database or information (decision) system, cannot faithfully serve as the basis for a decision support in a complex real life situation. Hence, the need for an idea of interactive information system (decision system) is quite prominent.

An interactive information system has both an abstract module and a real physical module. All the relevant information related to creating an information system available at a particular point of time, the window specifications of space, time, location for measuring an attribute, the reason behind selecting a particular set of attributes etc., are present at the abstract module. In the case of necessity, this abstract module can guide the real physical module to obtain information about the objects with respect to a required set of attributes by initiating real physical interactions. Thus, interactive information system is dynamic in nature and with the progress of time it develops an information system by perceiving the concerned real physical fragment through real physical interactions.

So, the first point is to bring in a notion of continuous interaction with the physical world so that the labels or values for the objects, with respect to certain set of attributes, remain associated with the window specifications explaining where and how they are obtained.

Based on IGrC, an interactive information system IS is defined relative to a given c-granule g as follows. IS is a dynamical system characterised by the information associated to a collection of states, i.e., configurations of ic-granules lying in the scope of g, and a mechanism for updating information based on the transition between states conducted by the control of g. These states are not presented as pure mathematical entities as information about them at a point of time is perceived partially using the control of g, which includes performance of an implementational module. So, given a c-granule g and a time point t, among all the configurations of ic-granules that can be potentially generated by the control of g, if we select a particular configuration say N_t of ic-granules and its corresponding information then that truncated part can be represented as some rows of an interactive information system. That is, given a particular name $nam_{is} \in Nam$, $Cont(g)$ can chunk out the respective information INF_{is} from INF available at that time point t. From INF_{is} a relevant set F_{att} of attributes, matching to the information perceived about N_t, can also be obtained. Moreover, based on F_{att} a set of compound attributes can also be defined where each such compound attribute $Prop$ is a function from a subset A of F_{att}, which assigns a given $inf \in INF_{is}$ a value from a suitable set V_{Prop} of values.

Definition 2. *An interactive information system IS in the context of a given c-granule g is a tuple of the form $(nam_{is}, INF_{is}, V_{is}, F_{att}, Upd_{is}, Cont(g))$, where*

(i) *$nam_{is} \in Nam$ is a distinguished name, obtained from the component INF of $Cont(g)$, which can be interpreted as an address specifying where the information system is stored,*

(ii) *INF_{is} is a selected subset of INF, obtained by some constraints filtering entries relevant to nam_{is} from INF,[6]*

(iii) *the set V_{is} is a collection $\{(U, A) : U \subseteq INF_{is}, A \subseteq F_{att}\}$ of information systems in the classical sense [12–14],*

(iv) *$Upd_{is} : (\cup_{N_t \in IG} Inf(N_t)) \times V_{is} \rightarrow V_{is}$ is a function updating a given information system to a new one based on the perceived properties of the objects from the relevant configuration of ic-granules.*

(v) *$Cont(g)$ is the control of the c-granule g.*

The update function Upd_{is} may have different strategies for updating an information system. It may simply match pairs of information from a selected information system with a currently perceived situation; however, in the context where such simple matching strategies do not fit to a situation, updating a

---

[6] The filtration is done with the help of $Cont(g)$ which for a given $nam_{is} \in Nam$ generates information associated with that label.

given information system with new values may depend on some real physical interactions conducted by $Cont(g)$. A discussion in regard to the formal structure of the update function may be included in our future work. For now we restrict ourselves to an intuitive illustration of the notion of update in the section below.

4 Example Illustrating Interactive Information System

In this sequel we present an illustrative example describing the idea of interactive information system and its dynamic nature of updating new entries to an existing information system with the help of the control of a c-granule. Updating an information system with time, in the context of a given c-granule g, depends on the changes in the configurations of ic-granules in the scope of g. It is already discussed that how from the changes of the configuration of the ic-granules the control can extract information about the current configuration, and thus the current state of the respective information system is learned. The state of an information system is updated using the currently perceived information stored in $Inf(N_t)$, the information layers of the current configuration of ic-granule. The process of updating an information system may be exemplified as follows.

Let us consider an IS related to a collection of configurations of ic-granules that can be potentially generated by the control of a c-granule g. So, with the help of $Cont(g)$ the respective address/name, say nam_{is} is selected. Once nam_{is} is selected INF_{is}, the corresponding information label is also selected by $Cont(g)$. Let from INF_{is} a finite set of attributes, say $F_{att} = \{a_1, a_2, \ldots, a_n\}$ is extracted; each a_k is associated to a set of windows $\{w_{k1}, w_{k2}, \ldots, w_{km}\}$ specifying different possible specifications for space, time, and process of measuring the respective attribute. Given INF_{is} and F_{att}, a family of information systems V_{is} (in classical sense) is obtained based on particular window chosen for an attribute a_k.

Now a particular $v_{is} \in V_{is}$ represents the state of a configuration of the ic-granules, say N_t from the collection of configurations of ic-granules IG of $Cont(g)$. From the notion of ic-granules we already know that every ic-granule corresponds to three sets of physical objects. For the objects, which belong to the soft_suit of an ic-granule, there is already some information available in the information layer of that ic-granule. That is, given an object u from the soft_suit of an ic-granule $g_x \in N_t$ and some attribute a_k, $1 \leq k \leq n$, we can expect some information about u with respect to a_k from the perspective of some window specification, say w_{kl}, where $1 \leq l \leq m$. This documentation is available in INF_{is}, and we can consider all such information together as the information signatures of the objects belonging to the soft_suit of the concerned ic-granule g_x. So, for the objects lying in the soft_suit, the particular information layer inf_{g_x} of g_x may contain information in the form of an usual information system (see Table 1).

The tuple $(a_1, w_{11}, inf_{a_1, w_{11}})$ denotes that $inf_{a_1, w_{11}}$ represents information of certain object with respect to the attribute a_1 based on the window specification w_{11}. Moreover, based on such tuples each situation/object, represented in a row, has ceratin properties such as $Prop_1, Prop_2, Prop_3, \ldots Prop_r$ where each

Table 1. Information layer of an ic-granule containing information about the objects in its soft_suit where the entries in the left-most column represent particular inf related to the objects in the soft_suit and each $Prop_i$ represents a compound concept defined based on a subset of attributes

	$Prop_1$	$Prop_2$	$Prop_3$	...
$(a_1, w_{11}, inf_{a_1,w_{11}}), \ldots, (a_1, w_{1m}, inf_{a_1,w_{1m}})$	:	:	:	...
$(a_2, w_{21}, inf_{a_2,w_{21}}), \ldots, (a_2, w_{2n}, inf_{a_2,w_{2n}})$	:	:	:	...
:	:	:	:	:

such $Prop_j$ can be considered as a function defined on some subset of the set of such tuples. This information is also available in the information layer inf_{g_x} of the ic-granule g_x. That is, inf_{g_x} contains the lower level perceptual information with respect to different window specifications as well as the higher level abstract information derived based on the lower level perceptual information.

The rows corresponding to the situations where the objects are partially perceived or not perceived yet, the tuples in the left most column are of the form $(a_k, w_{km}, ?)$. Generally, these can be considered as the objects belonging to the link_suit or hard_suit of the ic-granule g_x. In order to perceive information about such objects, relative to some attributes, $Cont(g)$ needs to initiate some real physical interactions on the respective configuration N_t, to which g_x belongs.

Let us assume that y is an object, lying in the hard_suit of g_x, for which no information yet has been perceived. So, as described before there might be a chain of objects connecting the already perceived objects in the soft_suit and the not yet perceived object y in the hard_suit. The reason behind such an assumption is that g_x represents a fragment of the real physical world where objects with their internal dynamics of interactions co-exist in the environment; some of them are in the direct reach of some perception/measurement process, and for some in order to gather perceptual information either some form of reasoning is required or some real physical interactions are required. These real physical actions are performed by the implementational module Mod_{imp} of $Cont(g)$.

For instance, let $x, x_1, x_2, \ldots, x_n, y$ be a chain of objects of which x lies in the soft_suit of g_x, $x_1, x_2, \ldots, x_n$ lie in the link_suit of g_x, and y lies in the hard_suit of g_x. That is, x is in the direct reach of the analysis of the control; information about $x_1, x_2, \ldots x_n$ may be to some extent available already in the information layer of g_x or they may be derived by some reasoning scheme available at INF_{is}. Gathering the available information for the objects $x, x_1, x_2, \ldots x_n$ some action plans for measuring/accessing y can be planned (using $\mathcal{R}_{inf}$ and $\mathcal{R}_{imp}$), and that plan needs to be implemented on the respective real physical configuration through real physical interactions. For simplicity, let for x all the informational tuples be present, for $x_1, x_2, \ldots, x_n$ some of the informational tuples be present, and for y no informational tuple be present. Here to be noted that even for the objects in the soft_suit, at a particular point of time, the information layer of the respective ic-granule may not have all the information. At a particular point of time it may have information with respect to a set of attributes and respective window specifications; both the set of attributes and the associated window specifications may change at a further point of time, and thus those objects which are regarded to be in the soft_suit of an ic-granule g_x at some point t_0 may become objects of the link_suit or hard_suit of the same ic-granule at some further point of time t_1.

Given the information available for x with respect to the given set of attributes and respective window specifications, as well as the partial documentations available for some of $x_1, x_2, \ldots x_n$, $Cont(g)$ generates a plan for gathering information about y.

1. As mentioned in Definition 1, with the help of $\mathcal{R}_{inf}$ from the given information of the objects $x, x_1, x_2, \ldots, x_n$ available at the inf_{g_x}, an abstract plan of actions is generated in Lan_{upp}.
2. This abstract plan is translated in Lan_{low}, the language of actuators.
3. Finally the *implementation module* Mod_{imp} embeds the plan of actions specified in Lan_{low} on some real physical objects.
4. After a while[7] the result of the changed configuration of the ic-granules is stored in the respective information layer inf_{g_x} of $Cont(g)$. The process may repeat finitely many times from step 1 as long as the informational tuples for the required objects $x_1, x_2, \ldots, x_n, y$ are not sufficient to derive information about y. In the case of sufficient information, the process stops, and new information related to the yet unknown objects are stored in INF_{is}.

Thus, the update function works with the help of $Cont(g)$ to change an existing information system to a new one. The functionalities of Mod_{imp}, mentioned at step 3, are out of the realm of IGrC. It only assumes that in order to perceive information about some objects e.g. y the specification of the plan of actions generated in Lan_{low} should be implemented through some real physical interactions on the concerned real physical configuration of the world. How the actions are in reality performed that is not the concern of the model of interactive information system. However, after such initiation of real physical actions the model is supposed to gather data from the real physical world. To do so, after a time lapse the process from step 1 again continues. A threshold for time lapse can be specified in the action specification generated in Lan_{low}; this threshold is learned from the already recorded data at INF_{is}, $\mathcal{R}_{inf}$ and $\mathcal{R}_{imp}$. After a passage of time based on new available cases this threshold may get updated.

At step 4, we have used a phrase, namely 'sufficient information'. Let us give an intuitive example regarding how to determine, or rather how the control will determine when there is sufficient information to discover information about an yet unknown object y. At step 1, to generate an abstract plan of action for learning information about y, the control matches the available tuples of $x, x_1, x_2, \ldots, x_n$ from inf_{g_x} with their respective values for the functions, denoted by $P_1, P_2, \ldots, P_r$ (see Fig. 3). For simplicity let us assume these values as 1 and 0 representing respectively a certain property P_i applies to a window

---

[7] The model assumes a time lapse between step 3 and step 4, specification of which is also available in the plan of actions generated by $\mathcal{R}_{imp}$.

specification and does not. Now based on the property which applies to most of the chain of the objects, the rule base of $\mathcal{R}_{inf}$ may generate the plan for determining whether that certain property applies to the objects, for which the related information is missing. As a particular property is a function of a set of tuples, the target of the control would be to gather information related to all those concerned tuples for each object in the above mentioned chain of objects. When the information related to all such tuples corresponding to all the objects in the chain will be complete, it can be considered as sufficient. However, we should remember that this interpretation for 'sufficient information' is only a simple example. Based on the task environment of the interactive information system, the interpretation for 'sufficient information' can vary and may have different approximated connotations.

The diagram in Fig. 3 may illustrate the whole process of learning information about situations/objects with respect to a certain set of attributes and respective window specifications based on some real physical interactions conducted by the control of a c-granule. In Fig. 3, g_x is a c-granule the scope of which contains a real physical object x. N_t is the network of all such g_x and $Inf(N_t)$ refers to the collection of all the respective information layers, that are shown in the form of an information table. For the objects belonging to the soft_suit (S) of a c-granule already some information is stored in the respective information layer. For the objects lying in the link_suit (L) and the hard_suit (H) some tuples in the left hand side of the table are missing. Based on the requirement of gathering particular information control of this granule generate a plan of interaction which via Lan_{upp} and Lan_{low} is embedded into the scope of the granule.

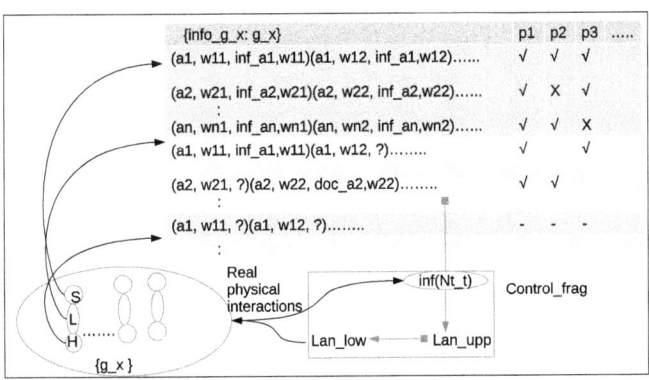

Fig. 3. Interactive information system

5 Conclusion

The needs of artificial intelligence simulating a complex real physical system or phenomenon focus not only to use static knowledge but also to acquire the ability of dynamically learning new information and accordingly updating the reasoning strategies based on the changes of the real physical environment. The existing practices of designing such models lack in having a component of continuous interactions with the real physical world and updating the information based on the changes perceived through such interactions. Thus, in the context of reasoning with complex system or phenomenon we need a new computing model. The desired model should be able to (i) continuously monitor (through some interactions) some basic properties of the respective real physical configuration associated to the complex system, (ii) learn and predict, based on already stored knowledge base and observations, more compound properties or rules for the seen and possibly unseen cases (iii) control the interaction process in order to reach to a desired goal, and (iv) update new information in the knowledge base.

Information systems and decision systems are widely used by machine learning communities in designing artificial decision support systems. However, they are far away from the ability of learning and updating based on continuous real physical interactions. According to IGrC, a suitable model for the above mentioned tasks should be endowed with an abstract module as well as an implementational module so that abstract module can store all the relevant information and implementational module can verify, learn and update through real physical interactions the relevance of all the stored properties in the future points of time. The reflection of this point of view can be observed in the definition of control of a c-granule (cf. Definition 1) and in the definition of interactive information system respective to a c-granule (cf. Definition 2).

Formal representation of such a construct having both mathematically equipped reasoning environment and technically equipped action environment is not very easy. This paper is an initial attempt to put forward the idea of such a model in a (semi) formal manner. In order to make the idea behind constructing such a model clear, in the present paper, we need to consider simplified versions of different aspects of such modeling. The aspects, which could not be covered in this paper, remain as our future directions of exploration.

References

1. Brooks Jr., F.P.: The Mythical Man-Month: Essays on Software Engineering, Anniversary Edition. Addison-Wesley (1995)
2. Campagner, A., Ciucci, D., Dorigatti, V.: Uncertainty representation in dynamical systems using rough set theory. Theor. Comput. Sci. **908**, 28–42 (2022). https://doi.org/10.1016/j.tcs.2021.11.009
3. Dong, L., Wang, R., Chen, D.: Incremental feature selection with fuzzy rough sets for dynamic data sets. Fuzzy Sets Syst. **467**, 108503 (2023). https://doi.org/10.1016/j.fss.2023.03.006
4. Dourish, P.: Where the Action Is. The Foundations of Embodied Interaction. The MIT Press, Cambridge, London (2004)
5. Dutta, S., Skowron, A.: Interactive granular computing connecting abstract and physical worlds: an example. In: Schlingloff, H., Vogel, T. (eds.) Proceedings of the 29th International Workshop on Concurrency, Specification and Programming (CS&P 2021), Berlin, Germany, 27–28 September 2021. CEUR Workshop Proceedings, vol. 2951, pp. 46–59. CEUR-WS.org (2021). ceur-ws.org/Vol-2951/paper18.pdf

6. Dutta, S., Skowron, A.: Interactive granular computing model for intelligent systems. In: Shi, Z., Chakraborty, M., Kar, S. (eds.) ICIS 2021. IAICT, vol. 623, pp. 37–48. Springer, Cham (2021). https://doi.org/10.1007/978-3-030-74826-5_4
7. Dutta, S., Skowron, A.: Toward a computing model dealing with complex phenomena: interactive granular computing. In: Nguyen, N.T., Iliadis, L., Maglogiannis, I., Trawinski, B. (eds.) ICCCI 2021. LNCS, vol. 12876, pp. 199–214. Springer, Cham (2021). https://doi.org/10.1007/978-3-030-88081-1_15
8. Gershenson, C., Heylighen, F.: How can we think the complex? In: Richardson, K. (ed.) Managing Organizational Complexity: Philosophy, Theory and Application, pp. 47–61. Information Age Publishing (2005)
9. Heller, M.: The Ontology of Physical Objects. Four Dimensional Hunks of Matter. Cambridge Studies in Philosophy. Cambridge University Press, Cambridge (1990)
10. Jankowski, A.: Interactive Granular Computations in Networks and Systems Engineering: A Practical Perspective. Springer, Heidelberg (2017). https://doi.org/10.1007/978-3-319-57627-5
11. Ortiz, C.L., Jr.: Why we need a physically embodied Turing test and what it might look like. AI Mag. **37**, 55–62 (2016). https://doi.org/10.1609/aimag.v37i1.2645
12. Pawlak, Z., Skowron, A.: Rudiments of rough sets. Inf. Sci. **177**(1), 3–27 (2007). https://doi.org/10.1016/j.ins.2006.06.003
13. Pawlak, Z.: Rough sets. Int. J. Comput. Inf. Sci. **11**, 341–356 (1982). https://doi.org/10.1007/BF01001956
14. Pawlak, Z.: Rough Sets: Theoretical Aspects of Reasoning about Data, System Theory, Knowledge Engineering and Problem Solving, vol. 9. Kluwer Academic Publishers, Dordrecht (1991). https://doi.org/10.1007/978-94-011-3534-4
15. Skowron, A., Dutta, S.: From information systems to interactive information systems. In: Wang, G., Skowron, A., Yao, Y., Ślęzak, D., Polkowski, L. (eds.) Thriving Rough Sets. SCI, vol. 708, pp. 207–223. Springer, Cham (2017). https://doi.org/10.1007/978-3-319-54966-8_10
16. Skowron, A., Jankowski, A., Dutta, S.: Interactive granular computing. Granular Comput. **1**, 95–113 (2016). https://doi.org/10.1007/s41066-015-0002-1
17. Skowron, A., Dutta, S.: Rough sets and fuzzy sets in interactive granular computing. In: Yao, J., Fujita, H., Yue, X., Miao, D., Grzymala-Busse, J.W., Li, F. (eds.) IJCRS 2022. LNCS, vol. 13633, pp. 19–29. Springer, Cham (2022). https://doi.org/10.1007/978-3-031-21244-4_2
18. Skowron, A., Jankowski, A.: Rough sets and interactive granular computing. Fundam. Inform. **147**(2–3), 371–385 (2016). https://doi.org/10.3233/FI-2016-1413
19. Skowron, A., Slezak, D.: Rough sets turn 40: from information systems to intelligent systems. In: Ganzha, M., Maciaszek, L.A., Paprzycki, M., Slezak, D. (eds.) Proceedings of the 17th Conference on Computer Science and Intelligence Systems, FedCSIS 2022, Sofia, Bulgaria, 4–7 September 2022. Annals of Computer Science and Information Systems, vol. 30, pp. 23–34 (2022). https://doi.org/10.15439/2022F310
20. Wolski, M.: Rough sets in terms of discrete dynamical systems. In: Szczuka, M., Kryszkiewicz, M., Ramanna, S., Jensen, R., Hu, Q. (eds.) RSCTC 2010. LNCS (LNAI), vol. 6086, pp. 237–246. Springer, Heidelberg (2010). https://doi.org/10.1007/978-3-642-13529-3_26

GBTWSVM: Granular-Ball Twin Support Vector Machine

Lixi Zhao[1], Zhifei Zhang[2], Wenjun Liu[1], and Guangming Lang[1(✉)]

[1] School of Mathematics and Statistics, Changsha University of Science and Technology, Changsha 410114, Hunan, China
`langguangming1984@126.com`
[2] Department of Computer Science and Technology, Tongji University, Shanghai 201804, China

Abstract. Twin Support Vector Machine (TWSVM) has gained popularity as a machine learning tool due to its low computational complexity. However, it may not be the most adaptable to all types of datasets containing noise and outliers. Granular-ball computing (GBC), known for its higher robustness to noise and outliers, provides a potential solution. In this paper, we introduce the Granular-ball Twin Support Vector Machine (GBTWSVM), which combines the strengths of GBC and TWSVM. We optimize TWSVM by using coarse-grained granular-balls as inputs instead of individual samples. Subsequently, we develop the dual model of GBTWSVM and design an algorithm for classifications. Through numerical experiments conducted on seventeen benchmark datasets, we demonstrate that GBTWSVM outperforms TWSVM, Intuitionistic Fuzzy Twin Support Vector Machines (IFTSVM) and Granular-ball Support Vector Machine (GBSVM) in terms of running time, accuracy, precision, and recall.

Keywords: Granular-ball · Twin support vector machine · Granular-ball computing · Classification

1 Introduction

Support Vector Machine (SVM), proposed by Vapnik in 1995 [1], is an effective method for classification. It is based on statistical learning theory and structural risk minimization theory. After its introduction, SVM and its variants demonstrated unique advantages in analyzing small sample, high-dimensional, and nonlinear data [2–5]. Among these variants, TWSVM is a new machine learning method based on SVM [6–8]. In comparison with SVM, the central idea of TWSVM is to generate two non-parallel hyperplanes instead of a single hyperplane. This ensures that each sample point is close to one of the two hyperplanes and far away from the other hyperplane [9]. For classification problems, a new sample point is assigned to the nearest hyperplane class. Unlike SVM, TWSVM determines the classification hyperplanes by solving two small-scale quadratic programming problems instead of one large-scale quadratic programming problem. This not only retains the advantages of SVM in handling small sample, high-dimensional, and nonlinear classification and regression problems but also achieves a training speed theoretically four times faster than SVM [10]. However, noise and outliers may still significantly impact TWSVM due to the outlier sensitivity shortcoming of the squared loss function [11]. To enhance the robustness of TWSVM, various improvement methods have been proposed [12–16]. For instance, Rezvani et al. combined the intuitionistic fuzzy sets with TWSVM, and presented the Intuitionistic Fuzzy Twin Support Vector Machine (IFTSVM), which not only reduces the impact of noise but also distinguishes support vectors from noise [13].

Granular computing is an efficient mathematical tool for managing uncertain and imprecise information. Initially, Zadeh proposed the concept of fuzzy information granulation in 1979 [17,18]. It leverages the inaccuracy, incompleteness, uncertainty, and compatibility of vast information to achieve easily processed, robust, low-cost intelligent systems, and intelligent controllers. Building upon these characteristics, many scholars have integrated granular computing with machine learning [19–24]. In recent years, GBC, as a robust and interpretable adaptive multi-particle representation and calculation method [25–30], has garnered increasing attention. For example, Xia et al. introduced a new approach for classification, namely, GBSVM, by combining GBS with SVM [25]. Unlike traditional methods that use samples as inputs, GBSVM employs granular-balls as data inputs, and represents a groundbreaking non-point input approach with remarkable robustness and effectiveness. Leveraging GBC's resilience to noise and outliers, GBSVM outperforms traditional methods in terms of both running time and accuracy.

To the best of our knowledge, TWSVM, which determines two hyperplanes instead of one, demonstrates superior performance over SVM in running time. Motivated by the robustness of GBC and the lower time complexity of TWSVM, our objective is to introduce GBTWSVM by combining GBC and TWSVM, which replaces fine-grained points with coarse-grained granular-balls as inputs, and offers a promising solution for robust and efficient classification tasks. Additionally, we present the dual model of GBTWSVM. Subsequently, we develop a corresponding algorithm tailored for classification problems with two classes. Finally, experimental evaluations are conducted to showcase the classification performance of GBTWSVM.

The remainder of this paper is structured as follows: Sect. 2 provides a review of GBC and TWSVM. Section 3 introduces GBTWSVM. Section 4 presents the experimental results on benchmark datasets. Finally, Sect. 5 concludes the paper.

2 Preliminaries

In this section, we review some concepts of GBC [25] and TWSVM [6].

2.1 Granular-Ball Computing

Unlike most existing classifiers, the core idea of GBC is to use granular-balls instead of sample points as input data and take several granular-balls to cover the original data samples. Given a dataset $X = \{x_k \mid k = 1, 2, ..., n\}$, which contains n data samples, the dataset X is covered by granular-ball $\mathcal{GB} = \{GB_i \mid i = 1, 2, ..., m\}$, $c_i = \frac{1}{l} \sum_{j=1}^{l} x_{ij}$ represents the centre of the i-th granular-ball GB_i, where x_{ij} and l represent the data samples and the number of data samples in the i-th granular-ball GB_i, respectively. There are two main methods to calculate the radius r_i of the granular-ball GB_i, namely, the maximum distance and average distance. The maximum distance is the maximum distance of all covered samples to the center of the granular-ball, which can be expressed by the formula: $r_i = \max |x_{ij} - c_i|$. Using the maximum distance ensures that the resulting granular-ball covers all sample points. The average distance is the average distance of all covered samples to the center of the granular-ball, which can be expressed by the formula: $r_i = \frac{1}{l}\sum_{j=1}^{l}|x_{ij} - c_i|$. The average distance can be used to reduce the size of the particle sphere affected by the outlying samples. To eliminate the effect of noisy data within each granular-ball and be more robust, the overall label y_i of the granular-ball GB_i is the label that appears the most times in that granular-ball.

Assuming that the coverage sample number of all granular-balls is $|\mathcal{GB}|$, its coverage can be expressed as $\frac{|\mathcal{GB}|}{n}$, and the objective function of granular-ball generation is as follows:

$$\min \quad \lambda_1 \times \frac{n}{|\mathcal{GB}|} + \lambda_2 \times m \quad (1)$$
$$\text{s.t.} \quad \text{quality}(GB_i) \geq T,$$

where λ_1 and λ_2 correspond to the weight coefficients, m is the number of granular-balls, and quality(GB_i) is the proportion of the majority of samples with the same label in the granular-ball GB_i. When the number of granular-balls is larger, the representation of granular-balls to data set X is more accurate. When the coverage is larger, less information is lost. Therefore, the granular-ball generation process is actually to adjust the parameters λ_1 and λ_2 to find the optimal granular-ball generation result to achieve the highest coverage with the minimum number of granular-balls.

2.2 TWSVM

TWSVM looks for a pair of non-parallel hyperplanes, such as $f_1(x) : x^T\omega_1 + b_1 = 0$ and $f_2(x) : x^T\omega_2 + b_2 = 0$, and positions each hyperplane closer to homogeneous data samples and somewhat far away from heterogeneous data samples. A new data sample $x \in \mathcal{R}^n$ is assigned to class +1 or −1, depending on which hyperplane it is closest to. A pair of non-parallel hyperplanes is obtained by solving the following Quadratic Programming Problems (QPPs):

$$\min_{\omega_1,b_1,\xi_2} \frac{1}{2}(x_A\omega_1 + e_1 b_1)^T(x_A\omega_1 + e_1 b_1) + C_1 e_2^T \xi_2; \quad (2)$$
$$\text{s.t.} \quad -(x_B\omega_1 + e_2 b_1) + \xi_2 \geq e_2, \xi_2 \geq 0,$$

and

$$\min_{\omega_2,b_2,\xi_1} \frac{1}{2}(x_B\omega_2 + e_2 b_2)^T(x_B\omega_2 + e_2 b_2) + C_2 e_1^T \xi_1; \quad (3)$$
$$\text{s.t.} \quad (x_A\omega_2 + e_1 b_2) + \xi_1 \geq e_1, \xi_1 \geq 0.$$

The samples of classes +1 and −1 are represented by matrices x_A and x_B, respectively. C_1 and C_2 are penalty parameters, e_1 and e_2 are vectors of ones of suitable dimensions, and ξ_1 and ξ_2 are slack variables.

The dual model of TWSVM is:

$$\max_{\alpha} \alpha^T e_2 - \frac{1}{2}\alpha^T G(H^T H)^{-1} G^T \alpha; \quad (4)$$
$$\text{s.t.} \quad 0 \le \alpha \le C_1,$$

and

$$\max_{\gamma} \gamma^T e_1 - \frac{1}{2}\gamma^T P(Q^T Q)^{-1} P^T \gamma; \quad (5)$$
$$\text{s.t.} \quad 0 \le \gamma \le C_2,$$

where $H = [x_A \ e_1]$, $G = [x_B \ e_2]$, $P = [x_A \ e_1]$, $Q = [x_B \ e_2]$, $u = \begin{bmatrix} \omega_1 \\ b_1 \end{bmatrix} = (H^T H)^{-1} G^T \alpha$, and $v = \begin{bmatrix} \omega_2 \\ b_2 \end{bmatrix} = (Q^T Q)^{-1} P^T \gamma$.

3 The model of Granular-Ball Twin Support Vector Machine

In this section, we give the principle of GBTWSVM and the dual model of GBTWSVM.

3.1 Granular-Ball Twin Support Vector Machine

We define two non-parallel hyperplanes $f_1(x) : x^T \omega_1 + b_1 = 0$ and $f_2(x) : x^T \omega_2 + b_2 = 0$, where $f_1(x)$ represents the hyperplane close to positive-class granular balls ($y_i = +1$), and $f_2(x)$ represents the hyperplane close to negative-class granular balls ($y_i = -1$). ω_t and b_t denote the normal vector and bias of $f_t(x)$, where $t \in \{1, 2\}$. These two hyperplanes are constrained by two rules: the first rule is that the hyperplanes should be as close as possible to the centers of the granular-balls belonging to the same class. The second rule is that the distance between the hyperplanes and the surface of granular-balls belonging to the other class should be as far as possible. For binary classification problems, assuming that m_1 granular-balls of class $+1$ and m_2 granular-balls of class -1 are generated, the centers of the granular-balls of classes $+1$ and -1 are represented by the matrices c_A and c_B, respectively, and the radii of the granular-balls of classes $+1$ and -1 are represented by the matrices r_A and r_B, respectively. The schematic diagram of GBTWSVM is shown in Fig. 1, where the red circles and blue circles represent granular-balls of class $+1$ and class -1, respectively.

The basic model can be written as:

$$\min_{\omega_1, b_1, \xi_2} \frac{1}{2}(c_A \omega_1 + e_1 b_1)^T (c_A \omega_1 + e_1 b_1) + C_1 e_2^T \xi_2; \quad (6)$$
$$\text{s.t.} \quad -(c_B \omega_1 + e_2 b_1) - r_B + \xi_2 \ge e_2, \xi_2 \ge 0,$$

and

$$\min_{\omega_2, b_2, \xi_1} \frac{1}{2}(c_B \omega_2 + e_2 b_2)^T (c_B \omega_2 + e_2 b_2) + C_2 e_1^T \xi_1; \quad (7)$$
$$\text{s.t.} \quad (c_A \omega_2 + e_1 b_2) - r_A + \xi_1 \ge e_1, \xi_1 \ge 0,$$

where C_1 and C_2 are constants and both are greater than 0, and e_1 and e_2 are unit vectors of the appropriate dimension.

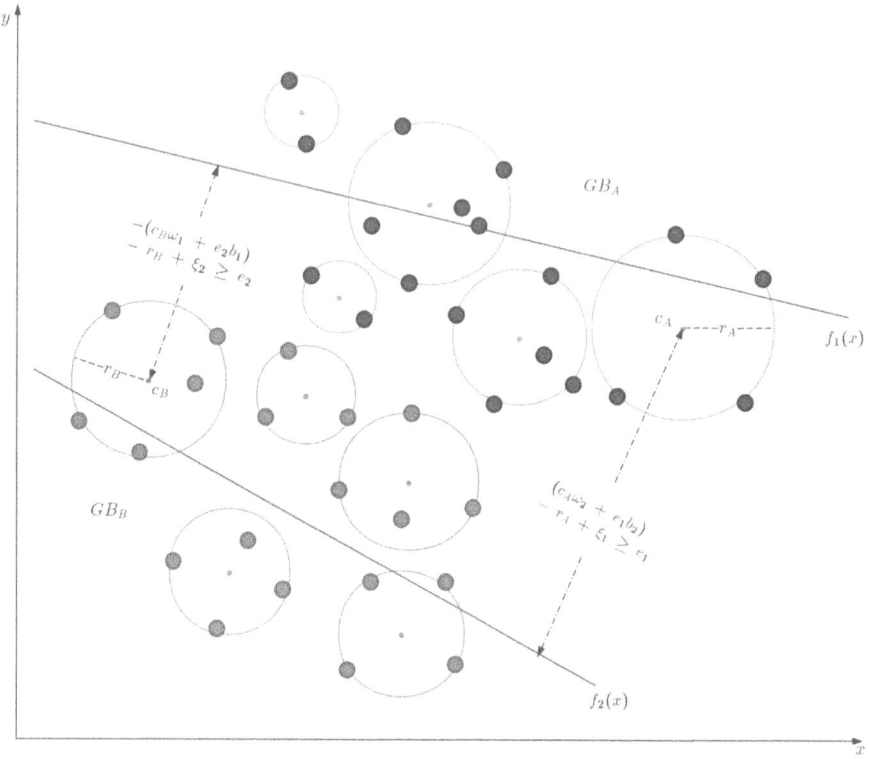

Fig. 1. Schematic diagram of GBTWSVM. (Color figure online)

3.2 The Dual Model of GBTWSVM

GBTWSVM minimizes structural risk by adding the regularization term to the margin maximization objective. This pair of quadratic programming problems can be achieved by constructing the following Lagrange function:

$$L(\omega_1, b_1, \xi_2, \alpha, \beta) = \frac{1}{2}\|c_A\omega_1 + e_1 b_1\|^2 + C_1 e_2^T \xi_2 - \alpha^T(-(c_B\omega_1 + e_2 b_1) + \xi_2 - r_B - e_2) - \beta^T \xi_2, \qquad(8)$$

where α and β are Lagrangian multipliers.

According to KKT conditions, we get:

$$\frac{\partial L}{\partial \omega_1} = c_A^T(c_A\omega_1 + e_1 b_1) + c_B^T \alpha = 0; \qquad(9)$$

$$\frac{\partial L}{\partial b_1} = e_1^T(c_A\omega_1 + e_1 b_1) + e_2^T \alpha = 0. \qquad(10)$$

By combining with Eqs. (9) and (10), we obtain:

$$\begin{bmatrix} c_A^T \\ e_1^T \end{bmatrix} \begin{bmatrix} c_A & e_1 \end{bmatrix} \begin{bmatrix} \omega_1 \\ b_1 \end{bmatrix} + \begin{bmatrix} c_B^T \\ e_2^T \end{bmatrix} \alpha = 0. \tag{11}$$

Then we define $E = [c_A \ e_1]$, $F = [c_B \ e_2]$ and $u = [\omega_1 \ b_1]^T$, and (11) can be changed as:

$$E^T E u + F^T \alpha = 0. \tag{12}$$

To enhance the generalization ability, we add the regularization item, and get the expression of u:

$$u = -(E^T E + \varepsilon I)^{-1} F^T \alpha. \tag{13}$$

Similarly, by defining $R = [c_A \ e_1]$, $S = [c_B \ e_2]$ and $v = [\omega_2 \ b_2]^T$, we get an expression of v:

$$v = (S^T S + \varepsilon I)^{-1} R^T \gamma. \tag{14}$$

The final dual model is solved as follows:

$$\max_{\alpha} \quad \alpha^T (e_2 + r_B) - \frac{1}{2} \alpha^T F (E^T E)^{-1} F^T \alpha; \tag{15}$$
$$\text{s.t.} \quad 0 \leq \alpha \leq C_1,$$

and

$$\max_{\gamma} \quad \gamma^T (e_1 + r_A) - \frac{1}{2} \gamma^T R (S^T S)^{-1} R^T \gamma; \tag{16}$$
$$\text{s.t.} \quad 0 \leq \gamma \leq C_2.$$

A new input data $x \in \mathcal{R}^n$ can be labeled as class $t \in \{+1, -1\}$ depending on which of the two hyperplanes is closer:

$$\text{Class } t = \arg \min_{t \in \{1, -1\}} \frac{|\langle \omega_t, x \rangle + b_t|}{\|\omega_t\|}. \tag{17}$$

3.3 Algorithm Design

We give an algorithm of GBTWSVM as follows:

Algorithm 1. GBTWSVM.

1: **Input:** A dataset $X = \{x_k \mid k = 1, 2, ..., n\}$, the corresponding labels $Y = \{y_k \mid k = 1, 2, ..., n\}$, the quality threshold T.
2: **Output:** $\omega_1, \omega_2, b_1, b_2$.
3: Generate a granular-ball set $\mathcal{GB} = \{GB_i \mid i = 1, 2, ..., m\}$ on X, and get the overall label y_i of each granular-ball GB_i;
4: Initialize c_A, c_B, r_A, and r_B as empty matrices.
5: **for** each GB_i **do**
6:   **if** $y_i = 1$ **then**
7:     add the center and radius of GB_i to matrix c_A and r_A;
8:   **else**
9:     add the center and radius of GB_i to matrix c_B and r_B;
10:   **end if**
11: **end for**
12: According to Equations (15) and (16), define the objective function for optimization;
13: Perform L-BFGS-B optimization for α and γ;
14: According to Equations (13), calculate ω_1 and b_1;
15: According to Equations (14), calculate ω_2 and b_2.
16: **return** $\omega_1, \omega_2, b_1, b_2$.

4 Experiment Analysis

In this section, we compare GBTWSVM with TWSVM, IFTSVM and GBSVM. All methods are implemented using Python 3.11 on a desktop with an Intel® Core^{TM} i7-10700 CPU @ 2.90 GHz with 16 GB RAM.

4.1 Penalty Parameters Selection

We perform a grid search on the penalty parameters that appear in different methods, and the best results obtained are the final results reported. More specifically, the grids used in the experiments are as follows:

$$TWSVM : C_1, C_2 \in \{2^i : i = -5, ..., 5\},$$
$$IFTSVM : C_1, C_2 \in \{2^i : i = -5, ..., 5\},$$
$$GBSVM : C \in \{2^i : i = -5, ..., 5\},$$
$$GBTWSVM : C_1, C_2 \in \{2^i : i = -5, ..., 5\}.$$

4.2 Evaluation Metrics

We use four evaluation criteria to assess the classifier's performance, where TP represents the number of correctly classified positive samples, TN represents the number of correctly classified negative samples, FP represents the number of negative samples incorrectly classified, and FN represents the number of positive samples incorrectly classified. *Accuracy (Acc)* represents the proportion of samples that are correctly classified, where $Acc = \frac{(TP+TN)}{(TP+FP+TN+FN)} \times 100\%$. *Precision* represents the probability of all the predicted positive samples actually being positive, where $Precision = \frac{TP}{(TP+FP)} \times 100\%$. *Recall* represents the probability that a sample that is actually positive is predicted to be positive, where $Recall = \frac{TP}{(TP+FN)} \times 100\%$.

4.3 Experimental Datasets

For experiments, we download seventeen benchmark datasets from the UCI machine learning repository [31], namely, Australian, Breast-cancer, Conn-bench-sonar-mines-rocks, Credit-approval, Diabetes, Diabetes-upload, Electrical, Fourclass, German.numer, Heart, Ionosphere, Liver-disorders, Messidor-features, Spambase, Tic-tac-toe, WDBC, and Wholesale-customers, which are binary classification problems from different topics. The statistics of these benchmark datasets are shown in Table 1.

For each dataset, we use ten-fold cross-validation to evaluate the performance of the three methods. This means that for all datasets, 90% of the samples are used for training, and 10% for testing. This process is repeated ten times, and the average of the ten test results is used as the performance measure. In order to fairly compare the efficiency, the four models are solved using the optimize library in Python under the same parameter settings.

Table 1. Datasets information.

Datasets	Numbers of samples	Numbers of attributes
Australian	690	13
Breast-cancer	277	9
Conn-bench-sonar-mines-rocks	208	60
Credit-approval	653	15
Diabetes	768	7
Diabetes-upload	520	16
Electrical	10000	13
Fourclass	682	2
German.numer	1000	23
Heart	270	12
Ionosphere	351	34
Liver-disorders	145	4
Messidor-features	1151	19
Spambase	4601	57
Tic-tac-toe	958	8
WDBC	569	30
Wholesale-customers	440	7

4.4 Experimental Results

GBTWSVM was compared with TWSVM, IFTSVM and GBSVM in terms of accuracy, precision, recall, and the standard deviation of accuracy (referred to as Sd), across 17 benchmark datasets without added noise. The findings are summarized in Table 2. Among the 17 datasets, GBTWSVM achieved the highest levels of accuracy, precision, and recall on 14 datasets. Furthermore, GBTWSVM exhibited the smallest standard deviation in accuracy, which indicates higher stability in performance. Additionally, GBTWSVM demonstrated superior classification performance. In conclusion, GBTWSVM not only excels in classification performance but also show higher stability.

Table 2. The performance of TWSVM, IFTSVM, GBSVM, and GBTWSVM on UCI datasets without added noise.

Dataset	TWSVM Acc ± Sd (Prec,Rec)	IFTSVM Acc ± Sd (Prec,Rec)	GBSVM Acc ± Sd (Prec,Rec)	GBTWSVM Acc ± Sd (Prec,Rec)
Australian	0.90 ± 0.03 (0.90,0.90)	0.90 ± 0.01 (0.90,0.90)	0.89 ± 0.04 (0.89,0.89)	**0.91 ± 0.03 (0.91,0.91)**
Breast-cancer	0.84 ± 0.05 (0.83,0.84)	0.86 ± 0.06 (0.85,0.86)	0.86 ± 0.07 (0.73,0.86)	**0.91 ± 0.04 (0.91,0.91)**
Conn-bench-sonar-mines-rocks	0.85 ± 0.04 (0.86,0.85)	0.79 ± 0.05 (0.80,0.79)	0.74 ± 0.09 (0.74,0.74)	**0.93 ± 0.03 (0.94,0.93)**
Credit-approval	**0.91 ± 0.03 (0.92,0.91)**	0.89 ± 0.02 (0.89,0.89)	0.86 ± 0.03 (0.87,0.86)	0.89 ± 0.02 (0.90,0.89)
Diabetes	0.81 ± 0.03 (0.82,0.81)	0.76 ± 0.02 (0.76,0.76)	0.79 ± 0.03 (0.78,0.79)	**0.83 ± 0.02 (0.83,0.83)**
Diabetes-upload	0.96 ± 0.03 (0.96,0.96)	0.94 ± 0.04 (0.94,0.94)	0.84 ± 0.15 (0.86,0.84)	**0.96 ± 0.02 (0.96,0.96)**
Electrical	0.99 ± 0.00 (0.99,0.99)	0.98 ± 0.01 (0.98,0.98)	0.75 ± 0.03 (0.76,0.75)	**1.00 ± 0.00 (1.00,1.00)**
Fourclass	0.79 ± 0.04 (0.81,0.79)	0.82 ± 0.03 (0.83,0.82)	0.83 ± 0.04 (0.84,0.83)	**0.83 ± 0.03 (0.85,0.83)**
German.numer	0.77 ± 0.03 (0.77,0.77)	0.76 ± 0.03 (0.76,0.76)	0.80 ± 0.04 (0.79,0.80)	**0.81 ± 0.04 (0.80,0.81)**
Heart	0.88 ± 0.06 (0.89,0.88)	0.90 ± 0.04 (0.90,0.90)	0.85 ± 0.04 (0.86,0.85)	**0.91 ± 0.05 (0.91,0.91)**
Ionosphere	0.93 ± 0.04 (0.93,0.93)	**0.95 ± 0.03 (0.95,0.95)**	0.80 ± 0.04 (0.83,0.80)	0.93 ± 0.02 (0.94,0.93)
Liver-disorders	0.87 ± 0.12 (0.86,0.87)	0.85 ± 0.08 (0.87,0.85)	0.86 ± 0.12 (0.88,0.86)	**0.90 ± 0.05 (0.91,0.90)**
Messidor-features	0.80 ± 0.03 (0.82,0.80)	0.76 ± 0.02 (0.78,0.76)	0.70 ± 0.05 (0.70,0.70)	**0.82 ± 0.03 (0.82,0.82)**
Spambase	0.85 ± 0.02 (0.87,0.85)	0.88 ± 0.02 (0.89,0.88)	0.66 ± 0.02 (0.67,0.66)	**0.94 ± 0.01 (0.94,0.94)**
Tic-tac-toe	0.70 ± 0.04 (0.68,0.70)	0.70 ± 0.04 (0.73,0.70)	0.67 ± 0.06 (0.68,0.67)	**0.75 ± 0.05 (0.76,0.75)**
WDBC	0.98 ± 0.01 (0.98,0.98)	**1.00 ± 0.01 (1.00,1.00)**	0.96 ± 0.03 (0.96,0.96)	0.96 ± 0.03 (0.97,0.96)
Wholesale-customers	0.95 ± 0.04 (0.95,0.95)	0.94 ± 0.03 (0.94,0.94)	0.85 ± 0.05 (0.85,0.85)	**0.95 ± 0.04 (0.95,0.95)**
Average Acc	0.87	0.86	0.81	**0.90**

Table 3. The running time of TWSVM, IFTSVM, GBSVM, and GBTWSVM.

Dataset	TWSVM	IFTSVM	GBSVM	GBTWSVM
Australian	2.309	5.640	8.758	**0.172**
Breast-cancer	0.289	1.495	1.217	**0.067**
Conn-bench-sonar-mines-rocks	0.862	0.451	1.067	**0.032**
Credit-approval	2.099	5.195	9.206	**0.090**
Diabetes	0.762	4.202	5.867	**0.194**
Diabetes-upload	1.541	4.215	0.208	**0.045**
Electrical	536.677	2055.804	347.536	**3.126**
Fourclass	1.233	7.034	0.156	**0.034**
German.numer	1.555	6.120	36.314	**1.038**
Heart	1.327	1.565	0.402	**0.064**
Ionosphere	1.465	2.963	0.745	**0.075**
Liver-disorders	0.120	0.137	0.065	**0.027**
Messidor-features	2.120	6.854	37.686	**0.940**
Spambase	99.972	651.741	210.924	**3.129**
Tic-tac-toe	0.730	4.421	38.297	**0.692**
WDBC	1.649	4.709	0.204	**0.040**
Wholesale-customers	1.499	3.284	0.366	**0.107**

For scenarios without added noise, we compared the training times of models across various datasets in Table 3. To illustrate computational efficiency more clearly, we created Fig. 2 based on the results in Table 3. Due to significant differences in training times across different datasets, we utilized relative time for visualization, where the training time of each model was divided by the maximum training time among different models on the same dataset. As depicted in the figure, GBTWSVM exhibits notably shorter training times compared to TWSVM, IFTSVM and GBSVM across all datasets, even being several times faster on certain datasets. This is primarily attributed to two factors: firstly, Table 4 illustrates the number of granular-balls generated by GBSVM and GBTWSVM across datasets without noise. We can observe that the number of granular-balls involved in GBSVM and GBTWSVM is significantly smaller than the number of samples in the dataset, and greatly reduces the complexity of model computation; secondly, GBTWSVM addresses two small-scale quadratic programming problems rather than one large-scale problem.

To evaluate the robustness of GBTWSVM against noise, we examined the accuracies of TWSVM, IFTSVM and GBTWSVM across 15 datasets under the influence of label noise ranging from 1% to 10%. The findings are depicted in Fig. 3. Upon scrutiny of the box plots, it is evident that GBTWSVM exhibits shorter box lengths and fewer outliers, and suggests a more stable prediction performance and enhanced resilience to noise compared to TWSVM and IFTSVM. The improved robustness of GBTWSVM may stem from two aspects: firstly, the relatively coarse granularity of the granules,

Table 4. The number of granular-balls of GBSVM and GBTWSVM.

Dataset	GBSVM	GBTWSVM
Australian	85.5	86.9
Breast-cancer	41.8	38.7
Conn-bench-sonar-mines-rocks	31.5	29.8
Credit-approval	88.5	55.8
Diabetes	87.3	90.3
Diabetes-upload	45.8	42.7
Electrical	1032.4	872.3
Fourclass	28.6	23.6
German.numer	148.1	131
Heart	42.8	37.3
Ionosphere	26.7	29.2
Liver-disorders	20.9	20.4
Messidor-features	188.8	159.2
Spambase	385	455.3
Tic-tac-toe	156	123.8
WDBC	24.8	31.6
Wholesale-customers	49.3	54.5

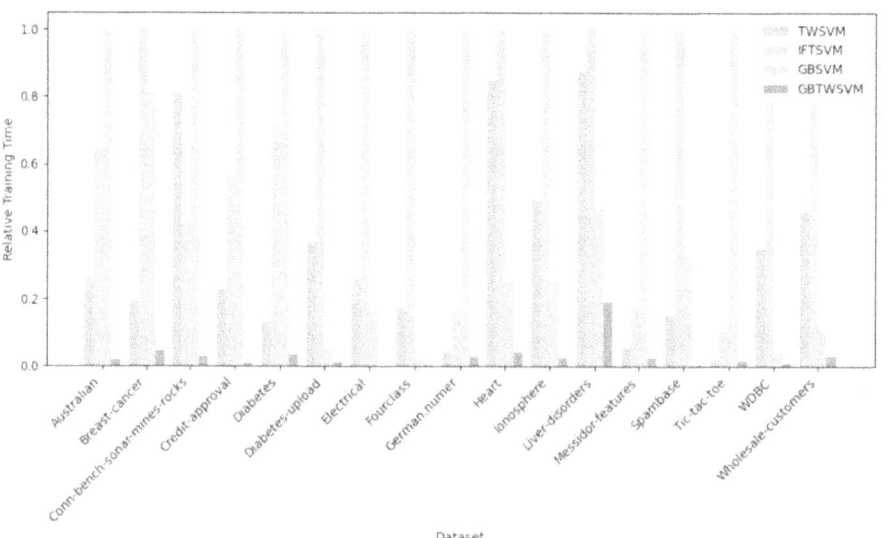

Fig. 2. Comparison of Relative Training Time Across Datasets and Models.

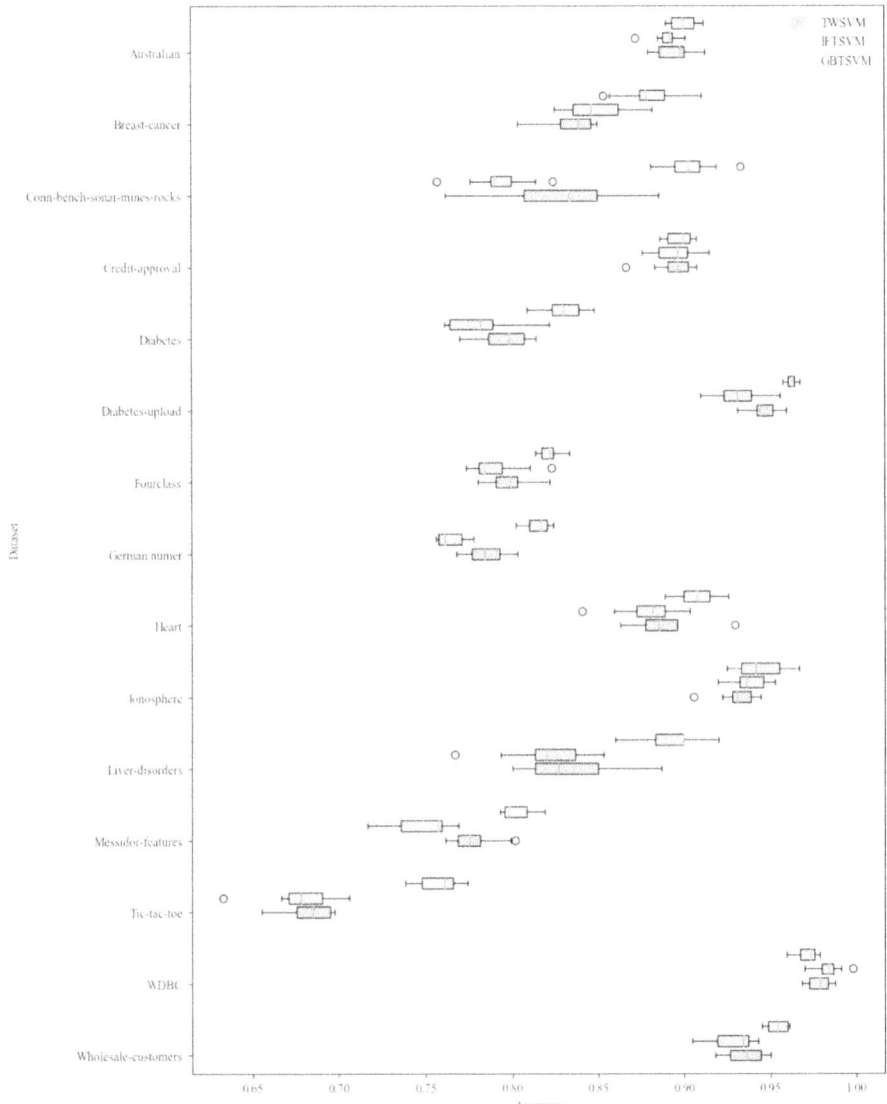

Fig. 3. Comparison of Model Accuracy Stability under Noisy Conditions

where labeling the granules as majority point labels can reduce the impact of noise points; secondly, GBTWSVM adopts two symmetric support vectors based on granules. Through symmetry, it can better resist noise and outliers in the data, and thus enhance the model's generalization ability.

5 Conclusions

Inspired by GBC and TWSVM, we proposed GBTWSVM to solve binary classification problems, which introduces granular-balls as input into TWSVM classification calculations, and provides a scalable, efficient, and robust data processing framework. Then we developed a dual model of GBTWSVM and designed an algorithm for classification. Finally, the experimental results showed that GBTWSVM performs better than TWSVM, IFTSVM and GBSVM on datasets without and with noise.

In real-world applications, there are numerous instances of nonlinear classification problems and multi-classification challenges. Our forthcoming research will be dedicated to proficient algorithms and classifiers tailored to address the array of nonlinear classification problems and multi-classification scenarios. This pursuit will entail the integration of the principles underlying GBC into our methodologies.

Acknowledgements. This work is supported by the National Natural Science Foundation of China (No. 62076040), the Scientific Research Fund of Hunan Provincial Education Department (No. 22A0233), the Scientific Research Fund of Chongqing Key Laboratory of Computational Intelligence (No. 2020FF04), the Postgraduate Scientific Research Innovation Project of Changsha University of Science and Technology (No. CSLGCX23106).

References

1. Vapnik, V.N.: Statistical Learning Theory, vol. 1. Wiley, New York (1998)
2. Heisele, B., Ho, P., Poggio, T.: Face recognition with support vector machines: global versus component-based approach. In: Proceedings Eighth IEEE International Conference on Computer Vision, vol. 2, pp. 688–694 (2001)
3. Mangasarian, O.L., Wild, E.W.: Multisurface proximal support vector machine classification via generalized eigenvalues. IEEE Trans. Pattern Anal. Mach. Intell. **28**(1), 69–74 (2006)
4. Deng, N.Y., Tian, Y.J., Zhang, C.H.: Support Vector Machines: Optimization Based Theory, Algorithms, and Extensions. CRC Press, Boca Raton (2012)
5. Vapnik, V.N., Izmailov, R.: Knowledge transfer in SVM and neural networks. Ann. Math. Artif. Intell. **81**, 3–19 (2017)
6. Jayadeva, Khemchandani, S., Chandra, S.: Twin support vector machines for pattern classification. IEEE Trans. Pattern Anal. Mach. Intell. **29**(5), 905–910 (2007)
7. Khemchandani, R., Jayadeva, Chandra, S.: Optimal kernel selection in twin support vector machines. Optim. Lett. **3**(1), 77–88 (2009)
8. Qi, Z., Tian, Y., Shi, Y.: Robust twin support vector machine for pattern classification. Pattern Recogn. **46**(1), 305–316 (2013)
9. Xie, X.J., Sun, S.L.: PAC-Bayes bounds for twin support vector machines. Neurocomputing **234**, 137–143 (2017)
10. Peng, X.: A v-twin support vector machine (v-TSVM) classifier and its geometric algorithms. Inf. Sci. **180**, 3863–3875 (2010)
11. Tikhonov, A.N., Arsenin, V.Y.: Solutions of Ill-Posed Problems. Wiley, New York (1977)
12. Tanveer, M.: Newton method for implicit Lagrangian twin support vector machines. Int. J. Mach. Learn. Cybern. **6**, 1029–1040 (2015)
13. Rezvani, S., Wang, X., Pourpanah, F.: Intuitionistic fuzzy twin support vector machines. IEEE Trans. Fuzzy Syst. **27**(11), 2140–2151 (2019)

14. Mei, B.S., Xu, Y.T.: Multi-task v-twin support vector machines. Neural Comput. Appl. **32**(15), 11329–11342 (2020)
15. An, R., Xu, Y., Liu, X.: A rough margin-based multi-task v-twin support vector machine for pattern classification. Appl. Soft Comput. **112**, 107769 (2021)
16. Richhariya, B., Tanveer, M.: Universum least squares twin parametric-margin support vector machine. In: 2020 International Joint Conference on Neural Networks (IJCNN), pp. 1–8, September 2020
17. Zadeh, L.A.: Fuzzy sets and information granularity. In: Fuzzy Sets, Fuzzy Logic, and Fuzzy Systems: Selected Papers, pp. 433–448, July 1979
18. Zadeh, L.A.: Toward a theory of fuzzy information granulation and its centrality in human reasoning and fuzzy logic. Fuzzy Sets Syst. **90**(2), 111–127 (1997)
19. Yao, Y.Y.: Interpreting concept learning in cognitive informatics and granular computing. IEEE Trans. Syst. Man Cybern. **39**(4), 855–866 (2009)
20. Skowron, A., Stepaniuk, J., Swiniarski, R.: Modeling rough granular computing based on approximation spaces. Inf. Sci. **184**(1), 20–43 (2012)
21. Ding, S., Huang, H., Yu, J., Zhao, H.: Research on the hybrid models of granular computing and support vector machine. Artif. Intell. Rev. **43**, 565–577 (2015)
22. Liu, H., Cocea, M.: Granular computing-based approach for classification towards reduction of bias in ensemble learning. Granul. Comput. **2**, 131–139 (2016)
23. Butenkov, S., Zhukov, A., Nagorov, A., Krivsha, N.: Granular computing models and methods based on the spatial granulation. Procedia Comput. Sci. **103**, 295–302 (2017)
24. Chen, L., Zhao, L., Xiao, Z., Liu, Y., Wang, J.: A granular computing based classification method from algebraic granule structure. IEEE Access **9**, 68118–68126 (2021)
25. Xia, S.Y., Liu, Y.S., Ding, X., Wang, G.Y., Yu, H., Luo, Y.G.: Granular ball computing classifiers for efficient, scalable and robust learning. Inf. Sci. **483**, 136–152 (2019)
26. Xia, S.Y., Zhang, H., Li, W.H., Wang, G.Y., Giem, E., Chen, Z.Z.: GBNRS: a novel rough set algorithm for fast adaptive attribute reduction in classification. IEEE Trans. Knowl. Data Eng. **34**(3), 1231–1242 (2022)
27. Xia, S.Y., Dai, X.C., Wang, G.Y., Gao, X.B., Giem, E.: An efficient and adaptive granular-ball generation method in classification problem. IEEE Trans. Neural Netw. Learn. Syst. **184**(1), 20–43 (2022)
28. Xia, S.Y., Zheng, S.Y., Wang, G.Y., Gao, X.B., Wang, B.G.: Granular ball sampling for noisy label classification or imbalanced classification. IEEE Trans. Neural Netw. Learn. Syst. **34**(4), 2144–2155 (2023)
29. Xie, J., Kong, W.Y., Xia, S.Y., Wang, G.Y., Gao, X.B.: An efficient spectral clustering algorithm based on granular-ball. IEEE Trans. Knowl. Data Eng. **35**(9), 9743–9753 (2023)
30. Qian, W.B., Xu, F.K., Qian, J., Shu, W.H., Ding, W.P.: Multi-label feature selection based on rough granular-ball and label distribution. Inf. Sci. **650**, 119698 (2023)
31. Kelly, M., Longjohn, R., Nottingham, K.: The UCI Machine Learning Repository (2023). https://archive.ics.uci.edu

Fuzzy Granular-Balls Based Spectral Clustering

Yueyang Li, Siheng Chen, and Guangming Lang(✉)

School of Mathematics and Statistics, Changsha University of Science and Technology,
Changsha 410114, Hunan, China
langguangming1984@126.com

Abstract. In real-world scenarios, datasets characterized by nonlinear separability and fuzziness are quite common. Although fuzzy granular-balls generated by fuzzy c-means can capture the fuzziness of the datasets, it exhibits limitations in effectively handling nonlinearly separable datasets. Due to the advantage of spectral clustering in handling nonlinearly separable datasets, we fully leverage this capability to propose a novel clustering model, namely, fuzzy granular-balls based spectral clustering (FGBSC). It enhances spectral clustering by introducing fuzzy granular-balls as inputs, and effectively addresses the nonlinear separability of the datasets on the basis of capturing the fuzziness of the datasets. We perform experiments to evaluate the effectiveness of the proposed method in handling nonlinearly separable datasets.

Keywords: Spectral clustering · Granular-ball · Information system · Three-way decisions

1 Introduction

Clustering is an unsupervised machine learning technique aimed at maximizing similarity within clusters and minimizing similarity between clusters [1,2]. It's categorized into hard and soft clustering based on how objects are assigned to clusters [3]. Spectral clustering (SC), which leverages graph theory, is widely used but lacks consistency analysis [4]. To address this limitation, some studies proposed methods to assess its consistency and introduce faster approximate algorithms [5,6]. Additionally, robustness issues were addressed through techniques like M-estimation [7]. Enhancing accuracy was achieved through the utilization of multiple data views with approaches like multi-view spectral clustering [8]. At the same time, an improvement involved a bipartite graph-based method for efficiency and ensemble methods for accuracy [9]. Another advancement incorporated a self-learning mechanism to utilize manifold structure for consensus graph partitioning [10]. A dataset is called nonlinearly separable dataset if it contains at least one cluster with a non-convex shape boundary. Based on the distribution of clusters, clustering can be divided into linearly separable clustering such as K-means [11] and nonlinearly separable clustering such as spectral clustering. The essential difference between them is that the linearly separable clustering can express a cluster through a center, while the nonlinearly separable clustering cannot express a cluster through a single center.

Human cognition exhibits a distinctive characteristic of large-scale priority. The human brain's thinking process tends to move from macro to micro or from coarse to fine [12]. Granular computing, a method acknowledged for its scalability, efficiency, and robustness, closely mirrors the cognitive patterns observed in the human brain [13, 14]. The concept of using granular-balls as grains introduced granular-balls computing [15]. A novel metric to assess the quality of granular-balls had significantly contributed to the fusion of clustering and granular-ball computing [16]. Spectral clustering with granular-ball computing (GBSC) had shown promise in substantially reducing computation time [16]. Density-based spatial clustering (DBSCAN), while effective, necessitated distance calculations between all data points. The integration of granular-ball computing into DBSCAN to form GB-DBSCAN [17] has the potential to markedly enhance the efficiency of the DBSCAN. The GBMST [18] technique which combines granular-balls and minimum spanning tree (MST) effectively handled outliers and speeds up MST construction. Afterward, Xie introduced the concept of fuzzy granular-balls [19], which utilizes fuzzy c-means (FCM) for granular-balls generation and exploits k-means for accelerated implementation of granular-balls clustering. These methods highlight the potential of integrating granular-ball computing with different clustering methods, and improve efficiency and performance in various clustering scenarios.

Datasets include nonlinearly separable and linearly separable datasets based on the nature of their relationships. Linearly separable datasets permit exploration of relationships through linear models; however, nonlinear models can uncover deeper associations within these datasets. Conversely, nonlinearly separable datasets pose limitations for linear models. For fuzzy granular-balls, k-means (FGBK) for clustering performs in shorter runtime, but it does not handle nonlinearly separable datasets well. GBSC uses k-means to divide granular-balls. Although it can accelerate spectral clustering, it has limitations in dealing with fuzzy datasets. This study combines the advantages of fuzzy granular-balls and spectral clustering to propose FGBSC, which can effectively handle nonlinearly separable datasets. We demonstrate the performance of FGBSC in handling nonlinearly separable datasets through visualization of clustering results on six artificial datasets. Furthermore, we calculate Adjusted Rand Index (ARI) [20], Normalized Mutual Information (NMI) [21] and Fowlkes-Mallows Index (FMI) [22] on nine other datasets to validate the superiority of FGBSC.

The rest of this paper is organized as follows: Sect. 2 reviews fuzzy c-means, spectral clustering and granular-ball computing. Section 3 presents the method of generating granular balls by FCM and the fusion of granular-balls by spectral clustering. Section 4 conducts experiments on datasets. Section 5 summarizes this paper and gives some future directions.

2 Preliminaries

2.1 Fuzzy C-Means

In 1982, Pawlak [23] provided the concept of an information system.

Definition 1. *An information system is represented as a tetrad $S = (X, A, V, f)$, where $X = \{x_1, x_2, \ldots, x_n\}$ is a nonempty finite set of objects, $A = \{a_1, a_2, \ldots, a_d\}$ stands for a nonempty finite set of attributes, $V = \bigcup_{a \in A} V_a$ is a family of sets of attribute values, V_a denotes the set of attribute values for all objects pertaining to attribute a, and $f : X \times A \to V$ is a function.*

In 2002, Zhang, Hall, and Goldgof [24] introduced fuzzy c-means for clustering, wherein they partitioned a set of objects into c fuzzy subsets.

Definition 2. Let $S = (X, A, V, f)$ be an information system, where $X = \{x_1, x_2, ..., x_n\}$ and $A = \{a_1, a_2, ..., a_d\}$, the objective function of fuzzy c-means (FCM) is defined by:

$$J_m(\mathcal{U}, \mathcal{V}) = \sum_{j=1}^{c} \sum_{i=1}^{n} (u_{ji})^m \|f(x_i, A) - v_j\|^2,$$

where $0 \leq u_{ji} \leq 1$, $\sum_{j=1}^{c} u_{ji} = 1$, c is the number of clusters, and m controls clustering fuzziness. Here, $\mathcal{U} = [u_{ji}]_{c \times n}$ is the membership matrix, where u_{ji} denotes the membership degree of x_i to the j-th cluster, $\mathcal{V} = [v_j]_{c \times 1}$ is the set of cluster centers, and $\| * \|$ represents the Euclidean norm of the vector $*$.

Next, we apply the Lagrange multiplier method, as described in Bertsekas [25], to derive expressions of u_{ji} and v_j as follows:

$$u_{ji} = \frac{1}{\sum_{k=1}^{c} \left(\frac{\|f(x_i, A) - v_j\|}{\|f(x_i, A) - v_k\|} \right)^{\frac{2}{m-1}}},$$

$$v_j = \frac{\sum_{i=1}^{n} u_{ij}^m f(x_i, A)}{\sum_{i=1}^{n} u_{ji}^m}.$$

Fuzzy c-means excels in handling fuzzy or uncertain data by providing a membership degree for objects within each cluster. Despite its effectiveness in specific scenarios, it is essential to acknowledge that fuzzy c-means comes with increased complexity, elevated computational costs, and necessitates careful parameter adjustments. Setting $m = 2$ is the most common choice.

2.2 Spectral Clustering

Spectral clustering is a method based on graph theory that constructs a graph G using the similarity matrix of the objects and performs clustering using the eigenvectors of the graph. The main idea is to treat the objects as nodes in the graph, with the similarity between nodes represented by the weight matrix W of the edges. Firstly, construct the similarity matrix of the objects. Then, the Laplacian matrix L of the graph is computed, which is given by $L = D - W$, where D is the degree matrix of G. Finally, the objects are clustered based on the eigenvectors of matrix L. The framework of the spectral clustering is shown in Fig. 1.

For the similarity matrix $W = [W_{pq}]_{n \times n}$, there are various types such as the Gaussian Similarity Matrix, K-Nearest Neighbors Similarity Matrix, etc. In this article, we choose the Gaussian Similarity Matrix. W_{pq} represents the similarity between object p and object q, σ is the parameter of the gaussian kernel function:

$$W_{pq} = \exp\left(-\frac{\|f(x_p, A) - f(x_q, A)\|^2}{2\sigma^2}\right).$$

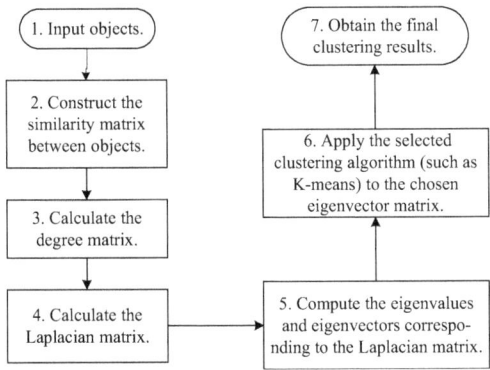

Fig. 1. The framework of spectral clustering.

2.3 Granular-Ball Computing

In granular-ball computing [15], a family of granular-balls serve as inputs to the classifier instead of individual object. This significantly diminishes the scale of input objects and consequently minimizes the overall algorithmic runtime. The symmetry of granular-balls allows the granular-ball to be succinctly described by its center and radius, and introduces only a limited number of parameters. In 2019, Xia, Liu, Ding, et al. [15] defined a specific model for granular-ball computing.

Definition 3. *Let $S = (X, A, V, f)$ be an information system, where $X = \{x_1, x_2, ..., x_n\}$, $A = \{a_1, a_2, ..., a_d\}$, a granular-ball is defined by $GB = \{x_{i_1}, x_{i_2}, ..., x_{i_t}\}$ with the center $O(GB)$ and radius $r(GB)$:*

$$O(GB) = \frac{1}{|GB|}(\sum_{x \in GB} f(x, a_1), \sum_{x \in GB} f(x, a_2), ..., \sum_{x \in GB} f(x, a_d)),$$
$$r(GB) = \max_{x \in GB}(\|f(x, A) - O(GB)\|).$$

Describing granular-balls through their centers and radiuses proves to be a dependable method. It provides a representation that effectively captures the relevant information of the involved objects.

Definition 4. *Let $S = (X, A, V, f)$ be an information system, and $\mathcal{GB} = \{GB_1, GB_2, ..., GB_k\}$ a family of granular-balls. The optimization function of granular-ball generation is given by:*

$$\min \lambda_1 * \frac{|X|}{\sum_{GB \in \mathcal{GB}} |GB|} + \lambda_2 * |\mathcal{GB}|, \quad s.t.\ quality(GB) \geq T,$$

where λ_1 and λ_2 represent two weight parameters to balance object coverage and the number of granular-ball, $|X|$ denotes the cardinality of X, and T represents the quality threshold.

In granular-ball computing, there exists a relationship between the reciprocal of granular-ball coverage and the quantity of granular-balls. It is anticipated that the coverage of granular-balls increases when the count of granular-balls decreases.

3 Fuzzy Granular-Balls Based Spectral Clustering

In this section, we explore the spectral clustering approach using fuzzy granular-balls.

In 2023, Xie, Kong, Xia, et al. [16] defined the distributed measure for clustering by granular-balls and specified the similarity measure between granular-balls.

Definition 5. *Let $S = (X, A, V, f)$ be an information system, the distributed measure DM to assess the quality of a granular-ball is defined by:*

$$DM(GB) = \frac{1}{|GB|} \sum_{x \in GB} \|f(x, A) - O(GB)\|.$$

Due to the inherently unsupervised nature of clustering problems, the metric DM is employed to assess the quality of granular-balls. A granular-ball with a smaller size suggests greater similarity among the samples it encompasses. In essence, a diminished distributed measure means the superior quality for the granular-ball.

Assuming the division of a granular-ball GB results in two distinct granular-balls GB_i and GB_j, the original granular-ball GB is referred to as the mother-ball, while GB_i and GB_j are considered as the sub-balls of GB.

Definition 6. *Let $S = (X, A, V, f)$ be an information system, the weighted distributed measure WD for the granular-ball GB with respect to the two granular-balls GB_i and GB_j is defined by:*

$$WD(GB) = \frac{|GB_i|}{|GB|} * DM(GB_i) + \frac{|GB_j|}{|GB|} * DM(GB_j).$$

If the weighted distributed measure $WM(GB)$ surpasses the distributed measure $DM(GB)$, it indicates that the quality of the two sub-balls GB_i and GB_j is relatively inferior compared to that of the mother-ball GB.

Once the dataset X is segmented into a set of granular-balls using FCM, these granular-balls can be effectively characterized by their respective centers and radiuses.

Definition 7. *For an information system $S = (X, A, V, f)$, the similarity between granular-balls GB_p and GB_q is defined as follows:*

$$W_{pq} = \begin{cases} \exp\left(-\frac{\|e_p - e_q\|^2}{2\sigma^2}\right), & \|e_p - e_q\| \geq 0, \\ 1, & \|e_p - e_q\| < 0, \end{cases}$$

where $\|e_p - e_q\| = \|O(GB_p) - O(GB_q)\| - r(GB_p) - r(GB_q)$ represents the distance between the centers of the granular-balls minus the radius of the granular-balls.

In fuzzy granular-balls based spectral clustering, the similarity matrix is: $W = [W_{pq}]_{k \times k}$. The dimension of the similarity matrix is $k \times k$ due to the existence of k granular-balls. For an individual granular-ball, there is a high probability that the objects within the granular-ball belong to the same cluster. When intersecting granular-balls share common objects, all objects within the intersecting granular-balls should also belong to the same cluster. The similarity between two granular-balls in this case would be 1. For non-intersecting granular-balls, their similarity can be expressed using an exponential function. The farther apart two granular-balls are, the lower the similarity; conversely, the closer they are, the higher the similarity.

The objective function of spectral clustering is RatioCut$(O(GB)) = \sum_{i=1}^{c} \frac{\text{cut}(C_i, \overline{C_i})}{|C_i|}$, which aims to minimize the sum of edge weights between clusters while ensuring that the number of samples in each cluster is moderate. C_i is a subgraph of $O(GB)$, and $\overline{C_i}$ is the complement subgraph of C_i. cut$(C_i, \overline{C_i})$ represents the sum of the weights of all edges connecting subgraph C_i and subgraph $\overline{C_i}$. Based on graph theory, we can conclude that minimizing the objective function of spectral clustering is equivalent to performing an eigendecomposition of L, where smaller eigenvalues correspond to more dispersed information in the corresponding eigenvectors.

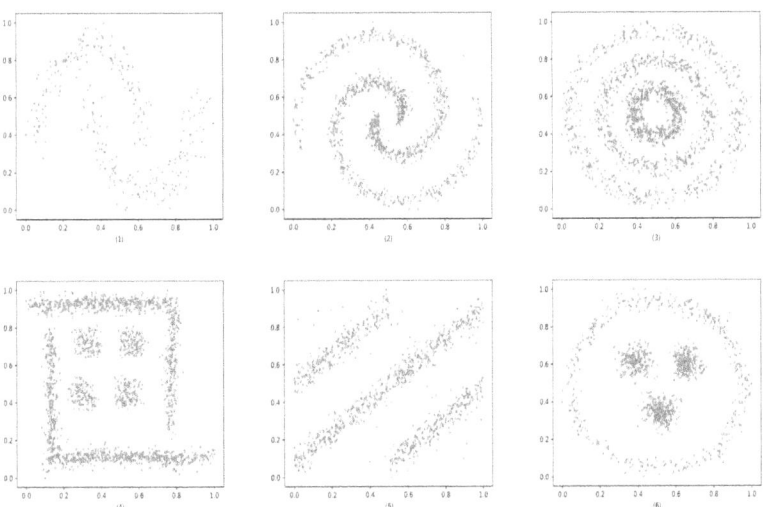

Fig. 2. Six artificially synthesized datasets.

Figure 2 displays the characteristics of FGBSC on six artificially synthesized datasets, each of which possesses relatively complex structures. To better handle the datasets with structures as shown in the Fig. 2, we perform spectral clustering based on fuzzy granular-balls. Because spectral clustering constructs a similarity graph rather than relying on the linearity of the data, it can effectively capture complex data structures. Additionally, by leveraging the accelerated clustering advantages of granular-ball computing and the fuzzy characteristics described by fuzzy granular-balls within datasets, we establish the similarity between these fuzzy granular-balls. Subsequently, we take spectral clustering to effectively merge these granular-balls, which not only accelerates spectral clustering but also handles nonlinearly separable datasets based on the description of fuzzy granular-balls for fuzzy datasets.

The process of FGBSC is illustrated in Algorithm 1. When constructing granular-balls, FCM is used, and its time complexity is $O(ndtT)$, where d is the dimension of objects, t is the number of iterations and T is the number of times the granular-balls are divided. The time complexity of spectral clustering is $O(n^2d + n^3)$, and the number of granular-balls is much smaller than n, the time complexity of FGBSC is significantly lower than $O(n^3 + n^2d + ndtT)$.

4 Experimental Analysis

In this section, we conduct the experiments to illustrate the effectiveness of the fuzzy granular-balls based spectral clustering.

4.1 Experimental Datasets and Metric Indexes

In the experiment, six artificially synthesized datasets and nine real datasets [26] are used. Furthermore, we have made a comparative experiment with GBSC, FGBK, SC, and FCM. The experimental equipment is a personal computer with Intel(R) Core(TM) i5-11500 @2.70 GHz, 2712 Mhz and Python 3.8.10.

Two-dimensional artificial datasets are more amenable to visualization analysis, allowing for the intuitive display of object distributions on a two-dimensional plane. To further validate the effectiveness of the FGBSC, we compared it with GBSC, FGBK, SC, and FCM on other nine datasets. The specific characteristics of the nine datasets are shown in Table 1. Iris comprises three distinct categories: setosa, versicolor, and virginica. Among these, the setosa category is linearly separable from the other two, whereas the versicolor and virginica categories cannot be linearly separated from each other. Despite the complexity of the data, it is impossible to determine whether the data is nonlinearly separable. However, even if the data is linear, FGBSC can still extract information from it.

We use Adjusted Rand Index (ARI) [20], Normalized Mutual Information (NMI) [21], and Fowlkes-Mallows Index (FMI) [22] to evaluate the clustering performance in the experiments. These metrics are defined as follows:

$$ARI = \frac{\sum \binom{n_{ij}}{2} - \left[\sum \binom{a_i}{2} \sum \binom{b_j}{2}\right] / \binom{n}{2}}{\frac{1}{2}\left[\sum \binom{a_i}{2} + \sum \binom{b_j}{2}\right] - \left[\sum \binom{a_i}{2} \sum \binom{b_j}{2}\right] / \binom{n}{2}}.$$

The value range of ARI is -1 to 1. Here, n_{ij} denotes the count of sample pairs presented in both cluster c_i and cluster c_j as defined in the true labels and predicted labels. Additionally, a_i represents the number of samples in cluster c_i, while b_i denotes the number of samples in cluster c_j.

$$NMI(X, C) = \frac{I(X; C)}{\sqrt{H(X) \cdot H(C)}}.$$

Algorithm 1. Fuzzy granular-balls spectral clustering.

1: **Input:** An information system $S = (X, A, V, f)$.
2: **Output:** The clustering result.
3: Initialize: $m = 2$, max_count = 1000, $center\_list = \emptyset$, $radius\_list = \emptyset$, $W = \emptyset$.
4: The set of GB in initialized with X, $\mathcal{GB} = \{X\}$;
5: **repeat**
6:     **for** each $GB \in \mathcal{GB}$ **do**
7:         Select the clustering center $\mathcal{V}$ with the farthest distance;
8:         **repeat**
9:             Compute membership matrix $\mathcal{U}$ based on $\mathcal{V}$ and GB;
10:            Update cluster center $\mathcal{V}$ based on $\mathcal{U}$ and GB;
11:         **until** $\mathcal{V}$ converges or reaches the maximum number of iterations max_count;
12:         Divide GB into GB_1, GB_2 according to the $\mathcal{U}$;
13:         Calculate $DM(GB)$ and $WM(GB)$ by Definitions 5 and 6;
14:         **if** $DM(GB) \geq WM(GB)$ **then**
15:             Take GB_1, GB_2 to replace GB;
16:         **else**
17:             $\mathcal{GB}$ unchanged;
18:         **end if**
19:     **end for**
20: **until** All GB do not meet step 14;
21: **for** each $GB \in \mathcal{GB}$ **do**
22:     **if** $r(GB) \geq 2 \cdot \max\{\text{mean}(r(GB)), \text{median}(r(GB))\}$ **then**
23:         Split GB;
24:     **end if**
25: **end for**
26: **for** each $GB \in \mathcal{GB}$ **do**
27:     Calculate the center and radius of GB;
28:     Add the center and radius of GB to $center\_list$ and $radius\_list$;
29: **end for**
30: Calculate the similarity matrix W based on Definition 7;
31: Compute the Laplacian matrix L;
32: Perform eigenvalue decomposition on the matrix L and select eigenvectors;
33: Use the selected eigenvectors as input for K-means to obtain the final clustering results;
34: **return** The clustering result.

The value range of NMI is 0 to 1. Here, $I(X;C)$ represents the mutual information between partition X and partition C, and $H(X)$ and $H(C)$ denote the entropy of X and C, respectively.

$$FMI = \frac{TP}{\sqrt{(TP+FP)\cdot(TP+FN)}}.$$

The value range of FMI is 0 to 1. Here, TP represents the number of data points correctly assigned to the same cluster in two different clustering results. FP is the number assigned to a certain cluster in the first clustering result and not assigned to the same cluster in another clustering result, and FN is the number assigned to a certain cluster in the second clustering result but not to the same cluster in the first clustering result.

Table 1. Basic Information of Experimental Datasets.

Dataset	Number of objects	Number of Attributes	Number of Clusters
Iris	150	4	3
Chart	600	60	6
Segment	2310	19	7
Pen	10992	16	10
Waveform	5000	21	3
Magic	19020	10	2
Satellite	6435	36	6
Twonorm	7400	20	2
Abalone	4177	8	3

4.2 The Experimental Results of the FGBSC

For artificially synthesized datasets, we can validate the experimental results based on visualization. For real datasets, we compare the results obtained by FGBSC with those of GBSC, FGBK, SC, and FCM.

Figure 3 shows the visualization of granular-balls boundaries for six artificial datasets. Most granular-balls are composed of multiple objects, and the number of granular-balls are much smaller than the sample size. Figure 4 displays the final clustering results, where the division between clusters is quite distinct.

Table 2 illustrates the counts of objects in the original datasets and the granular-balls derived using GBSC, FGBK and FGBSC. Specifically, the quantities of objects

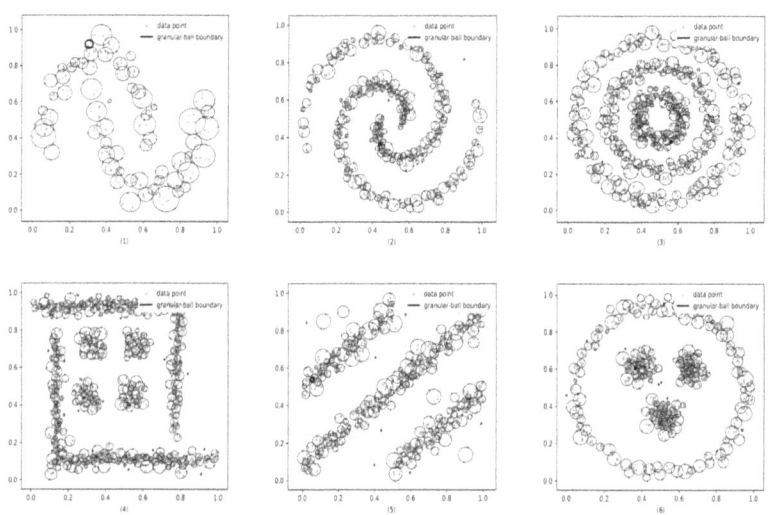

Fig. 3. Granular-ball boundaries obtained through FGBSC.

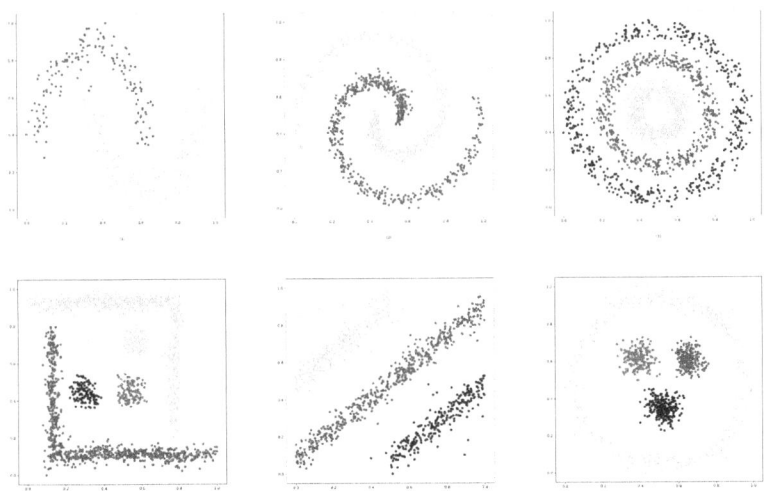

Fig. 4. Visualized clustering results obtained by FGBSC on six artificial datasets.

Table 2. The numbers of objects and granular-balls by GBSC, FGBK and FGBSC.

Dataset	Number of objects	GBSC	FGBK	FGBSC
Iris	150	35.0 ± 0.0	34.0 ± 0.8	34.0 ± 0.8
Chart	600	132.0 ± 0.0	125.7 ± 2.9	126.3 ± 1.9
Segment	2310	516.0 ± 0.0	496.3 ± 0.9	498.3 ± 0.5
Pen	10992	2477.0 ± 0.0	2405.0 ± 2.9	2407.0 ± 4.3
Waveform	5000	1104 ± 0.0	1061.3 ± 1.2	1063.7 ± 2.1
Magic	19020	4359.0 ± 29.7	4202.0 ± 5.4	4206.3 ± 2.4
Satellite	6435	1489 ± 0.0	1452.0 ± 1.6	1463.3 ± 2.6
Twonorm	7400	1673.2 ± 12.6	1631.3 ± 13.1	1620.3 ± 17.2
Abalone	4177	939.0 ± 0.0	937.3 ± 1.2	939.0 ± 2.2

Table 3. The running time of FGBSC, GBSC, FGBK, SC, and FCM.

Dataset	FGBSC	GBSC	FGBK	SC	FCM
Iris	0.1193	0.1165	0.1832	0.2580	0.0108
Chart	0.5434	0.5956	0.6371	1.3614	0.0345
Segment	5.0042	6.6226	4.4369	23.6532	0.1143
Pen	106.6175	129.8000	58.9572	475.2446	1.4957
Waveform	17.9307	26.3354	10.2035	86.5618	0.0655
Magic	233.0976	247.9335	38.1634	1475.9004	0.1393
Satellite	33.0721	47.8487	18.8474	144.7298	0.1601
Twonorm	38.9665	55.7694	20.6093	228.9936	0.0549
Abalone	14.0280	19.6759	8.6771	62.2959	0.0594

Table 4. The ARI of FGBSC, GBSC, FGBK, SC, and FCM on nine datasets.

Dataset	FGBSC	GBSC	FGBK	SC	FCM
Iris	**0.7404**	0.6498	0.7125	0.5805	0.7028
Chart	**0.6619**	0.6509	0.6204	0.6390	0.6106
Segment	0.4859	0.4273	0.4504	0.4092	**0.4902**
Pen	**0.4211**	0.4170	0.4073	0.3889	0.4042
Waveform	**0.2525**	0.2506	0.2509	0.2500	0.2485
Magic	0.1871	**0.2022**	0.0191	0.0366	0.0574
Satellite	**0.5856**	0.5776	0.3978	0.4831	0.5126
Twonorm	**0.9066**	0.9050	0.9063	0.9059	0.9059
Abalone	**0.1649**	0.1324	0.1529	0.1519	0.1492
Average	**0.4896**	0.4681	0.4353	0.4272	0.4525

Table 5. The NMI of FGBSC, GBSC, FGBK, SC, and FCM on nine datasets.

Dataset	FGBSC	GBSC	FGBK	SC	FCM
Iris	**0.7797**	0.7137	0.7431	0.6187	0.7277
Chart	**0.7785**	0.7678	0.7388	0.7500	0.7382
Segment	**0.6221**	0.6001	0.5945	0.5729	0.5996
Pen	**0.6026**	0.5994	0.5855	0.5607	0.5700
Waveform	**0.3705**	0.3695	0.3671	0.3692	0.3550
Magic	0.1080	**0.1196**	0.0091	0.0209	0.0317
Satellite	**0.6475**	0.6307	0.5082	0.5716	0.5856
Twonorm	**0.8372**	0.8356	0.8367	0.8362	0.8362
Abalone	**0.1480**	0.1242	0.1388	0.1417	0.1401
Average	**0.5438**	0.5290	0.5024	0.4935	0.5094

in the original datasets surpass those of granular-balls obtained through GBSC, FGBK and FGBSC. Furthermore, the number of granular-balls derived by FGBSC is nearly identical to that obtained by GBSC and FGBK.

In Table 3, we present the running times for FGBSC, GBSC, FGBK, SC, and FCM. We observe that the running times of FGBSC, GBSC, FGBK, and FCM are significantly lower than that of SC. The running times of FGBSC and GBSC are essentially the same, which may be due to factors such as the number of iterations. Meanwhile, we can observe that with the increase in object size, the running time of FGBK is relatively less compared to FGBSC and GBSC. Table 4 presents the Adjusted Rand Index (ARI) scores for FGBSC, GBSC, FGBK, SC, and FCM across nine datasets. Remarkably, FGBSC demonstrates higher ARI scores on the Iris, Chart, Pen, Waveform, Satellite, Twonorm, and Abalone datasets in comparison to other methods. On average, FGBSC outperforms other methods across all nine datasets based on ARI scores.

Table 6. The FMI of FGBSC, GBSC, FGBK, SC, and FCM on nine datasets.

Dataset	FGBSC	GBSC	FGBK	SC	FCM
Iris	**0.8278**	0.7712	0.8090	0.7194	0.8020
Chart	**0.7192**	0.7086	0.6870	0.7010	0.6788
Segment	**0.5779**	0.5291	0.5347	0.4994	0.5638
Pen	0.4938	0.4894	**0.4940**	0.4574	0.4684
Waveform	0.5023	0.5006	**0.5048**	0.5000	0.4997
Magic	0.6464	**0.6631**	0.6351	0.5429	0.5564
Satellite	**0.6705**	0.6675	0.5209	0.5820	0.6023
Twonorm	**0.9533**	0.9525	0.9531	0.9530	0.9530
Abalone	**0.4444**	0.4224	0.4394	0.4366	0.4340
Average	**0.6484**	0.6338	0.6087	0.5991	0.6176

Table 5 illustrates the Normalized Mutual Information (NMI) scores for FGBSC, GBSC, FGBK, SC, and FCM across the same nine datasets. Notably, FGBSC achieves higher NMI scores on the Iris, Chart, Segment, Pen, Waveform, Satellite, Twonorm, and Abalone datasets compared to other methods. On average, FGBSC demonstrates superior NMI performance across all nine datasets. In Table 6, the Fowlkes-Mallows Index (FMI) for FGBSC, GBSC, FGBK, SC, and FCM on nine datasets is presented. FGBSC exhibits higher FMI scores on the Iris, Chart, Segment, Satellite, Twonorm, and Abalone datasets compared to other methods. On average, FGBSC attains superior FMI scores across all nine datasets.

5 Conclusions

In this paper, we combined the advantages of fuzzy granular-balls and spectral clustering to propose FGBSC for handling nonlinearly separable datasets. This method enables us to capture the information of fuzzy datasets more precisely and addresses the challenges posed by nonlinear datasets while accelerating spectral clustering. We designed the algorithm of FGBSC for classification. The effectiveness of the proposed method was verified through experiments which provides significant support for practical applications in data analysis and mining.

In practice, there are a large number of dynamic datasets, and it is a challenge to deal with them. In the future, we will focus on designing effective clustering algorithms for dynamic datasets based on FGBSC to achieve superior clustering results.

Acknowledgements. This work is supported by the National Natural Science Foundation of China (No. 62076040), the Scientific Research Fund of Hunan Provincial Education Department (No. 22A0233), the Scientific Research Fund of Chongqing Key Laboratory of Computational Intelligence (No. 2020FF04), the Postgraduate Scientific Research Innovation Project of Changsha University of Science and Technology (No. CSLGCX23106).

References

1. Barlow, H.B.: Unsupervised learning. Neural Comput. **1**(3), 295–311 (1989)
2. Dubes, R.C.: Cluster Analysis and Related Issues. Handbook of Pattern Recognition and Computer Vision, pp. 3–32 (1999)
3. Kearns, M., Mansour, Y., Ng, A.Y.: An information-theoretic analysis of hard and soft assignment methods for clustering. In: Jordan, M.I. (eds.) Learning in Graphical Models. NATO ASI Series, vol. 89, pp. 495–520 (1998). Springer, Dordrecht. https://doi.org/10.1007/978-94-011-5014-9_18
4. Ng, A., Jordan, M., Weiss, Y.: On spectral clustering: analysis and an algorithm. In: Advances in Neural Information Processing systems, pp. 849–856 (2001)
5. Ulrike, V.L., Belkin, M., Bousquet, O.: Consistency of spectral clustering. Ann. Stat. **36**(2), 555–586 (2008)
6. Yan, D.H., Ling, H., Jordan, M.I.: Fast approximate spectral clustering. In: Proceedings of the 15th ACM SIGKDD International Conference on Knowledge Discovery and Data Mining, pp. 907–916 (2009)
7. Chang, H., Yeung, D.Y.: Robust path-based spectral clustering. Pattern Recogn. **41**(1), 191–203 (2008)
8. Kumar, A., Piyush, R., Hal, D.: Co-regularized multi-view spectral clustering. In: Advances in Neural Information Processing Systems, pp. 1413–1421 (2011)
9. Li, Y.Q., Huang, F.P., Huang, H., Huang, J.Z.: Large-scale multi-view spectral clustering via bipartite graph. In: Proceedings of the AAAI Conference on Artificial Intelligence, pp. 2750–2756 (2015)
10. Yue, G.L., Deng, A., Qu, Y.P., Cui, H., Wang, X.Y.: Stratified multi-density spectral clustering using Gaussian mixture model. Inf. Sci. **633**, 182–203 (2023)
11. MacQueen, J.: Some methods for classification and analysis of multivariate observations. In: Proceedings of the Fifth Berkeley Symposium on Mathematical Statistics and Probability, vol. 1, no. 14, pp. 281–297 (1967)
12. Chen, L.: Neural correlation of global-first object formation: anterior temporal lobe. Chin. Bull. Life Sci. **20**(5), 718–721 (2008)
13. Yao, J.T., Vasilakos, A.V., Pedrycz, W.: Granular computing: perspectives and challenges. IEEE Trans. Cybern. **43**(6), 1977–1989 (2013)
14. Bargiela, A., Pedrycz, W.: Granular Computing. Handbook on Computer Learning and Intelligence: Deep Learning, Intelligent Control and Evolutionary Computation, pp. 97–132 (2022)
15. Xia, S.Y., Liu, Y.S., Ding, X., Wang, G.Y., Yu, H., Luo, Y.G.: Granular ball computing classifiers for efficient, scalable and robust learning. Inf. Sci. **483**, 136–152 (2019)
16. Xie, J., Kong, W.Y., Xia, S.Y., Wang, G.Y., Gao, X.B.: An efficient spectral clustering algorithm based on granular-ball. IEEE Trans. Knowl. Data Eng. **35**(9), 9743–9753 (2023)
17. Cheng, D.D., Zhang, C., Li, Y., Xia, S.Y., Wang, G.Y.: GB-DBSCAN: a fast granular-ball based DBSCAN clustering algorithm. SSRN 4379714 (2023)
18. Xie, J., Xia, S.Y., Wang, G.Y., Gao, X.B.: GBMST: an efficient minimum spanning tree clustering based on granular-ball. arXiv preprint arXiv:2303.01082 (2023)
19. Xie, J., Deng, Q., Xia, S.Y., Zhao, Y.Z., Wang, G.Y., Gao, X.B.: Research on efficient fuzzy clustering method based on local fuzzy granular balls. arXiv preprint arXiv:2303.03590
20. Sundqvist, M., Chiquet, J., Rigaill, G.: Adjusting the adjusted rand index: a multinomial story. Comput. Stat. **38**(1), 327–347 (2023)
21. Vinh, L.T., Lee, S., Park, Y.T., Auriol, B.J.: A novel feature selection method based on normalized mutual information. Appl. Intell. **37**, 100–120 (2012)

22. Campello, R.G.: A fuzzy extension of the Rand index and other related indexes for clustering and classification assessment. Pattern Recogn. Lett. **28**(7), 833–841 (2007)
23. Pawlak, Z.: Rough sets. Int. J. Comput. Inf. Sci. **11**, 341–356 (1982)
24. Zhang, M.R., Hall, L.O., Goldgof, D.B.: A generic knowledge-guided image segmentation and labeling system using fuzzy clustering algorithms. IEEE Trans. Syst. Man Cybern. Part B (Cybern.) **32**(5), 571–582 (2002)
25. Bertsekas, D.P.: Constrained Optimization and Lagrange Multiplier Methods. Academic Press, New York (2014)
26. Sun, C., Yue, S.H., Li, Q.: Clustering characteristics of UCI dataset. In: Chinese Control Conference, pp. 6301–6306 (2020)

A Vector Is a Granule: A Novel Extension of the Variable Precision Rough Set Model

Hajime Okawa[1](✉), Yasuo Kudo[1](✉), and Tetsuya Murai[2]

[1] Muroran Institute of Technology, 27-1 Mizumoto, Muroran, Hokkaido 050-8585, Japan
{23096003,yask}@muroran-it.ac.jp
[2] Chitose Institute of Science and Technology, 758-65 Bibi, Chitose, Hokkaido 066-8655, Japan
t-murai@photon.chitose.ac.jp

Abstract. In the current extended variable precision rough set models (VPRS models), a granule is a set of objects. Examples of such models include the variable precision neighbourhood rough set model and the variable precision covering-based rough set model. These objects can be considered as real vectors when dealing with real-valued data, and, therefore, a granule is effectively a set of real vectors. In this paper, we show that a single vector can be considered as a granule, too. More precisely, we introduce the granule vector and the approximation vector, and show that the negativity of the inner product of these two vectors is equivalent to a granule being contained the lower (or upper) approximation. Based on this equivalence, we propose a novel extension of VPRS model which treats a single vector as a granule. In particular, this generalised model can deal with applications not covered by the existing models.

Keywords: variable precision rough set · inner product space · granule vector

1 Introduction

The rough set model [11] is commonly used in fields like data mining, feature selection, and machine learning. The approximation of any set of objects using granules is the fundamental concept of rough set model. Several extended rough set models have been proposed based on how granules are formed. For instance, the neighbourhood-based rough set model [6,7,10,17] uses neighbourhoods of objects as granules. The covering-based rough set model [12,13,20,22,25,26] considers each element of the covering as a granule. In the fuzzy rough set model [3,4,21,24], granules are fuzzy equivalence classes.

Another extension of rough set model, the variable precision rough set (for short VPRS) model [2,23,27] was proposed to deal with noisy and probabilistic data. To this end, the model uses the inclusion degree of a granule contained in a decision class as a criterion of approximation. The value of this degree can be varied only by the number of objects in the intersection of a granule and a decision class, which is considered to represent the essence of granules in VPRS models.

We introduce in this paper granule vectors and approximation vectors. Granule vectors are real vectors that represent the number of objects in the intersection of a granule and a decision class. The representation of the approximation can be achieved by calculating the inner product of a granule vector and an approximation vector. Based on this, we propose an extended VPRS model using the granule vectors and the approximation vectors. The approximation vectors are defined as the linear combinations of decision class vectors. The proposed model is capable of approximating a vector, and is therefore applicable to a variety of applications, including collaborative filtering for recommendation systems, where a decision class is represented by a vector.

2 Preliminaries

In this section, we review the notation of Pawlak's rough set model and Ziarko's variable precision rough set model. These models employ Pawlak's approximation space, where any set of objects can be approximated; however, in many practical applications [5,9,15,16], only decision classes are approximated. Therefore we restrict the presentation to notations referring to decision systems. Furthermore, we also review the notation of inner product spaces. The content of this section is mainly based on [1,11,27].

By $\mathbb{Z}$ and $\mathbb{N}_0$ denote the set of integers and the set of natural numbers containing 0, respectively.

2.1 Pawlak's Rough Set Model

A decision system DS is a tuple $(U, \mathcal{E}, \mathcal{D})$, where U is a finite non-empty set of objects, and $\mathcal{E}$ and $\mathcal{D}$ are partitions of U. The elements of $\mathcal{E}$ and elements of $\mathcal{D}$ are called *granules* and *decision classes*, respectively. In particular, $\mathcal{E}$ is called a *knowledge* about U.

For any decision class $D \in \mathcal{D}$, its *lower approximation* $\underline{\mathcal{E}}(D)$ and *upper approximation* $\overline{\mathcal{E}}(D)$ are defined as follows:

$$\underline{\mathcal{E}}(D) = \bigcup \{E \in \mathcal{E} \mid E \subseteq D\}, \tag{1}$$

$$\overline{\mathcal{E}}(D) = \bigcup \{E \in \mathcal{E} \mid E \cap D \neq \emptyset\} \tag{2}$$

$$= \bigcup \{E \in \mathcal{E} \mid E \nsubseteq U \setminus D\}. \tag{3}$$

The lower approximation of D is a set of objects that certainly belong to D by means of the knowledge $\mathcal{E}$. The upper approximation of D consists of objects

Table 1. An example of a table

Patient	Condition Attributes				Decision Attribute
	Headache	Sneeze	Temperature	Muscle-pain	Flu
p_1	no	no	very high	yes	yes
p_2	no	yes	high	yes	yes
p_3	yes	no	high	no	no
p_4	no	yes	high	yes	yes
p_5	no	no	very high	yes	no
p_6	yes	no	high	no	yes
p_7	no	no	very high	yes	no

for which the knowledge $\mathcal{E}$ does not allow us to exclude the possibility that they belong to D. The *boundary region* of $D \in \mathcal{D}$ and the *positive region* of $DS = (U, \mathcal{E}, \mathcal{D})$ are defined as follows:

$$BN\mathcal{E}(D) = \overline{\mathcal{E}}(D) \setminus \underline{\mathcal{E}}(D), \tag{4}$$

$$Pos_{\mathcal{E}}(\mathcal{D}) = \bigcup_{D \in \mathcal{D}} \underline{\mathcal{E}}(D). \tag{5}$$

Example 1. A table is an example of decision system. Consider Table 1, where the patients $p_1, \ldots, p_7$ are objects. The knowledge $\mathcal{E}$ and the set $\mathcal{D}$ of decision classes are built according to the values of the condition attributes and the decision attribute, respectively. Therefore

$$\mathcal{E} = \{E_1, E_2, E_3\}, \tag{6}$$
$$\mathcal{D} = \{D_{\text{yes}}, D_{\text{no}}\}, \tag{7}$$

where

$$E_1 = \{p_1, p_5, p_7\}, \quad E_2 = \{p_2, p_4\}, \quad E_3 = \{p_3, p_6\}, \tag{8}$$

$$D_{\text{yes}} = \{p_1, p_2, p_4, p_6\}, \quad D_{\text{no}} = \{p_3, p_5, p_7\}. \tag{9}$$

Then the lower and upper approximations of D_{yes} and D_{no} are

$$\underline{\mathcal{E}}(D_{\text{yes}}) = \{p_2, p_4\}, \quad\quad \underline{\mathcal{E}}(D_{\text{no}}) = \emptyset, \tag{10}$$
$$\overline{\mathcal{E}}(D_{\text{yes}}) = \{p_1, p_2, p_3, p_4, p_5, p_6, p_7\}, \quad \overline{\mathcal{E}}(D_{\text{no}}) = \{p_1, p_3, p_5, p_6, p_7\}. \tag{11}$$

2.2 Ziarko's Variable Precision Rough Set Model

The practical decision systems may contain stochastic data and noise. In such cases, Pawlak's rough set model is not useful because it prohibits any classification errors of objects in the lower approximation. A variable precision rough set (VPRS) model was proposed by Ziarko [27] to address these shortcomings. The heart of VPRS model is the extension of the standard set inclusion. The extended inclusion is called majority inclusion.

Majority Inclusion. Let X and Y be non-empty subsets of a finite universe U. We introduce the measure $c(X,Y)$ of the *relative degree of misclassification* of the set X with respect to the set Y defined as follows:

$$c(X,Y) = \begin{cases} 1 - \frac{|X \cap Y|}{|X|}, & \text{if } |X| > 0, \\ 0, & \text{otherwise,} \end{cases} \quad (12)$$

where $|\cdot|$ denotes the cardinality of a set. Then one has

$$X \subseteq Y \text{ iff } c(X,Y) = 0. \quad (13)$$

By setting a precision $\beta \in [0, 0.5)$, the β-*majority inclusion relation* is defined as

$$X \stackrel{\beta}{\subseteq} Y \text{ iff } c(X,Y) \leq \beta. \quad (14)$$

In order to maintain the classification errors a minority, the precision β is set within the range $0 \leq \beta < 0.5$.

Approximation in the VPRS Model. Since Pawlak's rough set model employs the set inclusion relation, the lower approximation is not suitable for treating noisy data and probabilistic data. To address this problem, the VPRS model replaces the set inclusion relation with the β-majority inclusion relation. Formally, for any decision class $D \in \mathcal{D}$ and precision $\beta \in [0, 0.5)$, the β-*lower approximation* $\underline{\mathcal{E}}_\beta(D)$ of D is defined as

$$\underline{\mathcal{E}}_\beta(D) = \bigcup \{E \in \mathcal{E} \mid E \stackrel{\beta}{\subseteq} D\}. \quad (15)$$

The β-*upper approximation* of the decision class $D \in \mathcal{D}$ is defined as

$$\overline{\mathcal{E}}_\beta(D) = \bigcup \{E \in \mathcal{E} \mid E \stackrel{\beta}{\not\subseteq} U \setminus D\} \quad (16)$$

$$= \bigcup \{E \in \mathcal{E} \mid c(E,D) < 1 - \beta\}, \quad (17)$$

and thus the boundary region of $D \in \mathcal{D}$ and positive region of DS is given by

$$BN\mathcal{E}_\beta(D) = \overline{\mathcal{E}}_\beta(D) \setminus \underline{\mathcal{E}}_\beta(D) \quad (18)$$

$$= \bigcup \{E \in \mathcal{E} \mid \beta < c(E,D) < 1 - \beta\}. \quad (19)$$

$$Pos_{\mathcal{E}}^{\beta}(\mathcal{D}) = \bigcup_{D \in \mathcal{D}} \underline{\mathcal{E}}_{\beta}(D). \tag{20}$$

Notice that if $\beta = 0$, then these mathematical objects are identical to their correspondents from Pawlak's rough set model.

We list below some properties of the VPRS model.

Proposition 1 ([8]). *Given a decision system* $DS = (U, \mathcal{E}, \mathcal{D})$ *where* $\mathcal{D} = \{D_1, \ldots, D_n\}$ *and a precision* $\beta \in [0, 0.5)$, *the following four properties hold:*

(i) $\underline{\mathcal{E}}_{\beta}(D) \subseteq \overline{\mathcal{E}}_{\beta}(D), \forall D \in \mathcal{D}$;

(ii) *If* $\frac{1}{n} > \beta$, *then* $U = \bigcup_{D \in \mathcal{D}} \overline{\mathcal{E}}_{\beta}(D)$;

(iii) $\overline{\mathcal{E}}_{\beta}(D) = \underline{\mathcal{E}}_{\beta}(D) \cup BN\mathcal{E}_{\beta}(D), \forall D \in \mathcal{D}$;

(iv) $\underline{\mathcal{E}}_{\beta}(D) = \overline{\mathcal{E}}_{\beta}(D) \setminus BN\mathcal{E}_{\beta}(D), \forall D \in \mathcal{D}$.

Notice that for $D \in \mathcal{D}$, the inclusions $\underline{\mathcal{E}}_{\beta}(D) \subseteq D$ and $D \subseteq \overline{\mathcal{E}}_{\beta}(D)$ do not always hold.

Example 2. In the case of Table 1, one obtains

$$\underline{\mathcal{E}}_{0.34}(D_{\text{yes}}) = \{p_2, p_4\}, \qquad \underline{\mathcal{E}}_{0.34}(D_{\text{no}}) = \{p_1, p_5, p_7\}, \tag{21}$$

$$\overline{\mathcal{E}}_{0.34}(D_{\text{yes}}) = \{p_2, p_3, p_4, p_6\}, \qquad \overline{\mathcal{E}}_{0.34}(D_{\text{no}}) = \{p_1, p_3, p_5, p_6, p_7\}. \tag{22}$$

2.3 Inner Product Spaces

By $\mathbb{R}$ and $\mathbf{0}$ denote the set of real numbers and a zero vector, respectively. In this paper, we exclusively employ the real vector spaces.

An inner product space is a real vector space V with an inner product $\langle \cdot, \cdot \rangle$. For two vectors $\mathbf{u}, \mathbf{v} \in V$, if $\langle \mathbf{u}, \mathbf{v} \rangle = 0$, then $\mathbf{u}, \mathbf{v}$ are said to be orthogonal. Given a subset $U \subseteq V$, then the orthogonal complement $U^{\perp}$ of U is a subspace of V defined as follows:

$$U^{\perp} = \{\mathbf{v} \in V \mid \langle \mathbf{u}, \mathbf{v} \rangle = 0, \forall \mathbf{u} \in U\}. \tag{23}$$

3 Granule Vectors and Approximation Vectors

For a decision system, we introduce two kinds of vectors called granule vectors and approximation vectors. We can examine whether a granule is contained in the approximation of a decision class by the inner product of the granule vector and the approximation vector. Before the description of the general case, we show the case of using the standard basis of a Cartesian coordinate system $\mathbb{R}^n$.

3.1 The Case of Using the Standard Basis

The granule vectors and the approximation vectors are introduced based on the following properties of Ziarko's VPRS model:

Proposition 2. *Let $DS = (U, \mathcal{E}, \mathcal{D})$ be a decision system, where $\mathcal{E} = \{E_1, \ldots, E_m\}$, $\mathcal{D} = \{D_1, \ldots, D_n\}$, and $\beta \in [0, 0.5)$ be a precision. Then, for any granule E_i ($i \in \{1, \ldots, m\}$) and decision class D_j ($j \in \{1, \ldots, n\}$), denoting by $\mathbf{e}_j \in \mathbb{R}^n$ the j-th standard basis vector, $j \in \{1, \ldots, n\}$, the following two properties hold:*

(i) $E_i \subseteq \underline{\mathcal{E}}_\beta(D_j)$ iff $\left\langle \left((1-\beta)\sum_{k=1}^n \mathbf{e}_k \right) - \mathbf{e}_j, \begin{bmatrix} |E_i \cap D_1| \\ |E_i \cap D_2| \\ \vdots \\ |E_i \cap D_n| \end{bmatrix} \right\rangle \leq 0;$

(ii) $E_i \subseteq \overline{\mathcal{E}}_\beta(D_j)$ iff $\left\langle \left(\beta \sum_{k=1}^n \mathbf{e}_k \right) - \mathbf{e}_j, \begin{bmatrix} |E_i \cap D_1| \\ |E_i \cap D_2| \\ \vdots \\ |E_i \cap D_n| \end{bmatrix} \right\rangle < 0.$

Proof. Let $i \in \{1, \ldots, m\}$ and $j \in \{1, \ldots, n\}$.

(i) Since $\mathcal{D}$ is a partition of U, then $|E_i| = \sum_{k=1}^n |E_i \cap D_k|$. From the definition of the β-lower approximation,

$$E_i \subseteq \underline{\mathcal{E}}_\beta(D_j) \Leftrightarrow E_i \overset{\beta}{\subseteq} D_j$$
$$\Leftrightarrow c(E_i, D_j) \leq \beta$$
$$\Leftrightarrow 1 - \frac{|E_i \cap D_j|}{|E_i|} \leq \beta$$
$$\Leftrightarrow 1 - \beta \leq \frac{|E_i \cap D_j|}{|E_i|} \quad (24)$$
$$\Leftrightarrow (1-\beta)|E_i| \leq |E_i \cap D_j|$$
$$\Leftrightarrow (1-\beta) \sum_{k=1}^n |E_i \cap D_k| \leq |E_i \cap D_j|$$
$$\Leftrightarrow (1-\beta) \left(\sum_{k=1}^n |E_i \cap D_k| \right) - |E_i \cap D_j| \leq 0.$$

On the other hand, since for $k \in \{1, \ldots, n\}$ one has

$$\left\langle \mathbf{e}_k, \begin{bmatrix} |E_i \cap D_1| \\ |E_i \cap D_2| \\ \vdots \\ |E_i \cap D_n| \end{bmatrix} \right\rangle = |E_i \cap D_k|, \quad (25)$$

from the linearity (additivity and homogeneity) of the inner product one gets

$$\left\langle \left((1-\beta)\sum_{k=1}^{n}\mathbf{e}_k\right) - \mathbf{e}_j, \begin{bmatrix} |E_i \cap D_1| \\ |E_i \cap D_2| \\ \vdots \\ |E_i \cap D_n| \end{bmatrix} \right\rangle \leq 0$$

$$\Leftrightarrow \left((1-\beta)\sum_{k=1}^{n}\left\langle \mathbf{e}_k, \begin{bmatrix} |E_i \cap D_1| \\ |E_i \cap D_2| \\ \vdots \\ |E_i \cap D_n| \end{bmatrix}\right\rangle \right) - \left\langle \mathbf{e}_j, \begin{bmatrix} |E_i \cap D_1| \\ |E_i \cap D_2| \\ \vdots \\ |E_i \cap D_n| \end{bmatrix} \right\rangle \leq 0 \quad (26)$$

$$\Leftrightarrow (1-\beta)\left(\sum_{k=1}^{n}|E_i \cap D_k|\right) - |E_i \cap D_j| \leq 0.$$

Hence, by the two equivalences (24) and (26), the property (i) holds.
(ii) The proof is similar to the one of (i). □

Proposition 2 leads to the definition of the granule vectors and the approximation vectors as follows.

Definition 1. *Let $DS = (U, \mathcal{E}, \mathcal{D})$ be a decision system where $\mathcal{E} = \{E_1, \ldots, E_m\}$, $\mathcal{D} = \{D_1, \ldots, D_n\}$, and $\beta \in [0, 0.5)$ be a precision. For DS and $i \in \{1, \ldots, m\}$, we define the granule vector $\mathbf{g}_i \in \mathbb{R}^n$ of E_i as*

$$\mathbf{g}_i = \sum_{j=1}^{n}|E_i \cap D_j|\mathbf{e}_j = \begin{bmatrix} |E_i \cap D_1| \\ |E_i \cap D_2| \\ \vdots \\ |E_i \cap D_n| \end{bmatrix}. \quad (27)$$

For $j \in \{1, \ldots, n\}$, the j-th standard vector $\mathbf{e}_j \in \mathbb{R}^n$ is called the decision class vector of D_j. For each decision class $D_j \in \mathcal{D}$, we introduce the β-lower and β-upper approximation vectors $\mathbf{l}_j^\beta, \mathbf{u}_j^\beta \in \mathbb{R}^n$ as:

$$\mathbf{l}_j^\beta = \left((1-\beta)\sum_{k=1}^{n}\mathbf{e}_k\right) - \mathbf{e}_j \quad (28)$$

$$\mathbf{u}_j^\beta = \left(\beta\sum_{k=1}^{n}\mathbf{e}_k\right) - \mathbf{e}_j. \quad (29)$$

Using the framework introduced in Definition 1, Proposition 2 can be rewritten as follows.

Theorem 1. *Let $DS = (U, \mathcal{E}, \mathcal{D})$ be a decision system where $\mathcal{E} = \{E_1, \ldots, E_m\}$, $\mathcal{D} = \{D_1, \ldots, D_n\}$, and $\beta \in [0, 0.5)$ be a precision. Then, for any granule $E_i \in \mathcal{E}$ ($i \in \{1, \ldots, m\}$) and decision class $D_j \in \mathcal{D}$ ($j \in \{1, \ldots, n\}$), the following two properties hold:*

(i) $E_i \subseteq \underline{\mathcal{E}}_\beta(D_j)$ iff $\left\langle \mathbf{l}_j^\beta, \mathbf{g}_i \right\rangle \leq 0$,

(ii) $E_i \subseteq \overline{\mathcal{E}}_\beta(D_j)$ iff $\left\langle \mathbf{u}_j^\beta, \mathbf{g}_i \right\rangle < 0$,

where $\mathbf{l}_j^\beta$ and $\mathbf{u}_j^\beta$ are the β-lower and β-upper approximation vectors of D_j, respectively, and $\mathbf{g}_i$ is a granule vector of E_i.

Proof. The proof is similar to the one of Proposition 2.

Example 3. In the case of Table 1, the granule vectors and the lower and upper approximation vectors are as follows:

$$\mathbf{g}_1 = \begin{bmatrix} 1 \\ 2 \end{bmatrix}, \mathbf{g}_2 = \begin{bmatrix} 2 \\ 0 \end{bmatrix}, \mathbf{g}_3 = \begin{bmatrix} 1 \\ 1 \end{bmatrix},$$

$$\mathbf{l}_1^\beta = \begin{bmatrix} -\beta \\ 1-\beta \end{bmatrix}, \mathbf{l}_2^\beta = \begin{bmatrix} 1-\beta \\ -\beta \end{bmatrix}, \quad \mathbf{u}_1^\beta = \begin{bmatrix} \beta-1 \\ \beta \end{bmatrix}, \mathbf{u}_2^\beta = \begin{bmatrix} \beta \\ \beta-1 \end{bmatrix}.$$

If $\beta = 0$, then

$\langle \mathbf{l}_1^0, \mathbf{g}_1 \rangle = 2 \not\leq 0, E_1 \not\subseteq \underline{\mathcal{E}}(D_1)$, $\langle \mathbf{l}_2^0, \mathbf{g}_1 \rangle = 1 \not\leq 0, E_1 \not\subseteq \underline{\mathcal{E}}(D_2)$,

$\langle \mathbf{u}_1^0, \mathbf{g}_1 \rangle = -1 < 0, E_1 \subseteq \overline{\mathcal{E}}(D_1)$, $\langle \mathbf{u}_2^0, \mathbf{g}_1 \rangle = -2 < 0, E_1 \subseteq \overline{\mathcal{E}}(D_2)$,

$\langle \mathbf{l}_1^0, \mathbf{g}_2 \rangle = 0 \leq 0, E_2 \subseteq \underline{\mathcal{E}}(D_1)$, $\langle \mathbf{l}_2^0, \mathbf{g}_2 \rangle = 2 \not\leq 0, E_2 \not\subseteq \underline{\mathcal{E}}(D_2)$,

$\langle \mathbf{u}_1^0, \mathbf{g}_2 \rangle = -2 < 0, E_2 \subseteq \overline{\mathcal{E}}(D_1)$, $\langle \mathbf{u}_2^0, \mathbf{g}_2 \rangle = 0 \not< 0, E_2 \not\subseteq \overline{\mathcal{E}}(D_2)$,

$\langle \mathbf{l}_1^0, \mathbf{g}_3 \rangle = 1 \not\leq 0, E_3 \not\subseteq \underline{\mathcal{E}}(D_1)$, $\langle \mathbf{l}_2^0, \mathbf{g}_3 \rangle = 1 \not\leq 0, E_3 \not\subseteq \underline{\mathcal{E}}(D_2)$,

$\langle \mathbf{u}_1^0, \mathbf{g}_3 \rangle = -1 < 0, E_3 \subseteq \overline{\mathcal{E}}(D_1)$, $\langle \mathbf{u}_2^0, \mathbf{g}_3 \rangle = -1 < 0, E_3 \subseteq \overline{\mathcal{E}}(D_2)$.

If $\beta = 0.34$, then

$\langle \mathbf{l}_1^{0.34}, \mathbf{g}_1 \rangle = 0.98 \not\leq 0, E_1 \not\subseteq \underline{\mathcal{E}}_{0.34}(D_1)$, $\langle \mathbf{l}_2^{0.34}, \mathbf{g}_1 \rangle = -0.02 \leq 0, E_1 \subseteq \underline{\mathcal{E}}_{0.34}(D_2)$,

$\langle \mathbf{u}_1^{0.34}, \mathbf{g}_1 \rangle = 0.02 \not< 0, E_1 \not\subseteq \overline{\mathcal{E}}_{0.34}(D_1)$, $\langle \mathbf{u}_2^{0.34}, \mathbf{g}_1 \rangle = -0.98 < 0, E_1 \subseteq \overline{\mathcal{E}}_{0.34}(D_2)$,

$\langle \mathbf{l}_1^{0.34}, \mathbf{g}_2 \rangle = -0.68 \leq 0, E_2 \subseteq \underline{\mathcal{E}}_{0.34}(D_1)$, $\langle \mathbf{l}_2^{0.34}, \mathbf{g}_2 \rangle = 1.32 \not\leq 0, E_2 \not\subseteq \underline{\mathcal{E}}_{0.34}(D_2)$,

$\langle \mathbf{u}_1^{0.34}, \mathbf{g}_2 \rangle = -1.32 < 0, E_2 \subseteq \overline{\mathcal{E}}_{0.34}(D_1)$, $\langle \mathbf{u}_2^{0.34}, \mathbf{g}_2 \rangle = 0.68 \not< 0, E_2 \not\subseteq \overline{\mathcal{E}}_{0.34}(D_2)$,

$\langle \mathbf{l}_1^{0.34}, \mathbf{g}_3 \rangle = 0.32 \not\leq 0, E_3 \not\subseteq \underline{\mathcal{E}}_{0.34}(D_1)$, $\langle \mathbf{l}_2^{0.34}, \mathbf{g}_3 \rangle = 0.32 \not\leq 0, E_3 \not\subseteq \underline{\mathcal{E}}_{0.34}(D_2)$,

$\langle \mathbf{u}_1^{0.34}, \mathbf{g}_3 \rangle = -0.32 < 0, E_3 \subseteq \overline{\mathcal{E}}_{0.34}(D_1)$, $\langle \mathbf{u}_2^{0.34}, \mathbf{g}_3 \rangle = -0.32 < 0, E_3 \subseteq \overline{\mathcal{E}}_{0.34}(D_2)$.

3.2 The Case of Using Linearly Independent Vectors

In Sect. 3.1, the granule vectors and the approximation vectors are defined as the linear combinations of the standard basis. Thus, each basis vector represents a decision class. The key property of a granule vector in the proof of Theorem 1 is

$$\langle \mathbf{e}_k, \mathbf{g}_i \rangle = |E_i \cap D_k|. \tag{30}$$

To construct vectors satisfying this property, it is sufficient to have n linearly independent vectors [1]. According to this, we redefine the granule vectors and the approximation vectors in the more general framework of inner product spaces of possibly infinite dimension.

Definition 2. *Let $DS = (U, \mathcal{E}, \mathcal{D})$ be a decision system where $\mathcal{E} = \{E_1, \ldots, E_m\}$, $\mathcal{D} = \{D_1, \ldots, D_n\}$, $\beta \in [0, 0.5)$ be a precision, V be an inner product space such that $n \leq \dim V$, and $\mathbf{d}_1, \mathbf{d}_2, \ldots, \mathbf{d}_n \in V$ be n linearly independent vectors (called decision class vectors). For each granule E_i ($i \in \{1, \ldots, m\}$), we redefine the granule vector $\mathbf{g}_i$ of E_i by a vector satisfying the condition*

$$\langle \mathbf{d}_j, \mathbf{g}_i \rangle = |E_i \cap D_j| \text{ for } j = 1, 2, \ldots, n. \tag{31}$$

(The existence of a vector with this property is guaranteed by [1]).

For each decision class $D_j \in \mathcal{D}$ ($j \in \{1, \ldots, n\}$), we also redefine the β-lower and β-upper approximation vectors $\mathbf{l}_j^\beta, \mathbf{u}_j^\beta \in \mathbb{R}^n$ as

$$\mathbf{l}_j^\beta = \left((1-\beta) \sum_{k=1}^n \mathbf{d}_k \right) - \mathbf{d}_j, \tag{32}$$

$$\mathbf{u}_j^\beta = \left(\beta \sum_{k=1}^n \mathbf{d}_k \right) - \mathbf{d}_j. \tag{33}$$

Note that the granule vectors are not uniquely determined in Definition 2.

Theorem 2. *Let $DS = (U, \mathcal{E}, \mathcal{D})$ be a decision system where $\mathcal{E} = \{E_1, \ldots, E_m\}$ and $\mathcal{D} = \{D_1, \ldots, D_n\}$, $\beta \in [0, 0.5)$ be a precision, V be an inner product space such that $n \leq \dim V$, and $\mathbf{d}_1, \mathbf{d}_2, \ldots, \mathbf{d}_n \in V$ decision class vectors (in particular, linearly independent vectors). Then, for any granule $E_i \in \mathcal{E}$ ($i \in \{1, \ldots, m\}$) and decision class $D_j \in \mathcal{D}$ ($j \in \{1, \ldots, n\}$), the following two properties hold:*

(i) $E_i \subseteq \underline{\mathcal{E}}_\beta(D_j)$ iff $\langle \mathbf{l}_j^\beta, \mathbf{g}_i \rangle \leq 0$;

(ii) $E_i \subseteq \overline{\mathcal{E}}_\beta(D_j)$ iff $\langle \mathbf{u}_j^\beta, \mathbf{g}_i \rangle < 0$,

where $\mathbf{l}_j^\beta$ and $\mathbf{u}_j^\beta$ are the β-lower and β-upper approximation vectors of D_j, respectively, and $\mathbf{g}_i$ is a granule vector of E_i, and these vectors generated from $\mathbf{d}_1, \mathbf{d}_2, \ldots, \mathbf{d}_n \in V$.

Proof. (i) By the proof of Theorem 1, for $i \in \{1, \ldots, m\}$ and $j \in \{1, \ldots, n\}$ one has

$$E_i \subseteq \underline{\mathcal{E}}_\beta(D_j) \Leftrightarrow (1-\beta) \left(\sum_{k=1}^n |E_i \cap D_k| \right) - |E_i \cap D_j| \leq 0. \tag{34}$$

On the other hand, from the definition of the β-lower approximation vector and the additivity and homogeneity of the inner product,

$$\langle \mathbf{l}_j^\beta, \mathbf{g}_i \rangle \leq 0 \Leftrightarrow \left\langle (1-\beta)\left(\sum_{k=1}^n \mathbf{d}_k\right) - \mathbf{d}_j, \mathbf{g}_i \right\rangle \leq 0$$

$$\Leftrightarrow \left((1-\beta)\sum_{k=1}^n \langle \mathbf{d}_k, \mathbf{g}_i \rangle\right) - \langle \mathbf{d}_j, \mathbf{g}_i \rangle \leq 0 \tag{35}$$

$$\Leftrightarrow (1-\beta)\left(\sum_{k=1}^n |E_i \cap D_k|\right) - |E_i \cap D_j| \leq 0.$$

Hence, by (34) and (35), $E_i \subseteq \underline{\mathcal{E}}_\beta(D_j)$ iff $\langle \mathbf{l}_j^\beta, \mathbf{g}_i \rangle \leq 0$.

(ii) The proof is similar to the one of (i). □

Example 4. In Table 1, given $\mathbf{d}_1, \mathbf{d}_2 \in \mathbb{R}^3$ as

$$\mathbf{d}_1 = \begin{bmatrix} 1 \\ 0 \\ -1 \end{bmatrix}, \mathbf{d}_2 = \begin{bmatrix} -1 \\ 2 \\ 0 \end{bmatrix}, \tag{36}$$

and put

$$\mathbf{g}_1 = \begin{bmatrix} 4 \\ 3 \\ 3 \end{bmatrix}, \mathbf{g}_2 = \begin{bmatrix} 4 \\ 2 \\ 2 \end{bmatrix}, \mathbf{g}_3 = \begin{bmatrix} 3 \\ 2 \\ 2 \end{bmatrix}, \tag{37}$$

then

$$\langle \mathbf{d}_1, \mathbf{g}_1 \rangle = |E_1 \cap D_1| = 1, \langle \mathbf{d}_1, \mathbf{g}_2 \rangle = |E_2 \cap D_1| = 2, \langle \mathbf{d}_1, \mathbf{g}_3 \rangle = |E_3 \cap D_1| = 1$$
$$\langle \mathbf{d}_2, \mathbf{g}_1 \rangle = |E_1 \cap D_2| = 2, \langle \mathbf{d}_2, \mathbf{g}_2 \rangle = |E_2 \cap D_2| = 0, \langle \mathbf{d}_2, \mathbf{g}_3 \rangle = |E_3 \cap D_2| = 1. \tag{38}$$

The β-lower and β-upper approximation vectors are

$$\mathbf{l}_1^\beta = \begin{bmatrix} -1 \\ 2-2\beta \\ \beta \end{bmatrix}, \mathbf{l}_2^\beta = \begin{bmatrix} 1 \\ -2\beta \\ \beta-1 \end{bmatrix}, \mathbf{u}_1^\beta = \begin{bmatrix} -1 \\ 2\beta \\ 1-\beta \end{bmatrix}, \mathbf{u}_2^\beta = \begin{bmatrix} 1 \\ 2\beta-2 \\ -\beta \end{bmatrix}, \tag{39}$$

and then $E_i \subseteq \underline{\mathcal{E}}_\beta(D_j)$ iff $\langle \mathbf{l}_j^\beta, \mathbf{g}_i \rangle \leq 0$, and $E_i \subseteq \overline{\mathcal{E}}_\beta(D_j)$ iff $\langle \mathbf{u}_j^\beta, \mathbf{g}_i \rangle < 0$.

Since the only hypothesis imposed on V is $n \leq \dim V$, Theorem 2 holds even for infinitely dimensional inner product spaces.

Example 5. $C[-\pi, \pi]$ is the inner product space of real-valued functions defined on $[-\pi, \pi]$, whose inner product is defined by

$$\langle \mathbf{x}, \mathbf{y} \rangle = \int_{-\pi}^\pi f(t)g(t)dt, \tag{40}$$

where $\mathbf{x} = f(t) \in C[-\pi, \pi]$ and $\mathbf{y} = g(t) \in C[-\pi, \pi]$.

In Table 1, let $\mathbf{d}_1, \mathbf{d}_2 \in C[-\pi, \pi]$ given as

$$\mathbf{d}_1 = \frac{1}{\sqrt{2\pi}}, \mathbf{d}_2 = \frac{1}{\sqrt{2\pi}} \cos t. \tag{41}$$

and set

$$\mathbf{g}_1 = \frac{1}{\sqrt{2\pi}} + \frac{2}{\sqrt{2\pi}} \cos t, \mathbf{g}_2 = \frac{2}{\sqrt{2\pi}}, \mathbf{g}_3 = \frac{1}{\sqrt{2\pi}} + \frac{1}{\sqrt{2\pi}} \cos t. \tag{42}$$

Then,

$$\langle \mathbf{d}_1, \mathbf{g}_1 \rangle = |E_1 \cap D_1| = 1, \langle \mathbf{d}_1, \mathbf{g}_2 \rangle = |E_2 \cap D_1| = 2, \langle \mathbf{d}_1, \mathbf{g}_3 \rangle = |E_3 \cap D_1| = 1,$$
$$\langle \mathbf{d}_2, \mathbf{g}_1 \rangle = |E_1 \cap D_2| = 2, \langle \mathbf{d}_2, \mathbf{g}_2 \rangle = |E_2 \cap D_2| = 0, \langle \mathbf{d}_2, \mathbf{g}_3 \rangle = |E_3 \cap D_2| = 1. \tag{43}$$

The β-lower and β-upper approximation vectors are

$$\mathbf{l}_1^\beta = -\frac{\beta}{\sqrt{2\pi}} + \frac{1-\beta}{\sqrt{2\pi}} \cos t, \mathbf{l}_2^\beta = \frac{1-\beta}{\sqrt{2\pi}} - \frac{\beta}{\sqrt{2\pi}} \cos t, \tag{44}$$

$$\mathbf{u}_1^\beta = \frac{\beta-1}{\sqrt{2\pi}} + \frac{\beta}{\sqrt{2\pi}} \cos t, \mathbf{u}_2^\beta = \frac{\beta}{\sqrt{2\pi}} + \frac{\beta-1}{\sqrt{2\pi}} \cos t, \tag{45}$$

and then $E_i \subseteq \underline{\mathcal{E}}_\beta(D_j)$ iff $\langle \mathbf{l}_j^\beta, \mathbf{g}_i \rangle \leq 0$, and $E_i \subseteq \overline{\mathcal{E}}_\beta(D_j)$ iff $\langle \mathbf{u}_j^\beta, \mathbf{g}_i \rangle < 0$.

4 An Extended VPRS Model

In Sect. 3, we generated the granule vectors and the approximation vectors from a given decision system. Moreover, the reverse is also possible.

Theorem 3. *Given an inner product space V such that $\dim V \geq n$, n linearly independent vectors $\mathbf{d}_1, \ldots, \mathbf{d}_n \in V$ (called decision class vectors), and $\mathbf{g}_1, \ldots, \mathbf{g}_m \in V$ satisfying $\langle \mathbf{d}_j, \mathbf{g}_i \rangle \in \mathbb{N}_0$ for all $\mathbf{g}_i$ and $\mathbf{d}_j$, and taken such that for every $\mathbf{g}_i$ there exists $\mathbf{d}_j$ such that $\langle \mathbf{d}_j, \mathbf{g}_i \rangle \neq 0$, then if a decision system $DS = (U, \mathcal{E}, \mathcal{D})$ where $\mathcal{E} = \{E_1, \ldots, E_m\}, \mathcal{D} = \{D_1, \ldots, D_n\}$ satisfies*

$$|E_i \cap D_j| = \langle \mathbf{d}_j, \mathbf{g}_i \rangle, \tag{46}$$

then, for any $E_i \in \mathcal{E}$ and $D_j \in \mathcal{D}$, the following two properties hold:

(i) $E_i \subseteq \underline{\mathcal{E}}_\beta(D_j)$ *iff* $\langle \mathbf{l}_j^\beta, \mathbf{g}_i \rangle \leq 0$;
(ii) $E_i \subseteq \overline{\mathcal{E}}_\beta(D_j)$ *iff* $\langle \mathbf{u}_j^\beta, \mathbf{g}_i \rangle < 0$,

where $\mathbf{l}_j^\beta, \mathbf{u}_j^\beta$ are given in Definition 2.

Proof. The proof is similar to the one of Theorem 2. □

Notice that $\emptyset \notin \mathcal{E}$ derives a sufficient condition to ensure that for every $\mathbf{g}_i$, there exists $\mathbf{d}_j$ such that $\langle \mathbf{d}_j, \mathbf{g}_i \rangle \neq 0$.

Theorem 3 does not allow the situation $\langle \mathbf{d}_j, \mathbf{g}_i \rangle < 0$; however, to handle negative values, the additive inverse of the decision class vectors can be used to generate the corresponding decision system.

Definition 3. *Given an inner product space V such that $\dim V \geq n$, n linearly independent vectors $\mathbf{d}_1, \ldots, \mathbf{d}_n \in V$, and $\mathbf{g}_1, \ldots, \mathbf{g}_m \in V$, then we define for $i \in \{1, \ldots, m\}$ and $j \in \{1, \ldots, n\}$ the decision class vector $\mathbf{d}_{ji}$ of $\mathbf{d}_j$ for $\mathbf{g}_i$ as*

$$\mathbf{d}_{ji} = \begin{cases} \mathbf{d}_j, & \text{if } \langle \mathbf{d}_j, \mathbf{g}_i \rangle \geq 0, \\ -\mathbf{d}_j, & \text{otherwise.} \end{cases} \tag{47}$$

Given a precision $\beta \in [0, 0.5)$, then the β-lower approximation vector $\mathbf{l}^\beta_{ji}$ and the β-upper approximation vector $\mathbf{u}^\beta_{ji}$ of $\mathbf{d}_j$ for $\mathbf{g}_i$ are defined as follows:

$$\mathbf{l}^\beta_{ji} = \left((1 - \beta) \sum_{k=1}^n \mathbf{d}_{ki} \right) - \mathbf{d}_{ji}, \tag{48}$$

$$\mathbf{u}^\beta_{ji} = \left(\beta \sum_{k=1}^n \mathbf{d}_{ki} \right) - \mathbf{d}_{ji}. \tag{49}$$

Theorem 4. *Given an inner product space V such that $\dim V \geq n$, n linearly independent vectors $\mathbf{d}_1, \ldots, \mathbf{d}_n \in V$, and $\mathbf{g}_1, \ldots, \mathbf{g}_m \in V$ satisfying $\langle \mathbf{d}_j, \mathbf{g}_i \rangle \in \mathbb{Z}$ for all $\mathbf{g}_i$ and $\mathbf{d}_j$, and for every $\mathbf{g}_i$, there exists $\mathbf{d}_j$ such that $\langle \mathbf{d}_j, \mathbf{g}_i \rangle \neq 0$, then if a decision system $DS = (U, \mathcal{E}, \mathcal{D})$ where $\mathcal{E} = \{E_1, \ldots, E_m\}, \mathcal{D} = \{D_1, \ldots, D_n\}$ such that*

$$|E_i \cap D_j| = \langle \mathbf{d}_{ji}, \mathbf{g}_i \rangle \in \mathbb{N}_0, \tag{50}$$

then, for any $E_i \in \mathcal{E}$ and $D_j \in \mathcal{D}$, the following two properties hold:

(i) $E_i \subseteq \underline{\mathcal{E}}_\beta(D_j)$ iff $\langle \mathbf{l}^\beta_{ji}, \mathbf{g}_i \rangle \leq 0$;

(ii) $E_i \subseteq \overline{\mathcal{E}}_\beta(D_j)$ iff $\langle \mathbf{u}^\beta_{ji}, \mathbf{g}_i \rangle < 0$,

where $\mathbf{d}_{ji}, \mathbf{l}^\beta_{ji}$, and $\mathbf{u}^\beta_{ji}$ are introduced in Definition 3.

Proof. The proof is similar to the one of Theorem 2. □

Theorem 4 states that n linearly independent vectors $\mathbf{d}_1, \ldots, \mathbf{d}_n \in V$ and $\mathbf{g}_1, \ldots, \mathbf{g}_m \in V$ can represent a decision system if $\langle \mathbf{d}_j, \mathbf{g}_i \rangle \in \mathbb{Z}$ and for all $\mathbf{g}_i$, there exists $\mathbf{d}_j$ such that $\langle \mathbf{d}_j, \mathbf{g}_i \rangle \neq 0$. This leads us to the generalisation of a decision system using the vectors satisfying $\langle \mathbf{d}_j, \mathbf{g}_i \rangle \in \mathbb{R}$ and that for every $\mathbf{g}_i$ there exists $\mathbf{d}_j$ such that $\langle \mathbf{d}_j, \mathbf{g}_i \rangle \neq 0$.

Definition 4. *A vectorial decision system VDS is a tuple (V, G, D), where V is an inner product space such that $\dim V \geq n$, $G = (\mathbf{g}_1, \ldots, \mathbf{g}_m)$ is a tuple of vectors in V, $D = (\mathbf{d}_1, \ldots, \mathbf{d}_n)$ is a tuple of linearly independent vectors in V, and for each $\mathbf{g}_i$ there exists $\mathbf{d}_j$ such that $\langle \mathbf{d}_j, \mathbf{g}_i \rangle \neq 0$, i.e., $\mathbf{g}_i \notin (\mathrm{span}(D))^{\perp}$. The vectors $\mathbf{g}_i$ and $\mathbf{d}_j$ are called a granule vector and a decision class vector, respectively. For a decision class vector $\mathbf{d}_j$ and a precision $\beta \in [0, 0.5)$, we define the β-lower approximation $\underline{G}_\beta(\mathbf{d}_j)$ and β-upper approximation $\overline{G}_\beta(\mathbf{d}_j)$ of $\mathbf{d}_j$ as*

$$\underline{G}_\beta(\mathbf{d}_j) = \left\{ \mathbf{g}_i \in G \mid \langle \mathbf{l}^\beta_{ji}, \mathbf{g}_i \rangle \leq 0 \right\}, \tag{51}$$

$$\overline{G}_\beta(\mathbf{d}_j) = \left\{ \mathbf{g}_i \in G \mid \langle \mathbf{u}^\beta_{ji}, \mathbf{g}_i \rangle < 0 \right\}, \tag{52}$$

where $\mathbf{l}^\beta_{ji}$ and $\mathbf{u}^\beta_{ji}$ are given in Definition 3. The boundary region of $\mathbf{d}_j$ and the positive region of VDS are defined by

$$BNG_\beta(\mathbf{d}_j) = \overline{G}_\beta(\mathbf{d}_j) \setminus \underline{G}_\beta(\mathbf{d}_j), \tag{53}$$

$$Pos^\beta_G(D) = \bigcup_{\mathbf{d}_j \in D} \underline{G}_\beta(\mathbf{d}_j). \tag{54}$$

As previously mentioned, if $\langle \mathbf{d}_j, \mathbf{g}_i \rangle \in \mathbb{Z}$ for all decision class vectors $\mathbf{d}_j$ and all granule vectors $\mathbf{g}_i$, then the vectorial decision system can represent a decision system with respect to the β-lower and β-upper approximations.

We have the following properties extending the ones exhibited in Proposition 1 to vectorial decision systems.

Proposition 3. *Given a vectorial decision system $VDS = (V, G, D)$ where $G = (\mathbf{g}_1, \ldots, \mathbf{g}_m)$ and $D = (\mathbf{d}_1, \ldots, \mathbf{d}_n)$ and a precision $\beta \in [0, 0.5)$, the following four properties hold:*

(i) $\underline{G}_\beta(\mathbf{d}_j) \subseteq \overline{G}_\beta(\mathbf{d}_j), \forall j = 1, \ldots, n$;
(ii) If $\frac{1}{n} > \beta$, then $\{ \mathbf{g}_i \mid i = 1, \ldots, m \} = \bigcup_{j=1,\ldots,n} \overline{G}_\beta(\mathbf{d}_j)$;
(iii) $\overline{G}_\beta(\mathbf{d}_j) = \underline{G}_\beta(\mathbf{d}_j) \cup BNG_\beta(\mathbf{d}_j), \forall j = 1, \ldots, n$;
(iv) $\underline{G}_\beta(\mathbf{d}_j) = \overline{G}_\beta(\mathbf{d}_j) \setminus BNG_\beta(\mathbf{d}_j), \forall j = 1, \ldots, n$.

Proof. (i) Let $\mathbf{g}_i \in \underline{G}_\beta(\mathbf{d}_j)$. Then, one has

$$(1 - \beta) \sum_{k=1}^{n} \langle \mathbf{d}_{ki}, \mathbf{g}_i \rangle - \langle \mathbf{d}_{ji}, \mathbf{g}_i \rangle \leq 0. \tag{55}$$

Since $0 \leq \beta < 0.5$ and $\forall \mathbf{g}_i \exists \mathbf{d}_j \langle \mathbf{d}_j, \mathbf{g}_i \rangle \neq 0$, then

$$(1 - 2\beta) \sum_{k=1}^{n} \langle \mathbf{d}_{ki}, \mathbf{g}_i \rangle > 0. \tag{56}$$

Therefore

$$(1-\beta)\sum_{k=1}^{n}\langle \mathbf{d}_{ki}, \mathbf{g}_i\rangle - \langle \mathbf{d}_{ji}, \mathbf{g}_i\rangle < (1-2\beta)\sum_{k=1}^{n}\langle \mathbf{d}_{ki}, \mathbf{g}_i\rangle \tag{57}$$

$$\Leftrightarrow \beta \sum_{k=1}^{n}\langle \mathbf{d}_{ki}, \mathbf{g}_i\rangle - \langle \mathbf{d}_{ji}, \mathbf{g}_i\rangle < 0, \tag{58}$$

and this implies $\mathbf{g}_i \in \overline{G}_\beta(\mathbf{d}_j)$.

(ii) For any granule vector $\mathbf{g}_i$, take

$$j = \arg\max_{k=1,\ldots,n}\langle \mathbf{d}_{ki}, \mathbf{g}_i\rangle, \tag{59}$$

then

$$n\langle \mathbf{d}_{ji}, \mathbf{g}_i\rangle \geq \sum_{k=1}^{n}\langle \mathbf{d}_{ki}, \mathbf{g}_i\rangle. \tag{60}$$

Since, from $\dfrac{1}{n} > \beta$, $\dfrac{1}{\beta}\langle \mathbf{d}_{ji}, \mathbf{g}_i\rangle > n\langle \mathbf{d}_{ji}, \mathbf{g}_i\rangle$, then

$$\frac{1}{\beta}\langle \mathbf{d}_{ji}, \mathbf{g}_i\rangle > \sum_{k=1}^{n}\langle \mathbf{d}_{ki}, \mathbf{g}_i\rangle \Leftrightarrow \beta\sum_{k=1}^{n}\langle \mathbf{d}_{ki}, \mathbf{g}_i\rangle - \langle \mathbf{d}_{ji}, \mathbf{g}_i\rangle < 0, \tag{61}$$

and this implies $\mathbf{g}_i \in \overline{G}_\beta(\mathbf{d}_j)$.

(iii) and (iv) are obvious from Definition 4. □

Example 6. Figures 1, 2 and 3 show which region a granule vector must locate in order to be contained in the lower or upper approximation. In Fig. 1, $VDS = (\mathbb{R}^2, G, D)$ and $D = \{\mathbf{d}_1, \mathbf{d}_2\}$, where $\mathbf{d}_1 = \mathbf{e}_1, \mathbf{d}_2 = \mathbf{e}_2$. In Fig. 2, $VDS = (\mathbb{R}^2, G, D)$ and $D = \{\mathbf{d}_1, \mathbf{d}_2\}$, where $\mathbf{d}_1 = {}^t(2,1), \mathbf{d}_2 = {}^t(1,2) \in \mathbb{R}^2$ In Fig. 3, $VDS = (\mathbb{R}^2, G, D)$ and $D = \{\mathbf{d}_1, \mathbf{d}_2\}$, where $\mathbf{d}_1 = {}^t(3,-1), \mathbf{d}_2 = {}^t(-2,2) \in \mathbb{R}^2$.

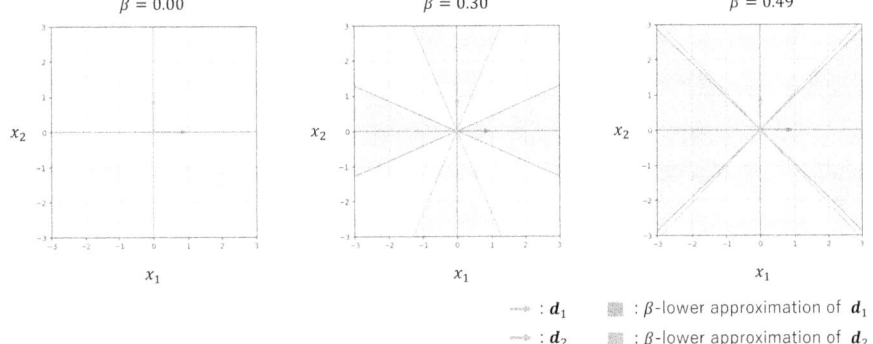

Fig. 1. The region of lower approximation using the standard basis

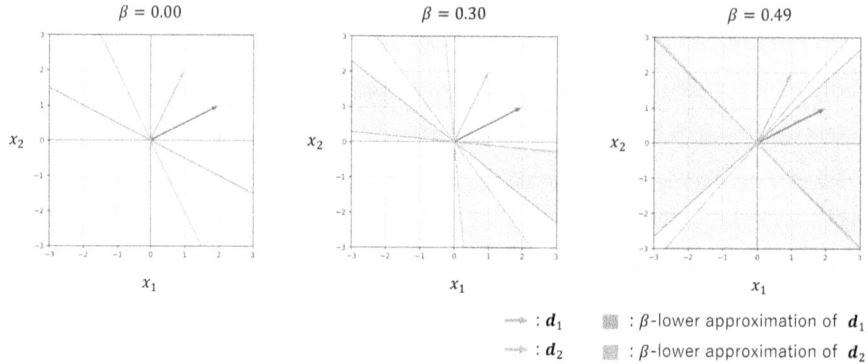

Fig. 2. The region of lower approximation using the two vectors such that the angle between them is acute

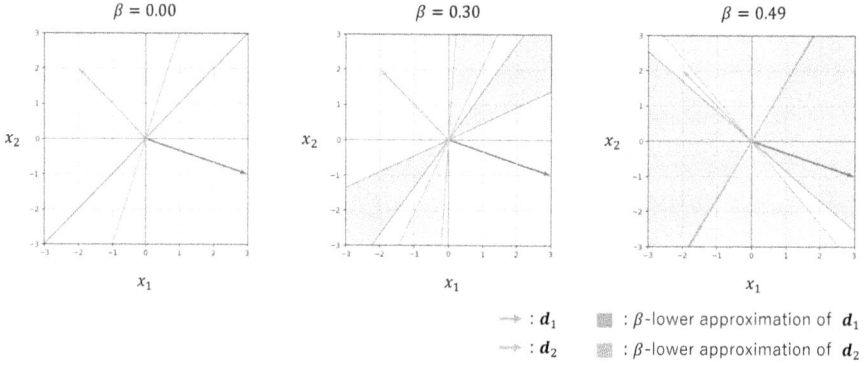

Fig. 3. The region of the lower approximation using the two vectors such that the angle between them is obtuse

5 Conclusion

Motivated by applications not covered by the existing VPRS model and its precursors, we propose an extended version of it, which is based on a vectorial decision system and approximation operators. The examination of these approximation operators reveals that they inherit the properties of their counterparts from Ziarko's VPRS model, which thus makes the proposed model an organic generalization of the latter. In particular, the newly proposed model allows a single vector to be considered as a granule, too, which is a feature not available within the framework of its existing counterparts from the literature. Therefore, this model can deal with the approximation of a vector. Our claims are illustrated by simple examples.

The region of vectors in the approximation is constructed based on the inner product. Therefore, it can be transformed into a more complex one by a feature map utilized in kernel methods.

The proposed methodology is readily applicable to probabilistic rough set models, including asymmetric variable precision rough set models [18] and decision-theoretic rough set models [14,19]. This is achieved by independently setting the thresholds for the lower and upper approximations.

In the future, our research will concentrate on the specific properties of vectorial decision systems.

Acknowledgement. This work was supported by JST SPRING, Grant Number JPMJSP2153.

References

1. Axler, S.: Linear Algebra Done Right. Springer, Heidelberg (2015)
2. Chen, Y., Chen, Y.: Feature subset selection based on variable precision neighborhood rough sets. Int. J. Comput. Intell. Syst. **14**, 572–581 (2021)
3. Dubois, D., Prade, H.: Rough fuzzy sets and fuzzy rough sets. Int. J. Gen. Syst. **17**, 191–209 (1990)
4. Dubois, D., Prade, H.: Putting rough sets and fuzzy sets together. In: Słowiński, R. (ed.) Intelligent Decision Support, pp. 203–232. Springer, Dordrecht (1992). https://doi.org/10.1007/978-94-015-7975-9_14
5. Heng, Y.P., Lee, H.Y., Chong, J.W., Tan, R.R., Aviso, K.B., Chemmangattuvalappil, N.G.: Incorporating machine learning in computer-aided molecular design for fragrance molecules. Processes **10**(9) (2022)
6. Hu, M., Tsang, E.C., Guo, Y., Chen, D., Xu, W.: A novel approach to attribute reduction based on weighted neighborhood rough sets. Knowl.-Based Syst. **220**, 106908 (2021)
7. Hu, Q., Yu, D., Liu, J., Wu, C.: Neighborhood rough set based heterogeneous feature subset selection. Inf. Sci. **178**(18), 3577–3594 (2008)
8. Inuiguchi, M.: Structure-based attribute reduction in variable precision rough set models. J. Adv. Comput. Intell. Intell. Inform. **10**(5), 657–665 (2006)
9. Lei, L., Chen, W., Wu, B., Chen, C., Liu, W.: A building energy consumption prediction model based on rough set theory and deep learning algorithms. Energy Build. **240**, 110886 (2021)
10. Luo, S., Miao, D., Zhang, Z., Zhang, Y., Hu, S.: A neighborhood rough set model with nominal metric embedding. Inf. Sci. **520**, 373–388 (2020)
11. Pawlak, Z.: Rough Sets: Theoretical Aspects of Reasoning About Data. Theory and Decision Library. Kluwer Academic Publishers (1991)
12. Pomykała, J.: On definability in the nondeterministic information system. Bull. Pol. Acad. Sci. Math. **36**(3–4), 193–210 (1988)
13. Pomykala, J.A.: Approximation operations in approximation space. Bull. Pol. Acad. Sci **35**(9–10), 653–662 (1987)
14. Suo, M., et al.: Single-parameter decision-theoretic rough set. Inf. Sci. **539**, 49–80 (2020)
15. Wang, J., Guo, J.: Research on rock mass quality classification based on an improved rough set-cloud model. IEEE Access **7**, 123710–123724 (2019)
16. Yan, C., et al.: River pattern discriminant method based on rough set theory. J. Hydrol.: Regional Stud. **45**, 101285 (2023)

17. Yang, X., Chen, H., Li, T., Wan, J., Sang, B.: Neighborhood rough sets with distance metric learning for feature selection. Knowl.-Based Syst. **224**, 107076 (2021)
18. Yao, Y.: Probabilistic rough set approximations. Int. J. Approximate Reasoning **49**(2), 255–271 (2008)
19. Yao, Y., Wong, S., Lingras, P.: A decision-theoretic rough set model. In: Proceedings of the 5th International Symposium on Methodologies for Intelligent Systems, pp. 17–25 (1990)
20. Yao, Y., Yao, B.: Covering based rough set approximations. Inf. Sci. **200**, 91–107 (2012)
21. Ye, J., Zhan, J., Ding, W., Fujita, H.: A novel fuzzy rough set model with fuzzy neighborhood operators. Inf. Sci. **544**, 266–297 (2021)
22. Zakowski, W.: Approximations in the space (u, π). Demonstratio Math. **16**(3), 761–770 (1983)
23. Zhan, J., Jiang, H., Yao, Y.: Covering-based variable precision fuzzy rough sets with PROMETHEE-EDAS methods. Inf. Sci. **538**, 314–336 (2020)
24. Zhang, K., Zhan, J., Wu, W.Z.: On multicriteria decision-making method based on a fuzzy rough set model with fuzzy α-neighborhoods. IEEE Trans. Fuzzy Syst. **29**(9), 2491–2505 (2021)
25. Zhu, W.: Relationship among basic concepts in covering-based rough sets. Inf. Sci. **179**(14), 2478–2486 (2009)
26. Zhu, W., Wang, F.Y.: On three types of covering-based rough sets. IEEE Trans. Knowl. Data Eng. **19**(8), 1131–1144 (2007)
27. Ziarko, W.: Variable precision rough set model. J. Comput. Syst. Sci. **46**(1), 39–59 (1993)

Rough Set Applications

Cross-Weighting Knowledge Distillation for Object Detection

Zhaoyi Li[1,2], Zihao Li[1], and Xiaodong Yue[1,2,3(✉)]

[1] School of Computer Engineering and Science, Shanghai University, Shanghai, China
{zhaoyili,zihao,yswantfly}@shu.edu.cn
[2] Artificial Intelligence Institute of Shanghai University, Shanghai, China
[3] VLN Lab, NAVI MedTech Co., Ltd., Shanghai, China

Abstract. Knowledge distillation(KD) has been widely utilized for compressing object detection models and enhancing their accuracy. However, most current knowledge distillation methods have not adequately addressed common issues in object detection, including the severe imbalance between foreground and background samples, and the discrepancy in integrating classification scores and IoU between the detector's training and testing phases. In this paper, we propose the Cross-Weighting Knowledge Distillation(CWKD) method, where localization and classification are jointly considered during the distillation training phase. Within this framework, we introduce two specific losses, namely, the IoU-aware Classification KD Loss and the Class-aware Localization KD Loss. We also propose the Valuable Distillation Object(VDO) selection module, aimed at discovering additional objects beyond the ground truth objects that are worth distillation, thereby further expanding positive samples to address the imbalance issue. Utilizing the logit-based knowledge distillation approach, our method can seamlessly integrate into both single-stage and multi-stage object detection models. Through extensive experiments on the challenging object detection dataset COCO, we demonstrate our effectiveness and surpass other logit-based and feature-based knowledge distillation methods.

Keywords: Knowledge Distillation · Object Detection · Cross-Weighting · Valuable Distillation Object (VDO)

1 Introduction

In recent years, the effectiveness of object detection models has significantly improved. However, due to the limited computational resources of edge devices, model compression techniques such as model pruning and knowledge distillation have emerged. Knowledge distillation, as expounded by Hinton et al. [10], assumes a pivotal role among these techniques. It facilitates model light-weighting by transferring the knowledge of proficient teacher models to student models, thereby reducing the size of the latter while enhancing their accuracy.

Discrepancies or inconsistencies in the field of object detection between large and small models can be categorized into three main aspects: response inconsistency, feature inconsistency, and relation inconsistency. Addressing feature inconsistency, methods attempt to mimic intermediate hidden layer features [24] or attention features at different scales in the neck [9,28,29]. Furthermore, in the domain of 3D object detection, [14] prescribe methods to compel the student model to mimic the intermediate voxel feature maps of the teacher. Relation inconsistency arises because object detection models focus on instances, and related works using graph [13] models or non-local attention modules [32] to model relation information and transfer such relation knowledge. Regarding response inconsistency, some models aim to distill through logits mimicking [5,10,34], imitating the outputs of teacher models.

Given the differences in feature levels between heterogeneous teacher-student models and the variations in scale brought about by homogeneous teacher-student model configurations, as well as the complexity of modeling the relation between instances, this paper elects to focus on addressing response inconsistency. The proposed method presents a straightforward and efficient knowledge distillation approach, providing flexibility for implementation in current heterogeneous or homogeneous teacher-student object detection frameworks.

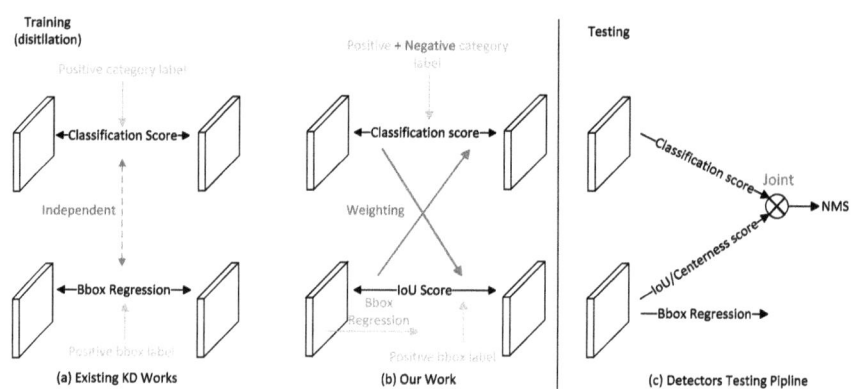

Fig. 1. Comparisons between existing works proposed separate classification and localization distillation and joint representation of classification and localization quality estimation. (a): Current works [5,9,10,28,29,34,36] using the separate usage of the quality branch to distill during training. (b): Our cross-weighting classification and localization branch enables high consistency between training and testing. (c): The existing pipeline for the testing phase of the object detection model.

As shown in Fig. 1 the classification score and IoU score are considered simultaneously in the NMS operation during the testing phase. However, in most existing object detection models and distillation methods, separate localization and classification branches during the training phase, leading to a gap. To address the aforementioned issue, we propose a cross-weighting method to enhance the coupling between the two.

Moreover, the imbalance between foreground and background samples is common in object detection. Existing knowledge distillation methods address this problem by distilling only foreground classes or masking [9,28] Ground Truth(GT) regions to distill foreground and background directly. The paper argues against segregating foreground and background samples because there might still be some objects worth distilling as foreground in the background, perhaps because they belong to unknown classes or are difficult to recognize. We naturally draw inspiration from soft sampling [16] and hard sampling [17,25] methodologies, employing a weighted approach and introducing the concept of Valuable Distillation Objects (VDO) through a selection module to tackle the imbalance issue. The VDO selection module can identify objects beyond the ground truth (GT) that are most valuable for distillation, thereby augmenting the pool of positive samples. The main contributions of this paper are:

1. Introduction of the cross-weighting knowledge distillation method aims to bridge the gap between the training and testing phases, leading to joint consideration of both classification and localization quality.
2. Introduction of the VDO concept and selection module to address sample imbalance issues.
3. Our method has validated the effectiveness of knowledge distillation methods on the COCO dataset while maintaining simplicity. Additionally, compared with existing feature-based and logit-based methods, our approach achieves state-of-the-art results.

2 Related Work

2.1 Object Detection

Object detection has always been an important challenge in computer vision tasks. In recent years, several proposed methodologies' performance has been greatly improved. In the early days, the object detection task was divided into two mainstream categories, namely, the two-stage method [15,22,33] and the one-stage method [16,18]. The two-stage methods usually include heuristic methods such as an RPN network or Selective Search. Generate the preset anchors into the region proposals of the object. In the second stage, the generated proposals are mapped onto scale features utilizing techniques such as RoI Align, followed by refinement of the final results. In one-stage methods, the object detection models directly utilize feature maps in a single stage, thereby eliminating the need for heuristic class-agnostic object detectors to generate proposals.

In recent years, anchor-free and anchor-based methods have been mentioned to distinguish most of the existing object detection methods. Anchor-based detection models [33] employed predefined anchor boxes or anchor points and further refined the object detection results. However, the anchors have randomness in manual design and need to rely on differences in the distribution of the dataset. Based on the aforementioned issues, anchor-free methods [8,11,17,18] have been proposed, which do not require predefined anchors, and most of such kinds of methods' bounding box(B-Box) regression tasks directly through feature points on the feature map.

For object detection tasks, the problem of imbalance between the number of background and foreground samples has always been one of the goals solved by many methods. Hard sampling tackles imbalance by selecting a subset of positive and negative examples from labeled B-Boxes using heuristic methods, with non-selected examples ignored for the current iteration. Techniques like Online Hard Example Mining (OHEM) [25] consider the loss values from both positive and negative examples to find hard samples. On the other hand, soft sampling adjusts the contribution of each example based on its relative importance to the training process. Unlike hard sampling, no sample is discarded, and the entire dataset is utilized for updating the parameters. For instance, Focal loss [16] dynamically assigns more weight to hard examples, while Gradient Harmonizing Mechanism [12] penalizes the loss of a sample if many samples have similar gradients. Drawing inspiration from sampling methods, our objective is to improve existing knowledge distillation methods by concentrating on the selection of hard samples and introducing weighting techniques tailored to hard samples.

2.2 Knowledge Distillation

Given the performance gap between small and large models and the need for model compression to meet application requirements, knowledge distillation(KD) was first mentioned by Hinton et al. [10]. Knowledge distillation is essentially a transfer learning idea. There are two components involved: a teacher model, pre-trained on either the same or different large-scale datasets and a student model, which is a small-scale model requiring training. During the training phase of the student model, knowledge transfer can be used to accelerate model convergence and enhance performance. To pursue better effect and accuracy in different visual tasks, the complexity and scale of the vision model have increased [7,21]. Therefore, knowledge distillation is widely employed in various visual tasks, such as Classification [2,5,20,34], Object Detection [13,28,29,32,36], Segmentation [31], Pose Recognition [1] and so on.

In the knowledge distillation pipeline, the focus mainly revolves around two levels: response-based and feature-based distillation methods. Feature-based methods employ attention modules to extract global features from the features of the neck. Specifically, attention is directed towards instances and relation information in object detection tasks. These methods employ graph models [19] or introduce generic object representations [6,20]. Response-based methods primarily concentrate on the model's outputs. Object detection has been widely proposed in relevant works including [10,34,36]. In addition, the [36] provides an approach to describe the bonding box(B-Box) using distribution and distilling positioning results by KL divergence. Response-based methods are deemed more amenable for implementation across homogeneous or heterogeneous detectors than feature-based methods.

3 Method

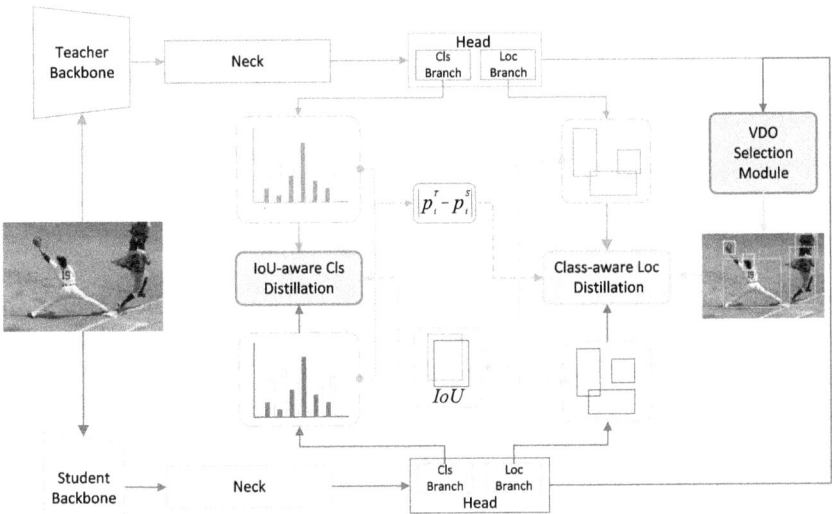

Fig. 2. An illustration of the proposed CWKD framework.

The framework of the proposed cross-weighting method consists of two losses, IoU-aware Classification, and Class-aware Localization knowledge distillation losses, as well as a Valuable Distillation Objects (VDO) selection module. As shown in Fig. 2, the classification and regression heads of both the teacher and student models yield corresponding classification scores and B-Boxes. We calculate the difference in classification scores between the teacher and student models as weights to participate in localization distillation calculations. Similarly, we use the IoU of B-Boxes between the teacher and student models as weights to participate in classification distillation loss calculations. Furthermore, by inputting the detection outputs of the teacher and student models into the selection module, VDOs highlighted in yellow boxes in the image are obtained. Our proposed framework can be easily applied to detectors with different backbones or frameworks.

3.1 Cross-Weighting Knowledge Distillation

In this subsection, we proposed adopting a cross-weighting knowledge distillation strategy for the classification and localization branches. As shown in Fig. 1, existing methods mostly concentrate on optimizing each of the two branches separately. However, during the testing phase or the common NMS operation, it can be observed that IoU and classification scores are coupled together. This coupling, along with the disparity in decoupling during the training phase, leads to sub-optimal results.

IoU-Aware Classification Knowledge Distillation Loss. During the training phase of detectors, there is typically an imbalance between foreground and background samples. Existing methods address the issue of sample imbalance during the training phase through techniques such as Hard Sampling [17,25], which involves the direct selection of meaningful samples and discarding irrelevant ones, Soft Sampling methods [3,12,16] that weight samples that directly generate samples to mitigate sample imbalance.

Due to the extreme imbalance issue during the knowledge distillation, noise can be introduced, leading to sub-optimal performance of the model. Our approach is inspired by the idea of Focal Loss [16] and aims to solve the issue concisely. Next, we will briefly introduce Focal Loss and some preliminary information. Focal Loss adjusts the weights between foreground and background samples to address the imbalance issue. It is defined as follows:

$$\mathcal{FL}(p, y) = \begin{cases} -\alpha(1-p)^\gamma log(p), & \text{if } y = 1; \\ -(1-\alpha)p^\gamma log(1-p), & \text{if } y = 0. \end{cases} \quad (1)$$

In the definition of Focal Loss, y represents the binary class supervision information with $y \in \{\pm 1\}$, where p denotes the probability of the foreground class. The modulating factor $(1-p)^\gamma$ and p^γ are used to down-weight easy samples and up-weight hard samples, and γ is a hyperparameter. α is used to balance foreground and background samples.

Following the classic classification distillation method [10], as shown in Eq. 2, the Kullback-Leibler(KL) Divergence is employed as the knowledge distillation loss, enabling the student to learn the class probability distribution obtained from the teacher classifier. The classification logit values z^t from the teacher model and z^s from the student model are utilized. Here, p^t represents the class probability derived from the teacher model output logit value z^t after being computed through $softmax(., \tau)$, where τ is the temperature coefficient, defaults to 1. A higher value for τ produces a softer class probability distribution.

$$\mathcal{L}_{VanillaKD} = \mathcal{KL}(p^t \parallel p^s), \quad (2)$$

$$p^t = \frac{exp(z^t/\tau)}{\sum_j exp(z_j^t/\tau)}. \quad (3)$$

Focal Loss is derived from the Binary Cross-Entropy (BCE) loss function. Instead, consider using BCE to replace the original class KL divergence loss. This enables the handling of binary classification problems and the balancing of foreground and background samples through weighted means, distinguishing it from Focal Loss, which treats foreground and background samples consistently. Hence, the following loss function is proposed:

$$\mathcal{L}_{cls}^{dis}(x) = \begin{cases} -w^{iou}((1-p^t) \cdot log(1-p^s) + p^t \cdot log(p^s)) & \text{if } x \in pos\_set \\ -\alpha(p^s)^\gamma log(1-p^s) & \text{otherwise} \end{cases}, \quad (4)$$

$$w^{iou} = IoU(\mathcal{B}^t, \mathcal{B}^s). \quad (5)$$

Inspired by [27], for foreground samples $x \in pos\_set$, the IoU score computed from the B-Box output by both teacher and student models serves as weights in the Binary Cross-Entropy (BCE) computation. This facilitates the classification loss for foreground samples to reflect the consistency in localization capability between the teacher and student models. In contrast to the Focal loss, a scaling factor of $(p^s)^\gamma$, with a hyperparameter γ, is employed to adjust the loss, thereby reducing the contribution from negative examples. As Eq. 5 shows, $\mathcal{B}^t$ denotes the predicted B-Box by the teacher, and $\mathcal{B}^s$ denotes the predicted B-Box by the student. For anchor-based models, the bounding box (B-Box) is obtained using the $decode(x, a)$ function with anchor a, whereas, for anchor-free models, it is obtained using $decode(x)$.

Class-Aware Localization Knowledge Distillation Loss. Transferring localization knowledge in the knowledge distillation of object detection models has gradually gained attention. Existing methods often focus on designing separate regression heads to handle localization capability, and l_n-norm loss functions are widely used for box regression.

In [15], it is proposed to partition the continuous box regression values into n sub-intervals, and the localization head can predict n logits value, corresponding to the endpoints of the sub-intervals. Above that, each edge of the given B-Box can be represented as a probability distribution. In [36], the KL Divergence loss is directly employed to enable the student to learn the distribution of edges.

However, these regression methods have several issues: Firstly, as mentioned earlier, the difference in the usage of IoU in regression training and evaluation introduces a gap, and regression methods are sensitive to variant scales. Additionally, the approach used in [36] requires the design of specific detection heads to obtain and distill distributions.

Given the aforementioned issues, in [30], IoU calculation was first used as a localization loss, and it was further developed in subsequent works such as [23,35]. To address the performance loss resulting from the decoupling of IoU and classification scores during training, IoU series loss functions are adopted to facilitate localization knowledge transfer. Furthermore, to ensure equal treatment of classification and regression during training, classification scores are utilized to weight the localization knowledge distillation. Therefore, the Class-aware Localization Knowledge Distillation loss is proposed:

$$\mathcal{L}_{loc}^{dis}(x) = \sum_{i=1}^{N_p} w_i \cdot f(\mathcal{B}_i^t, \mathcal{B}_i^s), \tag{6}$$

$$w_i = \max_{1 \leq k \leq C} \left\{ |p_{i,k}^t - p_{i,k}^s| \right\}. \tag{7}$$

$\mathcal{B}_i^t$ and $\mathcal{B}_i^s$ represent the predicted B-Boxes by the teacher and student models, respectively. N_p indicates the number of positive samples. The function f denotes the GIoU Loss is used to transfer localization knowledge. Calculate the value of the k-th class with the greatest difference in category probability between the teacher output p_i^t and student output p_i^s as the weight w_i. This allows us to down-weight simple or extremely difficult samples, those that are challenging for both the teacher and student to recognize, or background samples while simultaneously up-weighting some important samples.

3.2 Valuable Distillation Object Selection Module

Most existing distillation methods focus on distilling positive samples, believing that negative samples do not contribute much to the distillation process. Methods like [9,28] use masks to decouple foreground and background in non-GT regions. However, our observation is that, during the initial stages of training, the student model indeed focuses more on learning knowledge from GT regions. However, as training progresses, the knowledge that GT can provide becomes limited. Therefore, this paper proposes the VDO selection module, aimed at obtaining some targets worth distilling besides those highly correlated with GT, to compensate for the lack of knowledge transfer in the later stages.

Algorithm 1 is used to obtain our VDOs. Firstly, we compute the VDO score P_i^{vdo}, in two-stage object detectors, p_i^t and p_i^s represent the object scores output by the RPN of the teacher and student models respectively, and in one-stage detectors, we use the predicted probabilities from the classification branch. We then further select B-Boxes with high category probabilities from either the teacher $\mathcal{B}_t$ or student $\mathcal{B}_s$ as VDOs' B-Box $\mathcal{B}^{vdo}$. This way, we obtain a set of pseudo-labels $\{P_i^{vdo}, \mathcal{B}_i^{vdo}\}$. Since ground truth (GT) is employed in the aforementioned distillation process, further filtering of VDOs is required. The IoU between the VDOs and the GT is computed, and objects with IoU < 0.5 are subsequently filtered out. To prevent redundant calculations, Non-Maximum Suppression (NMS) is employed with an IoU threshold of 0.3 to select the top k high-quality VDOs, where k represents a hyperparameter. Finally, the subsequent distillation step is undertaken.

It can be observed that these objects serve as supplements to the GT supervision information. Hence, additional loss functions are employed to complement them, as depicted in Eq. 8. For model simplicity, proposals are selected for anchor-based methods, while for anchor-free methods, a mask is utilized to determine which ones undergo distillation.

Algorithm 1. Valuable Distillation Object Selection Module

Require: Three sets of student's and teacher's predicted bounding-box and ground truth bounding-box $\mathcal{B}^s = \{\mathcal{B}_i^s\}$, $\mathcal{B}^t = \{\mathcal{B}_i^t\}$ and $\mathcal{B}^{gt} = \{\mathcal{B}_j^{gt}\}$, $1 \leq i \leq N, 1 \leq j \leq J$.
Ensure: $VDO = \{P_i, \mathcal{B}_i\}$
1: $i \leftarrow 1$
2: **while** $i < N$ **do**
3: $\quad P_i^{vdo} = \max\limits_{1 \leq k \leq C} \{|p_i^t - p_i^s|\}$
4: $\quad$ **if** $p_{i,k}^t > p_{i,k}^s$ **then**
5: $\quad\quad \mathcal{B}_i = \mathcal{B}_t$
6: $\quad$ **else**
7: $\quad\quad \mathcal{B}_i = \mathcal{B}_s$
8: $\quad$ **end if**
9: $\quad \mathcal{B}^{vdo} \leftarrow \{P_i^{vdo}, \mathcal{B}_i\}$
10: $\quad i \leftarrow i + 1$
11: **end while**
12: Compute IoU between $\mathcal{B}^{gt}$ and $\mathcal{B}^{vdo}$, and select VDOs with $IoU < 0.5$ to add to the final set $\mathcal{B}^{vdo}$.
13: VDOs $\leftarrow$ top k $NMS(P^{vdo}, B^{vdo})$
14: **return VDOs**

3.3 Overall Distillation Loss

After the discussions in the preceding sections, overall the total knowledge distillation loss function with the cross-weighting strategy is as follows:

$$\mathcal{L}^{dis} = \lambda_0 \mathcal{L}_{loc}^{dis} + \lambda_1 \mathcal{L}_{cls}^{dis} + I_{VDO}(\lambda_2 \mathcal{L}_{loc}^{dis} + \lambda_3 \mathcal{L}_{cls}^{dis}). \tag{8}$$

$\mathcal{L}_{loc}^{dis}$ and $\mathcal{L}_{cls}^{dis}$ are the weighted knowledge distillation loss functions mentioned above for localization and classification tasks, respectively. I_{VDO} is a binary indicator, taking values of $\{0,1\}$, indicating whether to conduct VDO distillation. And the weighting terms λ_0, λ_1, λ_2, λ_3 are set to 4.0, 1.0, 4.0, 1.0 respectively.

4 Experiment

4.1 Experiment Settings

To verify our knowledge distillation method, following other mainstream detectors, We also use *train2017* images in COCO for training and *val2017* images for testing and reporting ablation results. We compare our method with other knowledge distillation methods in the same detectors on Average Precision.

4.2 Implementation and Details

The Cross-weighting Strategy Knowledge Distillation method is implemented with *MMDetection* [4] framework. The initial learning rate is set as 0.01 and the linear scaling rule is employed with the 0.1 warm-up ratio. We use a single A100 GPU for training with a total batch size of 16 in all performance comparisons. All the object detection models and the baselines are trained to use an image scale of [800, 1333].

4.3 Main Results

Our method can be applied to different detection frameworks easily, so we initially conduct experiments on GFL [15], which can based on either a one-stage or two-stage detector. Also, we compare with LD [36], which also is a logit-based KD method. As indicated in Table 1, detectors with ResNet-50, ResNet-34, and ResNet-18 are selected as the students, respectively, while the detector with ResNet-101 serves as the teacher. All student models demonstrated improvements in average precision (AP). Furthermore, compared to the logit-based LD [36] method on different backbones, our method achieved improvements in mAP scores.

Table 1. Quantitative results of proposed methods for lightweight detectors based on GFL [15]. The results are reported on MS COCO *val2017*.

Method	Schedule	mAP	AP_{50}	AP_{75}	AP_S	AP_M	AP_L
GFL-Res101 (Teacher) [15]	2×	44.9	63.1	49.0	28.0	49.1	57.2
GFL-Res50 (Student)	1×	40.1	58.2	43.1	23.3	44.4	52.5
LD	1×	42.1	60.3	45.6	24.5	46.2	54.8
Ours	1×	**43.0**	**61.2**	**46.6**	**25.1**	**47.3**	**55.4**
GFL-Res34 (Student)	1×	38.9	56.6	42.2	21.5	42.8	51.4
LD	1×	41.0	58.6	44.6	23.2	45.0	54.2
Ours	1×	**41.9**	**59.6**	**45.5**	**24.1**	**46.2**	**54.6**
GFL-Res18 (Student)	1×	35.8	53.1	38.2	18.9	38.9	47.9
LD	1×	37.5	54.7	40.4	20.2	41.2	49.4
Ours	1×	**38.5**	**56.4**	**41.5**	**21.4**	**41.8**	**50.1**

As shown in Table 2, experiments are conducted to compare our proposed methods with several state-of-the-art feature-based methods, logit-based methods, and relation-based methods, demonstrating significant AP score gains achieved by ours.

4.4 Analysis

Sensitivity Study of Different Losses. In this paper, experiments are conducted on the IoU-aware Classification KD loss ($\mathcal{L}_{cls}^{dis}$) and Class-aware Localization KD loss ($\mathcal{L}_{loc}^{dis}$) to evaluate their effects on the student model (GFL-Res18). Additionally, Valuable Distillation Objects (VDOs) are introduced as positive samples added to the training phases. As illustrated in Table 3, the

Table 2. Quantitative results of the proposed method on various popular dense object detectors with logit-based, feature-based, and relation-based KD methods. The results are reported on MS COCO *val2017*.

Method	mAP	AP_{50}	AP_{75}	AP_S	AP_M	AP_L
GFL-Res101 (Teacher) [15]	44.9	63.1	49.0	28.0	49.1	57.2
GFL-Res50 (Student)	40.1	58.2	43.1	23.3	44.4	52.5
FitNets [24]	40.7	58.6	44.0	23.7	44.4	53.2
DeFeat [9]	40.8	58.6	44.2	24.3	44.6	53.7
Fine-Grained [26]	41.1	58.8	44.8	23.3	45.4	53.1
FGD [28]	41.3	58.8	44.8	24.5	45.6	53.0
GID [6]	41.5	59.6	45.2	24.3	45.7	53.6
LD [36]	42.1	60.3	45.6	24.5	46.2	54.8
Ours	**43.0**	**61.2**	**46.6**	**25.1**	**47.3**	**55.4**

combination of $\mathcal{L}_{cls}^{dis}$ and $\mathcal{L}_{loc}^{dis}$ demonstrates an improvement in average precision (AP). Moreover, the inclusion of VDOs contributes to improved performance, thus reinforcing the effectiveness of incorporating additional positive samples.

Table 3. Ablation study of the distillation loss.

Method	GFocal ResNet101-ResNet18						
	$\mathcal{L}_{cls}^{dis}$		$\mathcal{L}_{loc}^{dis}$		mAP	AP_{50}	AP_{75}
	GT	VDO	GT	VDO			
Baseline					35.8	53.1	38.2
Ours	✔				37.9	55.4	40.9
			✔		37.6	54.5	40.5
	✔		✔		38.2	56.1	40.9
	✔	✔	✔	✔	**38.5**	**56.4**	**41.5**

Visualization. As shown in Fig. 3, the visualization of detection results on *COCO test-dev*. Our method yields detections closer to the teacher model's, showing superior localization ability and classification predictions compared to the original student model.

Fig. 3. Visualization of detection results in the first column to the last column corresponds to the teacher (GFL-r101), ours (r18), and the student (GFL-r18), respectively, on *COCO test-dev*. The score threshold for visualization is 0.3.

5 Conclusion

In this paper, we present a novel framework for knowledge distillation in object detection. Our Cross-Weighting Knowledge Distillation method aims to mitigate challenges arising from the imbalance between foreground and background samples in object detection methods, as well as the discrepancy between the separation of IoU and classification score during distillation training phases, and their strong correlation during testing or evaluation stages. Additionally, to address the imbalance issue, we introduce a VDO selection module to identify objects more valuable for distillation. Experimental results on the COCO dataset demonstrate the effectiveness of our knowledge distillation method, significantly outperforming the baseline and several other knowledge distillation methods. In future research, we plan to explore the effectiveness of such cross-weighting schemes in detection frameworks based on the Vision Transformer (ViT).

Acknowledgment. This work was supported by the National Natural Science Foundation of China (Serial Nos. 61991410, 61976134), OpenProject Foundation of Intelligent Information Processing Key Laboratory of Shanxi Province, China (No. CICIP2021001), Natural Science Foundation of Shanghai (No. 21ZR1423900), and Shanghai Science and Technology Innovation Action Plan (22511101903).

References

1. Abouelnaga, Y., Bui, M., Ilic, S.: DistillPose: lightweight camera localization using auxiliary learning. In: 2021 IEEE/RSJ International Conference on Intelligent Robots and Systems (IROS), pp. 7919–7924. IEEE (2021)
2. Cao, W., Zhang, Y., Gao, J., Cheng, A., Cheng, K., Cheng, J.: PKD: general distillation framework for object detectors via Pearson correlation coefficient. Adv. Neural. Inf. Process. Syst. **35**, 15394–15406 (2022)
3. Cao, Y., Chen, K., Loy, C.C., Lin, D.: Prime sample attention in object detection. In: Proceedings of the IEEE/CVF Conference on Computer Vision and Pattern Recognition, pp. 11583–11591 (2020)
4. Chen, K., et al.: MMDetection: Open MMLab detection toolbox and benchmark. arXiv preprint arXiv:1906.07155 (2019)
5. Chen, P., Liu, S., Zhao, H., Jia, J.: Distilling knowledge via knowledge review. In: Proceedings of the IEEE/CVF Conference on Computer Vision and Pattern Recognition, pp. 5008–5017 (2021)
6. Dai, X., et al.: General instance distillation for object detection. In: Proceedings of the IEEE/CVF Conference on Computer Vision and Pattern Recognition, pp. 7842–7851 (2021)
7. Dosovitskiy, A., et al.: An image is worth 16×16 words: transformers for image recognition at scale. In: ICLR (2021)
8. Duan, K., Bai, S., Xie, L., Qi, H., Huang, Q., Tian, Q.: CenterNet: keypoint triplets for object detection. In: Proceedings of the IEEE/CVF International Conference on Computer Vision, pp. 6569–6578 (2019)
9. Guo, J., et al.: Distilling object detectors via decoupled features. In: Proceedings of the IEEE/CVF Conference on Computer Vision and Pattern Recognition, pp. 2154–2164 (2021)

10. Hinton, G., Vinyals, O., Dean, J.: Distilling the knowledge in a neural network. Comput. Sci. **14**(7), 38–39 (2015)
11. Law, H., Deng, J.: Cornernet: Detecting objects as paired keypoints. In: Proceedings of the European Conference on Computer Vision (ECCV), pp. 734–750 (2018)
12. Li, B., Liu, Y., Wang, X.: Gradient harmonized single-stage detector. In: Proceedings of the AAAI Conference on Artificial Intelligence, vol. 33, pp. 8577–8584 (2019)
13. Li, G., Li, X., Wang, Y., Zhang, S., Wu, Y., Liang, D.: Knowledge distillation for object detection via rank mimicking and prediction-guided feature imitation. In: Proceedings of the AAAI Conference on Artificial Intelligence, vol. 36, pp. 1306–1313 (2022)
14. Li, L., Yue, X., Xu, Z., Xie, S.: Multi-dimensional pruned sparse convolution for efficient 3D object detection. In: 2023 IEEE International Conference on Image Processing (ICIP), pp. 3190–3194. IEEE (2023)
15. Li, X., et al.: Generalized focal loss: learning qualified and distributed bounding boxes for dense object detection. Adv. Neural. Inf. Process. Syst. **33**, 21002–21012 (2020)
16. Lin, T.Y., Goyal, P., Girshick, R., He, K., Dollár, P.: Focal loss for dense object detection. In: Proceedings of the IEEE International Conference on Computer Vision, pp. 2980–2988 (2017)
17. Liu, W., et al.: SSD: single shot multibox detector. In: Leibe, B., Matas, J., Sebe, N., Welling, M. (eds.) ECCV 2016. LNCS, vol. 9905, pp. 21–37. Springer, Cham (2016). https://doi.org/10.1007/978-3-319-46448-0_2
18. Long, J., Shelhamer, E., Darrell, T.: Fully convolutional networks for semantic segmentation. In: Proceedings of the IEEE Conference on Computer Vision and Pattern Recognition, pp. 3431–3440 (2015)
19. Ni, Z.L., Yang, F., Wen, S., Zhang, G.: Dual relation knowledge distillation for object detection. In: Proceedings of the Thirty-Second International Joint Conference on Artificial Intelligence, pp. 1276–1284 (2023)
20. Park, W., Kim, D., Lu, Y., Cho, M.: Relational knowledge distillation. In: Proceedings of the IEEE/CVF Conference on Computer Vision and Pattern Recognition, pp. 3967–3976 (2019)
21. Redmon, J., Divvala, S., Girshick, R., Farhadi, A.: You only look once: unified, real-time object detection. In: Proceedings of the IEEE Conference on Computer Vision and Pattern Recognition, pp. 779–788 (2016)
22. Ren, S., He, K., Girshick, R., Sun, J.: Faster R-CNN: towards real-time object detection with region proposal networks. Adv. Neural Inf. Process. Syst. **28** (2015)
23. Rezatofighi, H., Tsoi, N., Gwak, J., Sadeghian, A., Reid, I., Savarese, S.: Generalized intersection over union: a metric and a loss for bounding box regression. In: Proceedings of the IEEE/CVF Conference on Computer Vision and Pattern Recognition, pp. 658–666 (2019)
24. Romero, A., Ballas, N., Kahou, Bengio, Y.: FitNets: hints for thin deep nets. In: ICLR (2015)
25. Shrivastava, A., Gupta, A., Girshick, R.: Training region-based object detectors with online hard example mining. In: Proceedings of the IEEE Conference on Computer Vision and Pattern Recognition, pp. 761–769 (2016)
26. Wang, T., Yuan, L., Zhang, X., Feng, J.: Distilling object detectors with fine-grained feature imitation. In: Proceedings of the IEEE/CVF Conference on Computer Vision and Pattern Recognition, pp. 4933–4942 (2019)
27. Wu, S., Yang, J., Wang, X., Li, X.: IoU-balanced loss functions for single-stage object detection. Pattern Recogn. Lett. **156**, 96–103 (2022)

28. Yang, Z., et al.: Focal and global knowledge distillation for detectors. In: Proceedings of the IEEE/CVF Conference on Computer Vision and Pattern Recognition, pp. 4643–4652 (2022)
29. Yang, Z., Li, Z., Shao, M., Shi, D., Yuan, Z., Yuan, C.: Masked generative distillation. In: Avidan, S., Brostow, G., Cissé, M., Farinella, G.M., Hassner, T. (eds.) ECCV 2022. LNCS, vol. 13671, pp. 53–69. Springer, Cham (2022). https://doi.org/10.1007/978-3-031-20083-0_4
30. Yu, J., Jiang, Y., Wang, Z., Cao, Z., Huang, T.: UnitBox: an advanced object detection network. In: Proceedings of the 24th ACM International Conference on Multimedia, pp. 516–520 (2016)
31. Yu, L., Wang, S., Li, X., Fu, C.-W., Heng, P.-A.: Uncertainty-aware self-ensembling model for semi-supervised 3D left atrium segmentation. In: Shen, D., et al. (eds.) MICCAI 2019. LNCS, vol. 11765, pp. 605–613. Springer, Cham (2019). https://doi.org/10.1007/978-3-030-32245-8_67
32. Zhang, L., Ma, K.: Improve object detection with feature-based knowledge distillation: towards accurate and efficient detectors. In: International Conference on Learning Representations (2020)
33. Zhang, S., Chi, C., Yao, Y., Lei, Z., Li, S.Z.: Bridging the gap between anchor-based and anchor-free detection via adaptive training sample selection. In: Proceedings of the IEEE/CVF Conference on Computer Vision and Pattern Recognition, pp. 9759–9768 (2020)
34. Zhao, B., Cui, Q., Song, R., Qiu, Y., Liang, J.: Decoupled knowledge distillation. In: Proceedings of the IEEE/CVF Conference on Computer Vision and Pattern Recognition, pp. 11953–11962 (2022)
35. Zheng, Z., Wang, P., Liu, W., Li, J., Ye, R., Ren, D.: Distance-IoU loss: faster and better learning for bounding box regression. In: Proceedings of the AAAI Conference on Artificial Intelligence, vol. 34, pp. 12993–13000 (2020)
36. Zheng, Z., Ye, R., Hou, Q., Ren, D., Wang, P., Zuo, W., Cheng, M.M.: Localization distillation for object detection. IEEE Trans. Pattern Anal. Mach. Intell. (2023)

A Method of Multi-USV Reward Design Using Fuzzy Control

Jianfeng Xiao, Qun Liu[(✉)], and Xin Huang

Chongqing Key Laboratory of Computational Intelligence, Chongqing University of Posts and Telecommunications, Chongqing 400065, People's Republic of China
{s220231105,s210231080}@stu.cqupt.edu.cn, liuqun@cqupt.edu.cn

Abstract. This paper investigates the path planning problem of multiple unmanned surface vehicles (USVs). Since the maritime environment is more complex and variable than the land environment, none of the currently known path planning methods can overcome the uncertainties involved with desirable results. Deep reinforcement learning methods offer great promise for solving maritime uncertainty problems. However, the design of reward functions is often very difficult when it comes to solve practical problems. To enhance the path planning capability of multiple USVs in the maritime environment, we analyzed how maritime uncertainty affected path planning and designed a fuzzy logic-based reward function. This function is capable of effectively guiding the training of USVs while possessing good robustness to adapt to environmental uncertainty. In this paper, simulations of the motion model of USVs and the maritime environment are conducted, and the method's effectiveness is verified through experiments. The results show that the model can demonstrate strong adaptability in the complex maritime environment.

Keywords: Reinforcement Learning · Multi-USV · Path Planning · Fuzzy Control · Cloud Model

1 Introduction

With the advancement of intelligent technology, unmanned surface vehicles (USVs) are increasingly utilized in maritime applications such as marine scientific investigation, maritime search and rescue, and water quality testing. This is owing to their advantages of high safety, extended operational time, and robust deployability.

At the same time, as task complexity increases, single USV is no longer sufficient to meet the demands of tasks, leading to a rise in the use of multiple USVs for task execution. Multi-agent path planning is a crucial technology for enabling cooperation among multiple USVs, with a key constraint being the agents' ability to simultaneously follow planned paths without conflict. Multi-agent path planning techniques have found widespread application in automated warehouses [11], self-driving cars [7], and robots [2]. However, different environments have varying performance requirements for algorithms. Unlike the general terrestrial environment, the maritime environment presents significant dynamic complexity due to factors such as wind, waves, and currents. As a result, USVs often struggle to adhere to predetermined paths, necessitating path planning algorithms with autonomy, real-time capabilities, and strong resistance to interference.

Traditional multi-agent path planning algorithms are often unsuitable for complex and dynamic maritime environments because they struggle to handle the interference caused by the environment. However, deep reinforcement learning, which has become increasingly popular, is being applied to the maritime field due to its strong learning and adaptation capabilities. For instance, there have been studies on obstacle avoidance of USVs in uncertain environments [12], and formation planning for multiple unmanned boats [14].

In the process of solving real-world problems using deep reinforcement learning, the reward function's design is paramount in achieving optimal performance, which is divided into two main categories: dense and sparse rewards. Dense rewards usually give a corresponding reward after every action made by an agent, so that the agent can understand the reasonableness of the action. However, designing an effective dense reward function is very challenging. It is essential to have a thorough understanding of the factors that enable a task to be executed perfectly, and some of the values of the reward function may require repeated attempts to achieve a relatively optimal result. Additionally, dense rewards exhibit enhanced susceptibility to environmental noise, and the noise can be further amplified by the Bellman equation [4]. On the other hand, sparse rewards refer to binary reward functions, where completing a task receives $+1$ as a reward, failure receives -1 as a penalty, and no reward is given in any other cases. A sparse reward function is less susceptible to environmental noise. Theoretically, the agent can learn the global optimal strategy that a sufficient number of successful trajectories are collected.

However, when using reinforcement learning to complete a path planning task in a highly uncertain maritime environment, designing a reward function that can effectively handle environmental uncertainty becomes a major challenge. The dense reward function cannot accurately assign the appropriate reward value to the agent at each step. And for path planning tasks that cover wide time and space spans, the sparse reward function often fails to enable the agent to develop a better strategy within a limited time. To address these issues, a reward function framework based on fuzzy control is introduced in this paper.

The main contributions of this paper can be summarized as follows:

- We use fuzzy control theory to design a fuzzy reward function that is capable of adapting to environmental uncertainty in a multi-USV environment.
- Combining cloud model theory to deal with information uncertainty in uncertain environments, and membership values obtained through the cloud model improves the model's ability to cope with uncertainty.
- The motion model of the USV is considered in the simulation environment, while the effect of uncertainty noise is added to simulate the maritime situation.

2 Related Work

For multi-USV path planning in maritime environments, the traditional algorithms are currently used relatively more often. Liu et al. [8] proposed an improved ant colony clustering algorithm that dynamically adjusted the search range based on the complexity of different environments. They also used USV maneuvering rules to regulate the dynamic search path. Liu et al. [9] proposed a fast marching-based method for path planning in different dynamic environments using a novel constrained FM method. Chen et al. [3] proposed an enhanced ant colony optimized artificial potential field method that searched for the globally optimal path of a USV, starting from a specified point and reaching the endpoint within a grid environment. They used an improved ant colony optimization method and then employed an improved artificial potential field algorithm to avoid unpredictable obstacles during the navigation of the USV. Zhang et al. [18] combined genetic algorithm and simulated annealing algorithm for USV path planning. This method improves the efficiency of population evolution and optimizes the paths through the Jung family insertion and deletion operators. In order to address the issues of slow search speed, low collision avoidance efficiency, and long paths encountered in existing path planning methods, Bai et al. [1] proposed a plant growth path algorithm. This algorithm utilizes the principle of phototropism in plant growth to guide the USV in avoiding obstacles and reaching the target by leveraging the intensity of light. However, these approaches rarely consider the dynamic complexity of the marine environment and cannot guarantee reliable performance in extreme real-life environments.

In recent decades, deep learning has experienced rapid development, deep reinforcement learning has also achieved good results in path planning for USV. In response to the usability problem arising from the complexity of the control law in traditional methods, Xu et al. [16] proposed a path planning method which leveraged the gradient of the deep deterministic policy to achieve superior performance compared to traditional approaches. Xing et al. [15] proposed an improved deep reinforcement learning method that superimposed a dangerous action matrix on the final output layer to mitigate the risk of USVs selecting hazardous actions during training. Wang et al. [13] proposed a greedy navigation and subtle obstacle avoidance algorithm based on actor-critic architecture, building a Markov process to fit the equations of USV kinematics and achieving the goal by designing a reward function with a behavioral prior. Zhao et al. [19] proposed a path planning algorithm using deep deterministic policy gradient (DDPG), which could obtain the optimal path faster and more accurately than the traditional A-star algorithm.

It is evident that most of the deep reinforcement learning methods can achieve the same or better results than traditional methods. However, the current research often neglects the impact of environmental uncertainty. Additionally, many reinforcement learning methods still rely on sparse reward function design, which is inefficient and leads to a significant waste of training resources. In contrast, this paper conducts simulation experiments in an environment that takes into account the USV motion model and uncertainty factors. It achieves high sample efficiency and robustness to uncertainty factors through a reward function design based on fuzzy logic and the incorporation of cloud model theory to address the interplay between randomness and ambiguity.

3 Preliminary

3.1 Reinforcement Learning

Standard reinforcement learning models typically include three fundamental concepts: state, action, and reward. At each time step, the agent observes the environment to acquire its own state information s_t, and determines the action a_t to be taken based on its policy $\pi(s_t)$. Upon execution of the action, the environment undergoes a change, and the agent will obtain a reward r according to the reward function. Subsequently, it updates its policy based on the obtained reward. The iterative process continues until the failure or success of the agent's tasks, at which point it proceeds to initiate training for the next episode.

Under the given strategy π, the actual Q value of action a taken in state s can be written as follows:

$$Q_\pi(s,a) = E[R_1 + \gamma R_2 + ... + \gamma^{n-1} R_n | S_0 = s, A_0 = a, \pi] \quad (1)$$

where $\gamma \in [0,1]$ is a discount factor that represents the discount effect of subsequent rewards on the current Q value, and R_n represents the reward at the final time step of one episode. In the field of deep reinforcement learning, neural networks are used to approximate Q-values, which helps fit the objective Q-function:

$$Y_t = R_{t+1} + \gamma Q(S_{t+1}, \underset{a}{argmax}\, Q(S_{t+1}, a; \theta_t); \theta_t) \quad (2)$$

where θ_t denotes the weight of Q-network at time step t. Using random gradient descent to update the current value $Q(S_t, A_t; \theta_t)$, it tends towards the objective function Y_t.

3.2 Fuzzy Control

The concept of fuzzy sets was initially introduced by Lotfi A. Zadeh in 1965 [17], and it has found extensive applications in dealing with problems involving uncertainty. Fuzzy control is an intelligent control approach that utilizes fuzzy set theory, fuzzy linguistic variables, and fuzzy logic reasoning to achieve precise and adaptive control. It is capable of handling problems with uncertainty, fuzziness, and complexity, and adapting to changes under different conditions.

The principle of fuzzy control emulates the fuzzy reasoning and decision-making processes in human cognition, which can be roughly divided into three steps: fuzzification, fuzzy inference, and defuzzification. Firstly, the data will be fuzzified, and then the fuzzified data is used to calculate a fuzzy output based on fuzzy rules set by prior human knowledge. Finally, the defuzzification process is applied to obtain the final output result. Fuzzification entails the utilization of membership functions to partition data into various fuzzy semantics and allocate membership values to each semantic category. On the other hand, fuzzy rules are formulated based on fuzzy semantics, using IF-THEN rules, typically in the following form:

$$\text{Rule } R_i : \text{ IF } x_1 \text{ is } g_1^i, x_2 \text{ is } g_2^i, ..., x_m \text{ is } g_m^i; \text{ THEN } y \text{ is } H^i, i = 1, 2, .., n$$

where $x_i, i = 1, 2, ..., m$ represents uncertain variables, y represents the resulting value affected by these variables, H^i and $g_j^i, j = 1, 2, ..., m$ are fuzzy sets, and n is the number of fuzzy rules.

3.3 Cloud Model

Cloud models are capable of capturing randomness and fuzziness [5], allowing our model to account for the uncertainty of the actual situation to some extent. Let X be a standard set, $X = \{x_0, x_1, x_2, ...\}$, which is referred to as the universe. A is a fuzzy set over the universe X. If there exists a random number $\mu_A(x)$ that exhibits a stable tendency for any element x (for example, a random number r in the interval $(3, 3.5)$), then $\mu_A(x)$ is known as the membership value of x to A. If the elements in the universe are not ordered, X can be mapped to another ordered universe X' based on some function f. When X' has only one x' corresponding to X, X' is referred to as a basis variable, and the distribution of the membership value on X' is referred to as a cloud. The strength of cloud model is its ability to effectively handle randomness and fuzziness. By combining conceptual and mathematical approaches, cloud model can accurately capture and address the diversity and incompleteness of information.

The normal cloud model enables fuzzy sets to overcome the limitations associated with precisely quantifying the membership value of an element within the strict range of 0 to 1. Therefore, it may be more suitable for describing linguistic concepts with uncertainty [6]. Unlike Gaussian distributions, normal cloud models incorporate hyper entropy to capture fuzziness to some extent. Hence, unlike ordinary Gaussian functions that are constructed using only expected value Ex and entropy En, we can add the extra parameter hyper entropy He to construct the membership function to enhance the model's robustness. Specifically, if x satisfies $x \sim N(Ex, En'^2)$, $En'^2 \sim N(En, He^2)$, then the membership value of x to A is denoted as

$$\mu(x) = exp[-\frac{(x - Ex)^2}{2(En'^2)}] \tag{3}$$

4 Proposed Method

This section provides a comprehensive overview of the proposed method, starting with an introduction to the deep reinforcement learning model and fuzzy control theory. It then goes on to explain the general framework of our model and the construction of our reward function.

4.1 Framework

As shown in Fig. 1, in order to consider the scalability of USVs, we adopt the independent learning framework in multi-agent reinforcement learning, most of the classical single-agent reinforcement learning algorithms can be directly combined with it. In this paper, we use the Double Deep Q-learning Network (DDQN) algorithm [10] as the training kernel for each USV. DDQN, as a variant of the value-based DQN, efficiently solves the transition estimation problem to prevent local optimum.

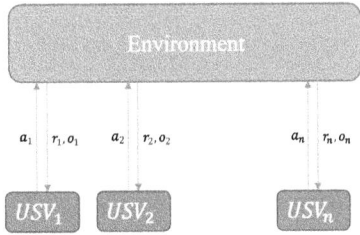

Fig. 1. The Independent Learning Framework allows each USV to interact with the environment independently, while also considering other agents as part of the environment. The framework is not theoretically convergent but has shown good results in practical applications.

In DDQN, there are two separate value functions utilized. During each update iteration, the first function is employed to identify the optimal strategy, while the second determines the value associated with the strategy. In contrast to classical DQN, the objective function of DDQN is as follows:

$$Y_t = R_{t+1} + \gamma Q(S_{t+1}, \underset{a}{argmax}\, Q(S_{t+1}, a; \theta_t); \theta'_t) \tag{4}$$

where θ'_t is the weight parameter of the second value function, the value of the current strategy is evaluated more fairly by using the first value function, and the two value functions are kept up to date by assigning values to both.

Our model framework is shown in Fig. 2. The USV with a built-in reinforcement learning algorithm generates the data $(s_t, a_t, r_t, s_{t+1}, d)$ by interacting with the environment, where s_t is the perceptual information of the USV, a_t is the action taken, r_t is the reward value obtained through the fuzzy controller, and d denotes the end state of the current round. The experience data is stored in the replay buffer, and the reinforcement learning algorithm samples from it to update the policy using the loss function equation:

$$Loss_t = |Y_t - Q(S_t, A_t; \theta)| \tag{5}$$

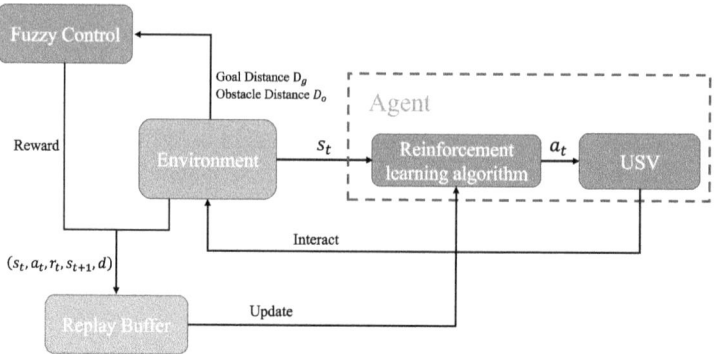

Fig. 2. System structure

4.2 Fuzzy Reward Design

On top of the general sparse reward function, we added a fuzzy reward function as an auxiliary reward value to guide the training of the USV. This includes collision penalties, rewards for reaching the target, and auxiliary fuzzy rewards. The overall function can be expressed as follows:

$$R(t) = \begin{cases} R_c, & \text{collision with obstacles} \\ R_r, & \text{reach the goal} \\ R_f, & \text{fuzzy auxiliary value} \end{cases} \quad (6)$$

As shown in Fig. 3, in the Auxiliary Fuzzy Reward, we design a two-input and single-output fuzzy control system to obtain fuzzy rewards. The inputs are D_o (the distance of the USV from the nearest obstacle or other USV within the observable range) and D_g (the straight line distance from the target). These two data obtained from the environment are transformed into membership values using the cloud model. Fuzzified reward values are then generated based on fuzzy rules, and finally, the determined reward values are obtained through defuzzification. The cloud model has a larger sampling range and enables greater robustness than using a Gaussian distribution.

Table 1. Cloud model parameters (Ex, En, He) of different fuzzy variables. D_o is the obstacles distance and D_g is the goal distance.

	D_o	D_g
close	(0, 1.5, 0.15)	(0, 0.15, 0.01)
middle	(5, 1.5, 0.1)	(0.5, 0.15, 0.01)
far	(10, 3, 0.16)	(1, 0.15, 0.01)

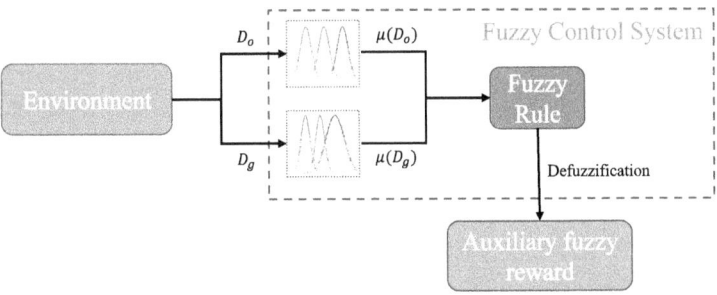

Fig. 3. The fuzzy control system for the auxiliary fuzzy rewards. D_o is the obstacles distance and D_g is the goal distance.

We divided the distance into three fuzzy variables: close, middle, and far, which can be visually represented as the shape of a cloud droplet, as shown in Fig. 4. We normalized the universe of obstacle distances to [0, 20] and target distances to [0, 1]. We determined the cloud model parameters in the form of (Ex, En, He) as shown in Table 1. The fuzzy rules are shown in Table 2. The fuzzy linguistic variable $\{low, medium, high\}$ denotes the fuzzy semantics of auxiliary rewards. The main purpose of the rules is to keep the USVs away from obstacles and other USVs, and to reduce the distance to the target.

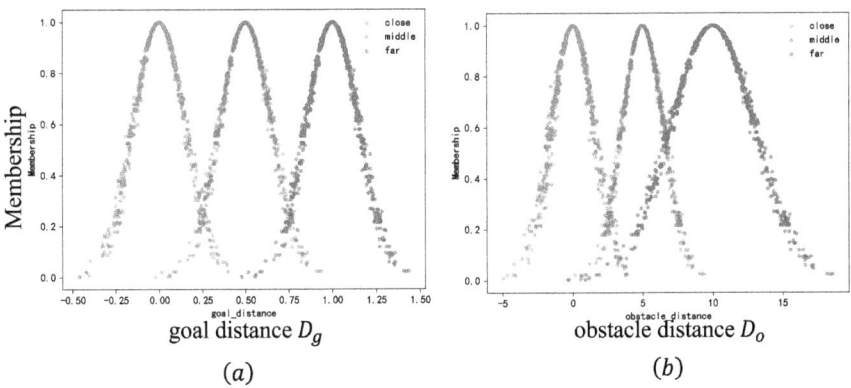

Fig. 4. Membership function.

5 Experiments

5.1 Simulation Environment

As shown in Fig. 5, we have constructed a 2D simulation environment using the OpenAI Gym framework. Blue represents obstacles, the center of gray indicates the USVs, gray represents the observed range of the USVs, and the remaining

Table 2. Fuzzy rules. D_o is the obstacles distance and D_g is the goal distance.

D_o \ D_g	close	middle	far
close	low	low	low
middle	high	medium	medium
far	high	high	medium

parts represent the target points of corresponding colors. The uncertainty of the maritime environment is simulated by having all objects in the environment move slowly in a random direction at each time step. As for the USVs, we simulated their real motion model by limiting their speed and turning angle, as well as restricting their field of view for observation. Specifically, the USVs had two actions: speed and turning angle. To reduce the dimensionality, we discretized the action space for speed as $\{10, 20\}$ and the action space for turning angle as $\{-45, -22.5, 0, 22.5, 45\}$. The USVs used radar to gather information about the surrounding environment. We simulated radar detection by setting up 36 perception lines spaced at 10° intervals centered on each USV. We included the target coordinates in the state space to assist the USVs in calculating their own relationship with the target, and the state space can be expressed as:

$$s(t) = \{p_1, p_2, ..., p_{36}, goal_x, goal_y\} \tag{7}$$

To simplify matters, we established boundaries that restricted the environment to a range of 700 × 700. The USVs could gather information about the borders through radar. If a USV collides with other USVs, obstacles, or reaches the boundary, the maneuver is deemed a failure. Conversely, if all USVs reach the target point successfully, the maneuver is considered to be a success.

5.2 Comparison Analysis

To assess the effectiveness of fuzzy reward functions in uncertain environments, we conducted comparative experiments using the following three kinds of reward functions: (i) sparse, (ii) dense, and (iii) fuzzy. We used these three reward functions in the same environment configuration, and the hyper-parameters of the algorithmic model are shown in Table 3.

In particular, α denotes the decay value of the exploration rate per 100 episodes. By decaying the exploration rate, we improve the sampling efficiency and gradually stabilize the strategy. γ denotes the importance of future rewards to USVs. The path planning task should not solely focus on immediate reward acquisition. It requires paying more attention to global paths, thus a higher value of γ needs to be set.

We conducted experiments in a high exploration rate configuration with two USVs and α value of 0.006. The results are shown in Fig. 6(a), in which dense reward fails to guide the USVs correctly, resulting in an almost 0 success rate. Consequently, the subsequent experiments will not consider the case of dense

Table 3. Parameter settings

hyper-parameter	definition	value
lr	Network learning rate	0.001
γ	Discount factor	0.98
ϵ	Exploration rate	0.9
α	Exploration decay rate	0.006
M	Size of replay buffer	100000
B	Number of samples per update	512

reward. The high uncertainty of the environment prevents the actions made by the USVs in the same state from generating results that match their own experience. For instance, let s_t denote a fixed state. Upon taking action a_t for the first time, a reward r_t is acquired. However, upon subsequent executions of the same action a_t, the reward might vary, possibly resulting in a distinct value r'_t due to environmental uncertainties. Deterministic reward models, particularly in uncertain environments, are prone to inducing distributional bias within the replay buffer, thereby preventing successful training outcomes. Conversely, sparse reward schemes exhibit enhanced robustness, demonstrating resilience against environmental uncertainties and yielding a superior success rate. Furthermore, fuzzy reward frameworks marginally outperform sparse rewards. Nevertheless, the shaded area in Fig. 6 indicates the fluctuation range of the training curve. Upon scrutinizing the shaded regions of the figure, it becomes evident that fuzzy rewards afford improved stability and show the advantage of dense reward structures.

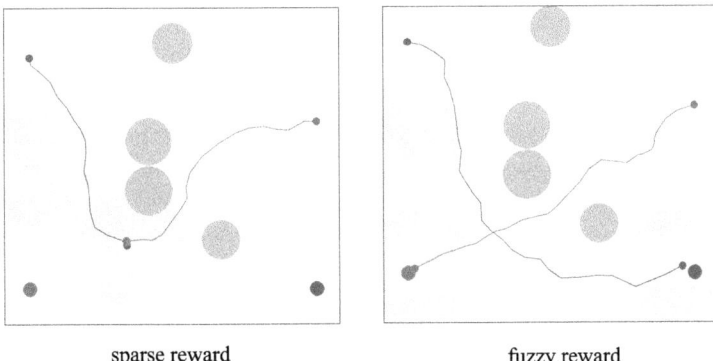

sparse reward   fuzzy reward

Fig. 5. Routes for different reward functions.

We visualized the planned path, and the results are shown in Fig. 5. The visual representation reveals that the trajectories of USVs trained using fuzzy rewards exhibits noticeable irregularities, and lacking smoothness. This is because the uncertain movement of obstacles can pose a threat to the USVs at any time. According to the fuzzy rule, the USVs will maintain a certain distance from the obstacles to ensure their safety. However, the paths generated by the USVs trained with sparse rewards are relatively smoother and less sensitive to dynamic obstacles. On the other hand, when considering the planned paths, the sparsely rewarded paths are clearly more dangerous. In contrast, the fuzzy rewarded paths are more spread out and offer higher levels of safety and feasibility.

To confirm the efficacy of fuzzy rewards in guiding training, we conducted experiments using a low exploration rate configuration with α value of 0.05. As illustrated in Fig. 6(b), it is evident that the significant reduction in sampling frequency due to the low exploration rate leads to a notable decrease in overall training efficiency. In such circumstances, sparse rewards are no longer effective. However, fuzzy rewards continue to demonstrate favorable training outcomes, albeit with reduced stability.

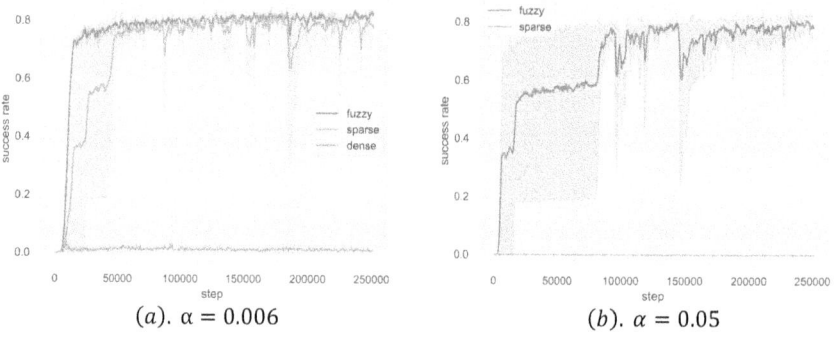

Fig. 6. Training results of different exploration rates.

To assess the effectiveness of the cloud model in this method, we conducted a comparative experiment by replacing the cloud model with an ordinary Gaussian function in the membership function. The experiment was carried out in an environment with three USVs. We compared the cumulative reward values and the success rate obtained by the two methods.

In Fig. 7(a), the use of the fuzzy reward with an ordinary Gaussian function as the membership function is comparable to the sparse reward, but the fuzzy reward using the cloud model as the membership function shows a higher success rate. In Fig. 7(b), we compared the cumulative reward value of different membership functions, and the result shows that the cloud model exhibits higher cumulative reward values. This also means that it is more adaptable to uncertain environments and is able to plan safer and more feasible paths.

Finally, we compare the performance of sparse rewards with fuzzy rewards for different numbers of USVs at normal high exploration rates. Due to the size of

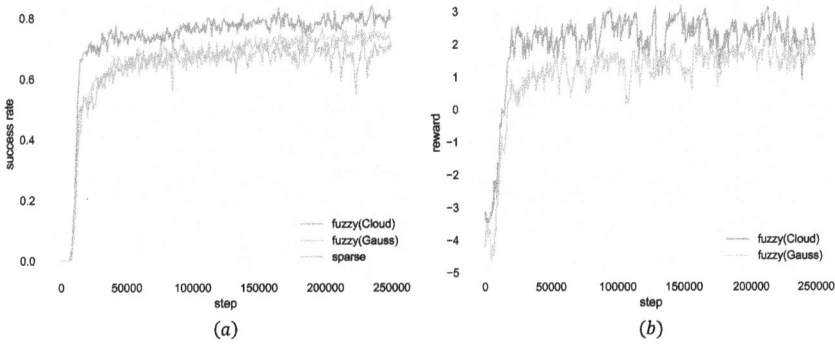

Fig. 7. Comparative experiments with different reward designs.

the restricted environment, increasing the number of USVs also implies a higher probability of collision. According to the comparison results shown in Fig. 8, the fuzzy reward function effectively guides the training process, enabling the achievement of superior results for different numbers of USVs.

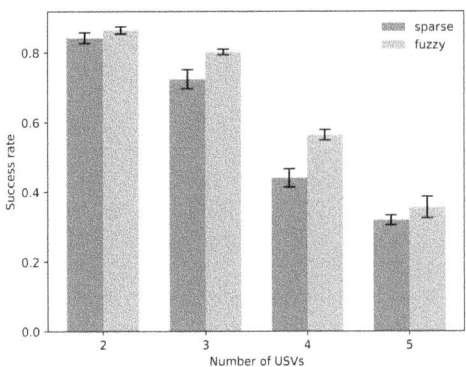

Fig. 8. Success rates at different numbers of USVs

6 Conclusion

In this paper, we study the problem of path planning for multiple USVs in uncertain marine environments. We use the cloud model to handle the uncertainty in the environment, and then design a reward function using fuzzy control theory that has a strong adaptability to the uncertain environment. We compare the training effect and the effect of various reward functions under different environment settings through experiments. The results show that this method is effective in adapting to unpredictable marine environments.

In the future, we plan to address the problem of communication uncertainty between different USVs within the independent learning framework using fuzzy rewards. Additionally, there is still a gap between our current experimental environment and the real environment, so we intend to test our model in real-world conditions.

References

1. Bai, X., Li, B., Xu, X., Xiao, Y.: USV path planning algorithm based on plant growth. Ocean Eng. **273**, 113965 (2023)
2. Barták, R., Švancara, J., Škopková, V., Nohejl, D.: Multi-agent path finding on real robots: first experience with ozobots. In: Simari, G.R., Fermé, E., Gutiérrez Segura, F., Rodríguez Melquiades, J.A. (eds.) IBERAMIA 2018. LNCS (LNAI), vol. 11238, pp. 290–301. Springer, Cham (2018). https://doi.org/10.1007/978-3-030-03928-8_24
3. Chen, Y., Bai, G., Zhan, Y., Hu, X., Liu, J.: Path planning and obstacle avoiding of the USV based on improved ACO-APF hybrid algorithm with adaptive early-warning. IEEE Access **9**, 40728–40742 (2021)
4. He, Q., Hou, X.: WD3: taming the estimation bias in deep reinforcement learning. In: 2020 IEEE 32nd International Conference on Tools with Artificial Intelligence (ICTAI), pp. 391–398. IEEE (2020)
5. Li, D., Cheung, D., Shi, X., Ng, V.: Uncertainty reasoning based on cloud models in controllers. Comput. Math. Appl. **35**(3), 99–123 (1998)
6. Li, D., Liu, C., Gan, W.: A new cognitive model: cloud model. Int. J. Intell. Syst. **24**(3), 357–375 (2009)
7. Li, J., Lin, E., Vu, H.L., Koenig, S., et al.: Intersection coordination with priority-based search for autonomous vehicles. In: Proceedings of the AAAI Conference on Artificial Intelligence, vol. 37, pp. 11578–11585 (2023)
8. Liu, X., Li, Y., Zhang, J., Zheng, J., Yang, C.: Self-adaptive dynamic obstacle avoidance and path planning for USV under complex maritime environment. IEEE Access **7**, 114945–114954 (2019)
9. Liu, Y., Bucknall, R.: Path planning algorithm for unmanned surface vehicle formations in a practical maritime environment. Ocean Eng. **97**, 126–144 (2015)
10. Van Hasselt, H., Guez, A., Silver, D.: Deep reinforcement learning with double Q-learning. In: Proceedings of the AAAI Conference on Artificial Intelligence, vol. 30 (2016)
11. Varambally, S., Li, J., Koenig, S.: Which MAPF model works best for automated warehousing? In: Proceedings of the International Symposium on Combinatorial Search, vol. 15, pp. 190–198 (2022)
12. Wang, P., Liu, R., Tian, X., Zhang, X., Qiao, L., Wang, Y.: Obstacle avoidance for environmentally-driven USVs based on deep reinforcement learning in large-scale uncertain environments. Ocean Eng. **270**, 113670 (2023)
13. Wang, X., Liu, X., Shen, T., Zhang, W.: A greedy navigation and subtle obstacle avoidance algorithm for USV using reinforcement learning. In: 2019 Chinese Automation Congress (CAC), pp. 770–775. IEEE (2019)
14. Wei, X., Wang, H., Tang, Y.: Deep hierarchical reinforcement learning based formation planning for multiple unmanned surface vehicles with experimental results. Ocean Eng. **286**, 115577 (2023)

15. Xing, B., Wang, X., Yang, L., Liu, Z., Wu, Q.: An Algorithm of Complete Coverage Path Planning for Unmanned Surface Vehicle Based on Reinforcement Learning. J. Marine Sci. Eng. **11**(3), 645 (2023)
16. Xu, H., Wang, N., Zhao, H., Zheng, Z.: Deep reinforcement learning-based path planning of underactuated surface vessels. Cyber-Phys. Syst. **5**(1), 1–17 (2019)
17. Zadeh, L.A., Klir, G.J., Yuan, B.: Fuzzy Sets, Fuzzy Logic, and Fuzzy Systems: Selected Papers, vol. 6. World Scientific (1996)
18. Zhang, W., Xu, Y., Xie, J.: Path planning of USV based on improved hybrid genetic algorithm. In: 2019 European Navigation Conference (ENC), pp. 1–7. IEEE (2019)
19. Zhao, J., Wang, P., Li, B., Bai, C.: A DDPG-based USV path-planning algorithm. Appl. Sci. **13**(19), 10567 (2023)

Hyp-DAN: Hyperbolic Distance-Aware Attention Networks

Fuchuan Xiang[1], Jianhang Tang[1], Shaobo Li[1], Guoyin Wang[2], and Ji Xu[1(✉)]

[1] State Key Laboratory of Public Big Data, College of Computer Science and Technology, Guizhou University, Guiyang 550025, China
jixu@gzu.edu.cn
[2] School of Computer Science and Technology, Chongqing University of Posts and Telecommunications, Chongqing 400065, China

Abstract. Hyperbolic embedding has advantages for hierarchically category images, hence it has been applied in few-shot learning and achieved significant results. However, treating all samples equally may not ensure that the learned hyperbolic embedding adequately considers various modalities within the same category. To address this issue, this paper proposes a novel hyperbolic attention mechanism for few-shot learning, which adjusts weights based on the hyperbolic distance of samples to the average position. This approach balances the typicality and diversity of labeled samples, aiding the model in a deeper understanding of data structures. After hyperbolic embedding, weights were redistributed through this attention mechanism for few-shot learning. Experiments were conducted on the CUB and miniImageNet datasets. The experiments demonstrate the superiority of the proposed attention mechanism when using hyperbolic embedding for hierarchically category data.

Keywords: Hyperbolic Embedding · Attention Mechanism · Few-Shot Learning

1 Introduction

Image embedding plays a crucial role in modern computer vision, but existing methods face certain limitations. In computer vision, learning high-dimensional embeddings is a fundamental and widespread task, aimed at grouping semantically similar images and separating dissimilar ones. Currently, there are numerous image classification and retrieval techniques available. For example, image classification networks [18,20] use linear operators (matrix multiplication) to map embeddings from the second-to-last layer to class logits. Therefore, the embeddings learned at the second-to-last layer by the model exist in Euclidean space. The same applies to image retrieval [32,38] and few-shot [23] learning using Euclidean distance. Alternatively, some few-shot learning and person re-identification methods learn spherical embeddings [25], to apply a spherical projection operator at the end of the network computing the embeddings. Cosine similarity (closely related to spherical geodesic distance) is then used by such architectures to match images. However, these methods often overlook

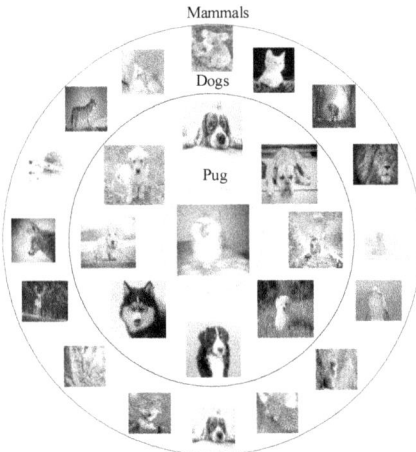

Fig. 1. In many computer vision tasks, our goal is to learn image embeddings that adhere to hierarchical constraints. For example, in recognition tasks, the hierarchy may stem from image degradation, where degraded images are inherently blurry, potentially corresponding to various identities/categories. The closer the images are to the center, the blurrier they become, containing more feature information. Hyperbolic space is more suitable for embedding data with such hierarchical structures.

the inherent non-linearity and hierarchical structure of image data, limiting the expressiveness of the embedding space.

Non-Euclidean embeddings, particularly those in hyperbolic space, offer significant advantages in handling hierarchically structured image data. Non-Euclidean spaces, especially hyperbolic spaces with their negative curvature, excel in embedding data with hierarchical structures. These spaces more naturally capture the complex relationships within hierarchical data, as exemplified in applications in the few-shot learning [23]. However, despite the superior performance of hyperbolic spaces in embedding hierarchical structures, existing research often overlooks the potential and possibility of applying attention mechanisms in such spaces.

In traditional domain relations, each data point is treated equally, and thus the optimization of target functions operating on them usually involves traversing all data points within the domain, with typical examples being the majority of machine learning algorithms such as k-means and EM. However, inspired by the Density Peak Clustering (DPC) method [10], we recognize that each sample, except for the root, is led to join the same class or cluster as its parent node, hence this type of tree is called a leading tree [48]. But due to the tree structure limitation where each node can only have one parent, the leading tree, being the connected graph with the fewest edges, can sometimes lead to learning results that are insensitive to parameters and not stable enough. Instead, We have defined a more general total order domain, which is not limited by the tree structure but still maintains the total order properties between nodes. This total order domain also takes into account the quantified importance or impact of samples or features, the deviation and representativeness of data points, as well as the degree of correlation between features.

Attention mechanisms have become a powerful tool for enhancing the performance of various models, particularly in tasks that require focused interpretation of complex data structures. Here, we argue that hierarchical relationships between images are also common in computer vision tasks (as shown in Fig. 1). Our work is inspired by recent studies that have demonstrated the advantages of learning hyperbolic embeddings for linguistic entities, such as classification categories, common words [3], phrases [45], and other NLP tasks (like neural machine translation [49]). Additionally, it is influenced by some research in image processing. Taking these into consideration, we propose a new hyperbolic distance-aware attention mechanism and conduct related experiments. The ambiguity between images can be viewed as containing different levels of information entropy [35], or as different hierarchies. In fact, the volume of hyperbolic space expands exponentially, while Euclidean space grows polynomially [7]. Therefore, the exponentially expanding hyperbolic space can capture the underlying hierarchies of visual data, which seems reasonable.

To construct deep learning models that operate on embeddings in hyperbolic space, we have utilized recent developments that build analogs of familiar layers (such as feed-forward layers or polynomial regression layers) in hyperbolic space. Experiments show that standard architectures used for image classification tasks, especially in few-shot learning settings, can be easily modified with hyperbolic operations [17], which usually brings performance improvement.

The main contributions of this paper are twofold:

- In the few-shot learning task, we considered the representativeness and deviation of the samples. The study found that the closer to the average hyperbolic distance, the more typical and representative the features of the samples are, and thus they should be assigned greater weight.
- We use the deviation from the average distance in hyperbolic space as a representation of the learning weight, we propose a novel distance-based perception attention mechanism in hyperbolic space and conduct multiple few-shot learning experiments. These experiments show that our method is effective.

2 Related Works

Hyperbolic Language Embeddings. In recent years, hyperbolic embeddings have been very successful in the field of natural language processing [26,47]. Their motivation lies in the inherent ability of hyperbolic space to embed hierarchical structures (such as dendrograms) with low distortion [7,36]. However, due to the discrete nature of data in NLP, these works typically use Riemannian optimization algorithms to learn embeddings of individual words into hyperbolic space. This approach is challenging to extend to visual data, where image representations are usually computed using CNNs [11].

Hyperbolic Image Embeddings. Another research direction more closely related to our current work involves imposing a hyperbolic structure on neural network activations [13,50]. However, the proposed architectures are mostly evaluated on various NLP tasks and accordingly modify traditional models, such as RNNs [39] or transformers [16]. We find that certain computer vision tasks, which extensively use graph embeddings, can also benefit from this kind of hyperbolic architecture.

Few-shot Learning. The task of few-shot learning focuses on the overall ability of a model to generalize to unknown data during training. Most existing few-shot learning models are based on metric learning approaches, utilizing distances between image representations computed by deep neural networks as a measure of similarity [4,6,37,41,44]. For example, Jeong's MAML [14], Lateko's Meta-Learner LSTM [19], and Baroudi's SNAIL [1] model, although these methods employ Euclidean or spherical geometry, they have not been extended to hyperbolic space. The recent work by Moreira [28] and others summarized research on graph embeddings in hyperbolic space. However, they did not make use of the distances after embedding, nor did they consider reallocating weights for the embedded points based on these distances. Beside, the weights of the nodes after embedding were not considered either. The recent proposal by Caglar [9] and others of a hyperbolic attention mechanism was limited to initially implementing the attention mechanism in Euclidean space, followed by hyperbolic embedding. They did not apply the attention mechanism after the hyperbolic embedding.

3 Preliminaries

3.1 Hyperbolic Geometry Initiation and Application

Hyperbolic space has been studied for a long time in differential geometry and, when considered under the five isometric models, specifically refers to manifolds with constant negative curvature in this context. Among them, the Poincaré ball model, Lorentz model, and Klein model are increasingly attracting attention in the machine learning community due to their appealing properties for modeling complex networks [33].

For each $n \geq 1$ and each $R > 0$ we will define a frame-homogeneous Riemannian manifold $\mathbb{H}^n(R)$, called *hyperbolic space of radius R*. There are three equivalent models of the hyperbolic spaces, each of which is useful in certain contexts. We introduce all of them and show that they are isometric.

Theorem 1 [21]. Let $n \geq 1$ ($n \in \mathbb{N}$). For each fixed $R>0$, the following Riemannian manifolds are all mutually isometric.

(a) (**hyperbolic model**). $\mathbb{H}^n(R)$ is the submanifold of Minkowski space $\mathbb{R}^{n,1}$ defined in standard coordinates $(\varepsilon^1, ..., \varepsilon^n, \tau)$ as the "upper sheet" ($\tau > 0$) of the two-sheeted hyperbolic $(\varepsilon^1)^2 + \cdots + (\varepsilon^n)^2 - \tau^2 = -R^2$, with the induced metric

$$\breve{g}_R^1 = \iota * \bar{q}, \qquad (1)$$

where $\iota: \mathbb{H}^n(R) \to \mathbb{R}^{n,1}$ is inclusion, and $\bar{q} = \bar{q}^{(n,1)}$ is the Minkowski, the formulas (2), (3), and (4), d represents differentiation, a standard notation in mathematics to describe the infinitesimal change of variables:

$$\bar{q} = (d\varepsilon^1)^2 + \cdots + (d\varepsilon^n)^2 - (d\tau)^2. \qquad (2)$$

(b) (**Klein model**). $\mathbb{K}^n(R)$ is the ball of radius R centered at the origin in $\mathbb{R}^n$, with the metric given in coordinates $(\omega^1, \ldots, \omega^n)$ by

$$\tilde{g}_R^2 = R^2 \frac{\sum_{i=1}^n (d\omega^i)^2}{R^2 - |\omega|^2} + R^2 \frac{\sum_{i=1}^n (\omega^i d\omega^i)^2}{(R^2 - |\omega|^2)^2}, \qquad (3)$$

where ω represents the coordinates of a point in space, while $\|w\|$ denotes the Euclidean distance from the origin to this point. $\|w\|^2 = (w^1)^2 + \cdots + (w^n)^2$, which is the square of the Euclidean norm of the point w.

(c) (**Poincaré ball model**). $\mathbb{B}^n(R)$ is the ball of radius R centered at the origin in $\mathbb{R}^n$, with the metric given in coordinates $(v^1, \ldots, v^n)$ by:

$$\tilde{g}_R^3 = 4R^4 \frac{(dv^1)^2 + \cdots + (dv^n)^2}{(R^2 - |v|^2)^2}. \qquad (4)$$

From above, we can find that in every model, a certain subset of Euclidean space is endowed with a metric, all these models are isomorphic to each other, and we may easily move from one to another base on which formulas are convenient to use. We follow majority works and use the Poincaré ball model.

The Poincaré ball model in our experiment is defined as a Riemannian manifold $\mathbb{P}^n = (\mathbb{B}^n, g^\mathbb{B})$, where $\mathbb{B}^n = \{\mathbf{x} \in \mathbb{R}^n : \|\mathbf{x}\| < 1\}$, and the metric tensor $g^\mathbb{P}(\mathbf{x}) = (\frac{2}{1-\|\mathbf{x}\|^2})^2 g^\varepsilon$, with the Euclidean metric tensor g^ε. The distance on this manifold is defined as

$$d_\mathbb{B}(\mathbf{x}, \mathbf{y}) = \operatorname{arcosh}\left(1 + 2 \frac{\|\mathbf{x} - \mathbf{y}\|^2}{(1 - \|\mathbf{x}\|)^2(1 - \|\mathbf{y}\|)^2}\right). \qquad (5)$$

Equation (4) defines the method of measuring length in hyperbolic space, while Eq. (5) is the specific application of this measurement to calculate the distance between two given points. The calculation of distance in Eq. (5) relies on the metric defined in Eq. (4), demonstrating how closely related they are: the metric defines the geometric structure of the space, and the distance is the measure of separation between two points within this structure.

To define the hyperbolic mean, we will utilize the Klein model of hyperbolic space, which is similar to the Poincaré model. Klein model is defined on $\mathbb{K}^n = \{\mathbf{x} \in \mathbb{R}^n : \|\mathbf{x}\| < 1\}$. The isomorphism between Klein model and Poincaré ball can be defined through a projection on or from the hemisphere model. One can get Klein points from Poincaré coordinates [17] by

$$\mathbb{B}^n \to \mathbb{K}^n : \text{B2K}(\mathbf{x}_\mathbb{K}) = \frac{2\mathbf{x}_\mathbb{B}}{1 + c\|\mathbf{x}_\mathbb{B}\|^2}, \qquad (6)$$

where $\mathbf{x}_\mathbb{B}$ represents a coordinate point in Poincaré ball model, and $\mathbf{x}_\mathbb{K}$ represents the corresponding coordinate point in the Klein model, c is the curvature of Poincaré ball.

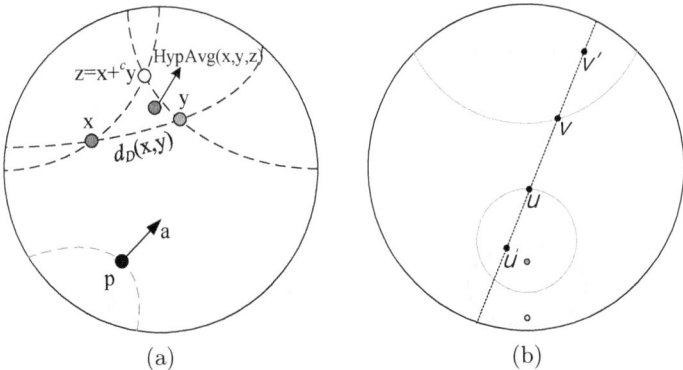

Fig. 2. In the left (a), visualization of a two-dimensional Poincaré ball, where the point **z** represents the Möbius addition of points **x** and **y**. HypAvg represents the hyperbolic average, and the gray lines represent averaging, the shortest curves connecting two points. In the right (b), the projection between the Poincaré model (red) and the Klein model (green). (i) In the Poincaré disk, geodesics are arcs of circles that are perpendicular to the boundary, whereas in the Klein model, geodesics are straight lines. (ii) Hyperbolic circles have their own hyperbolic radii. Circles in the Poincaré diskcorrespond to ellipses in the Klein model. However, the centers of both the circle and the ellipse do not lie at the Euclidean center. (Color figure online)

3.2 Hyperbolic Operations

Hyperbolic space is not a vector space in the traditional sense; standard operations like addition and multiplication cannot be used directly. To address this issue, we can use the formalism of the Möbius gyrovector space to extend many standard operations to hyperbolic space. Recently proposed hyperbolic neural networks adopt this formalism to define hyperbolic versions of feedforward networks, polynomial logistic regression, and recurrent neural networks [17]. We will discuss these networks and layers in detail and briefly summarize the various operations available in hyperbolic space in this section.

Similarly to the paper [17], we use an additional hyperparameter c which modifies the curvature of Poincaré ball; it is then defined as $\mathbb{B}_c^n = \{\mathbf{x} \in \mathbb{R}^n : c\|\mathbf{x}\|^2 < 1, c \geq 0\}$, the corresponding conformal factor now takes the form $\lambda_\mathbf{x}^c = \frac{2}{1-c\|\mathbf{x}\|^2}$. In practice, the choice of c allows one to balance between hyperbolic and Euclidean geometries, which is made precise by nothing that with $c \to 0$, all the formulas discussed below take their usual Euclidean form. The following operations are the main building blocks of hyperbolic networks.

Möbius Addition. For a pair $\mathbf{x}, \mathbf{y} \in \mathbb{B}_c^n$, the Möbius addition is defined as follows [17]:

$$\mathbf{x} \oplus^c \mathbf{y} := \frac{\alpha_{\mathbf{xy}} \mathbf{x} + \beta_{\mathbf{xy}} \mathbf{y}}{1 + 2c\mathbf{x}^\top \mathbf{y} + c^2 \|\mathbf{x}\|^2 \|\mathbf{y}\|^2}, \tag{7}$$

where $\alpha_{\mathbf{xy}} = 1 + 2c\mathbf{x}^\top \mathbf{y} + c\|\mathbf{y}\|^2$, and $\beta_{\mathbf{xy}} = 1 - c\|\mathbf{x}\|^2$.

Distance. The induced distance function is defined as follows:

$$d^c(\mathbf{x}, \mathbf{y}) := \frac{2}{\sqrt{c}} \operatorname{artanh}(\sqrt{c} \|-\mathbf{x} \oplus_c \mathbf{y}\|). \tag{8}$$

Note that with $c = 1$ one recovers the geodesic distance Eq. (5), while with $c \to 0$ we obtain the Euclidean distance $\lim_{c \to 0} d_c(\mathbf{x}, \mathbf{y}) = 2\|\mathbf{x} - \mathbf{y}\|$.

Hyp-DAN algorithm performs transformations between hyperbolic and Euclidean spaces through logarithmic mapping and exponential mapping. The logarithmic map $\log_\mathbf{x}(\mathbb{B} \to T_\mathbf{x}\mathbb{B})$ is used to map a point $\mathbf{x}$ from the Poincaré ball to the corresponding tangent vector from the tangent space, and the exponential map $\exp_\mathbf{x}(T_\mathbf{x}\mathbb{B} \to \mathbb{B})$ is used to map the tangent vector in the tangent space back to a point in the Poincaré ball. They are defined as follows:

$$\log_\mathbf{x}(\mathbf{y}) = \frac{2}{\lambda_\mathbf{x}} \operatorname{artanh}(\|-\mathbf{x} \oplus^c \mathbf{y}\|) \frac{-\mathbf{x} \oplus^c \mathbf{y}}{\|-\mathbf{x} \oplus^c \mathbf{y}\|}, \tag{9}$$

$$\exp_\mathbf{x}(v) = \mathbf{x} \oplus^c \left(\tanh\left(\frac{\lambda_\mathbf{x} \|v\|}{2}\right) \frac{v}{\|v\|} \right). \tag{10}$$

Both $\mathbf{x}$ and $\mathbf{y}$ are points in the Poincaré ball, and v is the tangent vector in the tangent space corresponding to $\mathbf{x}$. The relationship with Projection Poincaré model (red) and Klein model (green) is shown in Fig. 2.

Hyperbolic Averaging. In image processing, a common and important operation is the averaging of feature vectors. This is often used in few-shot learning with prototype networks [37]. In a Euclidean setting, the form of this operation is $\operatorname{Avg}(\mathbf{x}_1, \ldots, \mathbf{x}_N) \to \frac{1}{N} \sum_{i=1}^{N} \mathbf{x}_i$. Extension of this operation to hyperbolic spaces and takes the simple form in Klein coordinates [17], where $\lambda_i = \frac{1}{\sqrt{1 - c\|\mathbf{x}_i\|^2}}$ are the Lorentz factors:

$$\operatorname{HypAvg}(\mathbf{x}_1, \ldots, \mathbf{x}_N) = \sum_{i=1}^{N} \lambda_i \mathbf{x}_i / \sum_{i=1}^{N} \lambda_i. \tag{11}$$

Points in the Poincaré disk are mapped to the Klein model as seen in Eq. (6), and the formula for converting points from the Klein model to the Poincaré model is defined as follows [17]:

$$\mathbb{K}^n \to \mathbb{B}^n : \operatorname{K2B}(\mathbf{x}_\mathbb{B}) = \frac{\mathbf{x}_\mathbb{K}}{1 + \sqrt{1 - c\|\mathbf{x}_\mathbb{K}\|^2}}. \tag{12}$$

The meanings of $\mathbf{x}_\mathbb{K}$, $\mathbf{x}_\mathbb{B}$ and c are the same as in Eq. (6).

4 Networks

Hyp-DAN aims to emphasize more important features. If the hyperbolic distance of a sample is close to the average distance, it suggests that these samples might represent the *typical* features or patterns in the dataset. In classification tasks, emphasizing these typical features helps the model to learn more effectively the key features that differentiate various categories. At the same time, it enhances the internal consistency of the data. In hyperbolic space, samples close to the average distance may be more consistent in geometry or structure. This consistency aids the model in better understanding and learning the intrinsic structure of the data, especially when dealing with data that has hierarchical or non-Euclidean characteristics. It also can improve generalizability. By emphasizing more *universal* features, the model may generalize better to new data. This is particularly important in few-shot learning, where the model needs to learn patterns from a limited number of samples that can generalize. Finally, it can also reduce the impact of outliers. Samples that are far from the average embedding distance might be outliers or less representative samples. By assigning lower weights to these samples, their influence during the training process can be reduced, thereby increasing the model's sensitivity to typical data patterns.

The image shows hyperbolic activation, followed by an attention mechanism that performs partial order weight allocation, and then proceeds to downstream tasks. The overall scheme is as presented in Fig. 3.

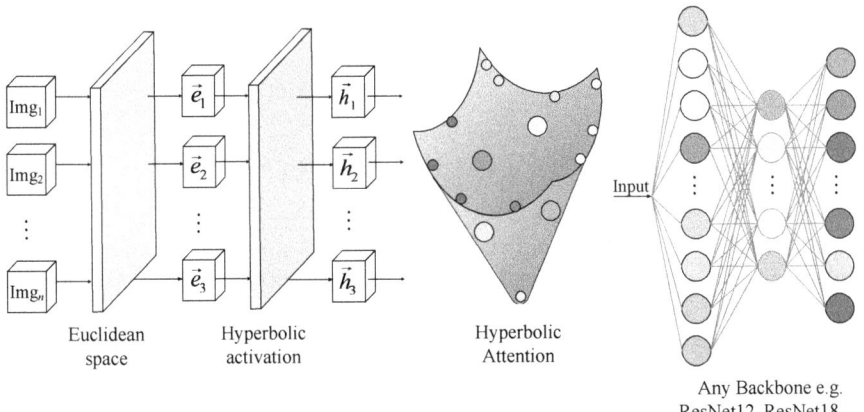

Fig. 3. Abstract structure of a distance-aware attention network in hyperbolic space. In the image, embeddings from traditional Euclidean space are activated into hyperbolic embeddings, and then weighted according to a distance-aware attention mechanism, before being applied to downstream tasks following the framework of a traditional network. The size of the filled circles implies different weights of the samples.

4.1 Hyperbolic Distance Attention

We now describe how to use a distance-aware attention mechanism to allocate individual weights.

As previously described, specific features within the images were activated in hyperbolic space through our network. The varying degrees of clarity or information entropy among these images clearly indicate that they occupy distinct levels within this space. Consequently, their contributions to the overall model differ significantly. Therefore, we need to develop a strategy to redistribute the weights.

Unlike the constant curvature of Euclidean space, the focus on design in hyperbolic space is not very apparent. Inspired by Tabaghi [42] and others, we consider using the distance from the points after hyperbolic embedding to the origin to design a kind of partial order weight distribution mechanism, and demonstrate a new interpretation in the context of image analysis.

Consider the Euclidean hidden state e, In order to utilize hyperbolic activation, we first take the transformation $e_i \to h_i$. The commonly used hyperbolic activation function is as shown in the previous Eq. (5). Then, to construct the attention for points, We utilize hyperbolic distance characteristics to measure the importance of each component. Our objective is to emphasize more important features, enhance the internal consistency of data, improve generalizability, and reduce the impact of outliers. Although training images are independently labeled, certain hidden states should play a dominant role. Furthermore, traditional image processing methods treat image labels as independent entities, but there can also be a hierarchical structure among the image labels. The weight is defined as follows:

$$\alpha_i = \frac{1}{\beta + (d^c{}_{(i,0)} - \sum_{i=1}^{N} d^c{}_{(i,0)}/N)^2}. \quad (13)$$

where α_i represents the weight of the i-th point, β is a parameter that, through training on the training set, can be adjusted to a suitable value, allowing the model to achieve better performance, $d^c{}_{(i,0)}$ is the distance from the i-th point to the origin in hyperbolic space, and N refers to the number of points in space.

5 Experiments

Experimental Setup. Our experiments are conducted using the Python programming language and model training is performed under the PyTorch framework. Under typical learning mechanisms, we conducted empirical evaluations on the few-shot recognition tasks of CUB200-2011 and MiniImageNet. It was found through experimentation that on the Poincaré disk, the overall accuracy of the model is better when $\beta = 1$. The dataset in this paper is experimented with under the condition of $\beta = 1$, the effect of β value range on accuracy is shown in Fig. 4. Our code is available at github[1].

MiniImageNet. The MiniImageNet dataset is a subset of the ImageNet dataset, consisting of 100 classes with 600 examples each. We divided the dataset according to the method in paper [17], where the training dataset consists of 64 classes, the validation dataset consists of 16 classes, and the remaining 20 classes form the test dataset. We tested our model on one-shot and five-shot classification tasks, each batch of query always contains 15 points. We used two different backbone CNN [4,37] models to test our method (represented in the table as "4 Conv") and ResNet12.

Table 1 and Table 2 displays the results obtained on the MiniImageNet dataset (along with other results from the literature). Interestingly, compared to the standard ProtoNet and common recent hyperbolic networks, our hyperbolic attention ProtoNet(termed as Hyp-DAN ProtoNet) significantly improved accuracy, especially in the one-shot setting. We observed that, in many cases, the accuracy values achieved sometimes surpassed those obtained by more advanced methods, even in the larger capacity hyperbolic architectures. This, to some extent, confirms our hypothesis that the perceptual distance attention mechanism proposed in this article is effective.

Table 1. Few-shot classification accuracy results on MiniImageNet on 1-shot 5-way and 5-shot 5-way tasks In 4 Conv Embedding Net. All accuracy results are reported with 95% confidence intervals.

Baselines	Embedding Net	1-Shot 5-way	5-shot 5-way
MatchingNet [44]	4 Conv	43.56±0.84%	55.31±0.73%
MAML [8]	4 Conv	48.70±1.84%	63.11±0.92%
RelationNet [41]	4 Conv	50.44±0.82%	65.32±0.70%
REPTILE [29]	4 Conv	49.97±0.32%	65.99±0.58%
ProtoNet [37]	4 Conv	49.92±0.78%	68.20±0.66%
Baseline [4]	4 Conv	41.08±0.70%	54.50±0.66%
Spotlearn [6]	4 Conv	51.03±0.78%	67.96±0.71%
DN4 [22]	4 Conv	51.24±0.74%	71.02±0.64%
Hyperbolic ProtoNet [17]	4 Conv	54.43±0.20%	72.67±0.15%
Hyp-DAN ProtoNet	4 Conv	**56.38 ± 0.43%**	**74.25 ± 0.52%**

Caltech-UCSD Birds. The CUB dataset contains 11,788 images of more than 200 species of birds and is commonly used for fine-grained classification. We use the segmentation strategy introduced in [46]: out of 200 classes, 100 are used for training, 50 for validation, and 50 for testing. In the 1-shot 5-way task, the learning rate is set to 0.001, with a step size of 50, and dim=1600. Due to the

[1] https://github.com/xiangfuchuan/Hyp-DAN.

Table 2. Few-shot classification accuracy results on MiniImageNet on 1-shot 5-way and 5-shot 5-way tasks In ResNets Embedding Net. All accuracy results are reported with 95% confidence intervals.

Baselines	Embedding Net	1-Shot 5-way	5-shot 5-way
SNAIL [27]	ResNet12	55.71 ± 0.99%	68.88 ± 0.92%
ProtoNet+ [37]	ResNet12	56.50 ± 0.40%	74.20 ± 0.20%
CAML [15]	ResNet12	59.23 ± 0.99%	72.35 ± 0.71%
TPN [24]	ResNet12	59.46%	75.65%
MTL [40]	ResNet12	61.20 ± 1.8%	75.50 ± 0.8%
DN4 [22]	ResNet12	54.37 ± 0.36%	74.44 ± 0.29%
TADAM [30]	ResNet12	58.50%	76.70%
Qiao-WRN [31]	ResNet28	59.60 ± 0.41%	73.74 ± 0.19%
LEO [34]	ResNet28	61.76 ± 0.08%	77.59 ± 0.12%
Dis.k-shot [2]	ResNet34	56.30 ± 0.40%	73.90 ± 0.30%
Self-Jig(SVM) [5]	ResNet50	58.80 ± 0.1.36%	76.71 ± 0.72%
Hyperbolic ProtoNet [17]	ResNet18	59.47 ± 0.20%	76.84 ± 0.14%
Hyp-DAN ProtoNet	ResNet12	**62.05 ± 0.45%**	**77.89 ± 0.44%**

dataset being relatively simple, we only considered a 4-Conv backbone, instead of modifying the training shot values like with MiniImageNet. Table 3 shows our research results. The study indicates that the attention mechanism we proposed is effective.

Table 3. Few-shot classification accuracy results on CUB on 1-shot 5-way and 5-shot 5-way tasks In ResNets Embedding Net. All accuracy results are reported with 95% confidence intervals.

Baselines	Embedding Net	1-Shot 5-way	5-shot 5-way
MatchingNet [43]	4 Conv	61.16 ± 0.89%	72.86 ± 0.70%
MAML [8]	4 Conv	55.92 ± 0.95%	72.09 ± 0.76%
ProtoNet [37]	4 Conv	51.31 ± 0.91%	70.77 ± 0.69%
MACO [12]	4 Conv	60.76%	74.96%
RelationNet [41]	4 Conv	62.45 ± 0.98%	76.11 ± 0.69%
Baseline++ [4]	4 Conv	60.53 ± 0.83%	79.34 ± 0.61%
DN4-DA [22]	4 Conv	53.15 ± 0.84%	81.90 ± 0.60%
Hyperbolic ProtoNet [17]	4 Conv	64.02 ± 0.24%	**82.53 ± 0.14%**
Hyp-DAN ProtoNet	4 Conv	**68.61 ± 0.63%**	82.13 ± 0.50%

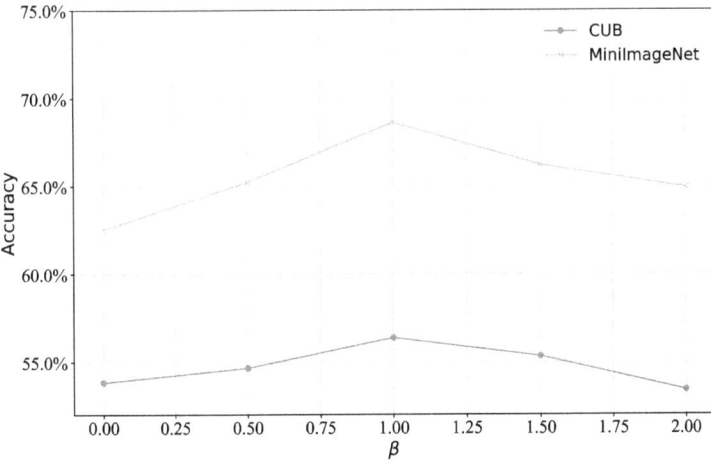

Fig. 4. Graph showing the impact of the values of parameter β on 1-shot 5-way accuracy in CUB and MiniImageNet datasets, where the horizontal axis represents the values of β, and the vertical axis represents the values of accuracy. Our experiment found that when $\beta=1$, the classification performance of the model is better for this type of dataset.

6 Discussion and Conclusion

We have studied the reassignment of weights to points on the Poincaré upper half-plane based on their distance from the origin after hyperbolic embedding. We trained and validated the model in the common Euclidean space, and at the end of the network, we used an exponential map to project from Euclidean to hyperbolic space. Therefore, the method we studied is compatible with existing backbone networks in Euclidean geometry.

Meanwhile, we have demonstrated that, in many tasks, especially in few-shot learning image classification tasks, the accuracy can be improved by using this method for data with hierarchical structures. Despite the observation in Table 2 that the LEO [34] baseline method shows a similar accuracy to the proposed method in both the 1-shot 5-way and 5-shot 5-way settings, it may be due to its use of larger convolutional layers. By contrast our method is based on Resnet12, which will also result in faster computation.

We hypothesize that after adopting hyperbolic embedding and mapping to the Poincaré disk, reassigning weights based on their distances can enhance the internal consistency of the data and emphasize typical features. If the hyperbolic distance of a sample is close to the average distance, it suggests that the sample may represent more typical features of its category.

Future work may include several potential modifications to this method. We have observed that adopting this hyperbolic attention mechanism is effective in certain tasks, particularly in few-shot learning tasks. However, it's important to note that the effect is more pronounced in 1-shot 5-way tasks, while relatively less obvious in 5-shot 5-way tasks. In the 1-shot scenarios, the scarcity of data presents a considerable challenge, however, it also underscores the potential impact of mechanisms adept at leveraging available data. The strategy of assigning weights based on distance is particularly effective in this context, as it maximizes the use of limited information, leading to noticeable improvements. Conversely, in 5-shot scenarios, the presence of additional samples per category (five in total) furnishes richer information. This abundance reduces the influence of individual sample characteristics and positions on the overall classification outcome, which may result in the benefits of distance-based weight allocation being less pronounced compared to 1-shot scenarios. These observations suggest that while our approach yields improvements in both scenarios, the underlying mechanisms of performance enhancement differ significantly between them. Moving forward, it is imperative to explore the design of a more robust attention mechanism, one that can precisely target and improve model accuracy across varying shot scenarios. This entails a deeper investigation into the nuanced dynamics of sample information utilization and its impact on classification performance, especially in more complex and less straightforward 5-shot scenarios.

Acknowledgement. This work has been supported by the National Key Research and Development Program of China under grant 2020YFB1713300, the National Natural Science Foundation of China under grants 62366008, 62221005 and 61966005.

References

1. Baroudi, F., Al Alam, J., Fajloun, Z., Millet, M.: Snail as sentinel organism for monitoring the environmental pollution; a review. Ecol. Indic. **113**, 106240 (2020)
2. Bauer, M., Rojas-Carulla, M., Świątkowski, J.B., Schölkopf, B., Turner, R.E.: Discriminative k-shot learning using probabilistic models. Cornell University - arXiv (2018)
3. Chen, B., Huang, X., Xiao, L., Cai, Z., Jing, L.: Hyperbolic interaction model for hierarchical multi-label classification. In: Proceedings of the AAAI Conference on Artificial Intelligence, vol. 34, pp. 7496–7503 (2020)
4. Chen, W.Y., Liu, Y.C., Kira, Z., Wang, Y.C.F., Huang, J.B.: A closer look at few-shot classification. In: International Conference on Learning Representations (2018)
5. Chen, Z., Yanwei, F., Chen, K., Jiang, Y.-G.: Image block augmentation for one-shot learning. In: Proceedings of the AAAI Conference on Artificial Intelligence, vol. 33, pp. 3379–3386 (2019)
6. Chu, W.H., Li, Y.J., Chang, J.C., Wang, Y.C.F.: Spot and learn: a maximum-entropy patch sampler for few-shot image classification. In: Proceedings of the IEEE/CVF Conference on Computer Vision and Pattern Recognition, pp. 6251–6260 (2019)
7. Fernando, J.F., Gamboa, J.M., Ueno, C.: Unbounded convex polyhedra as polynomial images of Euclidean spaces. Ann. Sci. Norm. Super. Pisa Cl. Sci. **19**(5), 509–565 (2019)
8. Finn, C., Abbeel, P., Levine, S.: Model-agnostic meta-learning for fast adaptation of deep networks. In: International Conference on Machine Learning, pp. 1126–1135. PMLR (2017)
9. Gulcehre, C., et al.: Hyperbolic attention networks. In: International Conference on Learning Representations (2019)

10. Guo, W., Wang, W., Zhao, S., Niu, Y., Zhang, Z., Liu, X.: Density peak clustering with connectivity estimation. Knowl.-Based Syst. **243**, 108501 (2022)
11. Gupta, J., Pathak, S., Kumar, G.: Deep learning (CNN) and transfer learning: a review. In: Journal of Physics: Conference Series, vol. 2273, p. 012029. IOP Publishing (2022)
12. Hilliard, N., Phillips, L., Howland, S., Yankov, A., Corley, C.D., Hodas, N.O.: Few-shot learning with metric-agnostic conditional embeddings. arXiv preprint arXiv:1802.04376 (2018)
13. Hong, J., Hayder, Z., Han, J., Fang, P., Harandi, M., Petersson, L.: Hyperbolic audio-visual zero-shot learning. In: Proceedings of the IEEE/CVF International Conference on Computer Vision, pp. 7873–7883 (2023)
14. Jeong, T., Kim, H.: OOD-MAML: meta-learning for few-shot out-of-distribution detection and classification. Adv. Neural. Inf. Process. Syst. **33**, 3907–3916 (2020)
15. Jiang, X., Havaei, M., Varno, F., Chartrand, G., Chapados, N., Matwin, S. : Learning to learn with conditional class dependencies. In: International Conference on Learning Representations (2018)
16. Khan, S., Naseer, M., Hayat, M., Zamir, S.W., Khan, F.S., Shah, M.: Transformers in vision: a survey. ACM Comput. Surv. (CSUR) **54**(10s), 1–41 (2022)
17. Khrulkov, V., Mirvakhabova, L., Ustinova, E., Oseledets, I., Lempitsky, V.: Hyperbolic image embeddings. In: Proceedings of the IEEE/CVF Conference on Computer Vision and Pattern Recognition, pp. 6418–6428 (2020)
18. Krizhevsky, A., Sutskever, I., Hinton, G.E.: ImageNet classification with deep convolutional neural networks. Commun. ACM **60**(6), 84–90 (2017)
19. Lateko, A.A.H., et al.: Stacking ensemble method with the RNN meta-learner for short-term PV power forecasting. Energies **14**(16), 4733 (2021)
20. LeCun, Y., Bottou, L., Bengio, Y., Haffner, P.: Gradient-based learning applied to document recognition. Proc. IEEE **86**(11), 2278–2324 (1998)
21. Lee, J.M.: Introduction to Riemannian Manifolds, vol. 2. Springer, Cham (2018). https://doi.org/10.1007/978-3-319-91755-9
22. Li, W., Wang, L., Xu, J., Huo, J., Gao, Y., Luo, J.: Revisiting local descriptor based image-to-class measure for few-shot learning. In: Proceedings of the IEEE/CVF Conference on Computer Vision and Pattern Recognition, pp. 7260–7268 (2019)
23. Liu, Q., Nickel, M., Kiela, D.: Hyperbolic graph neural networks. Adv. Neural Inf. Process. Syst. **32** (2019)
24. Liu, Y., et al.: Learning to propagate labels: transductive propagation network for few-shot learning. In: International Conference on Learning Representations (2018)
25. Yonggang, L., Liu, J., Zhu, L., Zhang, B., He, J.: 3D reconstruction from cryo-EM projection images using two spherical embeddings. Commun. Biol. **5**(1), 304 (2022)
26. Lyu, H., Sha, N., Qin, S., Yan, M., Xie, Y., Wang, R.: Advances in neural information processing systems. Adv. Neural Inf. Process. Syst. **32** (2019)
27. Mishra, N., Rohaninejad, M., Chen, X., Abbeel, P.: A simple neural attentive meta-learner. In: International Conference on Learning Representations (2018)
28. Moreira, G., Marques, M., Costeira, J.P., Hauptmann, A.: Hyperbolic vs Euclidean embeddings in few-shot learning: two sides of the same coin. In: Proceedings of the IEEE/CVF Winter Conference on Applications of Computer Vision, pp. 2082–2090 (2024)
29. Nichol, A., Achiam, J., Schulman, J.: On first-order meta-learning algorithms. arXiv preprint arXiv:1803.02999 (2018)
30. Oreshkin, B., Rodríguez López, P., Lacoste, A.: TADAM: task dependent adaptive metric for improved few-shot learning. Adv. Neural Inf. Process. Syst. **31** (2018)

31. Qiao, S., Liu, C., Shen, W., Yuille, A.L.: Few-shot image recognition by predicting parameters from activations. In: Proceedings of the IEEE Conference on Computer Vision and Pattern Recognition, pp. 7229–7238 (2018)
32. Redmon, J., Divvala, S., Girshick, R., Farhadi, A.: Proceedings of the IEEE conference on computer vision and pattern recognition. In: Proceedings of the IEEE Conference on Computer Vision and Pattern Recognition, pp. 779–788 (2016)
33. Robbin, J.W., Salamon, D.A.: Introduction to Differential Geometry. SSM, Springer, Heidelberg (2022). https://doi.org/10.1007/978-3-662-64340-2
34. Rusu, A.A., et al.: Meta-learning with latent embedding optimization. In: International Conference on Learning Representations (2018)
35. Sabirov, D.S., Shepelevich, I.S.: Information entropy in chemistry: an overview. Entropy **23**(10), 1240 (2021)
36. Sala, F., De Sa, C., Gu, A., Ré, C.: Representation tradeoffs for hyperbolic embeddings. In: International Conference on Machine Learning, pp. 4460–4469. PMLR (2018)
37. Snell, J., Swersky, K., Zemel, R.: Prototypical networks for few-shot learning. Adv. Neural Inf. Process. Syst. **30** (2017)
38. Sohn, K.: Improved deep metric learning with multi-class n-pair loss objective. Adv. Neural Inf. Process. Syst. **29** (2016)
39. Su, Y., Kuo, C.-C.J., et al.: Recurrent neural networks and their memory behavior: a survey. APSIPA Trans. Signal Inf. Process. **11**(1) (2022)
40. Sun, Q., Liu, Y., Chua, T.S., Schiele, B.: Meta-transfer learning for few-shot learning. In: Proceedings of the IEEE/CVF Conference on Computer Vision and Pattern Recognition, pp. 403–412 (2019)
41. Sung, F., Yang, Y., Zhang, L., Xiang, T., Torr, P.H., Hospedales, T.M.: Learning to compare: relation network for few-shot learning. In: Proceedings of the IEEE Conference on Computer Vision and Pattern Recognition, pp. 1199–1208 (2018)
42. Tabaghi, P., Dokmanić, I.: Hyperbolic distance matrices. In: Proceedings of the 26th ACM SIGKDD International Conference on Knowledge Discovery & Data Mining, pp. 1728–1738 (2020)
43. Ustinova, E., Lempitsky, V.: Learning deep embeddings with histogram loss. Adv. Neural Inf. Process. Syst. **29** (2016)
44. Vinyals, O., Blundell, C., Lillicrap, T., Wierstra, D., et al.: Matching networks for one shot learning. Adv. Neural Inf. Process. Syst. **29** (2016)
45. Vuorio, R., Sun, S.H., Hu, H., Lim, J.J.: Multimodal model-agnostic meta-learning via task-aware modulation. Adv. Neural Inf. Process. Syst. **32** (2019)
46. Wang, Z., Zhao, Y., Li, J., Tian, Y.: Cooperative bi-path metric for few-shot learning. In: Proceedings of the 28th ACM International Conference on Multimedia, pp. 1524–1532 (2020)
47. Canran, X., Ming, W.: Learning feature interactions with Lorentzian factorization machine. In: Proceedings of the AAAI Conference on Artificial Intelligence, vol. 34, pp. 6470–6477 (2020)
48. Ji, X., Wang, G., Deng, W.: DenPEHC: density peak based efficient hierarchical clustering. Inf. Sci. **373**, 200–218 (2016)
49. Zhang, Y., Wang, X., Shi, C., Jiang, X., Ye, Y.: Hyperbolic graph attention network. IEEE Trans. Big Data **8**(6), 1690–1701 (2021)
50. Zhou, M., Yang, M., Xiong, B., Xiong, H., King, I.: Hyperbolic graph neural networks: a tutorial on methods and applications. In: Proceedings of the 29th ACM SIGKDD Conference on Knowledge Discovery and Data Mining, pp. 5843–5844 (2023)

Optimizing Rough Set Flow Graph Inference

Jun Wang[1](✉) and Cory J. Butz[2]

[1] Xi'an Eurasia University, Xian, China
wangjun@eurasia.edu
[2] University of Regina, Regina, Canada
cory.butz@uregina.ca

Abstract. In this paper, we optimize Rough Set Flow Graph (RSFG) inference through several steps. Initially, we introduce the concept of barren variables and exploit them by safely removing them without necessitating any numeric computation in computer memory. Subsequently, we propose pruning independent-by-evidence variables and leveraging them in a similar manner as barren variables. Lastly, we eliminate all remaining variables by suggesting heuristics to determine effective elimination orderings for RSFG inference. Broadly, our heuristics are classified into domain heuristics and edge heuristics. Domain heuristics determine elimination orderings based on the domain cardinalities of the variables to be eliminated, while the graphical structure of RSFG with respect to edges serves as the primary factor for computing elimination orderings with edge heuristics. Our experimental analysis indicates that edge heuristics tend to produce favorable elimination orderings, whereas domain heuristics exhibit relatively less effectiveness.

Keywords: Rough Set Flow Graph(RSFG) · VE algorithm · BN inference

1 Introduction

Rough set flow graphs (RSFGs), founded by Pawlak [7,8], are proposed as a graphical framework for managing uncertainty. A RSFG extends traditional rough set theory [6,10] by organizing the rules obtained from decision tables as a DAG. Each edge is labeled with three coefficients, strength, certainty, and coverage, which have been shown to satisfy Bayes theorem [7,8]. Therefore, RSFGs provide a new perspective on Bayesian inference methodology [8].

RSFG inference [1] seeks to compute a binary RSFG on $\{A_i, A_j\}$, namely, a DAG on $\{A_i, A_j\}$ and the coefficient table, denoted $Ans(A_i, a_j)$, which involves strength, certainty, and coverage. The term query refers to any computation involving strength, certainty, or coverage.

This paper is dedicated to refining RSFG inference through a series of strategic steps. To begin, we introduce the notion of barren variables, eliminating them efficiently without requiring any numeric computations in computer memory. Following this, we propose the pruning of independent-by-evidence variables, employing a similar approach as with barren variables. Subsequently, we introduce heuristics aimed at determining optimal elimination orderings for RSFG inference. These heuristics are broadly categorized into two types: domain heuristics, which are based on the domain cardinalities of the variables to be eliminated, and edge heuristics, which rely on the graphical structure of RSFG with respect to its edges. Further, edge heuristics can be divided into incoming and outgoing categories. Our experimental results highlight the efficacy of edge heuristics in generating favorable elimination orderings, while domain heuristics demonstrate relatively lower effectiveness.

The remainder of this paper is organized as follows. Section 2 gives background information and reviews BN inference, RSFG inference. In Sect. 3, we optimize RSFG inference by pruning variables and propose elimination ordering heuristics based on domain heuristics and edge heuristics. In Sect. 4, we report our experiments and analysis, and discuss advantages and disadvantages. Section 5 draws conclusions and suggests future work.

2 Background

In this section, we briefly review the Baysian Network(BN) inference, RSFGs and RSFG inferences.

2.1 Bayesian Networks

A Bayesian Network (BN) [11] is a DAG on U together with CPTs $\{p(v_i|Pa_{(v_i)})| v_i \in U\}$, where $Pa_{(v_i)}$ denotes the parent set of variable v_i in the DAG.

The phrase probabilistic inference [3,12], also known as query processing, means the calculation of $p(X)$ or $p(X|Y)$, where $X, Y \in U$ and $X \cap Y = \phi$.

Variable elimination [5,14] (VE) is a well-known algorithm for answering probabilistic queries with respect to a Bayesian network [2]. VE answers a query $P(X|Y = y)$ in a BN $\mathcal{B}$ as follows: (i) remove all barren variables; (ii) remove all independent-by-evidence variables, giving $\mathcal{B}^s$; (iii) build a uniform distribution $1(v)$ for any root of $\mathcal{B}^s$ that is not a root of $\mathcal{B}$; (iv) set Y to Y=y in the CPTs of $\mathcal{B}^s$; (v) determine an elimination ordering σ form the moral graph $\mathcal{B}^s_m$; (vi) follow σ, eliminate variable v by multiplying all potentials involving v, and then summing v out of the product; (vii) multiply together all remaining potentials and normalize to obtain $P(X|Y = y)$.

A variable is leaf if it has no children. A variable is barren [14] with respect to query $P(X|Y = y)$, if it is a leaf and it is not in $X \cup Y$. Given a query $p(X|Y)$ posed to a BN, a variable Z is independent-by-evidence [14], if $I(Z, Y, X)$ holds in the BN.

Example 1. [14] Given query $p(e|b = 0)$ and BN $\mathcal{B}$ in Fig. 1, variables g and f are barren variables. After removing g, the updated BN $\mathcal{B}$ is (ii). Variable f is now

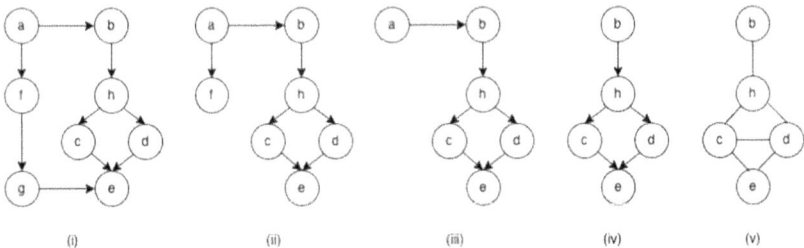

Fig. 1. Given query $p(e|b = 0)$ posted to BN $\mathcal{B}$ in (i); removing barren variable g in (ii); removing barren variable f in (iii); removing independent-by-evidence variable a in (iv), giving $\mathcal{B}^s$;

barren. Removing barren variable f gives (iii). Now variable a is independent-by-evidence because $I(a, b, e)$ holds in the BN, removing a in (iv). VE builds 1(b) and updates $p(h|b)$ as $p(h|b = 0)$. Then determine elimination ordering $\sigma = (c, d, h)$ from the moral graph $\mathcal{B}^s_m$ in (v).

The query p(e—b=0) can be computed as follows:

$$p(c, e|d, h) = p(c|h) \cdot p(e|c, d)$$
$$p(e|d, h) = \sum_c p(c, e|d, h)$$
$$p(e|h) = \sum_d p(d|h) \cdot p(e|d, h)$$
$$p(e|b = 0) = \sum_h p(h|b = 0) \cdot p(e|h)$$

2.2 Rough Set Flow Graphs

Rough Set Flow Graph(RSFGs) are built from decision tables. A decision table is a potential $\phi(C, D)$, where C is a set of conditioning attributes and D is a decision attribute. If a decision table is normalized, then we denote as $\phi(C, D)$.

Each decision table defines a binary RSFG. The set of nodes in the flow graph are $\{c_1, c_2, \ldots, c_k\} \cup \{d_1, d_2, \ldots, d_l\}$, where $\{c_1, c_2, \ldots, c_k\}$ and $\{d_1, d_2, \ldots, d_l\}$ are the values of C and D appearing in the decision table, respectively. For each row in the decision table, there is a directed edge (c_i, d_j) in the flow graph, where c_i is the value of C and d_j is the value of D. Clearly, the defined graphical structure is a directed acyclic graph (DAG). Each edge (c_i, d_j) is labelled with three coefficients, namely, strength, certainty, and coverage [7,8], which are introduced in decision theory [9]. The strength of (c_i, d_j) is $\phi(c_i, d_j)$ obtained from the decision table. From $\phi(c_i, d_j)$, we can compute the certainty $\phi(d_j|c_i)$ and the coverage $\phi(c_i|d_j)$.

The task of RSFG inference is to compute a binary RSFG on (A_i, A_j), namely, a DAG on $\{A_i, A_j\}$ and the coefficient table, denoted $Ans(A_i, A_j)$, which is a table with strength, certainty, and coverage columns. We use the term query to refer to any request involving strength, certainty or coverage.

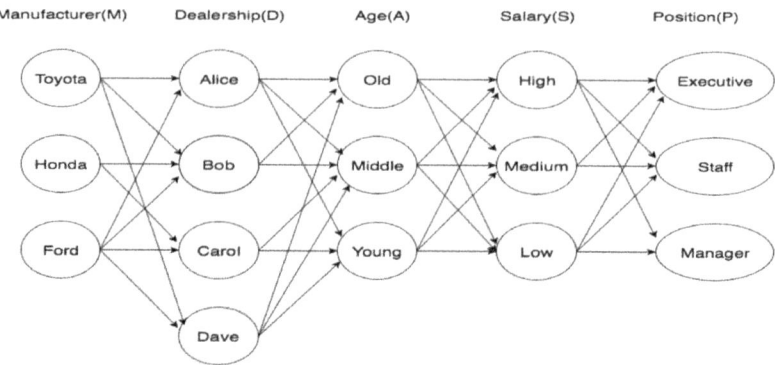

Fig. 2. A RSFG for $\{M, D, A, S, P\}$, where five attributes are Manufacturer (M), Dealership (D), Age (A), Salary (S), Position (P).

2.3 Marginalize All-at-once

Pawlak proposed algorithm [7,8] to answer queries in a RSFG. The main idea is to exploit the factorization to eliminate variables all-at-once.

Example 2. Given a query on $\{M, P\}$ posed to the RSFG in Fig. 2. Let us focus on M = "Ford" and P = "Staff", which we succinctly write as "Ford" and "Staff", respectively. The certainty $\phi(\text{"Staff"}-\text{"Ford"})$ is computed as:

$$\phi(\text{"Staff"}|\text{"Ford"}) = \sum_{D,A,S} \phi(\text{ D}|\text{"Ford"}) \cdot \phi(A|D) \cdot \phi(S|A) \cdot \phi(\text{"Staff"}|S).$$

The coverage $\phi(\text{"Ford"}-\text{"Staff"})$ is computed as:

$$\phi(\text{"Ford"}|\text{"Staff"}) = \sum_{D,A,S} \phi(\text{"Ford"}|D) \cdot \phi(D|A) \cdot \phi(A|S) \cdot \phi(S|\text{"Staff"}).$$

The strength $\phi(\text{"Ford"}, \text{"Staff"})$ is computed as:

$$\phi(\text{"Ford"}, \text{"Staff"}) = \phi(\text{"Ford"}) \cdot \phi(\text{"Staff"}|\text{"Ford"}).$$

In Example 2, computing coefficients $\phi(\text{"Ford"},\text{"Staff"})$, $\phi(\text{"Staff"}-\text{"Ford"})$ and $\phi(\text{"Ford"}-\text{"Staff"})$ in Ans(M, P) required 181 multiplications and 58 additions.

2.4 Marginalize One-by-one

The main idea is to exploit the factorization to eliminate variables one-by-one [1], instead of all-at-once.

Example 3. Recall Example 2. Again, we focus on the edge("Ford","Staff"). Variables $\{D, A, S\}$ need be eliminated. The certainty ϕ("Staff"—"Ford") is compute as

$$\phi(\text{"Staff"}|\text{"Ford"}) = \sum_D \sum_A \sum_S \phi(D|\text{"Ford"}) \cdot \phi(A|D) \cdot \phi(S|A) \cdot \phi(\text{"Staff"}|S),$$

while the coverage ϕ("Ford"—"Staff") is

$$\phi(\text{"Ford"}|\text{"Staff"}) = \sum_D \sum_A \sum_S \phi(\text{"Ford"}|D) \cdot \phi(D|A) \cdot \phi(A|S) \cdot \phi(S|\text{"Staff"}).$$

The strength ϕ("Ford","Staff") is computed as:

$$\phi(\text{"Ford"},\text{"Staff"}) = \phi(\text{"Ford"}) \cdot \phi(\text{"Staff"}|\text{"Ford"}).$$

In Example 3, computing coefficients ϕ("Ford","Staff"), ϕ("Staff"—"Ford") and ϕ("Ford"—"Staff") in Ans(M, P) required 45 multiplications and 30 additions.

3 Optimizing RSFG Inference

In this section, we follow the steps in answering Bayesian Network inference.

3.1 Prune Barren Variables

In a BN, a variable is barren if it is a leaf and not in query. Similarly, in RSFG, when answering a query, if the variable does not have a decision attribute, then the variable can be treated as leaf.

Definition 1. *A variable is barren in RSFG with refer to query $P(X|Y)$, if it is a leaf and it is not in $X \cup Y$.*

Example 4. Given the query $p(B, E)$ on $\{B, E\}$ posted to RSFG in Fig. 3, where attributes will be $\{A, B, C, D, E, F\}$. Variable F is a barren variable because it is a leaf and is not mentioned in the query $p(B, E)$. After removing F, the updated RSFG is in Fig. 4.

3.2 Prune Independent-by-evidence Variable

A variable is independent-by-evidence variable in BN, if a certain independence holds in a BN. Similarly, we try to find a certain independence in a RSFG. RSFGs make implicit independency assumptions regarding the problem domain [1]. Two tables $\phi_1(A_i, A_j)$ and $\phi_2(A_j, A_k)$ are pairwise consistent [4,13], if

$$\phi_1(A_j) = \phi_2(A_j).$$

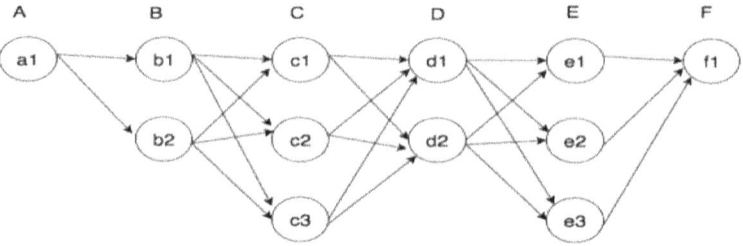

Fig. 3. A RSFG for $\{A, B, C, D, E, F\}$

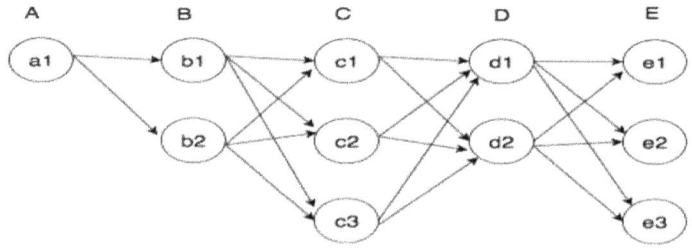

Fig. 4. Removing barren variable F form RSFG in Fig. 3

Consider $m-1$ potentials $\phi_1(A_1, A_2), \phi_2(A_2, A_3), \ldots, \phi_{m-1}(A_{m-1}, A_m)$, such that each consecutive pair is pairwise consistent, namely,

$$\phi_i(A_{i+1}) = \phi_{i+1}(A_{i+1}),$$

for $i = 1, 2, \ldots, m-2$. Dawid and Lauritzen [4] have shown that if a given set of potentials satisfies Equation $\phi_1(A_j) = \phi_2(A_j)$ and are defined over an acyclic hypergraph, then the potentials are marginals of a unique potential $\phi(A_1, A_2, \ldots, A_m)$, defined as:

$$\phi(A_1, A_2, \ldots, A_m) = \frac{\phi_1(A_1, A_2) \cdot \phi_2(A_2, A_3) \cdot \ldots \cdot \phi_{m-1}(A_{m-1}, A_m)}{\phi_1(A_2) \cdot \ldots \cdot \phi_{m-2}(A_{m-1})}.$$

We call these potentials collective potential [4]. The collective potential $\phi(A_1, A_2, \ldots, A_m)$ represents the problem domain from a rough set perspective. Normalizing [7,8] $\phi(A_1, A_2, \ldots, A_m)$ yields a jpd $p(A_1, A_2, \ldots, A_m)$ by multiplying $1/N$, where N denotes the number of all cases.

$$p(A_1, A_2, \ldots, A_m) = \frac{1}{N} \cdot \phi(A_1, A_2, \ldots, A_m)$$

$$= \frac{1}{N} \cdot \frac{\phi_1(A_1, A_2) \cdot \phi_2(A_2, A_3) \cdot \ldots \cdot \phi_{m-1}(A_{m-1}, A_m)}{\phi_1(A_2) \cdot \ldots \cdot \phi_{m-2}(A_{m-1})}$$

The following theorem shows that RSFGs make implicit independency assumptions regarding the problem domain.

Theorem 1. [1] *Consider a RSFG defined by $m-1$ decision tables $\phi_1(A_1, A_2) \cdot \phi_2(A_2, A_3) \cdot \ldots \cdot \phi_{m-1}(A_{m-1}, A_m)$. Then $m-2$ probabilistic independencies $I(A_1, A_2, A_3, \ldots, A_m), I(A_1, A_2, A_3, A_4, \ldots, A_m), \ldots, I(A_1, \ldots, A_{m-2}, A_{m-1}, A_m)$ are satisfied by the jpd $p(A_1, A_2, \ldots, A_m)$, where $p(A_1, A_2, \ldots, A_m)$ is the normalization of collective potential $\phi(A_1, A_2, \ldots, A_m)$ representing the problem domain.*

Then the definition of independent-by-evidence variable in RSFG will be as follows:

Definition 2. *Given a query on $\{X|Y\}$, a variable Z is independent-by-evidence variable, if $I(Z, Y, X)$ holds in a RSFG.*

Example 5. Recall Example 4. Given the query $p(B|E)$ on $\{B, E\}$ posted to RSFG in Fig. 3. Removing barren variable F in Fig. 4. Variable A is independent-by-evidence variable, and can be removed in Fig. 5.

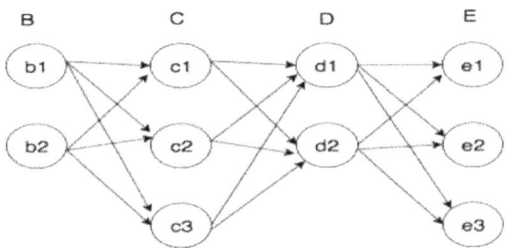

Fig. 5. Removing independent-by-evidence variable A form RSFG in Fig. 4

3.3 Elimination Orderings Heuristics

In a broad categorization, our heuristics fall into two main types: domain heuristics and edge heuristics. Domain heuristics dictate the order of elimination based on the cardinalities within the variable domains. Conversely, edge heuristics rely on the topological structure of RSFG to determine elimination orderings, emphasizing the significance of edges in this computational process.

3.3.1 Domain Heuristics
The main idea for domain heuristic is to eliminate variable one-by-one but follow the order given by heuristic.

Definition 3. *The domain for variable D is*

$$dom(D) = \{d_1, d_2, \ldots, d_n\}. \quad (1)$$

The value of domain will be cardinality $|dom(D)|$, i.e., n.

Example 6. The DAGs of the binary RSFGs are illustrated in Fig. 6. For each variable, we count its value of domain. In Fig. 6, domain of variable M will be

$$dom(M) = \{\text{``Toyota''}, \text{``Honda''}, \text{``Ford''}\},$$

Thus, the value of $dom(M)$ will be $|dom(M)| = 3$. Similarly, $|dom(D)| = 2$, $|dom(A)| = 4$, $|dom(S)| = 3$ and $|dom(P)| = 3$.

Given a query on $\{M, P\}$. Let us focus on M = "Ford" and P = "Staff" in the DAG in Fig. 6, variables $\{D, A, S\}$ need be eliminated. Consider the value of domain for each variable, we given increasing order D, S, A and decreasing order A, S, D to answer the query.

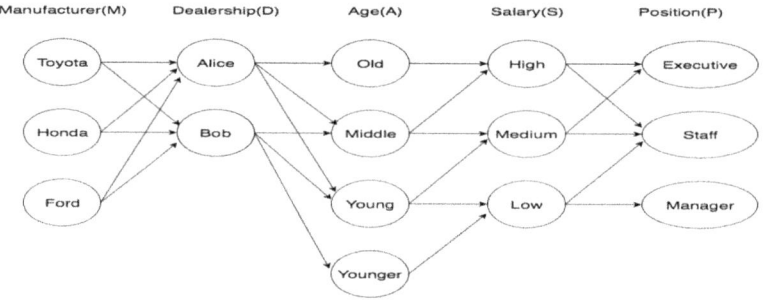

Fig. 6. Domains heuristics

Recall Example 6, we focus on same query $\{M, P\}$. The value of domain will be $|dom(D)| = 2$, $|dom(S)| = 3$, $|dom(A)| = 4$.

For the increasing order D, S, A based on number $\{2, 3, 4\}$. Computing coefficients $\phi(\text{``Ford''}, \text{``Staff''})$, $\phi(\text{``Staff''} - \text{``Ford''})$ and $\phi(\text{``Ford''} - \text{``Staff''})$ in Ans(M, P) required 37 multiplications and 16 additions.

For the decreasing order A, S, D based on $\{4, 3, 2\}$. Computing coefficients $\phi(\text{``Ford''}, \text{``Staff''})$, $\phi(\text{``Staff''} - \text{``Ford''})$ and $\phi(\text{``Ford''} - \text{``Staff''})$ in Ans(M, P) required 33 multiplications and 14 additions.

The distinction between these two orders is minimal.

3.3.2 Edge Heuristics The main idea for domain heuristic is to eliminate variable one-by-one but follow the order given by heuristic. For each variable, the edge can be count as incoming edges and outgoing edges.

First, we define the incoming edge.

Definition 4. *Incoming edges for variable D is*

$$inc(D) = \{(X, D) | X \text{ and } D \text{ are variables in RSFG}\}. \quad (2)$$

The value of incoming edge will be cardinality $|inc(D)|$.

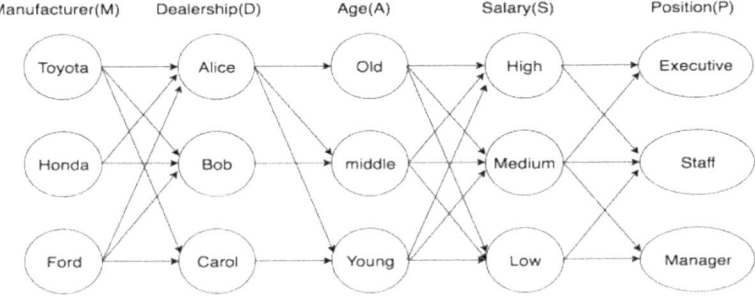

Fig. 7. Incoming edge heuristics

Example 7. The DAGs of the binary RSFGs are illustrated in Fig. 7. For each variable, we count the number of incoming edges. In Fig. 7, incoming edge of variable M will be $inc(M) = \{\}$ (empty set), Thus, the value of $inc(M)$ will be $|inc(M)| = 0$. Incoming edge of variable D will be

$$inc(D) = \{(\text{``Toyota''}, \text{``Alice''}), (\text{``Toyota''}, \text{``Bob''}), (\text{``Toyota''}, \text{``Carol''}),$$
$$(\text{``Honda''}, \text{``Alice''}), (\text{``Honda''}, \text{``Bob''}), (\text{``Honda''}, \text{``Carol''}),$$
$$(\text{``Ford''}, \text{``Bob''}), (\text{``Ford''}, \text{``Carol''})\}.$$

Thus, the value of $inc(D)$ will be $|inc(D)| = 8$. Similarly, $|inc(A)| = 5$, $|inc(S)| = 9$ and $|inc(P)| = 7$.

Given a query on $\{M, P\}$. Let us focus on M = "Ford" and P = "Staff" in the DAG in Fig. 7, variables $\{D, A, S\}$ need be eliminated. Consider the value of incoming edge for each variable, by increasing order we have A, D, S and by decreasing order S, D, A.

Recall Example 7, we focus on same query $\{M, P\}$. The value of incoming edge will be $|inc(S)| = 9$, $|inc(D)| = 8$, $|inc(A)| = 5$.

For the increasing order A, D, S based on number $\{5, 8, 9\}$. Computing coefficients $\phi(\text{``Ford''}, \text{``Staff''})$, $\phi(\text{``Staff''}-\text{``Ford''})$ and $\phi(\text{``Ford''}-\text{``Staff''})$ in $Ans(M, P)$ required 27 multiplications and 14 additions.

For the decreasing order S, D, A based on number $\{9, 8, 5\}$. Computing coefficients $\phi(\text{``Ford''}, \text{``Staff''})$, $\phi(\text{``Staff''}-\text{``Ford''})$ and $\phi(\text{``Ford''}-\text{``Staff''})$ in $Ans(M, P)$ required 17 multiplications and 10 additions.

This result demonstrates that utilizing a decreasing order of incoming edges entails less effort.

Second, we define the outgoing edge.

Definition 5. *Outgoing edges for variable D is*

$$outg(D) = \{(D, X) | D \text{ and } X \text{ are variables in } RSFG\}. \quad (3)$$

The value of outgoing edge will be cardinality $|outg(D)|$.

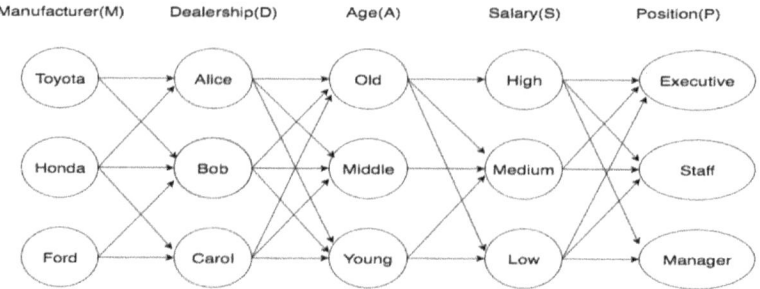

Fig. 8. Outgoing edge heuristics

Example 8. The DAGs of the binary RSFGs are illustrated in Fig. 8. For each variable, we count its value of outgoing edge. In Fig. 8, outgoing edge of variable M will be

$$outg(D) = \{(\text{``}Toyota\text{''}, \text{``}Alice\text{''}), (\text{``}Toyota\text{''}, \text{``}Bob\text{''}), (\text{``}Honda\text{''}, \text{``}Alice\text{''}),$$
$$(\text{``}Honda\text{''}, \text{``}Bob\text{''}), (\text{``}Honda\text{''}, \text{``}Carol\text{''}), (\text{``}Ford\text{''}, \text{``}Bob\text{''}), (\text{``}Ford\text{''}, \text{``}Carol\text{''})\},$$

Thus, the value of $outg(M)$ will be $|outg(M)| = 7$. Similarly, $|outg(D)| = 9$, $|outg(A)| = 6$, $|outg(S)| = 8$ and $|outg(P)| = 0$.

Given a query on $\{M, P\}$. Let us focus on M = "Ford" and P = "Staff" in the DAG in Fig. 8, variables $\{D, A, S\}$ need be eliminated. Consider to the domains of each attribute, we try both increasing order A, S, D and decreasing order D, S, A.

Recall Example 8, we focus on same query $\{M, P\}$. The value of outgoing edge will be $|outg(A)| = 6$, $|outg(S)| = 8$, $|outg(D)| = 9$.

For the increasing order A, S, D based on $\{6, 8, 9\}$. Computing coefficients $\phi(\text{``Ford ''}, \text{``Staff''})$, $\phi(\text{``Staff''} - \text{``Ford''})$ and $\phi(\text{``Ford''} - \text{``Staff''})$ in Ans(M, P) required 53 multiplications and 26 additions.

For the decreasing order D, S, A based on $\{9, 8, 6\}$. Computing coefficients $\phi(\text{``Ford''}, \text{``Staff''})$, $\phi(\text{``Staff''} - \text{``Ford''})$ and $\phi(\text{``Ford''} - \text{``Staff''})$ in Ans(M, P) required 29 multiplications and 14 additions.

This finding suggests that utilizing a decreasing order of outgoing edges results in less effort.

4   Experiments and Analysis

Recently, [15] discovered that conducting Rough Set tasks for various big data analyses within very large information systems poses challenges in processing time and memory utilization. Traditional Rough Set algorithms are no longer adequate for handling big data volumes. To address this issue, we conducted experiments to compares the effects of marginalizing one-by-one via random order, domain heuristic, and edge heuristic.

Example 9. Given a query on $\{A, E\}$. Let us focus on A = "a_2" and F = "e_1" in the DAG in Fig. 9. Variables $\{B, C, D\}$ need be eliminated. The work for all heuristics have been listed in Fig. 10.

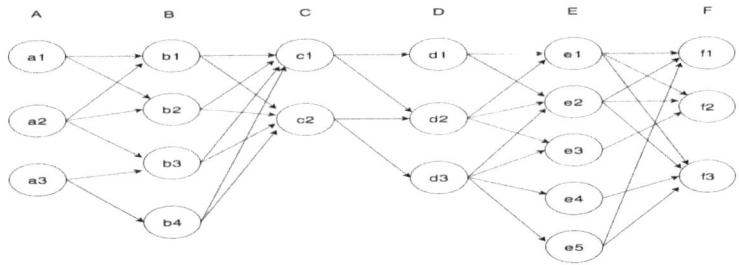

Fig. 9. Answering query Ans(A, E) in first experiment

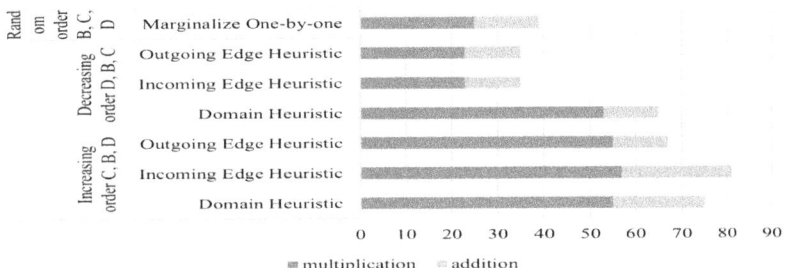

Fig. 10. Result for the experiment

Example 10. Given a query on $\{B, G\}$. Let us focus on B = "b_2" and G = "g_1" in the DAG in Fig. 11. Variables $\{C, D, E, F\}$ need be eliminated. The work for all heuristics have been listed in Fig. 12.

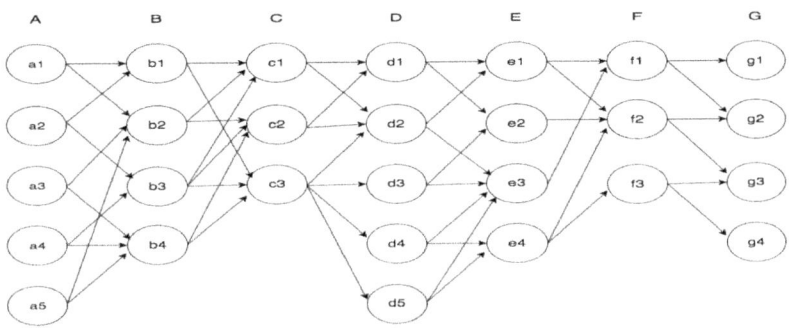

Fig. 11. Answering query Ans(B, G) in second experiment

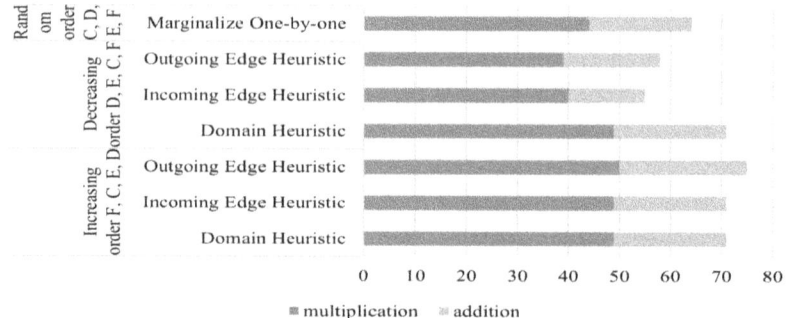

Fig. 12. Result for the experiment

Example 11. Given a query on $\{B, F\}$ in the DAG in Fig. 13. Variables $\{C, D, E\}$ need be eliminated. The work for all heuristics have been listed in Fig. 14.

When considering a query on $\{X, Y\}$ posed to an RSFG, employing a decreasing order of elimination proves to be more efficient. In this approach, variables are eliminated sequentially, starting from those associated with larger decision tables and progressing to smaller ones. This strategy minimizes computational complexity compared to an increasing order heuristic, which necessitates multiplying larger tables first. The significance of incoming edges becomes apparent in computation, with their number directly influencing query resolution. While domain attributes show minimal impact on computation, the consideration of incoming edges proves crucial. By prioritizing elimination based on decreasing order, tables encompassing more edges are addressed first, leading to marginalization into smaller tables before further multiplication. This optimized approach enhances computational efficiency and facilitates more streamlined query resolution within the RSFG framework. Moving forward, additional experiments will be conducted to further validate and refine these findings.

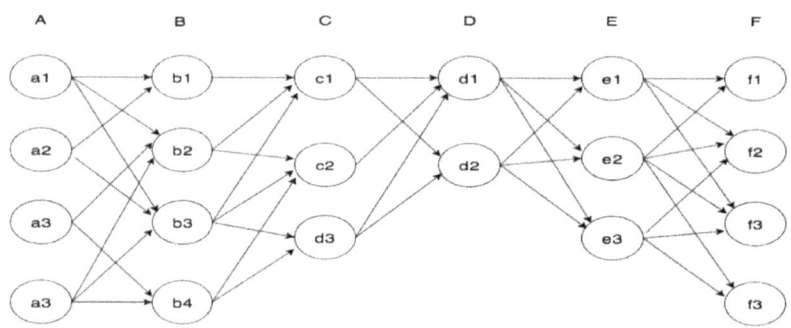

Fig. 13. Answering query Ans(B, F) in second experiment

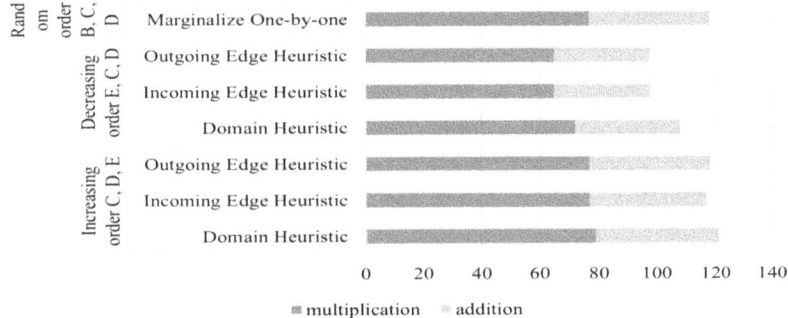

Fig. 14. Result for the experiment

5 Conclusion

Pawlak [7,8] introduced the concept of the rough set flow graph (RSFG) as a graphical framework for data reasoning. Additionally, Pawlak proposed an algorithm to address queries in RSFG, which involves the simultaneous elimination of variables, resulting in a significant computational load. Subsequently, a more efficient algorithm was proposed, which eliminates variables one by one. Notably, this algorithm never exceeds the computational workload of its predecessor.

In this paper, we optimize RSFG inference and make the following contributions. Firstly, we demonstrate the pruning of barren variables. Secondly, we propose the pruning of independent-by-evidence variables. Thirdly, we perform summation over all remaining variables. Furthermore, we propose heuristics to determine effective elimination orderings. Broadly categorized, our heuristics fall into two classes: domain heuristics and edge heuristics. Domain heuristics establish elimination orderings based on the domain cardinalities of the variables to be eliminated. Conversely, the topological structure of the RSFG with respect to edges drives the computation of elimination orderings with edge heuristics. Analysis of our experimental results suggests that the edge heuristic consistently produces favorable elimination orderings, while domain heuristics demonstrate relatively less effectiveness.

References

1. Butz, C.J., Yan, W., Yang, B.: An efficient algorithm for inference in rough set flow graphs. In: Peters, J.F., Skowron, A. (eds.) Transactions on Rough Sets V. LNCS, vol. 4100, pp. 102–122. Springer, Heidelberg (2006). https://doi.org/10.1007/11847465_5
2. MChavira, M., Darwiche, A.: Compiling Bayesian networks using variable elimination. In: IJCAI, pp. 2443–2449 (2007)
3. Cooper, G.F.: The computational complexity of probabilistic inference using Bayesian belief networks. Artif. Intell. **42**(2–3), 393–405 (1990)
4. Dawid, A.P., Lauritzen, S.L.: Hyper Markov laws in the statistical analysis of decomposable graphical models. Ann. Stat. **21**(3), 1272–1317 (1993)

5. Dechter, R.: Bucket elimination: a unifying framework for probabilistic inference. In: Jordan, M.I. (eds.) Learning in Graphical Models. NATO ASI Series, vol. 89, pp. 75–104. Springer, Dordrecht (1998). https://doi.org/10.1007/978-94-011-5014-9_4
6. Pawlak, Z.: Rough sets. Int. J. Comput. Inf. Sci. **11**(5), 341–356 (1982)
7. Pawlak, Z.: In pursuit of patterns in data reasoning from data - the rough set way. In: Alpigini, J.J., Peters, J.F., Skowron, A., Zhong, N. (eds.) RSCTC 2002. LNCS (LNAI), vol. 2475, pp. 1–9. Springer, Heidelberg (2002). https://doi.org/10.1007/3-540-45813-1_1
8. Pawlak, Z.: Flow graphs and decision algorithms. In: Wang, G., Liu, Q., Yao, Y., Skowron, A. (eds.) RSFDGrC 2003. LNCS (LNAI), vol. 2639, pp. 1–10. Springer, Heidelberg (2003). https://doi.org/10.1007/3-540-39205-X_1
9. Pawlak, Z.: Some issues on rough sets. In: Peters, J.F., Skowron, A., Grzymała-Busse, J.W., Kostek, B., Świniarski, R.W., Szczuka, M.S. (eds.) Transactions on Rough Sets I. LNCS, vol. 3100, pp. 1–58. Springer, Heidelberg (2004). https://doi.org/10.1007/978-3-540-27794-1_1
10. Pawlak, Z.: Rough Sets: Theoretical Aspects of Reasoning about Data, vol. 9. Springer Science & Business Media, Dordrecht (2012). https://doi.org/10.1007/978-94-011-3534-4
11. Pearl, J.: Probabilistic Reasoning in Intelligent Systems: Networks of Plausible Inference, 340 Pine Street. Elsevier, San Francisco (2014)
12. Shachter, R.D.: Probabilistic inference and influence diagrams. Oper. Res. **36**(4), 589–604 (1988)
13. Wong, S.M., Butz, C.J., Wu, D.: On the implication problem for probabilistic conditional independency. IEEE Trans. Syst. Man Cybern.-Part A Syst. Humans **30**(6), 785–805 (2000)
14. Zhang, N.L., Poole, D.: A simple approach to Bayesian network computations. In: Proceedings of the Biennial Conference-Canadian Society for Computational Studies of Intelligence, pp. 171–178 (1994)
15. Zhou, B., Cho, H., Zhang, X.: Scalable implementations of rough set algorithms: a survey. In: Mouhoub, M., Sadaoui, S., Ait Mohamed, O., Ali, M. (eds.) IEA/AIE 2018. LNCS (LNAI), vol. 10868, pp. 648–660. Springer, Cham (2018). https://doi.org/10.1007/978-3-319-92058-0_62

Multimodal Propaganda Detection in Memes with Tolerance-Based Soft Computing Method

Siddharth Kelkar[1], Srinivasa Ravi[1], Sheela Ramanna[2(✉)], and Anand Kumar Madasamy[1]

[1] Department of Information Technology, National Institute of Technology Karnataka, Surathkal, Mangaluru, India
{siddharthkelkar.211it067,srinivasar.211it070,m_anandkumar}@nitk.edu.in
[2] Department of Applied Computer Science, University of Winnipeg, Winnipeg, MB R3B 2E9, Canada
s.ramanna@uwinnipeg.ca

Abstract. This paper presents a tolerance-based near sets-based classifier applied to multimodal propaganda detection task using text and image data originating from Memes. Memes on the internet consist of an image superimposed with text and are very popular in social media. They are often used as a part of disinformation campaign whereby social media users are influenced via a number of rhetorical and psychological techniques known as persuasion techniques. The focus of this paper is on a subtask of the SemEval-2024 Multilingual Detection of Persuasion Techniques Competition in Memes to detect the presence or absence of a persuasion technique. We introduce a multimodal Tolerance Near Sets Classifier (MTNSC) trained on a combination of word embeddings (RoBERTa) and pre-trained image features (ResNet and ResNet-Memes) using the competition data. This work extends our earlier work in the Natural Language Processing domain where a tolerance-based near sets-based sentiment classifier was introduced. The proposed MTNSC achieves a macro F1 score of 70.15% and micro-F1 score of 75.33% on the test dataset demonstrating satisfactory performance of TNS-based classifiers in a multimodal setting. Our findings point to the model's effectiveness when compared to a few leading submissions based on deep learning techniques.

Keywords: Multimodal · Tolerance Near Sets · RoBERTa · ResNet · Memes · Persuasion Technique · Propaganda Detection

1 Introduction

Despite the many benefits that social media offer, more often than not, they are also used as a tool, by bots or human operators, to manipulate and to mislead unsuspecting users [3]. Propaganda is one such communication tool to influence the opinions and the actions of other people in order to achieve a predetermined goal using persuasive text [4]. Persuasive text is characterized by a specific use of language in order to influence readers (ex: justification, simplification, distraction and so on). Early work in detecting propaganda was reported in [14] which focused on detecting four classes (trusted, satire, hoax, and propaganda) at the document level in a case study using politicFac.com. In [2], the authors proposed 18 propaganda techniques and developed a corpus of annotated news articles at a more granular sentence level. In [13], a multilingual multifaceted annotated dataset of news articles was presented. The annotation was performed using a 2-level persuasion technique with a taxonomy of 23 fine-grained techniques grouped into 6 coarse categories.

Early propaganda techniques focused on text modality. With the availability of image modality, the focus has shifted to multimodal propaganda processing [7]. Internet memes are popular social media communication tools often used in disinformation campaigns [8]. Research on detecting the use of propaganda techniques in memes from a multimodal perspective was studied by the authors in [3]. Their focus was on annotating a multimodal corpus using 22 standardized propaganda techniques using pre-trained BERT transformer variants (SemEval-2021 task 6[1]). Curation of Arabic dataset from social media content for propaganda detection task was reported in [1]. In [5], various multimodal fusion approaches were applied to the task of detection of propaganda techniques in memes for the SemEval-2021 task 6.

In [9,10], we introduced a tolerance form of near sets-based soft computing algorithm (TSC1.0) for a supervised sentiment classification task. The TSC1.0 algorithm is a novel supervised learning method based on tolerance near sets [11,12] applied to text data. In near sets, the tolerance classes are directly induced from the feature vectors using a tolerance level parameter ε and a distance function. Recently, we introduced TSC2.0 algorithm which includes a tie-breaking component to address the labelling of prototype vectors in the presence of tolerance classes with equal or comparable number of elements having different labels [6]. This paper is extends our work from natural language domain to multimodal domain by focusing on a subtask of SemEval-2024 Multilingual Detection of Persuasion Techniques Competition. This competition includes subtask 1 which is a multilabel (text only) classification problem and subtask 2a which is multilabel, multimodal classification problem where the meme data includes both the textual and visual content[2]. In this paper, we focus on subtask 2b which is a multimodal binary classification task to identify whether a meme contains a persuasion technique or no technique. Our proposed MTNSC has demonstrated competitive performance when compared to a few leading submissions based on deep learning techniques.

In Sect. 2, we recall the definition of tolerance classes adapted to multimodal data, present the process flow of MTNSC, and briefly describe the dataset used in this paper. In Sect. 3, we discuss our findings in terms of the F1-scores measure

---

[1] https://semeval.github.io/SemEval2021/.
[2] https://propaganda.math.unipd.it/semeval2024task4/.

using text and image embeddings on the competition dataset followed by the concluding remarks in Sect. 4.

2 Methods and Materials

In this section, we briefly recall the definitions used to create tolerance classes.
Tolerance Classes - Preliminaries:
The formal model for the text-based tolerance relation is derived from [11]. We briefly review the notation.

Definition 1. Tolerance Relation $\cong_{\mathcal{T},\epsilon}$
Let T, F be a pair of nonempty sets where T is the set of objects and F is the feature set. Let $\mathcal{T} \subseteq F$ where $\mathcal{T}$ represents textual and visual features (embeddings). $\langle T, \cong_{\mathcal{T},\epsilon} \rangle$ defines a tolerance space with the relation $\cong_{\mathcal{T},\epsilon}$ defined as:

$$\cong_{\mathcal{T},\epsilon} = \{(t_i, t_j) \in T \times T : dist(t_i, t_j) \leq \epsilon\}, \quad (1)$$

where ϵ is a user-defined tolerance level, $dist$ is the cosine distance between objects (t_i, t_j) calculated as,

$$dist(\vec{t_i}, \vec{t_j}) = 1 - \frac{\vec{t_i} \cdot \vec{t_j}}{\|\vec{t_i}\| \|\vec{t_j}\|}, \quad (2)$$

where $\|\vec{a}\|$ represents the magnitude of any vector $\vec{a}$.

The tolerance relation $\cong_{\mathcal{T},\epsilon}$ which is a reflexive and symmetric relation induces a tolerance class. Each tolerance class represents an equivalence class where objects in a class are indistinguishable with respect to certain attributes within the given tolerance levels. MTNSC performs supervised meme binary classification using the cosine distance (defined in Eq. 2) using the feature vectors to compute tolerance classes based on a chosen value for ϵ.

Figure 1 gives the overview of the proposed MTNSC algorithm. The text and image embeddings are generated simultaneously from both text and visual content present in the memes. Then the two vectors are concatenated (early fusion) to create a final vector. Using the distance function from (1), a distance matrix is generated between each pair of vectors. Tolerance classes are created from the generated matrix followed by the computation of a prototype vector for each tolerance class by computing the mean of all the vectors within that class. Each prototype vector is then labelled using majority voting (with respect to the labels associated with the vectors within the corresponding tolerance class). Each test instance is labelled by determining the minimum distance to a prototype vector and assigning its label to the test instance. Details of the steps used in Fig. 1 are as follows:

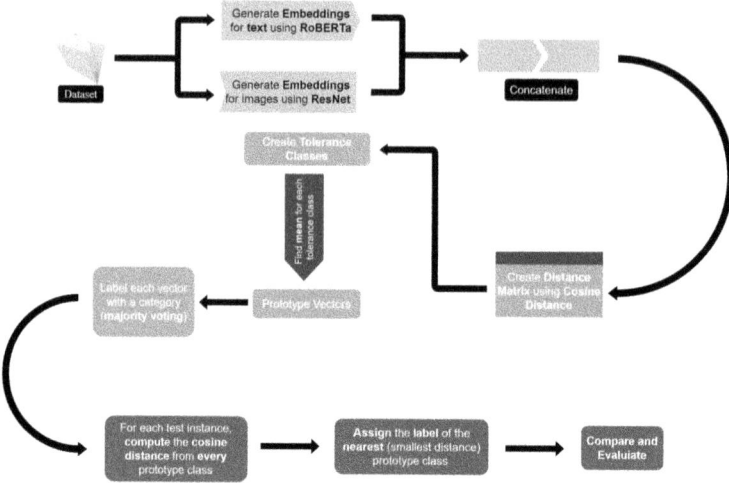

Fig. 1. MTNSC - High-level flow [9,10]

1° To generate the feature vectors for the text, a pretrained RoBERTa model fine-tuned on the text portion of the dataset was used. The last-but-one layer of the model was used to generate 768 dimensional word embeddings for each text instance.

2° Similarly, to generate the feature vectors for the images, a pretrained ResNet model fine-tuned on the memes portion of the dataset was used. The last-but-one layer of the model was used to generate 768 dimensional image embeddings (mapped from 2048 dimensions with the help of another layer) for each image instance. Here, two different ResNet models were used, one fine tuned on a large number of memes[3] and the other model on the ImageNet dataset.

3° The text and image embeddings were concatenated to form a single 1536 dimensional feature vector for each instance.

4° The multimodal feature vectors were used to generate the tolerance classes using the Tolerance-based Sentiment classifier (TSC) algorithm [9].

5° The threshold value (ϵ) was optimized to get the best accuracy for the model. The performance of the model was evaluated for threshold values between from 0.01 to 0.07. The model was iteratively tested on a wide range of values for the threshold, to get the approximate optimum range of threshold values.

6° The final performance of the model was evaluated using accuracy and the F_1-score.

---

[3] https://huggingface.co/jayanta/resnet-50-finetuned-memes-v2.

2.1 Materials

The provided meme dataset included 900 instances of training data and 450 instances of unlabelled development set. After the release of the test set, the labels for 450 instances of development set were released. Hence, the training data for MTNSC system consisted of a total of 1350 instances with 900 labelled as propagandistic and 450 labelled as non-propagandistic with a test set consisting of 600 instances (without annotations).

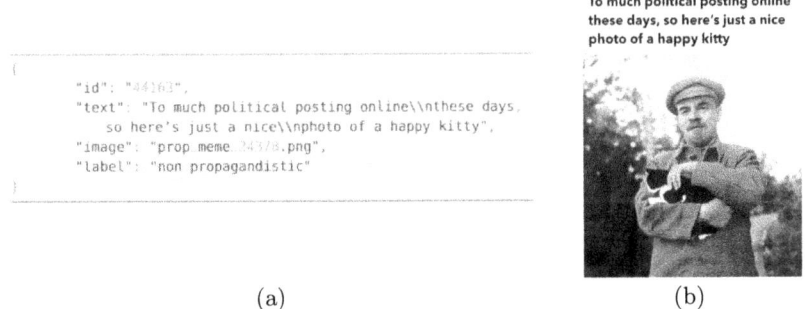

Fig. 2. Meme dataset as a JSON file with (a) text data, (b) corresponding image data

The raw data format given is in Fig. 2, where **id** is the unique identification for each meme, **text** is the text part of the data, **image** is the file name of the corresponding image and the **label** is either propagandistic or non-propagandistic representing the presence or absence of any of the persuasion techniques. These techniques include: Name Calling/Labeling; Exaggeration/Minimization; Appeal to fear/prejudice; Flag-Waving; Slogans; Repetition; Doubt; Reductio ad Hitlerum; Obfuscation/Intentional Vagueness/Confusion; Smears; Glittering Generalities; Causal Oversimplification; Black-and-White Fallacy; Appeal to Authority; Bandwagon; Red Herring; Whataboutism; Thought-terminating Cliches and Straw Men. These techniques were inspired by the document[4].

3 Results

The metric used for the evaluation of the models are variants of the F_1 score, defined as,

$$F_1 = 2 \times \frac{\text{Precision} \times \text{Recall}}{\text{Precision} + \text{Recall}}$$

---

[4] https://knowledge4policy.ec.europa.eu/text-mining/news-categorization-framing-persuasion-techniques-annotation-guidelines_e.

where *Precision* is the ratio of the number of true positive predictions to the number of all positive predictions and *Recall* is the ratio of the number of true positive predictions to the number of actual positive instances. The variants of the F_1 score used in the evaluation are the F_1-*Macro* and F_1-*Micro* scores. The F_1-Macro score is the average of the F_1 scores of each class and the F_1-Micro score is the F_1 score of the overall predictions.

Table 1 gives the performance of the MTNSC model with varying ε threshold values. The model was implemented in Python using in-built libraries. Since the dataset was small, no GPU computing capability was necessary. The optimum ϵ values were small compared to our previous work with just text data [6].

Table 1. Performance of MTNSC (using RoBERTa + ResNet Memes) with the threshold value (ϵ) varying from 0.01 to 0.07. # represents "number of".

ϵ	#(Tolerance Classes)	Accuracy	F_1-Macro	F_1-Micro
0.01	4	0.71	0.70	0.72
0.02	14	0.69	0.77	0.81
0.03	24	0.72	0.70	0.74
0.04	85	0.73	0.73	0.79
0.05	360	0.72	0.72	0.78
0.06	630	0.73	0.74	0.79
0.07	833	0.71	0.72	0.77

Figure 3 shows the values for accuracy and F_1-Macro scores for varying values of ε to determine the optimum threshold value for the model. The threshold value of 0.04 was chosen for the final prediction as it has a reasonable number of tolerance classes. From Table 1, one can observe that even though $\varepsilon = 0.02$ results in better F_1 scores overall, size of the tolerance class is small. Also higher ϵ values create more number of classes which could ultimately simplify down to individual similarity rather than that of clusters with a significant number of members.

Table 2 gives the F_1-Macro and F_1-Micro scores on the development set using the RoBERTa feature vectors for text embeddings and ResNet (Imagenet) for image embeddings as well as ResNet (Memes) for image embeddings. The image embeddings using ResNet (Memes) performed better which is expected due to the underlying task of meme classification.

Table 3 gives the F_1-Macro and F_1-Micro scores of the MTNSC model on the test set using the fine tuned RoBERTa feature vector for text embeddings and ResNet (Memes) for image embeddings combined together. It should be noted that the gold labels as well as the test set instances were not made available. Hence one could not judge the distribution of the test instances. However, the training instances were imbalanced.

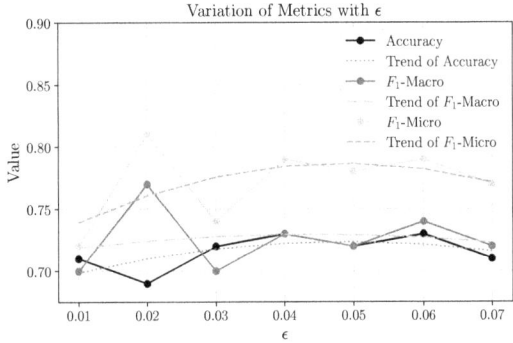

Fig. 3. Plot of the accuracy and F_1-Macro score with (RoBERTa + ResNet Memes) and their trends with respect to the threshold value

Table 2. Performance of the model (measured by F_1 scores) on the development set with different embeddings.

Embeddings	F_1-Macro	F_1-Micro
RoBERTa + ResNet (ImageNet)	0.7131	0.7596
RoBERTa + ResNet (Memes)	**0.7346**	**0.7966**

Table 3. Performance of the model (measured by F_1 scores) on the test set.

Embeddings	F_1-Macro	F_1-Micro
RoBERTa + ResNet (Memes)	**0.7015**	**0.7533**

Table 4. Comparative Snapshot of leaderboard of task 2b of Multilingual Detection of Persuasion Techniques in Memes.

Rank	Team Name	F_1-Macro	F_1-Micro
12	BERTastic	0.71582	0.76167
13	Hidetsune	0.71353	0.79000
14	**Scalar**	**0.70151**	**0.75333**
15	SheffieldVeraAI	0.64197	0.68667
16	HeirarchyEverywhere	0.56309	0.66167

Table 4 shows the ranking and the scores obtained by our team (Scalar) in the shared task[5] with 21 teams in the competition. The difference in scores of teams two places higher than our team is ≈ 0.02, indicative of our proposed model's ability to compete with other deep learning techniques. The best team on the leaderboard received a F_1-Macro score of 0.836.

---

[5] semeval_task4_leaderboard.

4 Conclusion and Future Work

The proposed TNS-based classifier for multimodal data (MTNSC) has demonstrated competitive performance which is comparable to a few leading submissions, indicating its effectiveness relative to deep learning techniques. Further refinement opportunities lie in systematic experimentation with tolerance values, embeddings, and architectural configurations. The model's adeptness in handling multimodal data and resilience to data imbalance underscore its practical utility. To optimize its performance, incorporating advanced multimodal embeddings, architectural refinements, and sophisticated threshold optimization techniques are proposed as future work. Expanding testing to diverse imbalanced multiclass datasets needs to be explored for comprehensive evaluation.

In addition to quantitative metrics, visualizing and interpreting the formed clusters offer qualitative insights into the model's decision boundaries, enriching the understanding of its inner workings. Evaluating performance on tasks beyond the current scope with imbalanced data enhances the model's generalizability and applicability in real-world scenarios.

Extending the model to address multilabel classification introduces a nuanced dimension to its capabilities, accommodating scenarios where data points may belong to multiple classes concurrently. Furthermore, an in-depth analysis of audio datasets provides a domain-specific perspective, highlighting the model's adaptability in tasks such as sound classification. In summary, our research outlines a pathway for refining the tolerance-based classifier by optimizing key parameters, incorporating qualitative insights through visualization, evaluating on diverse datasets, extending its capabilities to multilabel problems, and exploring domain-specific applications, contributing to the broader landscape of effective machine learning solutions.

Acknowledgements. Sheela Ramanna's work was supported by NSERC Discovery Grant # 194376.

References

1. Alam, F., Mubarak, H., Zaghouani, W., Martino, G.D.S., Nakov, P.: Overview of the WANLP 2022 shared task on propaganda detection in Arabic (2022). https://arxiv.org/abs/2211.10057
2. Da San Martino, G., Yu, S., Barrón-Cedeño, A., Petrov, R., Nakov, P.: Fine-grained analysis of propaganda in news article. In: Inui, K., Jiang, J., Ng, V., Wan, X. (eds.) Proceedings of the 2019 Conference on Empirical Methods in NLP and the 9th Intl. Joint Conference on NLP(EMNLP-IJCNLP), pp. 5636–5646. Association for Computational Linguistics, Hong Kong, China (2019). https://aclanthology.org/D19-1565
3. Dimitrov, D., et al.: Detecting propaganda techniques in memes (2021). https://arxiv.org/abs/2109.08013
4. Editor: Volume II of the publications of the institute for propaganda analys (1939)
5. Gundapu, S., Mamidi, R.: Detection of propaganda techniques in visuo-lingual metaphor in memes (2022). https://arxiv.org/abs/2205.02937

6. Hegde, T., Sanjay, K.S., Thomas, S.M., Kambhammettu, R., Anand Kumar, M., Ramanna, S.: Impact of vector embeddings on the performance of tolerance near sets-based sentiment classifier for text classification. Proc. Comput. Sci. KES **2023**(225), 645–654 (2023)
7. Ng, V., Li, S.: Multimodal propaganda processing (2023). https://arxiv.org/abs/2302.08709
8. Nieubuurt, J.T.: Internet memes: leaflet propaganda of the digital age. Front. Commun. **5** (2021). https://www.frontiersin.org/articles/10.3389/fcomm.2020.547065
9. Patel, V., Ramanna, S.: Tolerance-based short text sentiment classifier. In: Ramanna, S., Cornelis, C., Ciucci, D. (eds.) IJCRS 2021. LNCS (LNAI), vol. 12872, pp. 259–265. Springer, Cham (2021). https://doi.org/10.1007/978-3-030-87334-9_22
10. Patel, V., Ramanna, S., Kotecha, K., Walambe, R.: Short text classification with tolerance-based soft computing method. Algorithms **15**(8) (2022). https://www.mdpi.com/1999-4893/15/8/267
11. Peters, J.: Near sets. Special theory about nearness of objects. Fundamenta Informaticae **75**(1-4), 407–433 (2007)
12. Peters, J.: Tolerance near sets and image correspondence. Int. J. Bio-Inspired Comput. **1**(4), 239–245 (2009)
13. Piskorski, J., Stefanovitch, N., Nikolaidis, N., Martino, G.D.S., Nakov, P.: Multilingual multifaceted understanding of online news in terms of genre, framing, and persuasion techniques. In: Rogers, A., Boyd-Graber, J.L., Okazaki, N. (eds.) Proceedings of the 61st Annual Meeting of the Association for Computational Linguistics (Volume 1: Long Papers), ACL 2023, Toronto, Canada, July 9-14, 2023, pp. 3001–3022. Association for Computational Linguistics (2023). https://doi.org/10.18653/v1/2023.acl-long.169
14. Rashkin, H., Choi, E., Jang, J.Y., Volkova, S., Choi, Y.: Truth of varying shades: analyzing language in fake news and political fact-checking. In: Palmer, M., Hwa, R., Riedel, S. (eds.) Proc. of the 2017 Conference on Empirical Methods in NLP, pp. 2931–2937. Association for Computational Linguistics, Copenhagen, Denmark (2017)

Author Index

B
Bai, Zhongling II-14
Bao, Beining II-230
Bhutiyal, Tejasvi II-247
Butz, Cory J. I-329

C
Caleb, Jerry II-213
Chatterjee, Niladri I-134
Chen, Jiang II-14
Chen, Siheng I-252
Cornelis, Chris I-3
Cui, Xiangen II-230

D
Dai, Yihong II-230
De Domenico, Andrea I-67
Deja, Rafał I-105
Durdymyradov, Kerven I-188
Dutta, Soma I-34, I-223

F
Fujiwara, Takeshi I-157

G
Góra, Grzegorz I-201
Gupta, Rakshit II-247

H
Hou, Jialin II-47
Hu, Mengjun II-165
Huang, Xin I-300

K
Kandula, Venkata Sai Krishna Chaitanya II-310
Kelkar, Siddharth I-343
Koirala, Kritika II-213
Kudo, Yasuo I-266
Kumar, Amrit I-134

L
Lang, Guangming I-238, I-252, II-181
Li, Qiaoyi II-59, II-165
Li, Shaobo I-314
Li, Xiaonan II-196
Li, Yueyang I-252
Li, Zhaoyi I-285
Li, Zihao I-285
Liu, Hanyue II-280
Liu, Ping II-181
Liu, Qun I-300
Liu, Wenjun I-238
Liu, Zhuangzhuang II-29
Lu, Mingyu II-103
Luo, Nanfang II-86

M
Madasamy, Anand Kumar I-343
Mani, A. I-50
Manoorkar, Krishna B. I-67
Mehta, Neeyati II-213
Miao, Duoqian II-3, II-74
Moshkov, Mikhail I-93, I-173, I-188
Murai, Tetsuya I-266

N
Nakata, Michinori I-120, I-157
Naushad, Raoof II-247

O
Okawa, Hajime I-266
Ostonov, Azimkhon I-93, I-173

P
Palmigiano, Alessandra I-67
Pan, ZhengYue II-118
Pang, Kuo II-103
Patel, Sameer II-213
Prajapati, Vrushali II-247
Przybyła-Kasperek, Małgorzata I-105

Q
Qiu, ZiHeng   II-118

R
Ramanna, Sheela   I-343
Rao, Hong   II-3
Ravi, Srinivasa   I-343
Restrepo, Mauricio   I-3

S
Sakai, Hiroshi   I-120, I-157
Shaikh, Zaid   II-213
Shao, Shuangrun   II-230
Shi, Chengjun   II-59, II-165
Skowron, Andrzej   I-201, I-223
Ślęzak, Dominik   I-120, I-34
Ślęzak, Dominik   I-18
Su, Zuqiang   II-265
Sun, Shijie   II-280
Sun, XiaoJun   II-118
Suo, Langwangqing   II-149

T
Tang, Jianhang   I-314
Tu, Jing   II-3

W
Wang, Guoyin   I-314, II-86, II-265
Wang, Jun   I-329
Wang, Xiaoling   II-295
Wang, Ye   II-280
Wang, Zhen   II-103
Wang, Ziye   II-74
Wasilewski, Piotr   I-18
Watada, Junzo   I-120

X
Xia, Shuyin   II-29
Xiang, Fuchuan   I-314
Xiao, Jianfeng   I-300
Xiao, Qimei   II-181
Xie, Qin   II-86
Xu, Fanxin   II-265
Xu, Ji   I-314
Xu, Jianfeng   II-3
Xu, Taihua   II-29
Xu, Yi   II-118

Y
Yan, Yucong   II-196
Yang, Hai-Long   II-149
Yang, Han   II-59
Yang, Jie   II-29
Yao, Yiyu   II-47, II-59, II-135, II-149, II-165
Yu, Hong   II-280
Yu, Huiying   II-181
Yue, Wenjing   II-295
Yue, Xiaodong   I-285

Z
Zhang, Chuanlei   II-135
Zhang, Dingkai   II-295
Zhang, Qinghua   II-86
Zhang, Xianyong   II-14
Zhang, Yan   II-310
Zhang, Yuanjian   II-3
Zhang, Zhifei   I-238
Zhao, Bingxi   II-230
Zhao, Lixi   I-238
Zheng, Huanran   II-295
Zhu, QiSheng   II-118
Zou, Li   II-103

GPSR Compliance

The European Union's (EU) General Product Safety Regulation (GPSR) is a set of rules that requires consumer products to be safe and our obligations to ensure this.

If you have any concerns about our products, you can contact us on

ProductSafety@springernature.com

In case Publisher is established outside the EU, the EU authorized representative is:

Springer Nature Customer Service Center GmbH
Europaplatz 3
69115 Heidelberg, Germany

www.ingramcontent.com/pod-product-compliance
Ingram Content Group UK Ltd.
Pitfield, Milton Keynes, MK11 3LW, UK
UKHW022241230426
12048UKWH00018BA/1405